Christel Baier   Boudewijn R.
Holger Hermanns   Joost-Piete
Markus Siegle (Eds.)

Validation of Stochastic Systems

A Guide to Current Research

 Springer

Series Editors

Gerhard Goos, Karlsruhe University, Germany
Juris Hartmanis, Cornell University, NY, USA
Jan van Leeuwen, Utrecht University, The Netherlands

Volume Editors

Christel Baier
University of Bonn, Institute of Informatics I
Römerstr. 164, 53117 Bonn, Germany
E-mail: baier@cs.uni-bonn.de

Boudewijn R. Haverkort
University of Twente, Dept. of Computer Science and Electrical Engineering,
Design and Analysis of Communication Systems
P.O. Box 217, The Netherlands
E-mail: brh@cs.utwente.nl

Holger Hermanns
Saarland University, Dept. of Computer Science, Dependable Systems and Software
66123 Saarbrücken, Germany
E-mail: hermanns@cs.uni-sb.de

Joost-Pieter Katoen
University of Twente, Dept. of Computer Science, Formal Methods and Tools
P.O. Box 217, 7500 AE Enschede, The Netherlands
E-mail: katoen@cs.utwente.nl

Markus Siegle
University of Federal Armed Forces Munich, Dept. of Computer Science
85577 Neubiberg, Germany
E-mail: siegle@informatik.unibw-muenchen.de

Library of Congress Control Number: 2004110613

CR Subject Classification (1998): F.1, F.3, D.2, D.4, I.6, C.1

ISSN 0302-9743
ISBN 3-540-22265-0 Springer Berlin Heidelberg New York

Springer is a part of Springer Science+Business Media

springeronline.com

© Springer-Verlag Berlin Heidelberg 2004
Printed in Germany

Typesetting: Camera-ready by author, data conversion by Scientific Publishing Services, Chennai, India
Printed on acid-free paper   SPIN: 11015703   06/3142   5 4 3 2 1 0

Preface

It is with great pleasure that we present to you this tutorial volume entitled *Validation of Stochastic Systems*. It is one of the results of the Dutch-German bilateral cooperation project "Validation of Stochastic Systems" (VOSS), financed by NWO and DFG (the Dutch and German science foundations, respectively).

In the early days of 2002, the idea emerged to organize a seminar at Schloss Dagstuhl, not the usual Dagstuhl seminar with primarily invited participants, but a seminar aimed at young(er) people, and for which the organizers assign themes to be worked upon and presented on. Following an open call announced via the Internet in the spring of 2002, we received many applications for participation. After a selection procedure, we decided to assign (mostly) teams of two researchers to work on specific topics, roughly divided into the following four theme areas: "Modelling of Stochastic Systems," "Model Checking of Stochastic Systems," "Representing Large State Spaces," and "Deductive Verification of Stochastic Systems." These are the titles of the four parts of this volume.

The seminar was held in Schloss Dagstuhl during December 8–11, 2002 as part of the so-called GI/Research Seminar series. This series of seminars is financially supported by the *Gesellschaft für Informatik*, the German Computer Society. At that point in time the papers had already undergone a first review round. Each of the tutorial papers was presented in a one-hour session, and on the basis of the presentations we decided to bring together a selection of them into a book. A second review round was performed throughout 2003; at the end of 2003 all contributions were finished. We are glad that Springer-Verlag was willing to publish it in their well-established *Lecture Notes in Computer Science* series, in particular in the "green cover" Tutorial subseries.

To conclude this preface, we would like to thank NWO and DFG for making the VOSS bilateral cooperation project possible in the first place. Secondly, we would like to thank the *Gesellschaft für Informatik* for supporting the participants of the seminar. We would like to thank the whole team at Schloss Dagstuhl for their willingness to host us and for their hospitality. We also thank the authors of the tutorial papers as well as the reviewers for their efforts; without you, there would not have been a workshop! Finally, we would like to thank José Martínez (of the University of Twente) for his work on the editing of this volume.

Christel Baier
Boudewijn Haverkort
Holger Hermanns
Joost-Pieter Katoen
Markus Siegle

Table of Contents

VIII   Table of Contents

Probabilistic Automata: System Types, Parallel Composition and Comparison

Ana Sokolova[1] and Erik P. de Vink[2]

[1] Department of Mathematics and Computer Science,
TU/e, Eindhoven
a.sokolova@tue.nl
[2] LIACS, Leiden University
evink@win.tue.nl

Abstract. We survey various notions of probabilistic automata and probabilistic bisimulation, accumulating in an expressiveness hierarchy of probabilistic system types. The aim of this paper is twofold: On the one hand it provides an overview of existing types of probabilistic systems and, on the other hand, it explains the relationship between these models. We overview probabilistic systems with discrete probabilities only. The expressiveness order used to built the hierarchy is defined via the existence of mappings between the corresponding system types that preserve and reflect bisimilarity. Additionally, we discuss parallel composition for the presented types of systems, augmenting the map of probabilistic automata with closedness under this compositional operator.

Keywords: probabilistic automata (transition systems), probabilistic bisimulation, preservation and reflection of bisimulation, non-determinism, parallel composition.

1   Introduction

The notion of a state machine has proved useful in many modelling situations, amongst others, the area of validation of stochastic systems. In the literature up to now, a great variety of types of probabilistic automata has been proposed and many of these have been actually used for verification purposes. In this paper we discuss a number of probabilistic automata with discrete probability distributions. For continuous-time probabilistic systems the interested reader is referred to [11, 33, 32, 17, 45, 4]. Models of stochastic systems that are not represented by transition systems can also be found in [22] and [70].

Due to the variety of proposed models it is often the case that results have to be interpreted from one type of systems to another. Therefore we compare the considered types of probabilistic automata in terms of their expressiveness. The comparison is achieved by placing a partial order on the classes of such automata, where one class is less then another if each automaton in the class can be translated to an automaton of the other class such that translations both reflect and preserve the respective notions of bisimilarity. Hence, bisimulation and

C. Baier et al. (Eds.): Validation of Stochastic Systems, LNCS 2925, pp. 1–43, 2004.
© Springer-Verlag Berlin Heidelberg 2004

bisimilarity are central notions in this overview. Other comparison criteria are important as well, e.g. logical properties, logical characterization of bisimulation [61], complexity of algorithms for deciding bisimulation [9, 13, 31, 80] and so on. We choose the comparison criterion formulated in terms of strong bisimulation because of its simplicity and because we work with transition labelled systems, for which bisimulation semantics arises naturally from the step-by-step behavior.

A major distinction of probabilistic automata is that between fully probabilistic vs. non-deterministic ones. In a fully probabilistic automaton every choice is governed by a probability distribution (over set of states or states combined with actions). The probability distribution captures the uncertainty about the next state. If we abstract away from the actions in a fully probabilistic automaton, we are left with a discrete time Markov chain. Subsequently, standard techniques can be applied to analyze the resulting Markov chains. Sometimes, the incomplete knowledge about the system behavior can not be represented probabilistically. In these cases we should consider more than one transition possible. We speak in this case of a non-deterministic probabilistic automaton. Most of the models that we consider include some form of non-determinism and hence fall in the category of non-deterministic probabilistic automata. As pointed out by various authors, e.g. [47, 76, 3, 81] non-determinism is essential for modelling scheduling freedom, implementation freedom, the external environment and incomplete information. Furthermore, non-determinism is essential for the definition of an asynchronous parallel composition operator that allows interleaving. Often two kinds of non-deterministic choices are mentioned in the literature (see for e.g. [81]), *external* non-deterministic choices influenced by the environment, specified by having several transitions with different labels leaving from the same state, and *internal* non-determinism, exhibited by having several transitions with the same label leaving from a state. We use the term non-determinism for *full non-determinism* including both internal and external non-deterministic choices.

We introduce several classes of automata, ranging from the simplest models to more complex ones. The questions that we will address for each individual class are:

- the definition of the type of automaton and the respective notion of strong bisimulation;
- the relation of the model with other models;
- presence and form of non-determinism;
- the notion of a product or parallel composition in the model.

The set-up of the paper is as follows: Section 2 presents the necessary notions considering probability theory, automata (transition systems), and concurrency theory, in particular compositional operators. In section 3 we focus on the various definitions of probabilistic automata in isolation with their corresponding notions of bisimulation. In section 4 the operators of parallel composition are discussed. We address the interrelationship between the introduced types of automata in section 5. Section 6 wraps up with some conclusions.

Acknowledgements

We would like to thank Holger Hermanns for editorial support and for the plentitude of useful ideas and directions, as well as the other organizers of VOSS GI/Dagstuhl 2002 for initiating and organizing this nice event. We are in dept to the referees for various remarks. Special thanks go to Falk Bartels for his major contribution regarding the hierarchy of probabilistic systems, as well as for numerous comments, suggestions and his friendly cooperation.

2   Basic Ingredients

2.1   Probability Distributions

Let Ω be a set. A function $\mu\colon \Omega \to [0,1]$ is called a *discrete probability distribution*, or distribution for short, on Ω if $\{x \in \Omega|\ \mu(x) > 0\}$ is finite or countably infinite and $\sum_{x \in \Omega} \mu(x) = 1$. The set $\{x \in \Omega|\ \mu(x) > 0\}$ is called the *support* of μ and is denoted by $spt(\mu)$. If $x \in \Omega$, then μ_x^1 denotes the unique probability distribution with $\mu_x^1(x) = 1$, also known as the *Dirac distribution* for x. When μ is a distribution on Ω we use the notation $\mu[X]$ for $\sum_{x \in X} \mu(x)$ where $X \subseteq \Omega$. By $\mathcal{D}(\Omega)$ we denote the set of all discrete probability distributions on the set Ω. If μ is a distribution with finite support $\{s_1, \ldots, s_n\}$, we sometimes write $\{s_1 \mapsto \mu(s_1), \ldots, s_n \mapsto \mu(s_n)\}$. With this notation, $\mu_x^1 = \{x \mapsto 1\}$.

Let $\mu_1 \in \mathcal{D}(S)$ and $\mu_2 \in \mathcal{D}(T)$. The product $\mu_1 \times \mu_2$ of μ_1 and μ_2 is a distribution on $S \times T$ defined by $(\mu_1 \times \mu_2)(s,t) = \mu_1(s) \cdot \mu_2(t)$, for $\langle s, t \rangle \in S \times T$.

If $\mu \in \mathcal{D}(S \times T)$, we use the notation $\mu[s, T]$ for $\mu[\{s\} \times T]$ and $\mu[S, t]$ for $\mu[S \times \{t\}]$. We adopt from [51] the lifting of a relation between two sets to a relation between distributions on these sets.

Definition 1. *Let $R \subseteq S \times T$ be a relation between the sets S and T. Let $\mu \in \mathcal{D}(S)$ and $\mu' \in \mathcal{D}(T)$ be distributions. Define $\mu \equiv_R \mu'$ if and only if there exists a distribution $\nu \in \mathcal{D}(S \times T)$ such that*

1. *$\nu[s, T] = \mu(s)$ for any $s \in S$*
2. *$\nu[S, t] = \mu'(t)$ for any $t \in T$*
3. *$\nu(s, t) \neq 0$ if and only if $\langle s, t \rangle \in R$.*

The lifting of a relation R preserves the characteristic properties of preorders and equivalences (cf. [52]). For the special case of an equivalence relation there is a simpler way to define the lifting (cf. [52, 81, 9]).

Proposition 1. *Let R be an equivalence relation on the set S and let $\mu, \mu' \in \mathcal{D}(S)$. Then $\mu \equiv_R \mu'$ if and only if $\mu[C] = \mu'[C]$ for all equivalence classes $C \in S/R$.* □

Lifting of an equivalence relation on a set S to a relation $\equiv_{R,A}$ on the set $\mathcal{D}(A \times S)$, for a fixed set A, will also be needed.

Definition 2. *Let R be an equivalence relation on a set S, A a set, and let $\mu, \mu' \in \mathcal{D}(A \times S)$. Define*

$$\mu \equiv_{R,A} \mu' \iff \forall C \in S/R, \forall a \in A\colon \mu[a, C] = \mu'[a, C]$$

2.2   Non-probabilistic Automata, Markov Chains, Bisimilarity

Throughout the paper we will use the terms automaton, transition system or
just system as synonyms.

Non-probabilistic Automata

Definition 3. *A transition system, TS for short, is a pair* $\langle S, \alpha \rangle$ *where*

1. *S is a set of states*
2. *$\alpha : S \rightarrow \mathcal{P}(S)$ is a transition function, where $\mathcal{P}$ denotes the powerset of S.*

If $\langle S, \alpha \rangle$ is a transition system such that $s, s' \in S$ and $s' \in \alpha(s)$ we write
$s \rightarrow s'$ and call it a transition.

Often in the literature a TS is given as a triple, including besides the set of
states and the transition function also a subset of initial states, or a single initial
state. In this paper we will consider **no initial states** and therefore they are not
present in the definition. Instead of a transition function one could equivalently
consider a transition relation as a subset of $S \times S$. Our choice here is to always
present the transitions via a **transition function**.

A way of representing a TS is via its transition diagram. For example, the sys-
tem $\langle S, \alpha \rangle$ where $S = \{s_1, s_2, s_3, s_4\}$ and $\alpha(s_1) = \{s_2, s_3\}$, $\alpha(s_2) = \{s_4\}$, $\alpha(s_3) = \alpha(s_4) = \emptyset$, is represented as follows:

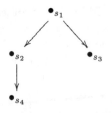

The states s_3 and s_4 are *terminating* states, with no outgoing transitions.

It is often of use to model the phenomenon that a change of state in a system
happens as a result of executing an *action*. Therefore, labelled transition systems
evolve from transition systems. There are two ways to incorporate labels in a
TS: by labelling the states (usually with some values of variables, or a set of
propositions true in a state), or by explicitly labelling the transitions with actions
or action names. In this paper we focus on **transition labelled systems**.

Definition 4. *A labelled transition system (LTS) (or a non-deterministic automaton) is a triple* $\langle S, A, \alpha \rangle$ *where*

1. *S is a set of states*
2. *A is a set of actions*
3. *$\alpha : S \rightarrow \mathcal{P}(A \times S)$ is a transition function.*

When $\langle S, A, \alpha \rangle$ is a LTS, then the transition function α can equivalently be
considered as a function from S to $\mathcal{P}(S)^A$, the collection of functions from A

to $\mathcal{P}(S)$. As in the case of TSs, for any state $s \in S$ of a LTS, every element $\langle a, s' \rangle \in \alpha(s)$ determines a transition which is denoted by $s \xrightarrow{a} s'$.

The class of non-deterministic automata (LTSs) is denoted by **NA**. Deterministic automata, given by the next definition, form a subclass of **NA**.

Definition 5. *A deterministic automaton is a triple $\langle S, A, \alpha \rangle$ where*

1. *S is a set of states*
2. *A is a set of actions*
3. *$\alpha : S \to (S+1)^A$ is a transition function.*

Notation 1 *We denote by $+$ the disjoint union of two sets. The set 1 is a singleton set containing the special element $*$, i.e. $1 = \{*\}$. We assume that $* \notin S$. The notation $(S+1)^A$ stands for the collection of all functions from A to $S+1$.*

The special set 1 and the disjoint union construction allow us to write partial functions as functions. Hence, in a deterministic automaton each state s is assigned a partial function $\alpha(s) : A \to S+1$ from the set of actions to the set of states, meaning that whenever $\alpha(s)(a) = s'$ for some $s' \in S$, i.e. $\alpha(s) \neq *$, then there is a transition $s \xrightarrow{a} s'$ enabled in S. We denote the class of all deterministic automata by **DA**.

We note that the class of automata **DA** exhibits external non-determinism, while in **NA** there is full non-determinism.

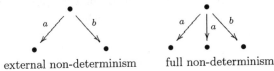

external non-determinism   full non-determinism

Markov Chains. The simplest class of fully probabilistic automata is the class of discrete time Markov chains. The theory of Markov chains is rich and huge (see, e.g., [57, 48, 16, 43]) and we only provide a simple definition of a discrete time Markov chain here.

Definition 6. *A Markov chain is a pair $\langle S, \alpha \rangle$ where*

1. *S is a set of states*
2. *$\alpha : S \to \mathcal{D}(S)$ is a transition function.*

Markov chains evolve from transition systems, when probability is added to each transition such that for any state the sum of the probabilities of all outgoing transitions equals 1. The class of all Markov chains is denoted by **MC**. If $s \in S$ and $\alpha(s) = \mu$ with $\mu(s') = p > 0$ then the Markov chain $\langle S, \alpha \rangle$ is said to go from a state s with probability p to a state s'. Notation: $s \rightsquigarrow \mu$ and $s \overset{p}{\rightsquigarrow} s'$.

Example 1.

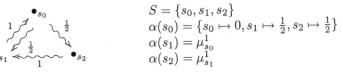

$S = \{s_0, s_1, s_2\}$
$\alpha(s_0) = \{s_0 \mapsto 0, s_1 \mapsto \frac{1}{2}, s_2 \mapsto \frac{1}{2}\}$
$\alpha(s_1) = \mu_{s_0}^1$
$\alpha(s_2) = \mu_{s_1}^1$

Bisimulation and Bisimilarity. Different semantics or notions of behavior can be given to labelled transition systems. We work with the bisimulation semantics (Milner [65, 66]) stating that two states in a system represented by LTSs are equivalent whenever there exists a bisimulation relation that relates them. A bisimulation relation compares the one-step behavior of two states and has a nice extension to the probabilistic case (as explored in [61]). In [54] probabilistic extensions of a number of other well known process equivalences have been studied like probability trace, completed trace, failure and ready equivalence. Other probabilistic process equivalences are probabilistic simulation and bisimulation by Segala and Lynch [78, 76], Yi and Larsen's testing equivalence [88], and CSP equivalences of Morgan et al. [67], Lowe [59] and Seidel [77]. An overview of several probabilistic process equivalences can be found in [58].

Definition 7. *Let $\langle S, A, \alpha \rangle$ and $\langle T, A, \alpha \rangle$ be two LTSs. A relation $R \subseteq S \times T$ is a bisimulation relation if for all $\langle s, t \rangle \in R$ and all $a \in A$ the following holds*

> *if $s \xrightarrow{a} s'$ then there exists $t' \in T$ such that $t \xrightarrow{a} t'$ and $\langle s', t' \rangle \in R$, and*
> *if $t \xrightarrow{a} t'$ then there exists $s' \in S$ such that $s \xrightarrow{a} s'$ and $\langle s', t' \rangle \in R$.*

Let $s \in S$ and $t \in T$. The states s and t are called bisimilar, *denoted by $s \approx t$ if there exists a bisimulation relation R with $\langle s, t \rangle \in R$.*

Example 2. For the following LTSs we have, for example, $s_0 \approx t_0$ since $R = \{\langle s_0, t_0 \rangle, \langle s_0, t_2 \rangle, \langle s_1, t_1 \rangle, \langle s_1, t_3 \rangle\}$ is a bisimulation.

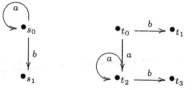

Remark 1. Instead of comparing states in two systems $\langle S, A, \alpha \rangle$ and $\langle T, A, \beta \rangle$ we can always consider one joined system $\langle S + T, A, \gamma \rangle$ with $\gamma(s) = \alpha(s)$ for $s \in S$ and $\gamma(t) = \beta(t)$ for $t \in T$. Therefore bisimulation can be defined as a relation on the set of states of a system. Furthermore, if $R \subseteq S \times S$ is a bisimulation, then it is reflexive and symmetric, and the transitive closure of R is also a bisimulation. Hence bisimilarity $\approx$ is not affected by the choice of defining bisimulation as an equivalence.

Definition 8. *An equivalence relation R on a set of states S of a LTS is an equivalence bisimulation if for all $\langle s, t \rangle \in R$ and all $a \in A$*

> *if $s \xrightarrow{a} s'$ then $\exists t' \in S: \ t \xrightarrow{a} t', \ \langle s', t' \rangle \in R$*

The states s and t are called bisimilar, *denoted by $s \approx_e t$ if there exists an equivalence bisimulation R with $\langle s, t \rangle \in R$.*

By Remark 1, the following proposition holds.

Proposition 2. *Let $\langle S, A, \alpha \rangle$ and $\langle T, A, \beta \rangle$ be two LTSs, and let $s \in S$, $t \in T$. Then $s \approx t$ if and only if $s \approx_e t$.*

$\square$

Bisimulation on **DA** is defined exactly the same as for **NA** i.e. with Definition 8.

The standard notion of probabilistic bisimulation is the one introduced by Larsen and Skou [61] originally formulated for reactive systems (see next subsection). An early reference to probabilistic bisimulation can be found in [23]. In the case of Markov chains, bisimulation corresponds to ordinary lumpability of Markov chains [57, 44, 27]. In [86, 85] it is shown that the concrete notion of bisimulation for Markov-chains coincides with a general coalgebraic notion of bisimulation [68, 53, 74, 64].

The idea behind probabilistic bisimulation is as follows. Since bisimilar states are considered "the same", it does not matter which element within a bisimulation class is reached. Hence, a bisimulation relation should compare the probability to reach an equivalence class and not the probability to reach a single state. In order to define bisimulation for Markov chains the lifting of a relation on a state S to a relation on $\mathcal{D}(S)$, as defined in Definition 1 and explained with Proposition 1, is used. Note that the comments of Remark 1 are in place here as well.

Definition 9. *An equivalence relation R on a set of states S of a Markov chain $\langle S, \alpha \rangle$ is a bisimulation if and only if for all $\langle s, t \rangle \in R$*

if $s \rightsquigarrow \mu$ then there is a transition $t \rightsquigarrow \mu'$ with $\mu \equiv_R \mu'$.

The states s and t are called bisimilar, *denoted by $s \approx t$, if there exists a bisimulation R with $\langle s, t \rangle \in R$.*

Definition 9 will be used, with some variations, for defining bisimulation relations for all types of probabilistic automata that we consider in this overview. However, note that in the case of Markov chains any two states of any two Markov chains are bisimilar, according to the given definition, since $\nabla = S \times S$ is a bisimulation on the state set of any Markov chain $\langle S, \alpha \rangle$. Namely, let $\langle S, \alpha \rangle$ be a Markov chain and $s, t \in S$, such that $\alpha(s) = \mu, \alpha(t) = \mu'$, i.e., $s \rightsquigarrow \mu, t \rightsquigarrow \mu'$. Then for the only equivalence class of ∇, S, we have $\mu[S] = 1 = \mu'[S]$ i.e. $\mu \equiv_R \mu'$ which makes $s \approx t$. This phenomenon can be explained with the fact that bisimilarity compares the observable behavior of two states in a system and the Markov chains are very simple systems in which there is not much to observe. Therefore the need comes to enrich Markov chains with actions or at least termination.

Notation. In Section 3 we will introduce ten other types of probabilistic automata, with corresponding notions of bisimulation. In order to avoid repetition we collect the following.

- A type of automata will always be a triple $\langle S, A, \alpha \rangle$ where S is a set of states, A is a set of actions and α is a transition function. The difference between the system types is expressed with the difference in the codomains of the corresponding transition functions.
- A bisimulation relation will always be defined as an equivalence on the set of states of a system. Depending on the type of systems the "transfer conditions" in the definition of bisimulation vary.
- For a particular type of system, the bisimilarity relation, denoted by $\approx$ is defined by: $s \approx t$ if and only if there exists a bisimulation R that relates s and t, i.e. $\langle s, t \rangle \in R$. Although we use the same notation $\approx$ for bisimilarity in different types of systems, it should be clear that for each type of systems, $\approx$ is a different relation.

2.3   Parallel Composition of LTSs and MCs

Compositional operators serve the need of modular specification and verification of systems. They arise from process calculi, such as CCS ([66]), CSP ([47]) and ACP ([19]), where process terms (models of processes) are built from atomic process terms with the use of compositional operators. Usually a model of a process calculi is a suitable class of transition systems. Therefore it is often the case that process terms are identified with their corresponding transition systems, and the compositional operators of the process calculus can be considered as operators for combining transition systems. In this overview we focus on the parallel composition operator. The definition of parallel composition varies a lot throughout different process calculi. In this section we consider the non-probabilistic case (LTSs) in order to explain variants of different parallel compositions, and the parallel composition of Markov chains in order to present the basics of probabilistic parallel composition.

Labelled Transition Systems. A major distinction between different parallel composition operators is whether they are *synchronous*, where the components are forced to synchronize whenever they can, or *asynchronous* where the components can either synchronize or act independently. Furthermore, different approaches for synchronization exist. The result of the parallel composition of two automata $\mathcal{A}_1 = \langle S_1, A, \alpha_1 \rangle$ and $\mathcal{A}_2 = \langle S_2, A, \alpha_2 \rangle$ is an automaton $\mathcal{A}_1 \| \mathcal{A}_2 = \langle S_1 \times S_2, A, \alpha \rangle$ where the definition of α varies. Instead of a pair $\langle s, t \rangle \in S_1 \times S_2$ we will write $s \| t$ for a state in the composed automaton. Throughout this subsection we will use as running example, the parallel composition of the following two automata.

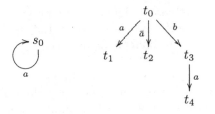

CCS style: The set of actions in this case contains compatible actions $a, \bar{a} \in A$ and a special idle or internal action $\tau \in A$. If one of the automata in state s can perform an action a changing to a state s' and the other one in state t can perform a's compatible action $\bar{a}$ moving to state t' then the composite automaton in state $s\|t$ can perform the idle action τ and move to state $s'\|t'$. Furthermore, independent behavior of each of the automata is possible within the composed automaton.

$s\|t \xrightarrow{a} s'\|t'$ if and only if

1. $s \xrightarrow{b} s', t \xrightarrow{\bar{b}} t', a = \tau$, for b and $\bar{b}$ compatible actions, or
2. $s \xrightarrow{a} s'$ and $t' = t$, or
3. $t \xrightarrow{a} t'$ and $s' = s$.

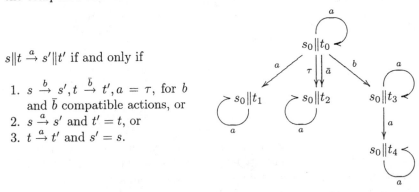

The presented CCS parallel composition is asynchronous. A synchronous variant (SCCS [65]) is defined by omitting clauses 2. and 3. in the definition above.

CSP style: Communication or synchronization in a CSP style parallel composition occurs on a set of synchronizing actions. Thus actions that are intended to synchronize are listed in a set $L \subseteq A$ and the rest of the actions can be performed independently.

$s\|_L t \xrightarrow{a} s'\|t'$ if and only if

1. $s \xrightarrow{a} s'$ and $t \xrightarrow{a} t'$ and $a \in L$, or
2. $s \xrightarrow{a} s', t = t'$ and $a \notin L$, or
3. $t \xrightarrow{a} t', s = s'$ and $a \notin L$.

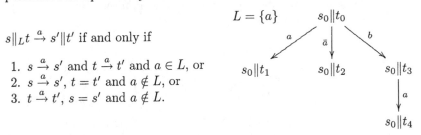

This type of parallel composition operator is synchronous for $L = A$, expresses only interleaving (shuffling) composition if $L = \emptyset$ and is never fully asynchronous with both independent behavior and communication allowed. An asynchronous CSP style parallel composition can be defined by omitting the clause "$a \notin L$"

in clauses 2. and 3. above. In case of different action sets A_1 and A_2, of the two component automata, L is taken to be a subset of $A_1 \cap A_2$. If $L = A_1 \cap A_2$ then we say that synchronization on common actions occurs.

ACP style: In ACP, parallel composition is fully asynchronous, allowing both interleaving (independent behavior) and synchronization via a communication function. A communication function is a commutative and associative partial function $\gamma : A \times A \hookrightarrow A$. Instead of $\gamma(a,b)$ we will write ab.

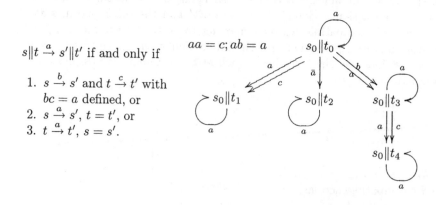

$s\|t \xrightarrow{a} s'\|t'$ if and only if

$aa = c; ab = a$

1. $s \xrightarrow{b} s'$ and $t \xrightarrow{c} t'$ with $bc = a$ defined, or
2. $s \xrightarrow{a} s'$, $t = t'$, or
3. $t \xrightarrow{a} t'$, $s = s'$.

Note that if A contains compatible actions and an idle action τ, and if $a\bar{a} = \tau$ for any compatible $a, \bar{a} \in A$ and undefined otherwise, then the ACP parallel composition operator specializes to the CCS parallel composition operator. On the other hand, for $aa = a$, ($a \in L \subseteq A$) we get the asynchronous variant of the CSP parallel composition operator. If clauses 2. and 3. are dropped from the definition, we get a synchronous variant of the ACP parallel composition operator called communication merge.

Markov Chains. Let $\mathcal{M}_1 = \langle S_1, \alpha_1 \rangle, \mathcal{M}_2 = \langle S_2, \alpha_2 \rangle$ be two Markov chains. Their parallel product is the Markov chain $\mathcal{M}_1\|\mathcal{M}_2 = \langle S_1 \times S_2, \alpha \rangle$, where $\alpha(s\|t) = \alpha_1(s) \times \alpha_2(t)$, $\times$ denoting the product of distributions. Hence $s\|t \rightsquigarrow \mu$ if and only if $s \rightsquigarrow \mu_1, t \rightsquigarrow \mu_2$ and $\mu = \mu_1 \times \mu_2$.

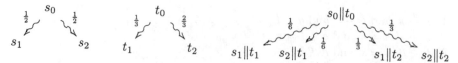

Note that the parallel composition of two Markov chains is synchronous, since each step in the composed automaton consists of independent steps performed by each of the components. The way of defining the product of two distributions goes in favor of the interpretation that when put in parallel, each of the automata independently chooses its transition that contributes to a transition in the composed automaton.

3   Probabilistic Models

This section defines the advanced types of probabilistic automata. The automata types are grouped in several subsections reflecting their common properties. Basically, every type of probabilistic automata arises from the plain definition of a transition system with or without labels. Probabilities can then be added either to every transition, or to transitions labelled with the same action, or there can be a distinction between probabilistic and ordinary (non-deterministic) states, where only the former ones include probabilistic information, or the transition function can be equipped with structure that provides both non-determinism and probability distributions.

Each kind of probabilistic automata comes equipped with a notion of bisimulation, and all these notions, frequently only subtly different, will also find their way in this section.

3.1   Reactive, Generative and I/O Probabilistic Automata

Two classical extensions of LTSs with probabilities are the reactive and the generative model. Throughout the years a large amount of research has been devoted to reactive and generative probabilistic systems. It is hard to note who introduced these systems first, but the reactive model was treated e.g. in [61, 62, 40, 39], the generative in e.g. [40, 39, 42, 50, 30, 29, 28], and the classification of these systems together with a so-called stratified model was proposed in [39, 40].

The way these models arise from LTSs, by changing the transition function, can be explained with the following figure, where α denotes the transition function of a LTS, α_r and α_g the transition function of a reactive and a generative system, respectively.

Definition 10. *A reactive probabilistic automaton is a triple $\langle S, A, \alpha \rangle$ where the transition function is given by*

$$\alpha : S \to (\mathcal{D}(S) + 1)^A.$$

If $s \in S$ and $\alpha(s)(a) = \mu_a$ then we write $s \xrightarrow{a} \mu_a$. More specifically, if $s' \in spt(\mu_a)$, $\mu_a(s') = p$ we write $s \xrightarrow{a[p]} s'$.

A *generative probabilistic automaton* is a triple $\langle S, A, \alpha \rangle$ with a transition function

$$\alpha : S \to \mathcal{D}(A \times S) + 1.$$

When $s \in S$ and $\alpha(s) = \mu \in \mathcal{D}(A \times S)$ then we write $s \rightsquigarrow \mu$. More particularly, if $\langle a, s' \rangle \in spt(\mu)$ with $\mu(\langle a, s' \rangle) = p$ we write $s \overset{a[p]}{\rightsquigarrow} s'$. We use $s \not\rightsquigarrow$ to denote that $\alpha(s) = *$.

Remark 2. In Definition 10 both uses of the special singleton set 1 appear. The first one, as in Definition 5 helps expressing partial functions. The second one, in the definition of generative transition function, expresses the possibility of termination. If s is a state in a generative system with $\alpha(s) = *$ then s is a terminating state allowing no transition. For LTSs, termination is allowed by the fact that $\emptyset \in \mathcal{P}(A \times S)$. Hence, when changing from subsets to distributions, $*$ is added to play the role of the $\emptyset$.

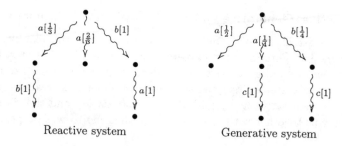

Reactive system   Generative system

In a reactive system probabilities are distributed over the outgoing transitions labelled with the *same action*, while in a generative system probabilities are distributed over *all* outgoing transitions from a state. A motivation for making this distinction is the different treatment of actions. In a reactive system actions are treated as *input* actions being provided by the environment. When a reactive system receives input from the environment then it acts probabilistically by choosing the next state according to a probability distribution assigned to this input. There are no probabilistic assumptions about the behavior of the environment. On the other hand, in a generative system, as the name suggests, actions are treated as *output* generated by the system. When a generative system is in a state s it chooses the next transition according to the probability distribution $\alpha(s)$ assigned to s. The transition being chosen, the system moves to another state while generating the output action which labels this transition. Note that in a generative system there is no non-determinism present, while in a reactive system there is only external non-determinism, as in **DA**. We denote by **React** and **Gen** the classes of reactive and generative probabilistic automata, respectively.

Definition 11. *An equivalence relation R on S is a bisimulation on the reactive probabilistic automaton $\langle S, A, \alpha \rangle$ if for all $\langle s, t \rangle \in R$ and for all actions $a \in A$:*

if $s \overset{a}{\rightsquigarrow} \mu$ then there exists a distribution μ' with $t \overset{a}{\rightsquigarrow} \mu'$ and $\mu \equiv_R \mu'$.

In order to state the definition of bisimulation for generative systems, the lifting from Definition 2 is used.

Definition 12. *An equivalence relation R on S is a bisimulation on the generative probabilistic automaton $\langle S, A, \alpha \rangle$ if for all $\langle s, t \rangle \in R$:*

if $s \rightsquigarrow \mu$ then there exists a distribution μ' with $t \rightsquigarrow \mu'$ and $\mu \equiv_{R,A} \mu'$.

Example 3. The equivalence relation R generated by the pairs $\langle C, D \rangle$, $\langle H, 1 \rangle$, $\langle H, 3 \rangle$, $\langle H, 5 \rangle$, $\langle T, 2 \rangle$, $\langle T, 4 \rangle$, $\langle T, 6 \rangle$ is a bisimulation for the probabilistic automaton given below. Hence, $C \approx D$. Note that this particular automaton belongs to both **React** and **Gen**.

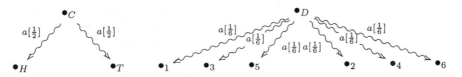

An intuitive interpretation of this example is obtained by adding meaning "flip" to the action a in the left sub-automaton and a meaning "roll" to the action a in the right sub-automaton. Then the state C represents flipping of a fair coin, and the state D represents rolling a fair dice. The bisimilarity of the states C and D shows that it is the same whether one flips a fair coin or rolls a fair dice being interested only in whether the outcome is odd or even.

I/O Probabilistic Automata. The model of input/output probabilistic automata, introduced by Wu, Smolka and Stark in [87], exploiting the input/output automata by Lynch and Tuttle, (cf. [63]), presents a combination of the reactive and the generative model.

Definition 13. *An input/output probabilistic automaton is a triple $\langle S, A, \alpha \rangle$ where*

1. *the set of actions A is divided into input and output actions, $A = A^{in} + A^{out}$;*
2. *$\alpha : S \rightarrow \mathcal{D}(S)^{A^{in}} \times (\mathcal{D}(A^{out} \times S) + 1) \times \mathbb{R}_{\geq 0}$ is the transition function.*

The third component in the transition function assigns an output delay rate to each state. If $s \in S$, then $\alpha(s) = \langle f^{in}, \mu^{out}, \delta_s \rangle$. We have that $\delta_s = 0$ iff $\mu^{out} = *$ i.e. delay is assigned only to states that generate output.

Denote the class of I/O automata by **IO**. We use a similar notation for transitions as in the reactive and the generative model. If $s \in S$ with $\alpha(s) = \langle f^{in}, \mu^{out}, \delta_s \rangle$ then

- if $a \in A^{in}$ with $f^{in}(a) = \mu_a$ we write $s \xrightarrow{a} \rightsquigarrow \mu_a$, furthermore, if $s' \in spt(\mu_a)$ with $\mu_a(s') = p$ we write $s \overset{a[p]}{\rightsquigarrow} s'$.
- if $\mu^{out} \neq *$ we write $s \rightsquigarrow \mu^{out}$ and if $\mu^{out}(a, s') = p > 0$ we write $s \overset{a[p]}{\rightsquigarrow} s'$.

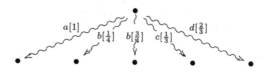

transitions from a state in an I/O probabilistic automaton
$$A^{in} = \{a, b\}, \ A^{out} = \{c, d\}$$

In an I/O automaton for every input action there is a reactive transition. Note that f^{in} is always a function and not a partial function as in the reactive model. Hence each input action is enabled in each state of an I/O probabilistic automaton. The output actions are treated generatively. At most one generative probabilistic transition gives the output behavior of each state. The delay rate parameter δ_s is an aspect from continuous-time systems, and its meaning will become clear in section 4 when we discuss compositions of I/O automata.

The I/O automata will not be compared and placed in the hierarchy of section 5 since they involve a continuous element. It is obvious that, when ignoring the 0 delays, for $A^{out} = \emptyset$ one gets the reactive model (with all actions enabled) and for $A^{in} = \emptyset$ one gets the generative model with a delay rate assigned to each state. A connection exists between I/O automata and some models with structured transition relation (section 3.3). Combined systems similar to I/O automata appear as models of process terms in the process algebra EMPA [14, 15].

Since we do not compare I/O automata in Section 4, we do not need a notion of bisimulation for them, although it can be defined by combining the transfer conditions for reactive and generative bisimulation, and taking care of the delay rate. In [87] no notion of bisimulation is introduced, instead a different notion of behavior of I/O automata is considered. A definition of bisumulation for I/O automata can be found in [75].

3.2   Automata with Distinction Between States

So far we have seen some types of automata that allow modelling of probabilistic behavior, but none of those has the capability of also modelling full non-determinism. The types of systems introduced in a minute allow full non-determinism while making a distinction between probabilistic states with outgoing probabilistic transitions, and non-deterministic states with action labelled transitions.

Stratified Probabilistic Automata. The simplest system with a distinction on states appears under the name of stratified probabilistic automaton, and is discussed in [39, 40, 79, 49]. Stratified automata do not yet allow any form of non-determinism although there is a distinction on states.

Definition 14. *A stratified probabilistic automaton is a triple* $\langle S, A, \alpha \rangle$ *where the transition function α is given by*

$$\alpha : S \to \mathcal{D}(S) + (A \times S) + 1$$

The class of all stratified automata we denote by **Str**. Due to the disjoint union in the codomain of the transition function, there are three types of states in a stratified automaton: probabilistic states consisting of $s \in S$ such that $\alpha(s) \in \mathcal{D}(S)$, deterministic states $s \in S$ for which $\alpha(s) = \langle a, s' \rangle$ allowing a single action labelled transition and terminating states $s \in S$ with $\alpha(s) = *$.

Definition 15. *An equivalence relation R on S is a bisimulation on the stratified probabilistic automaton $\langle S, A, \alpha \rangle$ if for all $\langle s, t \rangle \in R$:*

1. *if $s \rightsquigarrow \mu$ then there exists a distribution μ' with $t \rightsquigarrow \mu'$ and $\mu \equiv_R \mu'$;*
2. *if $s \xrightarrow{a} s'$ then there exists t' such that $t \xrightarrow{a} t'$ and $\langle s', t' \rangle \in R$.*

Vardi Probabilistic Automata. One of the earliest models of probabilistic automata was introduced by Vardi in [84] under the name *concurrent Markov chains*. The original definition of a concurrent Markov chain was given in terms of state labelled transition systems, for purposes of verification of logical properties. Therefore we slightly modify the definition, calling this class of automata Vardi probabilistic automata.

Definition 16. *A Vardi probabilistic automaton is a triple $\langle S, A, \alpha \rangle$ where the transition function α is given by*

$$\alpha : S \to \mathcal{D}(A \times S) \cup \mathcal{P}(A \times S)$$

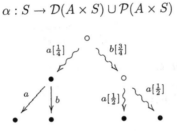

Vardi probabilistic automaton

Remark 3. Note that $\cup$ is used in Definition 16 rather than $+$. One could consider the union disjoint, but it is of more use to identify $\mu^1_{\langle a, s' \rangle}$ with the singleton $\{\langle a, s' \rangle\}$, i.e. a state with a transition $s \xrightarrow{a[1]} s'$ can be identified with a state allowing only one transition $s \xrightarrow{a} s'$.

In Vardi automata, the probabilistic states are of a generative kind, while the other states are non-deterministic with full non-determinism, as in an LTS. Therefore, the definition of bisimulation is a combination of Definition 8 and Definition 12.

Definition 17. *An equivalence relation R on S is a bisimulation on the Vardi probabilistic automaton $\langle S, A, \alpha \rangle$ if for all $\langle s, t \rangle \in R$:*

1. *if $s \rightsquigarrow \mu$ then there exists a distribution μ' with $t \rightsquigarrow \mu'$ and $\mu \equiv_{R,A} \mu'$;*
2. *if $s \xrightarrow{a} s'$ then there exists t' such that $t \xrightarrow{a} t'$ and $\langle s', t' \rangle \in R$.*

Remark 4. We note that in the literature, in particular in [84], there is no definition of bisimulation. However, the current understanding of probabilistic bisimulation, and the concept of a general coalgebraic definition of bisimulation allows us to state the previous definition.

We denote the class of Vardi probabilistic automata by **Var**.

The Alternating Models of Hansson. Another model that treats separately (purely) probabilistic and non-deterministic states is the alternating model introduced by Hansson, see for example [41, 46]. We present the class of alternating probabilistic automata **Alt**, its subclass of strictly alternating probabilistic automata **SA** and, in turn, two subclasses of **SA**, denoted by $\mathbf{SA}_n$ and $\mathbf{SA}_p$.

Definition 18. *An alternating probabilistic automaton is a triple $\langle S, A, \alpha \rangle$ where*

$$\alpha : S \to \mathcal{D}(S) + \mathcal{P}(A \times S).$$

The class of alternating automata is denoted by **Alt**. *Denote by N and P the subsets of S containing non-deterministic and probabilistic states, respectively.*

A strictly alternating automaton is an alternating automaton where for all $s \in S$ the following holds:

1. *if $s \in P$ with $\alpha(s) = \mu \in \mathcal{D}(S)$ then $\mathrm{spt}(\mu) \subseteq N$;*
2. *if $s \in N$ then for all $\langle a, s' \rangle \in \alpha(s)$, $s' \in P$.*

The class of all strictly alternating automata is denoted by **SA**.

An automaton of **SA** *belongs to* $\mathbf{SA}_n$ *if and only if*

$$\forall s \in S \colon (\forall s' \in S, \forall a \in A, \forall p \in [0,1] \colon s' \overset{p}{\not\rightarrow} s \wedge s' \overset{a}{\not\rightarrow} s) \Rightarrow s \in N. \qquad (1)$$

An automaton of **SA** *belongs to* $\mathbf{SA}_p$ *if and only if*

$$\forall s \in S \colon (\forall s' \in S, \forall a \in A, \forall p \in [0,1] \colon s' \overset{p}{\not\rightarrow} s \wedge s' \overset{a}{\not\rightarrow} s) \Rightarrow s \in P. \qquad (2)$$

The well known of these classes are the class **SA** [41, 46] and the class $\mathbf{SA}_n$ [5, 6], but we have chosen for presenting all these classes structurally. The class **Alt** is a slight generalization of the class **SA** and is very much similar to the stratified and Vardi models. Therefore it deserves its place in this overview. In an alternating automaton only a distinction on states is imposed. In the strictly alternating model it is required that all successors of a non-deterministic state are probabilistic states and vice versa. Furthermore, the two subclasses $\mathbf{SA}_n$ and $\mathbf{SA}_p$ take care that any "initial state" is non-deterministic (1) and probabilistic (2), respectively. We define the subclasses $\mathbf{SA}_n$ and $\mathbf{SA}_p$ in order to make a precise comparison of the class **SA** with some of the other models (section 5).

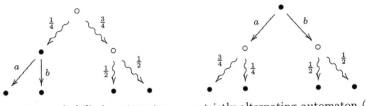

alternating probabilistic automaton   strictly alternating automaton ($\mathbf{SA}_n$)

One definition of bisimulation fits all the introduced classes of alternating automata, where the transfer conditions are exactly the same as for the stratified model, given in Definition 15.

3.3   Probabilistic Automata with Structured Transition Function

In this subsection we focus on three types of probabilistic automata that provide orthogonal coexistence of full non-determinism and probabilities without distinguishing between states.

Segala and Simple Segala Probabilistic Automata. Two types of probabilistic automata were introduced by Segala and Lynch in [78, 76]. We call them Segala probabilistic automata and simple Segala probabilistic automata. An extensive overview of the simple Segala model is given in [80, 81] and they have been used for verification purposes and developing theoretical results in several situations as reported in [82, 83, 24, 8, 13, 21, 20, 55, 56].

Definition 19. *A Segala probabilistic automaton is a triple $\langle S, A, \alpha \rangle$ where*

$$\alpha : S \to \mathcal{P}(\mathcal{D}(A \times S))$$

If $s \in S$ such that $\mu \in \alpha(s)$ we write $s \to\rightsquigarrow \mu$, and if $\langle a, s' \rangle \in spt(\mu)$ with $\mu(a, s') = p$ then we write $s \xrightarrow{a[p]}\rightsquigarrow s'$.

A simple Segala probabilistic automaton[1] is a triple $\langle S, A, \alpha \rangle$ for a transition function

$$\alpha : S \to \mathcal{P}(A \times \mathcal{D}(S))$$

If $s \in S$ with $\langle a, \mu \rangle \in \alpha(s)$ then we write $s \xrightarrow{a}\rightsquigarrow \mu$, and if $s' \in spt(\mu)$ we write $s \xrightarrow{a\ p}\rightsquigarrow s'$.

The simple Segala type of systems arise from **NA** by changing the target state with a distribution over possible target states. A transition in a simple Segala automaton and in a Segala automaton is shown in the next figure.

---

[1] Segala and Lynch call these models probabilistic automaton (PA) and simple probabilistic automaton, while Stoelinga calls them general PA and PA, respectively.

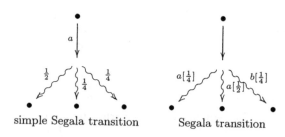

simple Segala transition   Segala transition

There can be more then one transition available in a state and that is where non-determinism occurs. Hence, the non-deterministic choices exist between transitions, while the probabilities are specified within a transition. In the original definition by Segala and Lynch distributions over an extended set $A \times S + 1$ (or over $S + 1$ in the simple case) were treated i.e. substochastic distributions, where the probability assigned to the special symbol $*$ was interpreted as the deadlock probability. We choose not to include this in the definition for two reasons: it disturbs the comparison (Section 4) since the other models do not include substochastic distributions, and it can be imitated by adding an extra deadlock state to a system.

We denote the class of Segala probabilistic automata by **Seg** and the class of simple Segala automata by **SSeg**.

The simple Segala automaton is a generalization towards full non-determinism of the reactive model and of the purely probabilistic automata of Rabin [73]. A deterministic version of the simple Segala automaton equivalent to the reactive model is known as *Markov decision process* ([34]), while the name *probabilistic transition system* is used for this model in [52] and for a state labelled version in [36, 37]. A comparison of $\mathbf{SA}_n$ and the simple Segala model can be found in [25].

Bisimulation for the simple Segala systems is defined with the same transfer conditions as for reactive systems given in Definition 11, while for the Segala systems the transfer conditions for bisimulation of Definition 12 for generative systems apply, when changing $\rightsquigarrow$ to $\longrightarrow\rightsquigarrow$.

A great novelty introduced with both types of Segala systems was the definition of a stronger probabilistic bisimulation relation that identifies states that have matching "combined transitions". For more information on this topic the interested reader is referred to [78, 76, 80, 81, 24].

Bundle Probabilistic Automata. Another way to include both non-determinism and probability is to consider distributions over sets of transitions as in the *bundle* model, introduced in [35]. (Recall that Segala systems have sets of distributions over transitions.)

Definition 20. *A bundle probabilistic automaton is a triple $\langle S, A, \alpha \rangle$ where*

$$\alpha : S \to \mathcal{D}(\mathcal{P}(A \times S)) + 1$$

When $s \in S$ and $\alpha(s) = \mu$ we write $s \rightsquigarrow \mu$, furthermore, if $T \subseteq A \times S$, $\mu(T) = p > 0$ we write $s \overset{p}{\rightsquigarrow} T$ and if $\langle a, t \rangle \in T$ then $s \overset{p}{\rightsquigarrow} \overset{a}{\rightarrow} t$.

The bundle model can be considered as generative, since probabilities are also distributed over actions. Therefore the bundle model offers a solution to the absence of non-determinism in the generative setting. Note that the original definition is even slightly more general, namely the codomain of the transition function is $\mathcal{D}(\mathcal{M}(A \times S))$ where $\mathcal{M}(X)$ denotes all the (finite) multi-subsets of a set X. Hence it is possible to have multiple transitions from one state to another with the same action within one bundle. Since it is not essential for the material presented here, we will not add multi-sets in the bundle model. The class of bundle probabilistic automata is denoted by **Bun**. A typical bundle probabilistic automaton is depicted below:

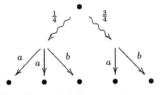

bundle probabilistic automaton

In the literature, in particular in [35], there is no definition of bisimulation on bundle probabilistic automata, instead they are transformed to generative systems and then compared with generative bisimulation. We give here a definition of bisimulation for the bundle probabilistic automata that is deduced from the general coalgebraic definition of bisimulation (cf. [53, 74, 68]). A justification for doing so is that all previously stated definitions of bisimulation which were based on the probabilistic bisimulation of Larsen and Skou [61] coincide with the general coalgebraic definition of bisimulation for the particular type of systems. In the non-probabilistic case this coincidence is well known (see e.g. [74]). For Markov chains it was proven in [86], for the Segala probabilistic automata in [26] and the same proof technique extends to all other cases.

Prior to stating the definition we need a way to lift a relation on a set S to a relation on the set $\mathcal{P}(A \times S)$.

Definition 21. *Let R be a relation on S and let X, $Y \in \mathcal{P}(A \times S)$. Define $X \equiv_{R,\mathcal{P}} Y$ if and only if for all $a \in A$:*

1. *if $\langle a, x \rangle \in X$ then there exists $\langle a, y \rangle \in Y$ with $\langle x, y \rangle \in R$;*
2. *if $\langle a, y \rangle \in Y$ then there exists $\langle a, x \rangle \in X$ with $\langle x, y \rangle \in R$.*

It holds that, if R is an equivalence on S, then $\equiv_{R,\mathcal{P}}$ is an equivalence on $\mathcal{P}(A \times S)$.

Definition 22. *An equivalence relation R is a bisimulation on the state set of a bundle probabilistic automaton $\langle S, A, \alpha \rangle$ if for all $\langle s, t \rangle \in R$ it holds*

$$\text{if } s \rightsquigarrow \mu \text{ then there exists } \mu' \text{ such that } t \rightsquigarrow \mu' \text{ and } \mu \equiv_{\equiv_{R,\mathcal{P}}} \mu'$$

where $\equiv_{\equiv_{R,\mathcal{P}}}$ denotes the lifting of the relation $\equiv_{R,\mathcal{P}}$ to distributions on $\mathcal{P}(A \times S)$ as defined by Definition 1.

3.4   Complex Models – Pnueli-Zuck and General Probabilistic Automata

An early model including probabilities and a structured transition relation was proposed by Pnueli and Zuck [71, 72] under the name *finite-state probabilistic programs* and later used in [7]. We call this type of automata Pnueli-Zuck probabilistic automata, and denote the class of all such by $\mathbf{PZ}^2$. The model of Pnueli and Zuck has the most complex transition function, it adds one more power set to the bundle model and so allows two types of non-determinism, both between the probabilistic transitions and inside the transitions. However, in order to get a top element for our hierarchy (section 5) we expand the model a bit further and define one most general type of probabilistic automata. The class of such will be denoted by $\mathbf{MG}$.

Definition 23. *A Pnueli-Zuck automaton is a triple $\langle S, A, \alpha \rangle$ where*

$$\alpha : S \to \mathcal{P}(\mathcal{D}(\mathcal{P}(A \times S)))$$

When $s \in S$ and $\mu \in \alpha(s)$ we write $s \to\!\!\rightsquigarrow \mu$, further on, if $T \subseteq A \times S$, $\mu(T) = p > 0$ we write $s \to\!\!\overset{p}{\rightsquigarrow} T$ and if $\langle a, t \rangle \in T$ then $s \to\!\!\overset{p}{\rightsquigarrow}\overset{a}{\to} t$. A general probabilistic automaton is a triple $\langle S, A, \alpha \rangle$ where

$$\alpha : S \to \mathcal{P}(\mathcal{D}(\mathcal{P}(A \times S + S)))$$

The notation for Pnueli-Zuck automata is also used for general automata. Furthermore, if $s \in S, \mu \in \alpha(s), T \subseteq A \times S + S$ with $\mu(T) = p > 0$ and $t \in T$, then we write $s \to\!\!\overset{p}{\rightsquigarrow} \to t$.

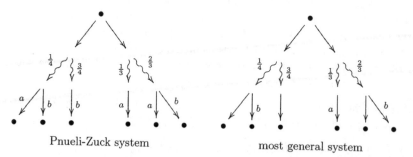

Pnueli-Zuck system   most general system

The unlabelled transitions which appear in the right figure (most general system) correspond to pure probabilistic transitions in Markov chains or alternating systems, where a change of state can happen with certain probability without performing an action.

As for bundle systems, there is no notion of bisimulation for Pnueli-Zuck systems in the literature. A bisimulation definition can be formulated out of the

---

[2] Like Vardi's model, these automata appear in the literature in a state labelled version for model checking purposes. Therefore we change the definition towards transition labels.

general coalgebraic definition, and it leads the same transfer conditions as in Definition 22 when changing $\leadsto$ to $\rightarrow\leadsto$. A small modification is needed for the general probabilistic automata.

4 Composing Probabilistic Systems in Parallel

Having introduced the probabilistic models, we consider possible definitions of the parallel composition operator for these extended systems. Lots of results on this topic exist in the literature. For a broad overview the reader is referred to [35, 9]. An overview of probabilistic process algebras covering other probabilistic operators as well is presented in [58].

For the classes **MC** and **IO** there is a unique parallel composition defined. In **MC** this operation is purely synchronous given by the product of distributions (cf. Section 2.3), whereas in **IO** the definition of the parallel composition operation strongly relies on the specific structure of the systems (cf. Section 4.3 below). For all other classes it is meaningful to consider various definitions of parallel composition. Such operations might be synchronous or asynchronous in nature and moreover might be based upon the styles CCS, CSP and ACP described in Section 2. The style CSP plays a special role in this respect since it is by its definition partly synchronous and partly asynchronous and hence gives rise to a somehow mixed variant of parallel composition.

The classes of (probabilistic) systems can be divided into three groups dependent on whether they show reactive, generative or alternating behavior. Classes belonging to the same of these groups allow in essence similar definition and investigation of parallel composition.

Instead of going through all, obviously quite numerous, variants of parallel composition for each single class of systems, we shall in the subsequent sections 4.1, 4.2 and 4.4. discuss a couple of instructive cases in some detail. However, let us give a complete scheme of possible (and/or already studied) definitions of parallel composition operator by means of a comprehensive table. In the table below each column is dedicated to one class of probabilistic automata, and each row to one of the introduced styles of parallel composition. In the intersecting cells a symbol representing the definability status of the corresponding parallel composition operator in the corresponding class is placed. Neighboring cells containing the same symbol within one column are merged. We use the following symbols:

$+$:   defined in the literature or straightforward
$+$:   definable but not carried out
$-$:   not definable
p: defined in the literature with parameters
p: parameterized version definable, but not carried out
n: normalized version definable, but not carried out
$+/$n: "$+$" for total communication function, "n" otherwise

Table 1 presents the overall situation concerning definability of parallel composition on probabilistic automata. A brief analysis of these summary results shows that allowing full non-determinism enables definition of any type of parallel composition.

Table 1. Definability of ∥

		DA	NA	React	SSeg	Gen	Seg	Bun	PZ	MG	Var	Alt, SA,..	Str
		- - - reactive - - -				- - - - - generative - - - - -						- - - alternating - - -	
sync	CCS	−	−			n	n				n		
	CSP	+	+	+	+	n	n	+	+	+	n	+	+
	ACP	−		−		+/n	+/n				+/n		
	CSP	−	+	−	+	p	p	+	+	+	p	+	−
async	CCS												
	CSP	−	+	−	+	p	p	+	+	+	p	+	−
	ACP												

4.1 Parallel Composition in the Reactive Setting

Systems with reactive behavior are systems in the classes **React** and **SSeg**, as well as **NA** and **DA** in the non-probabilistic case. Any parallel composition operator on LTS (Section 2.3.) nicely extends to the class **SSeg**. Let $\mathcal{A}_1 = \langle S_1, A, \alpha_1 \rangle, \mathcal{A}_2 = \langle S_2, A, \alpha_2 \rangle$ be in **SSeg**. Then $\mathcal{A}_1 \| \mathcal{A}_2 = \langle S_1 \times S_2, A, \alpha \rangle$ where α is defined as follows:

[*CCS style*]: $s\|t \xrightarrow{a}_{\rightsquigarrow} \mu$ if and only if

1. $a = \tau$, $s \xrightarrow{b}_{\rightsquigarrow} \mu_1$, $t \xrightarrow{\bar{b}}_{\rightsquigarrow} \mu_2$ and $\mu = \mu_1 \times \mu_2$, or
2. $s \xrightarrow{a}_{\rightsquigarrow} \mu_1$ and $\mu = \mu_1 \times \mu_t^1$, or
3. $t \xrightarrow{a}_{\rightsquigarrow} \mu_2$ and $\mu = \mu_s^1 \times \mu_2$.

[*CSP style*]: $s\|_L t \xrightarrow{a}_{\rightsquigarrow} \mu$ if and only if

1. $a \in L$, $s \xrightarrow{a}_{\rightsquigarrow} \mu_1$, $t \xrightarrow{a}_{\rightsquigarrow} \mu_2$ and $\mu = \mu_1 \times \mu_2$, or
2. $a \notin L$, $s \xrightarrow{a}_{\rightsquigarrow} \mu_1$ and $\mu = \mu_1 \times \mu_t^1$, or
3. $a \notin L$, $t \xrightarrow{a}_{\rightsquigarrow} \mu_2$ and $\mu = \mu_s^1 \times \mu_2$.

[*ACP style*]: $s\|t \xrightarrow{a}_{\rightsquigarrow} \mu$ if and only if

1. $a = bc$ defined, $s \xrightarrow{b}_{\rightsquigarrow} \mu_1$, $t \xrightarrow{c}_{\rightsquigarrow} \mu_2$ and $\mu = \mu_1 \times \mu_2$, or
2. $s \xrightarrow{a}_{\rightsquigarrow} \mu_1$ and $\mu = \mu_1 \times \mu_t^1$, or
3. $t \xrightarrow{a}_{\rightsquigarrow} \mu_2$ and $\mu = \mu_s^1 \times \mu_2$.

The definition of any of these operators is problematic for the class **React**. For $\mathcal{A}_1, \mathcal{A}_2 \in$ **React** it might happen that $\mathcal{A}_1 \| \mathcal{A}_2 \notin$ **React** in any variant of parallel composition. Even in the synchronous CCS style, multiple transitions labelled with τ may appear. In the CSP style, 2. and 3. may introduce internal non-determinism. However, if L contains all the common actions of $\mathcal{A}_1$ and $\mathcal{A}_2$, then this problem disappears. In case of ACP all of 1., 2. and 3. introduce internal non-determinism, hence **React** is not closed under this operator for any arbitrary communication function γ^3. For example, if $ab = ac = a$ then the ACP parallel product of the following two automata

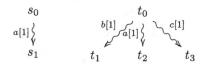

is not defined in **React**, since the definition yields: $s_0 \| t_0 \overset{a}{\rightarrow} \rightsquigarrow \mu_x^1$ for $x \in \{s_1 \| t_0, s_0 \| t_2, s_1 \| t_1, s_1 \| t_3\}$ i.e. more than one transition corresponds to the action a, which is prohibited in **React**.

An asynchronous parallel composition in CCS style on simple Segala systems was defined in [21], a synchronous parallel composition in CCS style on reactive systems was defined in [40, 39, 52], the last reference working with simple Segala systems. A synchronous CSP style parallel composition is defined for reactive systems in [55, 69], while an asynchronous CSP style parallel composition with synchronization on common actions is used in [78, 76, 80] for simple Segala systems.

4.2   Parallel Composition in the Generative Setting

Systems with generative behavior belong to the classes **Gen**, **Var**, **Seg**, **Bun**, **PZ** and **MG**. The Vardi systems express also alternating behavior and they will be discussed with the alternating systems. A common property of the generative systems is that always probability distributions over actions and states appear. This leads to difficulties in defining parallel composition operators (see [41, 30, 76, 35]), especially in the asynchronous case. Namely, a generative type system defines in each state a probability distribution over a set of enabled actions, offered by the environment. When two such systems are composed in parallel it is not clear how the common set of enabled actions should be defined, nor how the two probability distributions should be composed into one (cf. [52]). In this section we explain several approaches for solving this problem.

Let $\mathcal{A}_1 = \langle S_1, A, \alpha_1 \rangle, \mathcal{A}_2 = \langle S_2, A, \alpha_2 \rangle$ be two generative systems. Their parallel composition in all cases will be denoted by $\mathcal{A}_1 \| \mathcal{A}_2 = \langle S_1 \times S_2, A, \alpha \rangle$, possibly with parameters.

---

[3] The same problems arise in the class **DA** opposed to **NA**, namely parallel composition introduces internal non-determinism, and therefore **DA** is not closed under $\|$.

Synchronous CCS, CSP, ACP style parallel composition can be defined on generative systems, as done in [40, 39, 35] by:

$$s \overset{a[p]}{\rightsquigarrow} s', t \overset{b[q]}{\rightsquigarrow} t' \iff s\|t \overset{ab[pq]}{\rightsquigarrow} s'\|t'$$

where the set of actions is assumed to form a commutative semigroup (PCCS, [38]) and ab stands for the product of a and b in $A(\cdot)$.

The following figure presents an example of synchronous parallel composition of two generative systems.

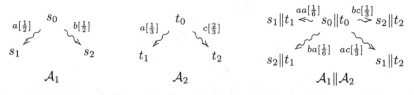

In order to capture possible asynchronous behavior, several parallel composition operators were defined in the literature that use *bias factors*. In most of the cases the composition is not symmetric. Namely, the main problem in defining asynchronous parallel composition is that any definition introduces non-determinism. In the proposed solutions, these non-deterministic choices are changed to probabilistic ones, by specifying parameters of the parallel composition.

An ACP style parallel composition operator for generative systems was defined in [10]. The definition follows the non-probabilistic definition of the ACP parallel operator, while changing the non-deterministic choices introduced by interleaving and/or communication into corresponding probabilistic choices. The operator is parameterized with two parameters σ and θ, denoted by $\mathcal{A}_1\|_{\sigma,\theta}\mathcal{A}_2$. In the product state $s\|_{\sigma,\theta}t$, synchronization between s and t can occur with probability $1 - \theta$ and an autonomous action of either s or t with probability θ. Furthermore, given that an autonomous move occurs, then it comes from s with probability σ and from t with probability $1 - \sigma$. For definability of $\mathcal{A}_1\|_{\sigma,\theta}\mathcal{A}_2$ it is necessary that the communication function is a total function.

We define $s\|_{\sigma,\theta}t \overset{a[P]}{\rightsquigarrow} s'\|_{\sigma,\theta}t'$ if and only if

1. $s \overset{b[p]}{\rightsquigarrow} s'$ and $t \overset{c[q]}{\rightsquigarrow} t'$, $bc = a$ and $P = (1 - \theta)pq$, or
2. $s \overset{a[p]}{\rightsquigarrow} s'$, $t = t'$, and $P = p\theta\sigma$, or
3. $t \overset{a[q]}{\rightsquigarrow} t'$, $s = s'$ and $P = q\theta(1 - \sigma)$, or
4. $s \overset{a[p]}{\rightsquigarrow} s'$, $t\not\rightsquigarrow$, $t = t'$ and $P = p$, or
5. $s\not\rightsquigarrow$, $t \overset{a[q]}{\rightsquigarrow} t'$, $s = s'$ and $P = q$.

Note that by this definition we might get two transitions $s\|_{\sigma,\theta}t \overset{a[p_1]}{\rightsquigarrow} s'\|_{\sigma,\theta}t'$ and $s\|_{\sigma,\theta}t \overset{a[p_2]}{\rightsquigarrow} s'\|_{\sigma,\theta}t'$, which then are replaced by one transition $s\|_{\sigma,\theta}t \overset{a[p_1+p_2]}{\rightsquigarrow} s'\|_{\sigma,\theta}t'$.

For $\mathcal{A}_1$ and $\mathcal{A}_2$ as in the previous figure, we get $\mathcal{A}_1\|_{\sigma,\theta}\mathcal{A}_2$ which looks like:

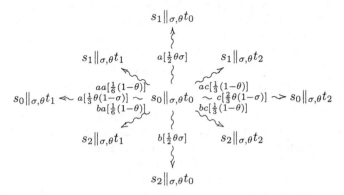

Another two biased, parameterized, parallel composition operators are defined in [35], one asynchronous CCS style operator, denoted by $\mathcal{A}_1^\theta\|^\sigma\mathcal{A}_2$ and one CSP-style operator, denoted by $\mathcal{A}_1\|_L^\sigma\mathcal{A}_2$. Denote by $s \overset{a[p]}{\rightsquigarrow}$ the clause $(\exists s')s \overset{a[p]}{\rightsquigarrow} s'$. We present the definition of $\mathcal{A}_1\|_L^\sigma\mathcal{A}_2$ first:

$s\|_L^\sigma t \overset{a[P]}{\rightsquigarrow} s'\|_L^\sigma t'$ if and only if one of the following is satisfied:

1. $s \overset{a[p]}{\rightsquigarrow} s'$, $t \overset{b[q]}{\rightsquigarrow}$, $a,b \notin L$, $t' = t$ and $P = \frac{pq\sigma}{\nu(s,t,L)}$;

2. $s \overset{b[p]}{\rightsquigarrow}$, $t \overset{a[q]}{\rightsquigarrow} t'$, $a,b \notin L$, $s' = s$ and $P = \frac{pq(1-\sigma)}{\nu(s,t,L)}$;

3. $s \overset{a[p]}{\rightsquigarrow} s'$, $t \overset{b[q]}{\rightsquigarrow}$, $a \notin L, b \in L$, $t' = t$ and $P = \frac{pq}{\nu(s,t,L)}$;

4. $s \overset{b[p]}{\rightsquigarrow}$, $t \overset{a[q]}{\rightsquigarrow} t'$, $a \notin L, b \in L$, $s' = s$ and $P = \frac{pq}{\nu(s,t,L)}$;

5. $s \overset{a[p]}{\rightsquigarrow} s'$, $t \not\rightsquigarrow$, $a \notin L$, $t' = t$ and $P = \frac{p}{\nu'(s,L)}$;

6. $s \not\rightsquigarrow$, $t \overset{a[q]}{\rightsquigarrow} t'$, $a \notin L$, $s' = s$ and $P = \frac{q}{\nu'(t,L)}$;

7. $s \overset{a[p]}{\rightsquigarrow} s'$, $t \overset{a[q]}{\rightsquigarrow} t'$, $a \in L$ and $P = \frac{pq}{\nu(s,t,L)}$.

Where the normalization factors are calculated by

$$\nu'(s,L) = 1 - \sum_{s\overset{a[p]}{\rightsquigarrow},\ a\in L} p, \quad \nu(s,t,L) = 1 - \sum_{s\overset{a[p]}{\rightsquigarrow},\ t\overset{b[q]}{\rightsquigarrow},\ a,b\in L,\ a\neq b} pq.$$

For this CSP style operator only one parameter is needed since the only nondeterminism occurs if both systems autonomously decide to perform actions not in the synchronizing set L. In $s\|_L^\sigma t$, the parameter σ denotes the probability that s performs an autonomous action, given that both s and t have decided not to synchronize. Furthermore, normalization factors are used to determine the actual probability of every transition. These normalization factors redistribute the probability mass that is due to autonomous decisions of both processes that would otherwise lead to deadlock.

For the asynchronous CCS parallel composition $\mathcal{A}_1^\theta\|^\sigma\mathcal{A}_2$ the interpretation of the probabilistic parameters $\theta, \sigma \in (0,1)$ is similar to the ACP approach. They provide the relevant information that an adversary needs in order to resolve non-determinism that arises when composing two systems. In $s^\theta\|^\sigma t$, σ denotes the probability that s performs an autonomous action given that both s and t do not want to synchronize, and θ denotes the probability that some autonomous action occurs, given that synchronization is possible. Hence, if synchronization is possible, it will take place with probability $1 - \theta$.

The earliest biased parallel composition operator for generative systems was defined in [30] and treated in detail in [58]. There the parallel composition $\mathcal{A}_1\|_\rho\mathcal{A}_2 = \langle S_1 \times S_2, A, \alpha \rangle$ uses one bias parameter ρ. A state $s\|_\rho t$ in the composed automaton can either do an action a with a certain probability if both s and t in their components can do an action a (CSP style), or can do a τ action if either s or t can do a τ action. However, whether a τ action from s or from t is chosen is biased with the bias factor ρ. The probability of the synchronous execution of a is calculated via a normalization function $\nu : S_1 \times S_2 \to [0,1]$. Basically, $\nu(s,t)$ sums up the probabilities of the possible outgoing transitions of the new state which would be obtained if asynchronous behavior introduced non-determinism. Then $\nu(s,t)$ is used to calculate the actual (conditional) probabilities of the distribution assigned to $s\|_\rho t$.

Finally, a completely different solution of the problem of defining a parallel composition operator in the generative setting is provided in [35], the introduction of the class of bundle systems **Bun**. The bundle systems possess non-determinism, which allows for an elegant definition of any asynchronous parallel composition operator, as follows.

Let $\mathcal{A}_1 = \langle S_1, A, \alpha_1 \rangle, \mathcal{A}_2 = \langle S_2, A, \alpha_2 \rangle \in$ **Bun**. Then $\mathcal{A}_1\|\mathcal{A}_2 = \langle S_1 \times S_2, A, \alpha \rangle$ where, for $s \in S_1, t \in S_2$,

$$s\|t \rightsquigarrow \mu = P(\mu_1, \mu_2) \iff s \rightsquigarrow \mu_1, t \rightsquigarrow \mu_2$$

and $P(\mu_1, \mu_2)$ denotes a specific product of distributions, defined as follows: For $\mu_1 \in \mathcal{D}(\mathcal{P}(A \times S_1)), \mu_2 \in \mathcal{D}(\mathcal{P}(A \times S_2))$, $\mu = P(\mu_1, \mu_2) \in \mathcal{D}(\mathcal{P}(A \times (S_1 \times S_2)))$ where for all $B_s \in spt(\mu_1), B_t \in spt(\mu_2)$, $\mu(B_s \otimes B_t) = \mu_1(B_s) \cdot \mu_2(B_t)$ and

$$B_s \otimes B_t =$$
$$\{\langle a, \langle s', t\rangle\rangle \mid \langle a, s'\rangle \in B_s\} \cup$$
$$\{\langle b, \langle s, t'\rangle\rangle \mid \langle b, t'\rangle \in B_t\} \cup$$
$$\{\langle ab, \langle s', t'\rangle\rangle \mid \langle a, s'\rangle \in B_s, \langle b, t'\rangle \in B_t\}$$

Furthermore, $P(\mu, *) = P(*, \mu) = \mu$ and $P(*, *) = *$.

Note that the defined parallel composition for bundle systems is ACP style. By a slight modification of the definition of $\otimes$, all the other variants can be obtained. In a similar manner asynchronous parallel composition can be defined on the classes **PZ** and **MG**. In the literature there is no definition of a parallel composition operator for the class **Seg**.

4.3   Parallel Composition in the I/O Setting

A rather clean solution to the problems in the generative setting is given for the class of I/O automata in [87]. The view taken there is that the actions are divided into input and output, and while there can be synchronization on input actions, as in the reactive setting, the sets of output actions in each of the components must be disjoint.

Let $\mathcal{A}_1 = \langle S_1, A_1, \alpha_1 \rangle$, $\mathcal{A}_2 = \langle S_2, A_2, \alpha_2 \rangle$ be two I/O automata. The automata $\mathcal{A}_1$ and $\mathcal{A}_2$ are *compatible* if and only if $A_1^{out} \cap A_2^{out} = \emptyset$. Parallel composition is only defined on compatible automata. Let $A = A_1 \cup A_2$. We use the following convention: if s is a state in $\mathcal{A}_1$ $(\mathcal{A}_2)$ and $a \in A \setminus A_1^{out}$ $(A \setminus A_2^{out})$, such that there is no transition from s involving a, then we consider that $s \overset{a[1]}{\rightsquigarrow} s$. This convention will enforce the "input always enabled" requirement for the composite automaton.

The parallel composition of $\mathcal{A}_1$ and $\mathcal{A}_2$ is the I/O automaton $\mathcal{A}_1 \| \mathcal{A}_2 = \langle S_1 \times S_2, A, \alpha \rangle$ where:

1. $A^{out} = A_1^{out} \cup A_2^{out}$,
2. $A^{in} = A \setminus A^{out} = (A_1^{in} \cup A_2^{in}) \setminus A^{out}$,
3. the transition function α is defined by the following:

   $s \| t \overset{a[P]}{\rightsquigarrow} s' \| t'$ if and only if one of the following holds

   a. $a \in A^{in}$, $s \overset{a[p]}{\rightsquigarrow} s'$, $t \overset{a[q]}{\rightsquigarrow} t'$ and $P = pq$;

   b. $a \in A_1^{out}$, $s \overset{a[p]}{\rightsquigarrow} s'$, $t \overset{a[q]}{\rightsquigarrow} t'$ and $P = \frac{\delta_1(s)}{\delta_1(s) + \delta_2(t)} pq$

   c. $a \in A_2^{out}$, $s \overset{a[p]}{\rightsquigarrow} s'$, $t \overset{a[q]}{\rightsquigarrow} t'$ and $P = \frac{\delta_2(t)}{\delta_1(s) + \delta_2(t)} pq$

   Hence, $\alpha(s\|t) = \langle f^{in}, \mu^{out}, \delta \rangle$ where $\delta(s\|t) = \delta_1(s) + \delta_2(t)$, $f^{in}(a) = \mu_a$ is determined by a., for $a \in A^{in}$, and μ^{out} is determined by b. and c., for $a \in A^{out}$.

Example 4. Let $\mathcal{A}_1 = \langle S_1, A_1, \alpha_1 \rangle$, $\mathcal{A}_2 = \langle S_2, A_2, \alpha_2 \rangle$ be two I/O automata, with $s \in S_1, t \in S_2$ and their corresponding transitions as in the following diagram.

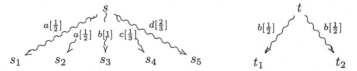

Take $A_1^{in} \supseteq \{a, b\}$, $A_1^{out} \supseteq \{c, d\}$, $A_2^{in} \supseteq \emptyset$, $A_2^{out} \supseteq \{b\}$. Assume the automata are compatible i.e. $A_1^{out} \cap A_2^{out} = \emptyset$ (clearly the states s and t are compatible). Then $A^{out} \supseteq \{b, c, d\}$ and $A^{in} \supseteq \{a\}$. Due to the convention we consider that $t \overset{a[1]}{\rightsquigarrow} t$, $t \overset{c[1]}{\rightsquigarrow} t$ and $t \overset{d[1]}{\rightsquigarrow} t$. The transitions from $s\|t$ are then given with the following diagram.

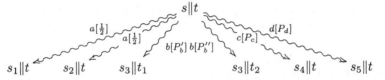

For $P_b' = P_b'' = \frac{\delta_2(t)}{\delta_1(s) + \delta_2(t)} \cdot \frac{1}{2}$, $P_c = \frac{\delta_1(s)}{\delta_1(s) + \delta_2(t)} \cdot \frac{1}{3}$ and $P_d = \frac{\delta_1(s)}{\delta_1(s) + \delta_2(t)} \cdot \frac{2}{3}$. Note that indeed $P_b' + P_b'' + P_c + P_d = 1$.

The proof that $\mathcal{A}_1 \| \mathcal{A}_2$ is well defined in the class **IO** can be found in [87]. Let us now informally explain the definition of parallel composition, and the role of the functions δ_1, δ_2 and δ. If s is a state of $\mathcal{A}_1$, then $\delta_1(s)$ is a positive real number corresponding to the delay rate in state s. It is a rate of an exponential distribution, determining the time that the automaton waits in state s until it generates one of its output actions. If no output actions are enabled in this state then $\delta_1(s) = 0$. When determining the distribution on output actions for $s \| t$, denoted by $\mu_{s\|t}^{out}$, the components' distributions μ_s^{out} and μ_t^{out} are joined in one such that any probability of μ_s^{out} is multiplied with normalization factor $\frac{\delta_1(s)}{\delta_1(s)+\delta_2(t)}$ and any probability of μ_t^{out} is multiplied with $\frac{\delta_2(t)}{\delta_1(s)+\delta_2(t)}$. Note that by the compatibility assumption, no action appears both in the support of μ_s^{out} and in the support of μ_t^{out}. The normalization factor models a racing policy between the states s and t for generating their own output actions. The value $\frac{\delta_1(s)}{\delta_1(s)+\delta_2(t)}$ is the probability that the state s has less waiting time left then the state t and therefore wins the race and generates one of its own output actions. On the other hand, synchronization occurs on all input actions, no autonomous behavior is allowed by the components on input actions, corresponding to the assumption that the input is provided by the environment and must be enabled in any state.

4.4   Parallel Composition in the Alternating Setting

In this section we focus on the classes **Str**, **Alt**, **SA** (**SA**$_n$, **SA**$_p$) and **Var** that exhibit alternating behavior i.e. make a distinction between probabilistic and non-deterministic states. In [40, 39] and in [41] a rather elegant parallel composition for the classes **Str** and **SA**, respectively, is defined.

We present the definition for the class **Alt** and discuss that the same definition can be restricted to the classes **Str**, **Alt**, **SA** (**SA**$_n$, **SA**$_p$). Let $\mathcal{A}_1 = \langle S_1, A, \alpha_1 \rangle$ and $\mathcal{A}_2 = \langle S_2, A, \alpha_2 \rangle$ be two alternating automata, with $S_1 = N_1 + P_1$ and $S_2 = N_2 + P_2$. Their parallel composition is the alternating automaton $\mathcal{A}_1 \| \mathcal{A}_2 = \langle S, A, \alpha \rangle$ where $S = S_1 \times S_2 = N + P$ for $N = N_1 \times N_2$ and $P = P_1 \times P_2 + N_1 \times P_2 + P_1 \times N_2$ and the transition function is defined as follows. Let $p_1 \in P_1, p_2 \in P_2, n_1 \in N_1, n_2 \in N_2$ and $s_1 \in S_1, s_2 \in S_2$. For the probabilistic states in the composed automaton, we have:

$$
\begin{aligned}
p_1 \| p_1 \rightsquigarrow \mu &\iff p_1 \rightsquigarrow \mu_1, p_2 \rightsquigarrow \mu_2, \mu = \mu_1 \times \mu_2 \\
p_1 \| n_2 \rightsquigarrow \mu &\iff p_1 \rightsquigarrow \mu_1, \mu = \mu_1 \times \mu_{n_2}^1 \\
n_1 \| p_2 \rightsquigarrow \mu &\iff p_2 \rightsquigarrow \mu_2, \mu = \mu_{n_1}^1 \times \mu_2.
\end{aligned}
$$

For the non-deterministic states in the composed automaton different variants (CCS, CSP or ACP style) can be chosen. We choose for the ACP style: $n_1 \| n_2 \xrightarrow{a} s_1 \| s_2$ if and only if

1. $s \xrightarrow{b} s'$, $t \xrightarrow{c} t'$ and $bc = a$ defined, or
2. $s \xrightarrow{a} s'$, $t = t'$, or
3. $t \xrightarrow{a} t'$, $s = s'$.

Hence, when composing a probabilistic state with any other state the result is a probabilistic state. If the other state is non-deterministic, then the composed state basically behaves as the probabilistic state and leaves the second component of the state unchanged, as in the following example.

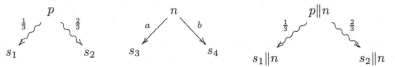

On the other hand, the composition of non-deterministic states is exactly the same as in the non-probabilistic case and therefore any possible parallel composition operator is definable here as in the case for LTSs.

By inspecting the definitions of the classes **SA**, $\mathbf{SA}_n$ and $\mathbf{SA}_p$ it is easy to see that the following statement is valid.

Proposition 3. *If $\mathcal{A}_1, \mathcal{A}_2 \in \mathbf{SA}$ (or $\mathbf{SA}_n$, or $\mathbf{SA}_p$), then $\mathcal{A}_1 \| \mathcal{A}_2 \in \mathbf{SA}$ (or $\mathbf{SA}_n$, or $\mathbf{SA}_p$), respectively.* $\qquad\square$

The definition of parallel composition for stratified systems is given in [40, 39] with synchronous behavior when composing two (non-) deterministic states. This is necessary in order to stay in the class **Str** when composing two such automata, since in the stratified model there is only a single action transition possible from a (non-) deterministic state. A parallel composition operator with no synchronization but only interleaving, for the stratified class of systems, is defined in [49]. In the original definition for strictly alternating systems of [41], non-deterministic states are composed in the CCS fashion.

Complications arise in the case of **Var** models, due to their generative probabilistic behavior. The behavior of the composite states $n_1 \| n_2$, $n_1 \| p_2$ and $p_1 \| n_2$ can be defined in the same way as above. However, there is no convenient way to define $p_1 \| p_2$, since this coincides with defining generative parallel composition. Any of the approaches described in section 4.2 can be used.

5   Comparing Classes

This section is based on the results of [26]. There a hierarchy of probabilistic system types is presented in a coalgebraic setting. The coalgebraic theory proved useful in providing a uniform framework and shortening the proofs. However, the results can be restated and explained with the machinery introduced so far.

An Expressiveness Criterion

Let $\mathbf{C}_1$ and $\mathbf{C}_2$ be two classes of probabilistic automata. We say that the class $\mathbf{C}_1$ is *included* or *embedded* in the class $\mathbf{C}_2$, i.e. the class $\mathbf{C}_2$ is *at least as expressive as* the class $\mathbf{C}_1$ (notation $\mathbf{C}_1 \to \mathbf{C}_2$) if and only if there exists a translation function $\mathcal{T}$ that maps each automaton of the first class to an automaton of the second class such that bisimilarity is both reflected and preserved. More explicitly, the translation function $\mathcal{T} : \mathbf{C}_1 \to \mathbf{C}_2$ should satisfy:

1. for $\mathcal{A} = \langle S, A, \alpha \rangle$ in $\mathbf{C}_1$, $\mathcal{T}(\mathcal{A}) = \langle S, A, \alpha' \rangle$ with the same set of states S,
2. the translation function $\mathcal{T}$ is injective, and
3. if $s, t \in S$, then $s \approx_{\mathcal{A}} t \Leftrightarrow s \approx_{\mathcal{T}(\mathcal{A})} t$, i.e. two states are bisimilar in the translated automaton (according to bisimilarity in the class $\mathbf{C}_2$) if and only if they were bisimilar in the original automaton (according to bisimilarity for the class $\mathbf{C}_1$).

The relation $\rightarrow$ between the classes of (probabilistic) automata is a preorder.

Basically our expressiveness criterion states that the class $\mathbf{C}_1$ is really embedded in the class $\mathbf{C}_2$, i.e. the translations are nothing else but "suitable copies" of the automata of the first class existing in the second class. Note that only preservation of bisimulation is not enough. For example, we could define a translation from reactive systems to LTSs that preserves bisimulation, by forgetting the probabilities, as in the following example.

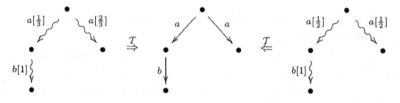

But we do not consider the class of LTSs more expressive than the class of reactive probabilistic automata, and the translation is by no means injective.

Another hierarchy result is the hierarchy of reactive, generative, stratified and non-deterministic automata of [40] and [39]. The expressiveness criterion used by van Glabbeek et al. is different. They consider a class $\mathbf{C}_2$ *more expressive* than a class $\mathbf{C}_1$ if there is an *abstraction* mapping that maps each automaton of the class $\mathbf{C}_2$ to an automaton of the class $\mathbf{C}_1$, such that the state set remains the same and bisimulation is preserved. The abstraction mappings are not injective by nature. An example of such an abstraction mapping is the translation that forgets the probabilities in the previous figure. Therefore, in their setting the class **React** is more expressive than the class **NA**.

The Hierarchy

Theorem 1. *[26] The class embeddings presented in Figure 1 hold among the probabilistic system types.*   □

The proof of Theorem 1 (except for the strictly alternating classes) is given in [26] using a technical result in terms of injective natural transformations. Due to the correspondence of concrete bisimulation with coalgebraic bisimulation for all the treated systems, the theorem remains valid without mentioning the coalgebraic theory behind it. Instead of presenting a proof here, we will explicitly state the translations for each arrow in Figure 1, give some examples and illustrate how preservation and reflection of bisimulation can be proven in one concrete case. We present the translations for the arrows of Figure 1 in several groups: simple arrows based on inclusion, arrows that show a change from external to

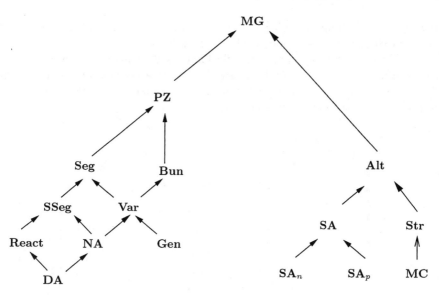

Fig. 1. Class embeddings

full non-determinism, arrows that change an element to a singleton, arrows that change an element to a corresponding Dirac distribution and more specific arrows. The translations are quite natural. It is the property of preservation and reflection of bisimilarity that adds justification to the translations.

Simple Arrows. Let $\mathbf{C}_1$ and $\mathbf{C}_2$ be two classes of probabilistic automata. If $\mathbf{C}_1 \subseteq \mathbf{C}_2$ then $\mathbf{C}_1 \to \mathbf{C}_2$. Therefore, the embeddings $\mathbf{SA} \to \mathbf{Alt}, \mathbf{SA}_n \to \mathbf{SA}$ and $\mathbf{SA}_p \to \mathbf{SA}$ hold. Furthermore, if $\mathbf{C}_1$ is defined with a transition function $\alpha_1 : S \to C_1(S)$ and $\mathbf{C}_2$ has a transition function of type $\alpha_2 : S \to C_2(S)$ such that for all S, $C_1(S) \subseteq C_2(S)$ then every automaton of the class $\mathbf{C}_1$ can be considered an automaton of the class $\mathbf{C}_2$, by only extending the codomain of the transition function. In this case also $\mathbf{C}_1 \to \mathbf{C}_2$. The following arrows of Figure 1 hold due to extending the codomain of the transition function: $\mathbf{MC} \to \mathbf{Str}$, $\mathbf{Gen} \to \mathbf{Var}$, $\mathbf{NA} \to \mathbf{Var}$, $\mathbf{PZ} \to \mathbf{MG}$. In each of these cases the translation of the automata is basically the identity mapping. For example, every generative automaton is a Vardi automaton without non-deterministic states, or every Markov chain is a stratified automaton that has no action-transitions, i.e. no deterministic states.

From External to Full Non-Determinism. Two of the embedding arrows of Figure 1, $\mathbf{DA} \to \mathbf{NA}$ and $\mathbf{React} \to \mathbf{SSeg}$, show that every system with only external non-determinism can be considered as a system with full non-determinism that never uses the full-nondeterminism option. Let $\mathcal{A} = \langle \mathcal{S}, \mathcal{A}, \alpha \rangle \in \mathbf{DA}$. Then $\mathcal{T}(\mathcal{A}) = \langle S, A, \alpha' \rangle \in \mathbf{NA}$ is given by

$$\alpha'(s) = \{\langle a, s' \rangle \mid \alpha(s)(a) = s' \in S\} \in \mathcal{P}(A \times S).$$

If we consider automata as diagrams, i.e. transition graphs, then this translation does not change the transition graph of a **DA** system.

For **React** → **SSeg** a similar translation is used, changing a partial function to its graph. Due to the notation used for reactive systems in Section 3, the diagram of a reactive system when considered a simple Segala system has to be re-drawn, as in the next example.

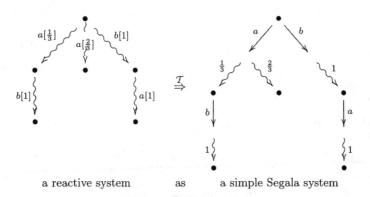

a reactive system   as   a simple Segala system

Singleton Arrows. In several cases the translation only changes an element (state, or pair of state and action, or distribution) into a singleton set containing this element.

Bun → **PZ**: Let $\mathcal{A} = \langle S, A, \alpha \rangle$ be a bundle probabilistic automaton, i.e. $\alpha : S \to \mathcal{D}(\mathcal{P}(A \times S)) + 1$, then the translation to a Pnueli-Zuck automaton is achieved by putting $\mathcal{T}(\mathcal{A}) = \langle S, A, \alpha' \rangle$ for $\alpha'(s) = \{\alpha(s)\}$ if $\alpha(s)$ is a distribution, and $\alpha'(s) = \emptyset$ if $\alpha(s) = *$.

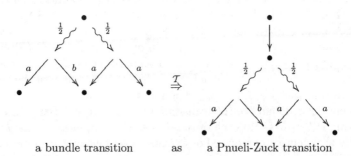

a bundle transition   as   a Pnueli-Zuck transition

Str → **Alt**: In this case $\mathcal{T}(\langle S, A, \alpha \rangle) = \langle S, A, \alpha' \rangle$ where $\alpha : S \to \mathcal{D}(S) + A \times S + 1$ and $\alpha' : S \to \mathcal{D}(S) + \mathcal{P}(A \times S)$ and

$$\alpha'(s) = \begin{cases} \{\alpha(s)\} & \text{if } \alpha(s) \in A \times S \\ \emptyset & \text{if } \alpha(s) = * \\ \alpha(s) & \text{otherwise} \end{cases}$$

The diagram of a stratified automaton when translated to alternating automaton stays the same.

Seg → PZ: Let $\mathcal{A} = \langle \mathcal{S}, \mathcal{A}, \alpha \rangle$ be a Segala automaton, $\alpha : S \to \mathcal{P}(\mathcal{D}(A \times S))$. Then $\mathcal{T}(\mathcal{A}) = \langle S, A, \alpha' \rangle$ where α' is determined from α in the following way:

$$\alpha'(s) = \{\mu' \mid \mu \in \alpha(s)\}$$

where μ' is constructed from μ by changing a distribution over pairs to distribution over singletons of pairs, i.e.

$$\mu' = \{\{\langle a, s'\rangle\} \mapsto \mu(a, s')\}.$$

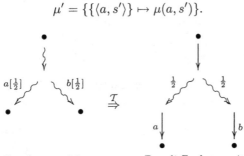

a Segala transition   as   a Pnueli-Zuck transition

Dirac Arrows. Sometimes a transformation from an element to a Dirac distribution for this element is needed. This kind of translation embeds the non-probabilistic automata, **DA** and **NA**, into the reactive and simple Segala automata, respectively.

DA → React: In this case every transition has to be changed to a probabilistic transition with probability 1.

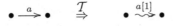

a **DA** transition as a reactive transition

For $\langle S, A, \alpha \rangle \in \mathbf{DA}$, the translation is $\mathcal{T}(\langle S, A, \alpha \rangle) = \langle S, A, \alpha' \rangle$, if $\alpha(s) \in (S + 1)^A$, we put $\alpha'(s) \in (\mathcal{D}(S) + 1)^A$ such that $\alpha(s)(a) = t \in S \iff \alpha'(s)(a) = \mu_t^1$.

NA → SSeg: For obtaining a simple Segala automaton out of a LTS, we change the next state of every transition to a Dirac distribution for this state.

a **NA** transition as a simple Segala transition

Formally, $\mathcal{T}(\langle S, A, \alpha \rangle) = \langle S, A, \alpha' \rangle$ such that $\alpha'(s) = \{\langle a, \mu_{s'}^1 \rangle \mid \langle a, s' \rangle \in \alpha(s)\}$.

Specific Arrows

Var → Seg: Let $\mathcal{A} = \langle S, A, \alpha \rangle$ be a Vardi automaton, with $\alpha(s) : S \to \mathcal{D}(A \times S) + \mathcal{P}(A \times S)$. The injective translation to a Segala automaton is given by $\mathcal{T}(\mathcal{A}) = \langle S, A, \alpha' \rangle$ for

$$\alpha'(s) = \begin{cases} \{\alpha(s)\} & \text{if } \alpha(s) \in \mathcal{D}(A \times S) \\ \{\mu_{\langle a, s'\rangle}^1 \mid \langle a, s'\rangle \in \alpha(s)\} & \text{if } \alpha(s) \in \mathcal{P}(A \times S) \end{cases}$$

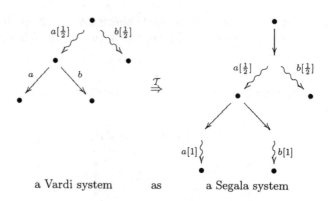

a Vardi system   as   a Segala system

Var → Bun: This translation is orthogonal to the one that gives us **Var → Seg.** For a Vardi automaton $\mathcal{A} = \langle \mathcal{S}, \mathcal{A}, \alpha \rangle$ we put $\mathcal{T}(\mathcal{A}) = \langle S, A, \alpha' \rangle$ where

$$\alpha'(s) = \begin{cases} \{\{\langle a, s' \rangle\} \mapsto \mu(a, s')\} & \text{if } \alpha(s) = \mu \in \mathcal{D}(A \times S) \\ \mu^1_{\alpha(s)} & \text{if } \alpha(s) \in \mathcal{P}(A \times S) \end{cases}$$

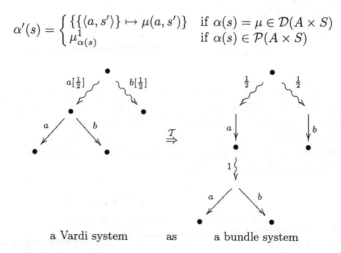

a Vardi system   as   a bundle system

Remark 5. Both in **Var → Seg** and in **Var → Bun** the translated transition function α' is well defined even when we consider $\mathcal{D}(A \times S) \cap \mathcal{P}(A \times S) \neq \emptyset$, i.e. we identify $\mu^1_{\langle a, s \rangle}$ with $\{\langle a, s \rangle\}$. Furthermore, this identification is needed to obtain injectivity of the translations.

Alt → MG: Similarly as when translating Vardi systems, with an extra singleton construction, an alternating automaton $\mathcal{A} = \langle \mathcal{S}, \mathcal{A}, \alpha \rangle$, $\alpha : S \to \mathcal{D}(S) + \mathcal{P}(A \times S)$ is translated into a general probabilistic automaton. We put $\mathcal{T}(\mathcal{A}) = \langle S, A, \alpha' \rangle$ where

$$\alpha'(s) = \begin{cases} \{\{\{\langle a, s' \rangle\} \mapsto \mu(a, s')\}\} & \text{if } \alpha(s) = \mu \in \mathcal{D}(S) \\ \{\mu^1_{\alpha(s)}\} & \text{if } \alpha(s) \in \mathcal{P}(A \times S) \end{cases}$$

SSeg → Seg: In order to change a transition of a simple Segala automaton to a transition of a Segala automaton it is enough to push the action label into the distribution.

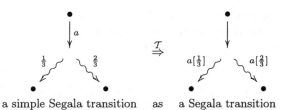

a simple Segala transition   as   a Segala transition

Formally, if $\mathcal{A} = \langle S, A, \alpha \rangle$, $\alpha : S \to \mathcal{P}(A \times \mathcal{D})$ then $\mathcal{T}(\mathcal{A}) = \langle S, A, \alpha' \rangle$ where

$$\langle a, \mu \rangle \in \alpha(s) \iff \mu_a \in \alpha'(s) \in \mathcal{P}(\mathcal{D}(A \times S))$$

and $\mu_a(a, s') = \mu(s')$ for all $s' \in S$.

For this last translation, as an illustration, we give the proof of preservation and reflection of bisimilarity. Let μ be a distribution on S and a an action in A. Denote by μ_a the distribution on $A \times S$ obtained from μ by $\mu_a(a, s) = \mu(s)$. Clearly, for any subset $X \subseteq S$ we have $\mu[X] = \mu_a[a, X]$, which yields $\mu \equiv_R \mu' \iff \mu_a \equiv_{A,R} \mu'_a$. By the translation, the following holds:

1. If $s \xrightarrow{a} \rightsquigarrow_{\mathcal{A}} \mu$, then $s \to \rightsquigarrow_{\mathcal{T}(\mathcal{A})} \mu_a$.
2. If $s \to \rightsquigarrow_{\mathcal{T}(\mathcal{A})} \mu$ then there exists $a \in A$ such that $\mu = \nu_a$ for some distribution ν on S and $s \xrightarrow{a} \rightsquigarrow_{\mathcal{A}} \nu$.

Now assume $s \approx_{\mathcal{A}} t$, i.e. there exists a bisimulation R (Definition 11) with $\langle s, t \rangle \in R$. We prove that R is a bisimulation (Definition 12) on the state space of $\mathcal{T}(\mathcal{A})$. Let $s \to \rightsquigarrow_{\mathcal{T}(\mathcal{A})} \mu$. By 2. we have that $\mu = \nu_a$ for some $a \in A$ and some $\nu \in \mathcal{D}(S)$, and $s \xrightarrow{a} \rightsquigarrow_{\mathcal{A}} \nu$. Since R is a bisimulation on the state space of $\mathcal{A}$ we have that there exists a distribution ν' such that $t \xrightarrow{a} \rightsquigarrow_{\mathcal{A}} \nu'$ and $\nu \equiv_R \nu'$. Now it follows by 1. that $s \to \rightsquigarrow_{\mathcal{T}(\mathcal{A})} \nu'_a$, and furthermore $\mu = \nu_a \equiv_{A,R} \nu'_a$. So, R is a bisimulation on the state set of $\mathcal{T}(\mathcal{A})$.

The opposite is analogous. If R is a bisimulation (Definition 12) with $\langle s, t \rangle \in R$ on the set of states of $\mathcal{T}(\mathcal{A})$, we prove that R is a bisimulation (Definition 11) on the set of states of $\mathcal{A}$. Assume $s \xrightarrow{a} \rightsquigarrow_{\mathcal{A}} \mu$. Then by 1., $s \to \rightsquigarrow_{\mathcal{T}(\mathcal{A})} \mu_a$. Since R is a bisimulation and $\langle s, t \rangle \in R$ we get that there exists $\nu \in \mathcal{D}(A \times S)$ such that $t \to \rightsquigarrow_{\mathcal{T}(\mathcal{A})} \nu$ and $\mu_a \equiv_{A,R} \nu$. Now, by 2., we get that there exists $a' \in A$ and a distribution μ' on S such that $\nu = \mu'_{a'}$, $t \xrightarrow{a'} \rightsquigarrow_{\mathcal{A}} \mu'$ and $\mu \equiv_R \mu'$. However, from $\mu_a \equiv_{A,R} \nu = \mu'_{a'}$ we get that $a' = a$ and hence $t \xrightarrow{a} \rightsquigarrow_{\mathcal{A}} \mu'$ and $\mu \equiv_R \mu'$, which completes the proof.

Remark 6. All the translations presented in this section are injective and preserve and reflect bisimilarity.

Strict Alternation vs. Complex Transition Function

A translation between simple Segala automata and automata of the class $\mathbf{SA}_n$ has been known for a long time, and has been recently justified in [25]. Similarly, the class $\mathbf{SA}_p$ can be compared with the class of bundle probabilistic automata. In order to carry out these comparisons in our framework we slightly change the comparison criterion itself so that it allows for translations that do not keep the same state set.

Relaxed Expressiveness Criteria. Let $\mathbf{C}_1$ and $\mathbf{C}_2$ be two classes of probabilistic automata and let $\mathcal{T} : \mathbf{C}_1 \to \mathbf{C}_2$ be a translation mapping, such that if $\mathcal{A} = \langle S, A, \alpha \rangle$ then $\mathcal{T}(\mathcal{A}) = \langle S', A, \alpha' \rangle$

We say that the class $\mathbf{C}_1$ *is embedded* in the class $\mathbf{C}_2$, by the translation $\mathcal{T}$, notation $\mathbf{C}_1 \twoheadrightarrow \mathbf{C}_2$ if the following conditions hold:

1. for any automaton of $\mathbf{C}_1$, $S \subseteq S'$ or for any automaton of $\mathbf{C}_1$, $S' \subseteq S$,
2. bisimilarity is preserved and reflected for common states, i.e.,
$$\forall s, t \in S \cap S' : s \approx_{\mathcal{A}} t \iff s \approx_{\mathcal{T}(\mathcal{A})} t, \text{ and}$$
3. the translation $\mathcal{T}$ is injective.

Furthermore, we say that the class $\mathbf{C}_1$ *is embedded up to irrelevant bisimilarity* in the class $\mathbf{C}_2$, by the translation $\mathcal{T}$, notation $\mathbf{C}_1 \longrightarrow_{/\approx} \mathbf{C}_2$ if the following conditions hold:

1. $S' \subseteq S$,
2. for any two common states $s, t \in S'$, $s \approx_{\mathcal{A}} t \iff s \approx_{\mathcal{T}(\mathcal{A})} t$, and
3. the translation $\mathcal{T}$ is injective up to bisimilarity of irrelevant states, i.e., if $\mathcal{A}_1 = \langle S_1, A, \alpha_1 \rangle$ and $\mathcal{A}_2 = \langle S_2, A, \alpha_2 \rangle$ are two automata of the first class such that $\mathcal{T}(\mathcal{A}_1) = \mathcal{T}(\mathcal{A}_2) = \langle S, A, \alpha \rangle$, then
$$s_1 \in S_1 \setminus S \Rightarrow \exists s_2 \in S_2 \setminus S : s_1 \approx s_2 \text{ and}$$
$$s_2 \in S_2 \setminus S \Rightarrow \exists s_1 \in S_1 \setminus S : s_1 \approx s_2.$$

Theorem 2. *The following embeddings compare the classes of strictly alternating, bundle and simple Segala probabilistic automata.*

Fig. 2. Strictly alternating models as models with a structured transition relation

Instead of a complete proof, we give the translations needed in each of the embeddings.

$\mathbf{SSeg} \twoheadrightarrow \mathbf{SA}_n$:

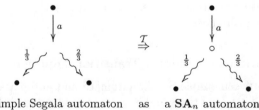

a simple Segala automaton   as   a $\mathbf{SA}_n$ automaton

Let $\mathcal{A} = \langle \mathcal{S}, A, \alpha \rangle$ be a simple Segala automaton. The translation function translates it to $\mathcal{T}(\mathcal{A}) = \langle S', A, \alpha' \rangle$ by inserting a new probabilistic state for each transition, i.e., by changing $s \xrightarrow{a} \mu$ to $s \xrightarrow{a} s_{s,a,\mu} \rightsquigarrow \mu$. Formally, $S' = S + S_p$ where S_p is a set of fresh probabilistic states, such that $S_p = \{s_{s,a,\mu} \mid s \in S, a \in A, \mu \in \mathcal{D}(S), \langle a, \mu \rangle \in \alpha(s)\}$. So, for every outgoing transition of a state there is a corresponding new state added to S_p. The transition function is defined as follows: if $s \in S$, such that $\langle a, \mu \rangle \in \alpha(s)$ and the corresponding new state is $s_{s,a,\mu}$ then and only then $\langle a, s_{s,a,\mu} \rangle \in \alpha'(s)$; if $s_{s,a,\mu} \in S_p$ corresponds to a transition $\langle a, \mu \rangle \in \alpha(s)$ then $\alpha'(s_{s,a,\mu}) = \mu$. The translation is injective and it preserves and reflects bisimilarity on S.

$\mathbf{SA}_n \longrightarrow_{/\approx} \mathbf{SSeg}$: The translation for this embedding, is the "inverse" of the case $\mathbf{SSeg} \twoheadrightarrow \mathbf{SA}_n$. If $\mathcal{A} = \langle S, A, \alpha \rangle \in \mathbf{SA}_n$ with a set of probabilistic states P and a set of non-deterministic states N, then $\mathcal{T}(\mathcal{A}) = \langle N, A, \alpha' \rangle$ where for every $s \in N$, $\alpha'(s) = \{\langle a, \mu \rangle \mid \langle a, s' \rangle \in \alpha(s)$ and $\alpha(s') = \mu\}$. Hence this translation forgets the probabilistic states, and turns two strictly alternating steps of the $\mathbf{SA}_n$ automaton into one transition of the corresponding Segala automaton. This translation is only injective up to bisimilarity of the disappearing probabilistic states. In fact it is even stricter than that, it is injective up to indistinguishable probabilistic states. The only cases of non-injectivity come as a consequence of having two probabilistic states s_{p1} and s_{p2} such that $\alpha(s_{p1}) = \alpha(s_{p2}) = \mu$. Then if s is a non-deterministic state with $s \xrightarrow{a} s_{p1}$ and $s \xrightarrow{a} s_{p2}$, the two "copies" $s \xrightarrow{a} s_{p1} \rightsquigarrow \mu$ and $s \xrightarrow{a} s_{p2} \rightsquigarrow \mu$ will be mapped to one single transition $s \xrightarrow{a} \mu$. For example, the following two different $\mathbf{SA}_n$ automata translate to one $\mathbf{SSeg}$ automaton.

$\mathbf{Bun} \twoheadrightarrow \mathbf{SA}_p$:

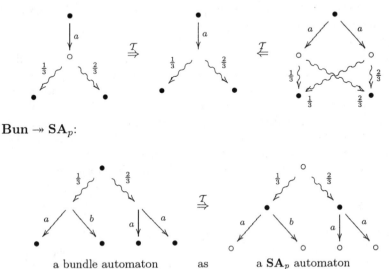

| a bundle automaton | as | a $\mathbf{SA}_p$ automaton |

The translation is as follows: $\mathcal{T}(\langle S, A, \alpha \rangle) = \langle S', A, \alpha' \rangle$ where $S' = S + S_n$, S_n containing the fresh non-deterministic states $S_n = \{s_{s,T} \mid s \in S, T \in spt(\alpha(s))\}$. The transition function is defined by: for $s \in S$ with $\alpha(s) = \mu$, we put $\alpha'(s) = \mu_s \in \mathcal{D}(S_n)$ where $\mu_s(s_{s,T}) = \mu(T)$, and for $s_{s,T} \in S_n$ we put $\alpha'(s_{s,T}) = T$.

$\mathbf{SA}_p \twoheadrightarrow \mathbf{Bun}$: The inverse of the translation $\mathbf{Bun} \twoheadrightarrow \mathbf{SA}_p$ gives us the translation in this case. We forget all the non-deterministic states in a $\mathbf{SA}_p$ automaton, being left with a bundle automaton. For a $\mathbf{SA}_p$ automaton $\mathcal{A} = \langle S, A, \alpha \rangle$ with a set of probabilistic states P and non-deterministic states N we put $\mathcal{T}(\mathcal{A}) = \langle P, A, \alpha' \rangle$ where for $s \in P$ with $\alpha(s) = \mu \in \mathcal{D}(N)$ we define $\alpha'(s) = \mu_s \in \mathcal{D}(\mathcal{P}(A \times S))$ such that $\mu_s(\alpha(s')) = \mu(s')$. Note that the phenomenon of losing "copies" as in $\mathbf{SA}_n \longrightarrow_{/\approx} \mathbf{SSeg}$ does not appear here. Even though states are lost, no transitions are identified, i.e. no arrow is lost.

Remark 7. Restricting to subclasses of the class $\mathbf{SA}$, i.e. considering the classes $\mathbf{SA}_n$ and $\mathbf{SA}_p$, is necessary to obtain injectivity (up to bisimilar irrelevant states) of the embedding mappings.

6   Conclusions

In this overview we have presented various types of probabilistic automata, including generative, reactive and stratified ones, strictly alternating and alternating ones, the simple Segala, Segala and Vardi type of probabilistic automata and the bundle, Pnueli-Zuck and general ones.

A major part of our work has been devoted to the comparison of the various classes of probabilistic automata, taking strong bisimilarity for these automata as a starting point, resulting in a hierarchy of probabilistic system types. Additionally, we have discussed the extent of non-determinism that can be modelled in the various types of automata and the operator of parallel composition for them. Classes positioned higher in the map of Figure 1 can be characterized as closures of the simpler classes under parallel composition, which clarifies the need for the more complex models.

The results obtained are briefly presented in Figure 1, Figure 2 and Table 1. From there various conclusions on probabilistic system modelling can be drawn. For instance, if presence of non-determinism and closedness under all variants of parallel composition are desired properties on the one hand, and having as simple a model as possible is needed on the other hand, then whether the choice is for input (reactive) or output (generative) type of systems, the best choice appears to be the simple Segala model and the bundle model, respectively. Different requirements lead to different choices, but we hope the map of probabilistic automata based models will prove to be useful in making a right decision.

References

1. R. Alur and T.A. Henzinger, *Reactive modules*, Formal Methods in System Design **15** (1999), 7–48, A preliminary version appeared in the Proceedings of the 11th Annual Symposium on Logic in Computer Science (LICS), IEEE Computer Society Press, 1996, pp. 207-218.
2. L. de Alfaro, T.A. Henzinger, and R. Jhala, *Compositional methods for probabilistic systems*, CONCUR 2001 - Concurrency Theory: 12th International Conference, Aalborg, Denmark, August 20-25, 2001, LNCS, vol. 2154, Springer-Verlag, 2001, pp. 351–365.

3. L. de Alfaro, *Formal verification of probabilistic systems*, Ph.D. thesis, Stanford University, 1997.

4. _____, *Stochastic transition systems*, International Conference on Concurrency Theory, CONCUR, vol. 1466, LNCS, 1998, pp. 423–438.

5. S. Andova, *Process algebra with probabilistic choice*, Proc. 5th International AMAST Workshop, ARTS'99, Bamberg, Germany (J.-P. Katoen, ed.), LNCS 1601, Springer-Verlag, 1999, pp. 111–129.

6. _____, *Probabilistic process algebra*, Ph.D. thesis, Eindhoven University of Technology, 2002.

7. A. Bianco and L. de Alfaro, *Model checking of probabilistic and nondeterministic systems*, Found. of Software Tech. and Theor. Comp. Sci., LNCS, vol. 1026, Springer-Verlag, 1995.

8. C. Baier, *Polynomial time algorithms for testing probabilistic bisimulation and simulation*, Proc. 8th International Conference on Computer Aided Verification (CAV'96), Lecture Notes in Computer Science, vol. 1102, 1996, pp. 38–49.

9. C. Baier, *On algorithmic verification methods for probabilistic systems*, Habilitationsschrift, FMI, Universitaet Mannheim, 1998.

10. J.C.M. Baeten, J.A. Bergstra, and S.A. Smolka, *Axiomatizing probabilistic processes: ACP with generative probabilities*, Information and Computation **121** (1995), no. 2, 234–255.

11. R. Blute, J. Desharnais, A. Edalat, and P. Panangaden, *Bisimulation for labelled Markov processes*, LICS'97, 1997, pp. 149–158.

12. C. Baier, P.R. D'Argenio, H. Hermanns, and J.-P. Katoen, *How to cook a probabilistic process calculus*, unpublished, 1999.

13. C. Baier, B. Engelen, and M. Majster-Cederbaum, *Deciding bisimilarity and similarity for probabilistic processes*, Journal of Computer and System Sciences **60** (1999), 187–231.

14. M. Bernardo, *Theory and application of extended Markovian process algebra*, Ph.D. thesis, University of Bologna, 1999.

15. M. Bernardo and R. Gorrieri, *A tutorial on EMPA: A theory of concurrent processes with nondeterminism, priorities, probabilities and time*, Theoretical Computer Science **202** (1998), no. 1, 1–54.

16. E. Brinksma and H. Hermanns, *Process algebra and Markov chains*, Lectures on Formal Methods and Performance Analysis, First EEF/Euro Summer School on Trends in Computer Science, Berg en Dal, The Netherlands, July 3–7, 2000 (E. Brinksma, H. Hermanns, and J.-P. Katoen, eds.), LNCS, vol. 2090, Springer-Verlag, 2001, pp. 183–232.

17. C. Baier, B. Haverkort, H. Hermanns, and J.-P. Katoen, *Model checking continuous-time Markov chains by transient analysis*, CAV 2000, vol. 1855, LNCS, Springer-Verlag, 2000, pp. 358–372.

18. S.D. Brookes, C.A.R. Hoare, and A.W. Roscoe, *A theory of communicating sequential processes*, Journal of the ACM **31** (1984), 560–599.

19. J.A. Bergstra and J.W. Klop, *Algebra of communicating processes with abstraction*, Theoretical Computer Science **37** (1985), 77–121.

20. C. Baier and M.Z. Kwiatkowska, *Domain equations for probabilistic processes*, 4th Workshop on Expressiveness in Concurrency (EXPRESS'97), Santa Margherita, vol. 7, Electronic Notes in Theoretical Computer Science, 1997.

21. _____, *Domain equations for probabilistic processes*, Mathematical Structures in Computer Science **10** (2000), 665–717.

22. A. Benveniste, B. C. Levy, E. Fabre, and P. Le Guernic, *A calculus of stochastic systems for the specification, simulation, and hidden state estimation of mixed stochastic/non-stochastic systems*, Theoretical Computer Science **152** (1995), 171–217.

23. B. Bloom and A. R. Meyer, *A remark on bisimulation between probabilistic processes*, Foundations of Software Technology and Theoretical Computer Science, LNCS, vol. 363, Springer-Verlag, 1989, pp. 26–40.

24. C. Baier and M.I.A. Stoelinga, *Norm fuctions for probabilistic bisimulations with delays*, Proceedings of 3rd International Conference on Foundations of Science and Computation Structures (FOSSACS), Berlin, Germany, March 2000 (J. Tiuryn, ed.), LNCS, vol. 1784, "Springer-Verlag", 2000, pp. 1–16.

25. E. Bandini and R. Segala, *Axiomatizations for probabilistic bisimulation*, Proceedings of the 28th International Colloquium on Automata, Languages and Programming (ICALP) 2001, Crete, LNCS 2076, 2001, pp. 370–381.

26. F. Bartels, A. Sokolova, and E.P. de Vink, *A hierarchy of probabilistic system types*, Electronic Notes in Theoretical Computer Science (H. Peter Gumm, ed.), vol. 82, Elsevier, 2003.

27. P. Buchholz, *Markovian process algebra: Composition and equivalence*, in Proc. of PAPM '94, Erlangen (Germany), 1994, pp. 11–30.

28. L. Christoff and I. Christoff, *Efficient algorithms for verification of equivalences for probabilistic processes*, Proc. Workshop on Computer Aided Verification 1991 (K. Larsen and A. Skou, eds.), LNCS, vol. 575, 1991.

29. I. Christoff, *Testing equivalences and fully abstract models for probabilistic processes*, Proceedings of CONCUR'90 (J.C.M. Baeten and J.W. Klop, eds.), LNCS 458, Springer-Verlag, 1990, pp. 126–140.

30. R. Cleaveland, S.A. Smolka, and A. Zwarico, *Testing preorders for probabilistic processes*, Automata, Languages and Programming (ICALP '92), Vienna, LNCS, vol. 623, Springer-Verlag, 1992, pp. 708–719.

31. C. Courcoubetis and M. Yannakakis, *The complexity of probabilistic verification*, Journal of the ACM (JACM) **42** (1995), 857–907.

32. P.R. D'Argenio, *Algebras and automata for timed and stochastic system*, Ph.D. thesis, University of Twente, 1999.

33. J. Desharnais, A. Edalat, and P. Panangaden, *A logical characterization of bisimulation for labeled Markov processes*, Proc. LICS'98 (Indianapolis), 1998, pp. 478–487.

34. C. Derman, *Finite state Markovian decision proceses*, Academic Press, 1970.

35. P. D'Argenio, H. Hermanns, and J.-P. Katoen, *On generative parallel composition*, Proc. PROBMIV'98, ENTCS 22, 1998, pp. 105–122.

36. P.R. D'Argenio, B. Jeannet, H.E. Jensen, and K.G. Larsen, *Reachability analysis of probabilistic systems by successive refinements*, PAPM-PROBMIV 2001, Aachen, Germany (L. de Alfaro and S. Gilmore, eds.), LNCS, vol. 2165, Springer-Verlag, 2001, pp. 29–56.

37. _____, *Reduction and refinement strategies for probabilistic analysis*, PAPM-PROBMIV 2002, Copenhagen, Denmark (H. Hermanns and R. Segala, eds.), LNCS, Springer-Verlag, 2002.

38. A. Giacalone, C. Jou, and S. Smolka, *Algebraic reasoning for probabilistic concurrent systems*, Proc. of the Working Conf. on Programming Concepts and Methods, 1990. (M. Broy and C.B. Jones, eds.), North Holland, 1990, pp. 443–458.

39. R.J. van Glabbeek, S.A. Smolka, and B. Steffen, *Reactive, generative, and stratified models of probabilistic processes*, Information and Computation **121** (1995), 59–80.

40. R. J. van Glabbeek, S. A. Smolka, B. Steffen, and C. M. N. Tofts, *Reactive, generative, and stratified models of probabilistic processes*, Logic in Computer Science, 1990, pp. 130–141.

41. H. A. Hansson, *Time and probability in formal design of distributed systems*, Ph.D. thesis, Uppsala University, Department of Computer Systems, 1991, Also appeared in Real-Time Safety Critical Systems, vol. 1, Elsevier, 1994.

42. J.I. den Hartog, *Probabilistic extensions of semantical models*, Ph.D. thesis, Vrije Universiteit Amsterdam, 2002.

43. B. R. Haverkort, *Markovian models for performance and dependability evaluation*, Lectures on Formal Methods and Performance Analysis, First EEF/Euro Summer School on Trends in Computer Science, Berg en Dal, The Netherlands, July 3–7, 2000 (E. Brinksma, H. Hermanns, and J.-P. Katoen, eds.), LNCS, vol. 2090, Springer-Verlag, 2001, pp. 38–84.

44. H. Hermanns, *Interactive Markov chains*, Ph.D. thesis, Universiät Erlangen-Nürnberg, 1998, Revised version appeared as Interactive Markov Chains And the Quest for Quantified Quality, LNCS 2428, 2002.

45. J. Hillston, *A compositional approach to performance modelling*, Ph.D. thesis, University of Edinburgh, 1994, Also appeared in the CPHC/BCS Distinguished Dissertation Series, Cambridge University Press, 1996.

46. H. Hansson and B. Jonsson, *A logic for reasoning about time and reliability*, Formal Aspects of Computing **6** (1994), 512–535.

47. C.A.R. Hoare, *Communicating Sequential Processes*, Prentice Hall, 1985.

48. R. A. Howard, *Dynamic probabilistic systems*, John Wiley & Sons, Inc., New York, 1971.

49. J.I. den Hartog and E.P. de Vink, *Mixing up nondeterminism and probability: A preliminary report*, Proc. PROBMIV'98 (C. Baier, M. Huth, M. Kwiatkowska, and M. Ryan, eds.), ENTCS 22, 1998.

50. _____, *Verifying probabilistic programs using a Hoare-like logic*, International Journal of Foundations of Computer Science **13** (2002), 315–340.

51. B. Jonsson and K.G. Larsen, *Specification and refinement of probabilistic processes*, Proceedings of Sixth Annual IEEE Symposium on Logic in Computer Science, 1991. LICS '91., IEEE, 1991.

52. B. Jonsson, K.G. Larsen, and W. Yi, *Probabilistic extensions of process algebras*, Handbook of Process Algebras, Elsevier, North Holland, 2001.

53. B.P.F. Jacobs and J.J.M.M. Rutten, *A tutorial on (co)algebras and (co)induction*, Bulletin of the EATCS **62** (1996), 222–259.

54. C.-C. Jou and S.A. Smolka, *Equivalences, congruences and complete axiomatizations for probabilistic processes*, Proceedings of CONCUR'90 (J.C.M. Baeten and J.W. Klop, eds.), Springer-Verlag, 1990, pp. 367–383.

55. B. Jonnson and W. Yi, *Testing preorders for probabilistic processes can be characterized by simulations*, Theoretical Computer Science **282** (2002), 33–51.

56. M.Z. Kwiatkowska and G.J. Norman, *A testing equivalence for reactive probabilistic processes*, EXPRESS '98 Fifth International Workshop on Expressiveness in Concurrency, ENTCS 16(2), 1998.

57. J. G. Kemeny and J. L. Snell, *Finite Markov Chains*, Springer-Verlag, New York, 1976.

58. N. López and M. Núñez, *An overview of probabilistic process algebras and their equivalences*, 2003, this volume.

59. G. Lowe, *Probabilistic and prioritized models of timed CSP*, Theoretical Computer Science **138** (1995), 315–352.

60. L. R. Lewis and C. H. Papadimitriou, *Elements of the theory of computation*, Prentice-Hall, Englewood Cliffs, NJ, 1981.
61. K. G. Larsen and A. Skou, *Bisimulation through probabilistic testing*, Information and Computation **94** (1991), 1–28.
62. K.G. Larsen and A. Skou, *Compositional verification of probabilistic processes*, CONCUR '92, Third International Conference on Concurrency Theory, Stony Brook, NY, USA (R. Cleaveland, ed.), LNCS, vol. 630, Springer-Verlag, 1992, pp. 456–471.
63. N. A. Lynch and M. Tuttle, *Hierarchical completeness proofs for distributed algorithms*, Proceedings of the 6th Annual ACM Symposium on Principles of Distributed Computing, 1987.
64. S. MacLane, *Categories for the working mathematician*, Springer-Verlag, 1971.
65. R. Milner, *Calculi for synchrony and asynchrony*, Theoretical Computer Science **25** (1983), 267–310.
66. _____, *Communication and Concurrency*, Prentice-Hall, 1989.
67. C. Morgan, A. McIver, K. Seidel, and J.W. Sanders, *Refinement oriented probability for CSP*, Formal aspects of computing **8** (1996), 617–647.
68. L.S. Moss, *Coalgebraic logic*, Annals of Pure and Applied Logic **96** (1999), 277–317.
69. G. Norman, *Metric semantics for reactive probabilistic processes*, Ph.D. thesis, School of Computer Science, University of Birmingham, 1997.
70. B. Plateau and K. Atif, *Stochastic automata network for modeling parallel systems*, IEEE Trans. on Software Engineering **17** (1991), 1093–1108.
71. A. Pnueli and L. Zuck, *Verification of multiprocess probabilistic protocols*, Distributed Computing **1** (1986), no. 1, 53–72.
72. _____, *Probabilistic verification*, Information and Computation **103** (1993), 1–29.
73. M.O. Rabin, *Probabilistic automata*, Information and Control **6** (1963), 230–245.
74. J.J.M.M. Rutten, *Universal coalgebra: A theory of systems*, Theoretical Computer Science **249** (2000), 3–80.
75. E.W. Stark, R. Cleaveland, and S.A. Smolka, *A process-algebraic language for probabilistic I/O automata*, Proc. CONCUR'03 (R. Amadio and D. Lugiez, eds.), LNCS, vol. 2761, Springer, 2003, pp. 193–207.
76. R. Segala, *Modeling and verification of randomized distributed real-time systems*, Ph.D. thesis, MIT, 1995.
77. K. Seidel, *Probabilistic communicating processes*, Theoretical Computer Science **152** (1995), 219–249.
78. R. Segala and N.A. Lynch, *Probabilistic simulations for probabilistic processes*, Proc. Concur'94, LNCS 836, 1994, pp. 481–496.
79. S. A. Smolka and B.U. Steffen, *Priority as extremal probability*, Proceedings of CONCUR'90 (J.C.M. Baeten and J.W. Klop, eds.), LNCS, vol. 458, Springer-Verlag, 1990, pp. 456–466.
80. M.I.A. Stoelinga, *Alea jacta est: verification of probabilistic, real-time and parametric systems*, Ph.D. thesis, University of Nijmegen, the Netherlands, 2002.
81. _____, *An introduction to probabilistic automata*, EATCS bulletin, vol. 78, 2002.
82. M.I.A. Stoelinga and F.W. Vaandrager, *Root contention in IEEE 1394*, Proc. 5th International AMAST Workshop, ARTS'99, Bamberg, Germany (J.-P. Katoen, ed.), LNCS, vol. 1601, Springer-Verlag, 1999, pp. 53–75.
83. _____, *A testing scenario for probabilistic automata*, Proceedings of the 30th International colloquium on automata, languages and programming (ICALP'03) Eindhoven, the Netherlands, June 2003, LNCS, vol. 2719, Springer-verlag, 2003, pp. 464–477.

84. M.Y. Vardi, *Automatic verification of probabilistic concurrent finite state programs*, Proc. FOCS'95 (Portland, Oregon), IEEE Computer Society Press, 1985, pp. 327–338.
85. E.P. de Vink, *On a functor for probabilistic bisimulation and the preservation of weak pullbacks*, Tech. Report IR–444, Vrije Universiteit Amsterdam, 1998.
86. E.P. de Vink and J.J.M.M. Rutten, *Bisimulation for probabilistic transition systems: a coalgebraic approach*, Theoretical Computer Science **221** (1999), 271–293.
87. S.-H. Wu, S. A. Smolka, and E. W. Stark, *Composition and behaviors of probabilistic I/O automata*, Theoretical Computer Science **176** (1997), 1–38.
88. W. Yi and K.G. Larsen, *Testing preorders for probabilistic and non-deterministic processes*, Protocol Specification, Testing and Verification (Florida, USA), vol. 12, 1992, pp. 47–61.

Tutte le Algebre Insieme:

Concepts, Discussions and Relations of Stochastic Process Algebras with General Distributions

Mario Bravetti[1] and Pedro R. D'Argenio[2]*

[1] Dip. di Scienze dell'Informazione, Università di Bologna,
Mura Anteo Zamboni 7, 40127 Bologna, Italy
bravetti@cs.unibo.it
[2] CONICET – FaMAF, Universidad Nacional de Córdoba,
Ciudad Universitaria, 5000 Córdoba, Argentina
dargenio@famaf.unc.edu.ar

Abstract. We report on the state of the art in the formal specification and analysis of concurrent systems whose activity duration depends on general probability distributions. First of all the basic notions and results introduced in the literature are explained and, on this basis, a conceptual classification of the different approaches is presented. We observe that most of the approaches agree on the fact that the specification of systems with general distributions has a three level structure: the process algebra level, the level of symbolic semantics and the level of concrete semantics. Based on such observations, a new very expressive model is introduced for representing timed systems with general distributions. We show that many of the approaches in the literature can be mapped into this model establishing therefore a formal framework to compare these approaches.

1 Introduction

The research community has widely recognized the importance of time aspects in the specification and analysis of concurrent systems and communication protocols (see e.g. [1, 3, 19] and the references therein). There are fundamentally two reasons behind this recognition: first of all, the (correct) behavior of certain systems/protocols often depends on real-time aspects; second, expressing time duration of system activities makes it possible to estimate system performance. In this paper we report on the state of the art in the formal specification and analysis of concurrent systems whose activity durations are random variables.

The random duration of an activity is represented by probability distribution functions. For example Fig. 1 depicts the probability of the duration of an activity. This can be any time value between two and four time units.

An important special case of duration distribution is the exponential distribution. Due to its nice mathematical properties (the so called "memoryless"

---

* Partially supported by the NWO visiting grant B-61-519.

C. Baier et al. (Eds.): Validation of Stochastic Systems, LNCS 2925, pp. 44–88, 2004.
© Springer-Verlag Berlin Heidelberg 2004

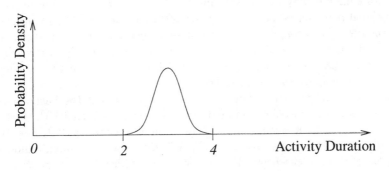

Fig. 1. Activity with a probabilistic duration

property),the problem of specifying systems which only make use of exponential distributions is easier than the general case and has been successfully studied (see e.g. [20, 19, 3, 5]). The approach obtained in this restricted setting is often referred to as the "Markovian" approach, in that system behavior turns out to be expressible by simple continuous time Markov chains (CTMCs). In spite of its advantages, the Markovian approach imposes a considerable limitation on the modeling of time aspects. Therefore, it is important to address the general case in which probability distributions can be arbitrary.

In order to better understand the power of general distributions, let us look again at the example of Fig. 1. The distribution of Fig. 1 expresses both *time bounds* for the activity (i.e. it will certainly last between 2 and 4 time units) and a *probabilistic quantification* over the possible duration values for the activity. This shows that, compared to the Markovian case where time bounds are not expressible, general distributions make it possible to express both the system aspects typical of real time modeling (where time bounds are represented) and the system aspects typical of stochastic modeling (where a probabilistic quantification of time values is expressed) in an integrated fashion. In particular, given a system specification with general distributions, it is possible both to validate its real time properties via, e.g., model checking, and to evaluate its performance.

The problem of specifying and analyzing systems with general distributions has been studied, e.g., in [11, 9, 36, 17, 32, 6, 4]. This paper focuses on such a general problem and presents the main concepts and results obtained in the literature. More precisely we proceed as follows. First of all we explain basic notions about the formal specification and analysis of systems with general distributions and, on this basis, we present a conceptual classification of the different approaches in the literature.

A result of such a conceptual step is the observation that much of the approaches in the literature are based on the common idea that the specification of systems with general distributions has a three level structure: the process algebra level, the level of symbolic semantics (where system timed behavior is represented in a symbolic way by clocks), and the level of concrete semantics (where system timed behavior is represented explicitly by numerical timed transitions). Based on such an observation, we introduce a new model to represent timed systems with general distributions. We show that many of the approaches in the literature can be mapped into this model establishing therefore a formal framework to compare such approaches.

The paper is organized as follows. In Sect. 2 we present the basic concepts about the formal specification and analysis of concurrent systems with stochastic time on the basis of the several approaches in the literature. In Sect. 3 we introduce "prioritized stochastic automata (PSA)": the common symbolic model for the representation of stochastic timed systems on which we will map such approaches. Sect. 4 introduces "probabilistic timed transition systems (PTTSs)": the concrete timed model used to define the semantics of PSA. Actually we consider two kind of semantics based on so-called "residual" (see [11]) and "spent" (see [4]) clock lifetimes. Sect. 5 defines symbolic bisimulation equivalence (over PSA) and concrete bisimulation equivalence (over PTTSs). In Sect. 6 we provide embeddings of the stochastic models presented in [4, 11, 9] into PSA. In Sect. 7, semantics for the different process algebras in [4, 11, 9] are given directly in terms of PSA and shown to be consistent with their original semantics. In Sect. 8 we report some notes and discussions concerning a detailed comparison of the approaches in [11, 9, 36, 17, 32, 6, 4]. Sect. 9 concludes the paper.

2   Concurrent Systems with Stochastic Time

In this section we explain how to formally represent and reason about concurrent systems whose behavior is specified by means of activities with random duration. As explained in Sect. 1, we focus on the general case, i.e., the expressiveness of our specification paradigms is not limited to "memoryless" exponential distributions. As we will see, abandoning such limitation changes the nature of the problem and increases its complexity. In order to better understand this it is important to glance some details of the Markovian approach.

2.1   Exponential Distributions Make Things Easy

Restricting to exponentially distributed durations gives the advantage of having activities for which the memoryless property holds. This property basically says that at each time point in which an activity has started but not terminated yet, the residual duration of the activity is still distributed as the entire duration of the activity. Such a property makes it possible to represent "timed" behavior of systems by a continuous time Markov chain (CTMC), i.e. a simple continuous time stochastic process where in each time point the future behavior of the process is completely independent of its past behavior and depends on its current state only (Markov property). In fact the memoryless nature of time in the Markovian approach makes it possible to avoid the explicit representation of time passage in system specifications. For instance, consider a simple example of

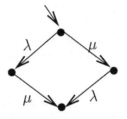

Fig. 2. Parallel of exponential delays

two exponentially timed activities with rates (the parameters of the exponential distribution) λ and μ executed in parallel. The resulting CTMC is the one in Fig. 2. Transitions in a CTMC represent exponentially distributed delays and choices in a state of a CTMC are resolved via the "race policy", i.e. the delays represented by the outgoing transitions are executed in parallel and the first delay that terminates determines the transition to be performed. Therefore, in the example, both delays count synchronously from the initial state and, when one of them terminates, the corresponding transition is executed. Such a transition leads to a state where the other delay counts its residual duration until it terminates as well. Note that, because of the memoryless property, the residual duration of the delay is also exponentially distributed and with the same rate. Hence the CTMC of Fig. 2 is an adequate representation of the parallel execution of the two considered activities.

If system behaviors in the form of CTMCs are obtained from system specifications expressed with a process algebra (see, e.g., [20, 19, 3, 5]), the "+" operator will be the natural choice to express the same kind of alternative given by transitions leaving a state of a CTMC. For example, $\lambda + \mu$ represents a choice between exponentially timed delays λ and μ and it is solved via the race policy explained above. As far as the "||" operator is concerned, the example of Fig. 2, should make clear that the simple standard interleaving semantics can be adopted: the semantics of $\lambda \,\|\, \mu$ is just that of $\lambda.\mu + \mu.\lambda$. This example also shows that having a race policy interpretation of operator "+" and an interleaving semantics for operator "||" yields an expansion law like $\lambda \,\|\, \mu = \lambda.\mu + \mu.\lambda$, which is crucial for building complete axiomatizations of process equivalence notions.

Regarding equivalences, it turns out that standard bisimulation equivalence (as opposed to other notions of equivalence) can be easily extended by following an approach similar to that of [26] for discrete probabilistic choices.

2.2   A Symbolic Model with Clocks

When the restriction to exponential distributions is abandoned, the behavior in a global system state *does* depend on the time partially spent by activities currently under execution. Therefore, we are forced to explicitly represent the passage of time. For instance, if the two activities in the example of Fig. 2 are *not* exponentially distributed, the time spent in a state reached after the termination of the first activity effectively depends on the time the second activity has already (partially) spent in execution.

A simple way of representing the execution time of an activity is to adopt a model with clocks in a similar manner timed automata do [1]. A timed automaton represents the behavior of a system in terms of a fixed set of clocks c_1, c_2, c_3,.... During the execution of the automaton, each clock has an associated time value. When the automaton sojourns in a state, clock time values increase in a synchronous manner. The transitions of the automaton are executed instantaneously: their execution may depend on some condition on clocks (e.g. $c_1 \geq t$ for some time value t) and may cause some clock to be set to some value (e.g. $c_3 := 0$). More precisely, labels of timed automata transitions are of three kinds:

- *actions*, used to express the occurrence of events and to synchronize system components (when composing in parallel several timed automata),
- *guards*, expressing a condition on clocks, and
- *clock setting* events.

Note that the representation of time passage in a timed automata is *symbolic* in the sense that the temporal behavior of the system is expressed by means of events like clock setting and clock constraints instead of *concrete* (real-valued) timed transitions. This makes it possible have a finite representation of system behavior, hence the chance of analyzing some of its properties in a computable way, even if the time domain is assumed to be (continuously) infinite. On the other hand the semantics of timed automata (the meaning of the symbolic representation) is usually defined in terms of an inherently infinite *concrete* semantic model which makes use of real-valued timed transitions.

It is easy to see that a simple probabilistic extension of the clocks of a timed automaton gives the possibility of representing generally distributed time.

Introducing Stochastic Clocks

In order to express generally distributed time it is sufficient to consider a variant of timed automata where:

- each clock has an associated *probability distribution* expressing its *duration*: $c_1 \mapsto f$, $c_2 \mapsto g$, $c_3 \mapsto h$,...;
- the three kinds of labels of a transitions are the following ones:
  - *actions* (represented with a, b,...), used to express the occurrence of events and to synchronize system components (as in timed automata),
  - *guards*, requiring all clocks in a certain set to be *terminated*, and
  - *clock setting* events representing the *start* of all clocks in a certain set;
- we possibly may express probabilistic and/or prioritized choices.

By using an approach like this, two parallel activities with random duration can be easily represented even if they are not exponentially distributed. Suppose that the duration of two activities are generally distributed according to distributions f and g. The behavior of $f \parallel g$ can be represented as in Fig. 3. In this figure,

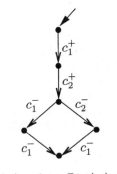

$$Dist(c_1) = f, \quad Dist(c_2) = g$$

Fig. 3. Parallel of generally distributed delays

the execution of these activities is represented by clocks c_1 and c_2, where their associated distribution functions are f and g, respectively. A c^+ label means that the clock c is set to be started, while a c^- label represents the clock c to be terminated. In this event-based representation, we assume that whenever a clock is not explicitly restarted it continues counting. Therefore, Fig. 3 is a correct representation of the parallel execution $f \parallel g$. (Compare to the exponential case in Fig. 2, in which a clock does not need to be explicitly started.)

Models to represent systems like this, which may execute generally distributed timed activities in parallel, are also used in probability theory. In particular the class of generalized semi-Markov processes (GSMPs) exploits a similar event-based symbolic representation where clocks are also called "elements". GSMPs also provide the capability of expressing probabilistic choices.

2.3   A Concrete Semantics

Similarly as for classic timed automata, the meaning of the symbolic representation given by an automata with probabilistic clocks can be formally defined in terms of a concrete model with real-valued timed transitions.

A Concrete Probabilistic Timed Model

In the case that a continuous time domain is considered (as we do in this paper), the concrete model is given by a transition system on an uncountably large state space where time passage is represented explicitly via transitions labeled with real numbers. More precisely, the concrete model must have at least the following three kinds of transitions:

- *actions* transitions,
- *time* transitions, labeled with a time $t \in \mathbb{R}_{\geq 0}$,
- *continuous probabilistic* transitions represented by probability spaces.

In particular, continuous probabilistic choices are used to represent the *sampling of time values* from distributions on durations.

The easiest way to understand the concrete model is to see it as a representation of the behavior of the symbolic probabilistic automata when it is "executed" (i.e. when actual time values are sampled from duration distributions). In the literature, there are two techniques to represent this execution (i.e. to define the semantics of the symbolic automata):

- keeping track of *residual lifetime* of clocks, i.e. the amount of time that a clock has to spend in execution from the current time instant until its termination.
- keeping track of *spent lifetime* of clocks, i.e. the amount of time that a clock has already spent in execution since it was started until the current time instant.

While the residual lifetime of a clock decreases as time passes, the spent lifetime of a clock increases. This last case is similar to the way clock values are treated in the semantics of timed automata.

Recording Residual Lifetimes

The technique based on clock residual lifetimes (see e.g. [11]) is the simplest one: the duration of a clock is decided once and for all when a clock is started (by sampling from its associated distribution) and the duration time is decreased until it gets to zero.

More precisely the concrete semantic model is derived from the stochastic automata as follows:

- When a clock is *started*, its residual lifetime is determined by sampling a value from its associated duration distribution.
- When a clock is under execution, its time to termination t_{TERM} is given by its residual lifetime t_R ($t_{TERM} = t_R$). This means that if we consider the situation of the system after a time period t (i.e., a t transition is performed):

  - if $t < t_{TERM}$ then the residual lifetime of the clock becomes $t_R - t$,
  - if $t \geq t_{TERM}$ then the clock is *terminated*.

The main advantage of this technique, which is commonly used in discrete event simulation, is its simplicity and applicability (see e.g. [27]). However, it may be argued that this approach is not adequate if non-deterministic choices need to be resolved by adversaries. This is due to the fact that the eventual duration of a clock is decided when it starts. Hence, an adversary can base its decisions on the knowledge of the future behavior of the system and therefore, play a more angelic (or more demonic) role than it is desired. For instance, if three clocks are being concurrently executed in a symbolic state, an adversary a priory may know not only which one of the three clocks will terminate first in such a state, but also which clock will be the next one to terminate, thus obtaining information about the future behavior of the system.

Recording Spent Lifetimes

The technique based on clock spent lifetimes (see [4]) is more complex but similar to that of timed automata: the spent lifetime of a clock is set to zero when it starts and it increases as time passes. At every (concrete) state the distribution of the time to termination of the clock is resampled from its associated duration distribution conditioned to the spent lifetime.

More precisely the concrete semantic model is derived from the automata with stochastic clocks as follows:

- When a clock is *started* its spent lifetime is set to 0.
- When a clock is under execution, its time to termination t_{TERM} is determined by sampling a value from its *residual duration distribution*. This is computed from:
  - the spent lifetime t_S of the clock and
  - the duration distribution associated to the clock.

  This means that if we consider the situation of the system after a time period t (i.e., a t transition is performed):
  - if $t < t_{TERM}$ then the spent lifetime of the clock becomes $t_S + t$,
  - if $t \geq t_{TERM}$ then the clock is *terminated*.

This technique presents an appropriate context for the resolution of non-determinism by adversaries. However, it cannot easily be taken into practice by means of discrete event simulation. This would require to compute the residual duration distribution of every clock in execution and sample a value from it, and this repeated at every state traversed by the automata.

2.4   Bisimulation Equivalences

Similarly to the Markovian case, we want to find some extension of standard bisimulation equivalence which makes it possible to reason about equivalence of systems with generally distributed durations.

It is important to have a definition of bisimulation equivalence both at the symbolic model and at the concrete model level. Equivalences at the symbolic level make it possible to actually decide the equivalence of two systems and to minimize the state space of a system in a computable way. Equivalences at the concrete level show that equivalence at the symbolic level is correct with respect to the adopted concrete semantics. Note that in general, at the concrete level, systems can be compared on the basis of actual time delays and probability spaces, while at the symbolic level we can only check correspondence of clock-events and correspondence of probability distributions. Therefore equivalence at the concrete level represents, somehow, the coarsest notion of equivalence that we can hope to gain with a symbolic equivalence (in fact the symbolic equivalence is typically much finer than the concrete one).

Symbolic Bisimulation

Symbolic bisimulation is an equivalence defined on the symbolic stochastic timed model. It is defined in such a way that:

- *action* transitions are matched as in standard bisimulation,
- *clock start* and *clock termination* events are matched if they refer to clocks with the same duration distribution.

Moreover if the symbolic model includes probabilistic choices they are matched as in standard probabilistic bisimulation [26].

Two approaches may be followed in order to match clock start and termination. The simplest one is to only match clocks if they have the same name, therefore ensuring that duration distributions are the same. The advantage of this approach is its simplicity; its disadvantage is that it hardly provides any stochastic insight.

The other approach relies on a *clock name association*, given by a function or a relation. This relation ensure that start and termination events of clocks with different names but with the same duration distribution are properly matched. This second type of symbolic bisimulation is coarser than the first one.

Examples of the first case appear in [14, 11]; examples of the second one can be found in [6, 11, 4].

A particularly distinct case appears in [8, 4] where symbolic models can be restricted to canonical one (in the sense that clocks are canonically named) for which building clock name associations is not needed (it works just like the first type of equivalence) and nonetheless it equates symbolic automata just like the second kind.

Concrete Bisimulation

Concrete bisimulation is an equivalence defined on the concrete timed model. In particular it is defined in such a way that:

- *action* transitions and *timed* transitions are matched as in standard bisimulation,
- *continuous probabilistic* transitions are matched according to an extension of probabilistic bisimulation [26] to continuous probability spaces.

2.5   Dealing with Composition: Process Algebra

The essence of designing a process algebra for modeling systems with generally distributed activities can be captured by simply considering the extension of a standard process algebra with a new prefix or guard "$f.P$". This new operation represents the execution of P after a random delay (sampled from the general distribution f) took place.

In order to obtain a symbolic model with clocks just like those previously described, the semantics of $f.P$ should take into account the following:

- it should represent the execution of the delay as the combination of a *start* and a *termination* event (e.g. denoted by f^+ and f^-, respectively).
- it must generate a (clock) *name* for the delay which keeps it distinguished from other delays being executed at the same time (e.g. in $f \parallel f$).

If we see f as being the "type" of the delay, a semantics which operates in this way is exactly the classical ST semantics of [15, 7]. Therefore, conceptually, the essence of the problem of representing symbolically general distributions is just like the classical problem of expressing ST semantics for actions in algebraic terms.

As in the standard case of ST semantics (see [7]), we can employ different kind of techniques for generating clock names from delay prefixes. In particular such techniques can be classified into *static* and *dynamic* techniques. Note that it is also possible (see [11]) to define the semantics of the process algebra by abstracting from the particular name generation mechanism. We can just say that names are generated by *arbitrary α-conversion*: any name can be chosen for a delay provided that it is not in use by another clock in execution. Terms are then considered to be equivalent up to (distribution preserving) clock associations.

On the other hand, choosing a particular technique (static or dynamic) may lead to smaller models or may make it easier to check symbolic bisimulation equivalence at the price of introducing some (often quite complex) unique name generation mechanism.

In static techniques, clock names are generated statically according to the *syntactic structure* of the process algebra term. For instance, in [6] clocks are named according to the syntactic position of the delay w.r.t. the parallel operators in the term (the *location* of the delay). A simple rule like this is enough to guarantee that two delays that may be executed at the same time always get different names. The main advantage of the static approach is simplicity and the size of the symbolic state space which is smaller than the one of the dynamic approach, due to the fact that a delay always gets the same name independently on when it is executed. The main drawback is that clock names must be explicitly associated in the symbolic bisimulation (see, e.g., [6, 11]) in order to capture equivalence of systems just based on duration distributions.

Example 1. In figure 4 we depict the behavior of "$f \parallel f$" (only the phase of clock starts) according to a static technique like that of [6] which assigns names to delays according to their position w.r.t. parallel operators. In the example of figure 4 the delay f to the left of the parallel operator is named "f_l", the delay f to the right "f_r". Note that, since names of delays are determined by their

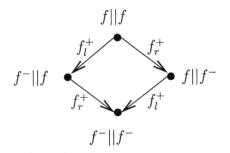

Fig. 4. Parallel of delays with the static technique (starting phase)

syntactical position in the current term, after an event of starting of a delay f, we just have to record that this happened by turning f prefix into f^- in the term (we do not have to record names assigned to starting delays).

In dynamic techniques, clock names are generated dynamically (at run-time) according to the order of delay execution, with a fixed rule. For instance, in the approach of [8] the clock name generated for a delay f is f_i, where i is the least $i \in \mathbb{N}$ which is not used in the name of any clock with the same distribution f already in execution. The main advantage of the dynamic approach is the fact that clock names do not need to be explicitly associated in the symbolic bisimulation. This is due to the fact that the method to compute new names is fixed: processes that perform equivalent computations generate the same "canonical" names for clocks. The main drawback is the complex mechanism to compositionally generate canonical names in operational semantics (we have to perform a so-called level wise renaming [8,4]) and the size of the state space which is larger than the one of the static approach since the same delay may get different names depending on the moment in which it is executed with respect to other delays.

Example 2. In figure 5 we depict the behavior of "$f \parallel f$" (only the phase of clock starts) according to a dynamic technique like that of [8] which assigns names to delays (indexes $i \in \mathbb{N}$) according to the rule of the minimum index not currently in use by delays with the same distribution. In the example of figure 5, depending on the order in which delays are started, either the lefthand one gets name 1 and the righthand one gets name 2, or vice-versa. Note that, since names of delays are determined by the order in which they are executed, after an f_i^+ event, we have to record, not only that this happened by turning f prefix into f^- in the term, but also the name i assigned to the delay f at the moment of delay start. As a consequence we produce an additional state w.r.t. the static naming case.

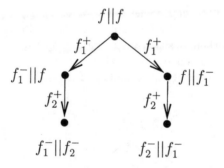

Fig. 5. Parallel of delays with the dynamic technique (starting phase)

It could be said that [9,23] also follow a dynamic approach. However, as event names are arbitrarily chosen, it does not have the advantages of using "canonically" chosen names. Other approaches do not generate a symbolic model as an intermediate semantics, therefore they are not concerned with clock name generation.

Basic Operators in the Case of Pure ST Semantics

As far as the choice operator "+" is concerned, due to the generation of start and termination events f^+ and f^- from delays f operated by ST semantics, we have that (intuitively)

$$f + g = f^+.f^- + g^+.g^-$$

i.e. choices between delays become *choices between starting events* in semantic models. Since start events are executed immediately this means that choices are solved by a *preselection policy*, where first we choose the delay to be performed and then we execute it, instead of the *race policy* used with exponential distributions. A possible justification of this fact (given in [8, 4]) is the following one. Preselection policy can be claimed to be more natural than race policy when the restriction to exponential distributions (justifying race policy) is abandoned. This because it causes $f + g$ to represent a real choice and not a form of parallel execution. In particular in the approach of [8, 4] probability information is attached to delays (by using "weight" values) and the choice between start events is resolved probabilistically. For instance, $<f, 1>.P + <g, 2>.Q$ means that $1/3$ of the time the delay f is chosen, after which P is executed, while the other $2/3$ the delay g followed by process Q is executed. This choice allows for a representation of the full GSMP model [10] (with the only restriction that we assume "decay rate" of delays to be always 1). On the other hand, adopting preselection policy for generally distributed delays is absolutely not mandatory (it is just a matter of taste) and as we will see, there are alternative technical solutions (e.g. that of [11]) which allow race policy for the "+" operator to be used.

As far as the parallel operator "||" is concerned, due to the generation of start and termination events f^+ and f^- from delays f operated by ST semantics, we have that (intuitively):

$$f \parallel g = f^+.f^- \parallel g^+.g^- = f^+.g^+.(f^-.g^- + g^-.f^-) + g^+.f^+.(f^-.g^- + g^-.f^-)$$

i.e. parallel of delays becomes *interleaving of events* in semantic models[1]. This shows that even in the case of general distributions it is possible to have an *expansion law at the level of events*. This makes it possible to produce complete axiomatizations for symbolic bisimulation equivalence (see [11, 4]). In the history of stochastic process algebra with general distributions obtaining an expansion law has been an important issue. First approaches failed to obtain expansion laws providing, at best, only partial decomposition (e.g. [17, 36, 32, 18]). Others pursued a true concurrency approach, dropping the idea of having an expansion law [9, 23]. Modern stochastic process algebras solved this problem and the solution depends on the splitting of the delay into start and termination *events*: parallel of delays becomes the *interleaving of events*.

---

[1] For the sake of simplicity we assume delay starts to have precedence over delay terminations of another process as in [8, 4], hence we do not generate the complete interleaving of events.

Technical Solutions Different to ST Semantics

The treatment of generally distributed durations in process algebra that we explained above is just intended to be conceptual and adheres completely to the approach of [8] only.

In the literature we can find approaches which essentially follows the concept that we explained, but adopt some different technical solution:

- We can have *event separation* already in the specification language [11]. In this case clock events C^+ and C^- are used directly in the initial system algebraic specification instead of delays f. This somehow gives more specification freedom, but forces the specifier to deal with clock names at the algebraic level.
- We can have that *clock start* events do *not resolve the choice* [11]. In this case the choice operator treats start events in a "special" way, i.e. in such a way that their execution does not resolve the choice. Therefore having that (intuitively):

$$(\{c_1^+\}.c_1^-) + (\{c_2^+\}.c_2^-) = \{c_1^+, c_2^+\}.(c_1^- + c_2^-)$$

  where $\{c_1^+, c_2^+, ...\}$ represents a set of clocks starting at the same time.
  In this way we can obtain *race policy* instead of *preselection policy* for choice even in the case of generally distributed durations.
- The possibility of dealing with *termination of multiple clocks* can be introduced [11, 6]. This basically means that we can specify systems behaving like this (intuitively):

$$\{c_1^+, c_2^+\}.(\{c_1^-\}.P + \{c_2^-\}.Q + \{c_1^-, c_2^-\}.R)$$

  where $\{c_1^-, c_2^-, \dots\}$ represents a set of terminated clocks.

2.6   Totally Different Approaches

In the following we briefly discuss approaches which deal with general distributions in a very different way with respect to the methodology above.

- The approach of [9] is based on a *truly concurrent* semantics for the process algebra. The semantics of process algebraic specifications is given in terms of *stochastic bundle event structures* instead of transitions over global system states. Therefore it does not represent "interleaved" execution of processes, but keeps instead a "local" representation of the behavior of each process. The advantage is that it does not need to represent residual duration distributions of generally distributed activities. However, it is not so clear how to use truly concurrent models for actual system analysis apart from discrete event simulation [24] and analysis of the first passage time of events [34] (a technique to approximate performance measures).
- The approach of [36, 18] is based on a *direct concrete semantics* for the process algebra. The semantics of process algebraic specifications is given directly in terms of a concrete model with explicit time. This is a fine approach

if we only expect to perform discrete event simulation of systems. The main advantage is that it is easy to deal with very expressive process algebras: we do not have to worry about how to develop symbolical representations. However, concrete models are mostly uncountably large which makes them unsuitable for system analysis not based on simulation. In a recent work [22] a methodology for obtaining finite semantic models from the algebra of [36, 18] is defined, which is based on symbolic operational semantics. Such semantics generates symbolical transition systems which abstract from time values by representing operations on values as symbolic expressions. In this way, for systems belonging to a certain class, it is possible to derive a GSMP via a (quite involved) procedure.

– The approach of [29] defines a testing theory for semi-Markov processes. This is done by using a process algebra which is similar to IGSMP [8, 4], but where the parallel operator is left out. Tests are processes which do not include probabilistic delays (just actions and discrete probabilistic choices).

3   A Common Model for Stochastic Timed Systems

The model we introduce in this section considers the ingredients discussed in the previous section: clock start, clock termination, execution of actions, and probabilistic jumps. All these ingredients are included in only one symbolic transition. The aim of this model is to encode many formalisms for modelling stochastic timed systems, hence having a common framework to formally compare existing approaches. In addition, this model also considers priorities because some frameworks implicitly include this feature and, moreover, they make it possible to represent maximal progress and urgency. This model is an extension of stochastic automata [14, 11] with priorities and probabilistic transitions.

With PDF we denote the set of all probability distributions functions and with $Prob(\Omega)$ we denote the set of all probability spaces with sample space in a subset of Ω. We let $Prob_d(\Omega)$ denote the subset of $Prob(\Omega)$ containing only discrete probability spaces. We use $\rho, \rho', \rho_i, \ldots$ to denote discrete probability spaces and $\pi, \pi', \pi_i, \ldots$ to denote probability spaces in general. In both cases, we will overload the notation and use these letters to represent the probability measure as well.

Definition 1. *A prioritized stochastic automata (PSA) is defined to be a structure* $(St, Ck, Distr, Act, \longrightarrow, s_0, C_0)$ *where*

– *St is a countable set of* control states *with $s_0 \in St$ being the* initial control state.
– *Ck is the set of* clock names *with $C_0 \subseteq Ck$ being the set of* clocks to be started at initialization. *For the sake of clarity in technical manipulation, we assume that Ck is totally ordered, and that if $C \subseteq Ck$, $\vec{C}$ is the vector induced by this order.*
– *$Distr : Ck \rightarrow PDF$ assigns a probability distribution function to each clock. If $f \in PDF$, we usually name a clock $c_f \in Ck$ to indicate that $Distr(c_f) = f$.*

- *Act is set of* actions *partitioned in the following sets:*
  - *Act_d, the set of* delayable *actions and*
  - *Act_u, the set of* urgent *actions*

  where Act_u is ordered according to a priority relation $\prec$ *which is a strict order with the* silent action $\tau \in Act_u$ *being the maximum element.*
- $\longrightarrow \subseteq St \times 2^{Ck} \times Act \times Prob_d(2^{Ck} \times St)$ *is the* control transition *relation.*

We write $s \xrightarrow{C,a} \rho$ if $(s, C, a, \rho) \in \longrightarrow$ and $s \xrightarrow{C,a}$ if there is a distribution ρ such that $s \xrightarrow{C,a} \rho$. If $\rho(C', s') = 1$ (i.e., ρ is trivial) then we write $s \xrightarrow{C,a,C'} s'$ instead of $s \xrightarrow{C,a} \rho$. The meaning of a control transition $s \xrightarrow{C,a} \rho$ is the following. To trigger the transition, all clocks in set C must terminate, that is, the transition cannot be executed as long as a clock in C is active. Transitions are labeled with actions. They can be delayable, meaning that they need to interact with the environment, or they can be urgent and they will not be allowed interact with the environment. Urgent actions impose maximal progress and therefore transitions labeled with actions of this type must be executed as soon as they are enabled. In addition, if a conflict between two enabled urgent transitions occurs, it may be solved according to the priority relation on the actions labelling the control transition provided it is defined (notice that the order may not be total). In addition, when a transition is executed, a probabilistic branching will take place according to the probability space ρ. $\rho(C', s')$ is the probability that all clocks in C' are started and the system reaches the control state s'.

Fig. 6 shows a system of three queues with two servers to process certain classes of job. Jobs arrive to the system according to an unknown rate and they queue in a buffer. Server A is intended to take jobs from this buffer a soon as it can and preprocess them. This preprocessing takes some (random) time. Moreover 1/3 of the outcome will result in a high priority job and the other 2/3 in a low priority one. As soon as the preprocessing finishes the job is queued in the high priority buffer or in the low priority one, depending whether it is high or low priority. Service B is intended to perform some postprocessing on these jobs. As soon as it can, it takes and process a job from the high priority queue provided there is any, otherwise it takes a job from the low priority queue. Processing

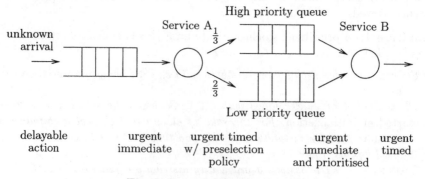

Fig. 6. A simple queuing network

takes a (random) time, and as soon as it finishes, the result is output. Notice that, since all ingredients above are present in control transitions $s \xrightarrow{C,a} \rho$, the PSA model allows for the modelling of systems like the queuing network depicted in Fig. 6.

Let *Jobs* be the possible set of jobs to be processed in the queuing network of Fig. 6. Suppose the serving time in A (resp. B) is distributed according to a distribution F_A (resp. F_B). Then, the queuing network can be modelled by the PSA defined as follows:

$$St = Jobs^* \times Jobs^* \times Jobs^* \times (Jobs \cup \{-\}) \times (Jobs \cup \{-\})$$
$$Ck = \{x_A, x_B\} \quad \text{with } Distr(x_A) = F_A \text{ and } Distr(x_B) = F_B$$
$$Act_d = \{input(j) \mid j \in Jobs\}$$
$$Act_u = \{get(j), put(j), getHigh(j), getLow(j), output(j) \mid j \in Jobs\}$$

where $getLow(j) \prec getHigh(j')$ for every $j, j' \in Jobs$ and the other actions in Act_u are not comparable.

A control state (q_I, q_H, q_L, A, B) saves in q_I, q_H, and q_L the contents of the input queue, the high priority queue, and the low priority queue, respectively, and A (resp. B) indicates if server A (resp. B) is processing a job $j \in Jobs$ or it is idling $(-)$. Initially, the system has all its queues empty and the servers are idling. Therefore $s_0 = (\epsilon, \epsilon, \epsilon, -, -)$. Moreover, initially the system is only waiting for a job to arrive so no timer needs to be set: $C_0 = \emptyset$. The control transitions are defined as follows (where $j \in Jobs$).

a job can be input at any time:
$$(q_I, q_H, q_L, A, B) \xrightarrow{\emptyset, input(j), \emptyset} (q_I j, q_H, q_L, A, B)$$
if server A is idling and q_I is not empty it must take a job and begin to process it (it starts clock x_A):
$$(jq_I, q_H, q_L, -, B) \xrightarrow{\emptyset, get(j), \{x_A\}} (q_I, q_H, q_L, j, B)$$
when clock x_A terminates, server A finishes processing its job and appends it to q_H with probability 1/3, or to q_L with 2/3:
$$(q_I, q_H, q_L, j, B) \xrightarrow{\{x_A\}, put(j)} \left\{ \begin{array}{l} \langle \emptyset, (q_I, q_H j, q_L, -, B) \rangle \mapsto \frac{1}{3}, \\ \langle \emptyset, (q_I, q_H, q_L j, -, B) \rangle \mapsto \frac{2}{3} \end{array} \right\}$$
if server B is idling and q_H is not empty it must take a job and begin to process it (it starts clock x_B), and similarly if q_L is not empty:
$$(q_I, jq_H, q_L, A, -) \xrightarrow{\emptyset, getHigh(j), \{x_B\}} (q_I, q_H, q_L, A, j)$$
$$(q_I, q_H, jq_L, A, -) \xrightarrow{\emptyset, getLow(j), \{x_B\}} (q_I, q_H, q_L, A, j)$$
when clock x_B terminates, server B finishes processing its job and outputs it:
$$(q_I, q_H, q_L, A, j) \xrightarrow{\{x_B\}, output(j), \emptyset} (q_I, q_H, q_L, A, -)$$

The server B prioritizes jobs from the high priority queue because $getLow(j) \prec getHigh(j')$ for every $j, j' \in Jobs$.

The PSA model is obtained by integrating the stochastic automata model [14, 11] with the IGSMP model [6, 8, 4] and it is based on the model used for the MoDeST language [13].

4   Semantics of PSA

The semantics of PSA is given in terms of probabilistic timed transition systems (PTTS). As discussed in Section 2.3, there are two possible interpretations. This is discussed in the following.

4.1   Probabilistic Timed Transition Systems

The PTTS model is an extension of Segala's simple probabilistic automata [35] with continuous probability spaces and time labelled transitions.

Definition 2. *A probabilistic timed transition system (PTTS) is a structure* $(\Sigma, Act \cup I\!\!R_{\geq 0}, \longrightarrow, \pi_0)$ *where*

- Σ *is a set of* states;
- Act *is a set of* actions *like in PSA;*
- $\longrightarrow \subseteq \Sigma \times (Act \cup I\!\!R_{\geq 0}) \times Prob(\Sigma)$ *is the transition relation; and*
- $\pi_0 \in Prob(\Sigma)$ *is the probability space that indicates how to select a probable initial state.*

In addition, the following requirements must hold:

1. *maximal progress:* $\forall \sigma \in \Sigma, a \in Act_u.\ \sigma \xrightarrow{a} \implies \not\exists t \in I\!\!R_{\geq 0}.\ \sigma \xrightarrow{t}$
2. *priority:* $\forall \sigma \in \Sigma, \{a, b\} \subseteq Act_u.\ (a \prec b \wedge \sigma \xrightarrow{a}) \implies \sigma \xrightarrow{b}\!\!\!\!/$

Above we use the following notation. $\sigma \xrightarrow{a} \pi$ *denotes* $(\sigma, a, \pi) \in \longrightarrow.$ *If there is a probability space* π *such that* $\sigma \xrightarrow{a} \pi$, *we write* $\sigma \xrightarrow{a};$ *if such* π *does not exists, we write* $\sigma \xrightarrow{a}\!\!\!\!/$ *Moreover, we write* $\sigma \xrightarrow{a} \sigma'$ *if* $\sigma \xrightarrow{a} \pi$ *and* π *is a trivial probability space where the atom* $\{\sigma'\}$ *has measure 1.*

Transition $\sigma \xrightarrow{a} \pi$ means that when the system is in state σ it may perform an action a and then move to some state with a probability determined by the probability space π. π may be a continuous probability space.

As an example, consider a metronome. A metronome is a device that marks the tempo of a piece of music (i.e. the speed at which it should be played). Thus, a metronome is a clock that ticks once each given interval of time. Suppose our metronome always plays *andante*; this would be a tick per second. As any clock a metronome may not be precise. Our metronome may skew up to 1 millisecond according to a uniform distribution. A possible modelling of the metronome's behaviour in terms of PTTS could be as follows (time is measured in milliseconds).

$\Sigma = \mathbb{R}$

$Act_d = \emptyset \qquad Act_u = \{tick\}$

$0 \xrightarrow{tick} \pi_U \qquad$ where π_U is the probability space in $\mathbb{R}$ with measure defined from a uniform in $[999, 1001]$

$t + t' \xrightarrow{t} t' \qquad$ with $t, t' > 0$

$\pi_0 = \pi_U$

Notice that the metronome satisfies maximal progress since $0 \xrightarrow{t}\!\!\!\!/\,$ for any real t. Maximal progress requires that when an urgent action becomes available, it must be executed without letting time pass. (See first requirement in Def. 2.)

To add an on/off button to the metronome, the previous PTTS should be modified as follows:

$\Sigma = \mathbb{R} \cup \{sys\_off\}$

$Act_d = \{on, off\}$

the transition relation is extended s.t. $sys\_off \xrightarrow{on} 0$,

$t \xrightarrow{off} sys\_off$, and $sys\_off \xrightarrow{t} sys\_off$ for all $t \geq 0$

with the rest of the components as before.

Our metronome is not as sophisticated as to include a priority scheme, but if it did, no state could show a non-deterministic choice between a high-priority labelled transition and a low-priority one. This is stated by the second requirement in Def. 2

We give two different interpretations to PSA in terms of PTTSes, one that observe the residual lifetime of the clocks to decide the enabling of a transition, and the other that observe the spent lifetime.

4.2   Residual Lifetime Semantics

In the *residual lifetime semantics*, when a clock c is started, its termination time is sampled according to $Distr(c)$. The clock c is *active* as long as it does not reach this termination time. Otherwise we say it is *terminated*. Therefore, in this context, a control transition $s \xrightarrow{C,a} \rho$ becomes enabled as soon as all clocks in C are terminated in the sense above. To carry the time value and the termination value of a clock, we use valuations. A *valuation* is a partial function v from Ck in the set $\mathbb{R}_{\geq 0}$ of non-negative real numbers. Let Val be the set of all valuations, and $TVal$ be the set of all total valuations on Ck, i.e. valuations such that the underlying function is total.

A state in the semantics is a triple (s, v, e) where s is a control state in the PSA, v is the *time valuation*, and e is the *enabling valuation*: both v and e are total valuations. Therefore, given a clock c, $v(c)$ registers the time that has passed since c was started, and $e(c)$ registers its termination time which was sampled when it was started. Notice that c is active if $v(c) \leq e(c)$. The difference $e(c) - v(c)$ is c's residual lifetime. As a consequence, a control transition $s \xrightarrow{C,a} \rho$ is enabled in state (s, v, e) whenever $v(c) \geq e(c)$ for all clocks in C. This is denoted by the predicate $enabled(s \xrightarrow{C,a} \rho, v, e)$,

Let $\mathcal{R}(f_1, \ldots, f_k)$ be the probability space in the k-dimensional real space with the unique probability measure induced by the probability distribution functions $f_1, \ldots, f_k$. For $C = \{c_{f_1}^1, \ldots, c_{f_k}^k\} \subseteq Ck$, $\mathcal{R}(Distr(\vec{C}))$ denotes the probability space $\mathcal{R}(f_1, \ldots, f_k)$.

Besides, we use the following notation. Given a probability space π and $p \in [0, 1]$, $p \cdot \pi$ is the measurable space obtained by multiplying p to the probability measure of π. Given a denumerable set of probability spaces π_i, $i \in I$, $\sum_{i \in I} \pi_i$ is the measurable space obtained by appropriately summing the measures of the different probability spaces. Given a probability space π and a function f defined on the domain of π, we take $f(\pi)$ has being the probability space on the range of f induced from f [4].

Definition 3. *Let* PSA $= (St, Ck, Distr, Act, \longrightarrow, s_0, C_0)$. *Its residual lifetime semantics is defined by the PTTS* $[\![PSA]\!]_r = (\Sigma, Act \cup \mathbb{R}_{\geq 0}, \longrightarrow, \pi_0)$ *where:*

- $\Sigma \stackrel{def}{=} St \times TVal \times TVal$
- $\longrightarrow$ *is defined by the following rules:*

$$\frac{enabled(s \xrightarrow{C,a} \rho, v, e)}{a \in Act_u \implies \not\exists b \in Act_u. \ a \prec b \wedge enabled(s \xrightarrow{C'',b} \rho'', v, e)}{(s, v, e) \xrightarrow{a} \sum_{\substack{s' \in St \\ C' \subseteq Ck}} \rho(C', s') \cdot sample_{v,e}^{s',C'}(\mathcal{R}(Distr(\vec{C'})))} \quad (1)$$

$$\frac{\forall t'. \ 0 \leq t' < t \implies \forall a \in Act_u. \ \neg enabled(s \xrightarrow{C,a} \rho, v + t', e)}{(s, v, e) \xrightarrow{t} (s, v + t, e)} \quad (2)$$

- $\pi_0 \stackrel{def}{=} sample_{v_0, v_0}^{s_0, C_0}(\mathcal{R}(Distr(\vec{C_0})))$

where

- $enabled(s \xrightarrow{C,a} \rho, v, e) \stackrel{def}{\Longleftrightarrow} \forall c \in C. \ v(c) \geq e(c);$
- $sample_{v,e}^{s,C}(\vec{t}) \stackrel{def}{=} (s, v[\vec{C}/0], e[\vec{C}/\vec{t}]);$ *and*
- *for all* $c \in Ck$, $v_0(c) \stackrel{def}{=} 0$, $(v + t)(c) \stackrel{def}{=} v(c) + t$, *and* $v[\vec{C}/\vec{t}](c) \stackrel{def}{=} \vec{t}(i)$ *whenever there is an index i such that $c = \vec{C}(i)$, otherwise* $v[\vec{C}/\vec{t}](c) \stackrel{def}{=} v(c)$.

Rule (1) defines the execution of a control transition $s \xrightarrow{C,a} \rho$ requiring, therefore, that it is enabled. Notice that if a is an urgent action, it is also necessary to check that no control transition with higher priority is also enabled. In addition, the postcondition of the concrete transition is a random selection of a control state together with the set of clocks to be started and a sample terminating value for this clocks. Function $sample_{v,e}^{s',C'}$ takes care of appropriately constructing the next state from a tuple of sampled time values. Note that $sample_{v,e}^{s',C'}$ is applied to a probability space on tuples of time values, thus yielding an induced probability space on states.

Rule (2) controls the passage of time. It states that the system is allowed to stay in the control state s as long as no urgent action becomes enabled. As a consequence, maximal progress on urgent action is ensured.

Besides, the initial state is determined by the initial control state together with a sample of terminating values for C_0 (clocks not in C_0 are considered to be terminated).

4.3   Spent Lifetime Semantics

In the *spent lifetime semantics* it is only important to keep track of the time value of the clock since its termination time is continuously resampled conditioned to the time that has passed since it was started. As before, a state is a triple (s, v, e) but now valuations are partial functions. A clock c is *active* whenever v is defined in c and, in this case, $v(c)$ is its spent lifetime. Otherwise we say it is *terminated*. $e(c)$ is only defined if c is sampled with the smallest time value among the active clocks and $e(c)$ is that value (note that this may hold true for several active clocks c). If time passes or an event occurs, this enabling value is resampled among the clocks which are still active, and function e changes according to this. Therefore, in this context, a control transition $s \xrightarrow{C,a} \rho$ becomes enabled at state (s, v, e), denoted $enabled(s \xrightarrow{C,a} \rho, v, e)$, if every clock in C is either terminated in the sense above or it has been sampled with value 0 (i.e., $e(c) = 0$).

Definition 4. *Let* $\mathsf{PSA} = (St, Ck, Distr, Act, \longrightarrow, s_0, C_0)$. *Its spent lifetime semantics is defined by the PTTS* $[\![\mathsf{PSA}]\!]_s = (\Sigma, Act \cup \mathbb{R}_{\geq 0}, \longrightarrow, \pi_0)$ *where:*

- $\Sigma \stackrel{\text{def}}{=} St \times Val \times Val$
- $\longrightarrow$ *is defined by the following rules:*

$$\frac{\begin{array}{c} enabled(s \xrightarrow{C,a} \rho, v, e) \\ a \in Act_u \implies \nexists b \in Act_u. \ a \prec b \wedge enabled(s \xrightarrow{C'',b} \rho'', v, e) \end{array}}{(s, v, e) \xrightarrow{a} \sum_{\substack{s' \in St \\ C' \subseteq Ck}} \rho(C', s') \cdot PSpace(s', (v - C)[\vec{C'}/0])} \quad (3)$$

$$\frac{\forall c \in \text{dom}(e). \ 0 \leq t < e(c) \qquad \forall a \in Act_u. \ \neg enabled(s \xrightarrow{C,a} \rho, v, e)}{(s, v, e) \xrightarrow{t} PSpace(s, v + t)} \quad (4)$$

$$\frac{\forall c \in \text{dom}(e). \ 0 \leq t = e(c) \qquad \forall a \in Act_u. \ \neg enabled(s \xrightarrow{C,a} \rho, v, e)}{(s, v, e) \xrightarrow{t} PSpace(s, (v + t) - \text{dom}(e))} \quad (5)$$

Notice that in rules (4) and (5), if $e = \emptyset$ (and $v = \emptyset$ as a consequence) then the precondition holds by emptiness and then $(s, v, \emptyset) \xrightarrow{t} PSpace(s, v + t)$ for any $t \in \mathbb{R}_{\geq 0}$.

$-\ \pi_0 \overset{\text{def}}{=} PSpace(s_0, \emptyset[\vec{C_0}/0])$

where

- $\text{enabled}(s \xrightarrow{C,a} \rho, v, e) \overset{\text{def}}{\Longleftrightarrow} \forall c \in C \cap \text{dom}(v).\ c \in \text{dom}(e) \wedge e(c) = 0;$
- $PSpace(s, v) \overset{\text{def}}{=} sample_v^s(\mathcal{R}([Distr(c_1) \mid v(c_1)], \ldots, [Distr(c_k) \mid v(c_k)]))$ provided $\text{dom}(v) = \{c_1, \ldots, c_k\}$ with $sample_v^s(t_1, \ldots, t_k) \overset{\text{def}}{=} (s, v, \{c_j \mapsto t_j \mid t_j = \min_i t_i\})^2;$ and
- $v - C$ *is undefined in* C *and whenever* v *is not defined, otherwise it takes the same values as* v.

Rule (3) defines the execution of a control transition $s \xrightarrow{C,a} \rho$ and the hypotheses are as before: it must be enabled and if a is urgent, no control transition with higher priority is also enabled. The postcondition of the concrete transition is a random selection of a control state together with the set of clocks to be started and a sample terminating value for these clocks. However, in this case, the sample is taken for all active clocks and conditioned to the time that has already passed. Function $sample_v^s$ takes care of appropriately constructing the next state ensuring that the enabling valuation e is only defined for the clocks whose sampled value is minimal among all the sampled values.

Rule (4) and (5) control the passage of time. Rule (4) defines the case in which no clock has reached its termination instant, while rule (5) considers the other case. Notice that for both cases it is required that no urgent action becomes enabled before letting time pass. Observe also that after letting time pass clock termination times are resampled. Besides, rule (5) ensures that terminated clock are removed from the domain of the time valuation v.

4.4   A Note on the Difference Between the Two Semantics

As it was already noted, residual lifetime semantics samples the clock termination time only once, namely, when the clock is started, while the spent lifetime semantics keeps resampling the termination value of the active clocks. This induces two main technical differences between the interpretations $[\![PSA]\!]_r$ and $[\![PSA]\!]_s$.

A first notorious difference is that timed transitions in $[\![PSA]\!]_r$, i.e. transitions labelled with a real number, are trivial (namely, of the form $\sigma \xrightarrow{t} \sigma'$), while this is not the case in $[\![PSA]\!]_s$. Here, transitions have the form $\sigma \xrightarrow{t} \rho$ where ρ is a probability space representing the resampling of all active clocks conditioned that t units of time have passed. To notice the difference compare rule (2) in Def. 3 with rules (4) and (5) in Def. 4.

---

^{2} $[f \mid t]$ is defined by $[f \mid t](t') \overset{\text{def}}{=} P(T \leq t + t' \mid T \geq t)$ where T is a random variable with distribution f.

The other difference lies in the form of the state, more precisely, in the valuations. Given a state (s, v, e) of $[\![PSA]\!]_s$, $\mathrm{dom}(v)$ contains exactly all active clocks at this state, while $\mathrm{dom}(e) \subseteq \mathrm{dom}(v)$ contains those clocks that have been sampled as the clocks next to terminate (which may vary if time passes). In $[\![PSA]\!]_r$, valuations are total functions. In this case, given (s, v, e), c is active if $e(c) - v(c) \geq 0$, i.e. if the elapsed time since c was activated did not reach the enabling value. As a consequence, predicate *enabled* has to be accommodated to this difference. Also rule (5) has to accommodate the fact that clocks in $\mathrm{dom}(e)$ are terminated.

5   Bisimulations

In this section, bisimulation relations are defined both on the symbolic model and the concrete model. We use bisimulation as our correctness criteria. At the end of this section, the relation between the bisimulation on PSA and the bisimulation on both its semantics is stated.

The definition of the symbolic bisimulation is a straightforward modification of probabilistic bisimulation [26] in order to fit clocks.

Definition 5. *Given a PSA, a relation $R \subseteq St \times St$ is a (symbolic) bisimulation on PSA if the following statements hold:*

1. *R is an equivalence relation, and*
2. *whenever $\langle s_1, s_2 \rangle \in R$ and $s_1 \xrightarrow{C',a} \rho_1$, there is a probability space ρ_2 such that*
   (a) *$s_2 \xrightarrow{C',a} \rho_2$ and*
   (b) *$\rho_1(S) = \rho_2(S)$ for every equivalence class $S \in (Ck \times St)/R_{Ck}$ induced by the relation $R_{Ck} \overset{\mathrm{def}}{=} \{\langle (C, s_1), (C, s_2) \rangle \mid C \subseteq Ck, \langle s_1, s_2 \rangle \in R\}$.*

Two control states s_1 and s_2 are bisimilar, notation $s_1 \sim s_2$, if there is a bisimulation R such that $\langle s_1, s_2 \rangle \in R$. Two PSA, PSA_1 and PSA_2 are bisimilar, notation $\mathsf{PSA}_1 \sim \mathsf{PSA}_2$, if $C_0^1 = C_0^2$ and $s_0^1 \sim s_0^2$ in the disjoint union of PSA_1 and PSA_2.

The definition of the concrete bisimulation is a little more involved since it must deal with continuous probability spaces (see e.g. [36, 11, 4]). In particular, we follow the definition given in [4]. We first give some necessary definitions.

Let $(\Omega, \mathcal{F}, \mu)$ and $(\Omega', \mathcal{F}', \mu')$ be two probability spaces. We say that they are *equivalent* (notation $(\Omega, \mathcal{F}, \mu) \approx (\Omega', \mathcal{F}', \mu')$) if (a) for all $A \in \mathcal{F}$, $A \cap \Omega' \in \mathcal{F}'$ and $\mu(A) = \mu'(A \cap \Omega')$, and (b) for all $A' \in \mathcal{F}'$, $A' \cap \Omega \in \mathcal{F}$ and $\mu'(A') = \mu(A' \cap \Omega)$.

Given an equivalence relation R on a set Σ and a set $I \subseteq \Sigma$, we define the function $EC_{I,R} : I \to \Sigma/R$ which maps each state $\sigma \in I$ into the corresponding equivalence class $[\sigma]_R$ in Σ.

Definition 6. *Given a PTTS, a relation $R \subseteq \Sigma \times \Sigma$ is a bisimulation on PTTS if the following statements hold:*

1. R is an equivalence relation, and
2. whenever $\langle \sigma_1, \sigma_2 \rangle \in R$ and $\sigma_1 \xrightarrow{a} \pi_1$, there is a probability space π_2 such that
   (a) $\sigma_2 \xrightarrow{a} \pi_2$ and
   (b) $EC_{\Sigma_1, R}(\pi_1) \approx EC_{\Sigma_2, R}(\pi_2)$ where Σ_i is the sample space of π_i.

Two states σ_1 and σ_2 are bisimilar, notation $\sigma_1 \sim \sigma_2$, if there is a bisimulation R such that $\langle \sigma_1, \sigma_2 \rangle \in R$. Two PTTS, PTTS_1 and PTTS_2 are bisimilar, notation $\mathsf{PTTS}_1 \sim \mathsf{PTTS}_2$, if $EC_{\Sigma_1, \sim}(\pi_0^1) \approx EC_{\Sigma_2, \sim}(\pi_0^2)$ in the disjoint union of PTTS_1 and PTTS_2.

Symbolic bisimulation on PSAs is preserved by both residual lifetime semantics and spent lifetime semantics in the following sense.

Theorem 1. Given two PSA, PSA_1 and PSA_2 such that $\mathsf{PSA}_1 \sim \mathsf{PSA}_2$, then $[\![\mathsf{PSA}_1]\!]_r \sim [\![\mathsf{PSA}_2]\!]_r$ and $[\![\mathsf{PSA}_1]\!]_s \sim [\![\mathsf{PSA}_2]\!]_s$.

Proof (Sketch). It is routine to prove that if R is a symbolic bisimulation, then $\{\langle (s_1, v, e), (s_2, v, e) \rangle \mid \langle s_1, s_2 \rangle \in R, v \in TVal, e \in TVal\}$ is a bisimulation in the residual lifetime semantics, and similarly, changing $TVal$ by Val, for the spent lifetime semantics.   □

6   Model Embeddings

In this section, three different stochastic formal models are encoded into PSA. Therefore, PSA is a reasonable framework that captures the characteristics of each one of them, and hence, this hints that those models are not so distant in expressiveness. Adequacy of these encodings are also given.

6.1   IGSMP and PSA

In [8, 4], an extension of generalized semi-Markov process [16] is presented in order to allow for compositional description of concurrent systems.

It considers three different types of transitions: standard action transitions, representing action execution, clock start transitions, representing the event in which a clock becomes active, and clock termination transition, representing the event of clock termination. This is different from PSA in which all these ingredients are integrated in only one control transition.

Clock start transitions allow for preselection policy, therefore they are also labelled with a weight. They have the form $s_1 \xrightarrow{<c,w>} s_2$ where c is a clock name and $w \in \mathbb{R}_{>0}$ is the weight of the transition. Therefore, if there is another clock start transition $s_1 \xrightarrow{<c',w'>} s_3$, the probability of taking the first one is $\frac{w}{w+w'}$. When $s_1 \xrightarrow{<c,w>} s_2$ is performed, the clock c starts and continues its execution in every state traversed by the IGSMP. Whenever the clock c terminates, the IGSMP executes a termination transition of the form $s_1' \xrightarrow{c^-} s_2'$. In particular, since each active clock c must continue its execution in each state traversed by the IGSMP, all such states must have an outgoing c^- transition. Like in PSA, simultaneously active clocks in an IGSMP must have different names so that the event of termination of a clock c^- is always related to a start event $<c, w>$ of the same clock (for some w). IGSMP treats this in a particular manner: clock names have the form $\langle f, i \rangle$ where f is its distribution and $i \in \mathbb{N}$ is a number to differentiate from other clocks $\langle f, j \rangle$ with the same distribution function. Besides, this natural number is used to dynamically name clocks in an ordered fashion.

An IGSMP has four different kind of states:

- *silent states*, enabling invisible action transitions τ and (possibly) visible action transitions only. In such states, the IGSMP performs a non-deterministic choice among the τ transitions and may interact with the environment through one of the visible actions but it is not allowed to idle.
- *probabilistic states*, enabling $<c, w>$ transitions and (possibly) visible action transitions only. In such states, the IGSMP performs a probabilistic choice among the clock start transitions and may interact with the environment through one of the visible actions but it is not allowed to idle.
- *timed states*, enabling c^- transitions and (possibly) visible action transitions only. In such states, the IGSMP is allowed to idle as long as no active clock terminates. The clock that terminates first determines the transition to be performed. Note that IGSMP is defined under the assumption that clocks cannot terminate at the same instant, therefore, only one clock terminates before the other ones. While the IGSMP sojourns in these states, it may interact with the environment through one of the outgoing visible action transitions.
- *waiting states* enabling standard visible action transitions only or no transition at all. In these states, the IGSMP sojourns indefinitely or, at any time, it may interact with the environment through one of the outgoing visible action transitions.

Formally, an IGSMP is defined as follows.

Definition 7. *An Interactive Generalized Semi-Markov Process is a structure* $\mathsf{IGSMP} = (St, Ck, Distr, Act_d \cup \{\tau\}, \longrightarrow, s_0)$ *where* St, s_0, Act_d, *and* τ *are as in Def. 1, and*

- $Ck = PDF \times \mathbb{N}$ *is the set of* clock *names;*
- $Distr : Ck \to PDF$ *assigns a probability distribution function to each clock such that for all* $\langle f, i \rangle \in Ck$, $Distr(\langle f, i \rangle) = f$; *and*
- $\longrightarrow \subseteq St \times (Ck^+ \cup Ck^- \cup Act_d \cup \{\tau\}) \times St$ *is the* control transition *relation, where*
    - $Ck^+ = Ck \times \mathbb{R}_{>0}$ *is the set of events indicating the start of a clock; for* $<c, w> \in Ck^+$, w *gives the* weight *that determines the probability of starting clock* c; *and*
    - $Ck^- = \{c^- \mid c \in Ck\}$ *is the set of events denoting the termination of a clock;*

  and satisfies

1. $\forall s \in St.\ s \xrightarrow{\tau} \implies \forall \theta \in Ck^+ \cup Ck^- : s \xrightarrow{\theta} \nrightarrow$

2. $\forall s \in St.\ (\exists \theta \in Ck^+.\ s \xrightarrow{\theta}) \implies \forall c^- \in Ck^-.\ s \xrightarrow{c^-} \nrightarrow$

3. *exists* $S : St \to 2^{Ck}$, *the* active clock function , *such that for all* $s \in St$,

   (a) $S(s_0) = \emptyset$

   (b) $\cdot\ \forall a \in Act_d \cup \{\tau\}.\ s \xrightarrow{a} s' \implies S(s') = S(s)$

   $\cdot\ \forall <c,w> \in Ck^+.\ s \xrightarrow{<c,w>} s' \implies S(s') = S(s) \cup \{c\}$

   $\cdot\ \forall c^- \in Ck^-.\ s \xrightarrow{c^-} s' \implies c^- \in S(s) \wedge S(s') = S(s) - \{c\}$

   (c) $\forall <\langle f,i\rangle, w> \in Ck^+$.

   $$s \xrightarrow{<\langle f,i\rangle, w>} \implies i = \min\{j \mid j \in \mathbb{N}, \langle f,j\rangle \in S(s)\}$$

   (d) $c \in S(s) \wedge s \xrightarrow{\tau} \nrightarrow \wedge (\forall \theta \in Ck^+.\ s \xrightarrow{\theta} \nrightarrow) \implies s \xrightarrow{c^-}$

4. $\forall s \in St.\ (\exists \theta \in Ck^+.\ s \xrightarrow{\theta} s') \implies act(s') \subseteq act(s)$ *(with* $act(s) = \{a \in Act_d \cup \{\tau\} \mid s \xrightarrow{a} \})$

The constraints over the transitions guarantee that each state in IGSMP belongs to one of the four kind of states mentioned above. In particular, the first requirement says that if a state can perform τ actions, it cannot perform clock starts or clock terminations. Such a property derives from the assumption of *maximal progress*. The second requirement, says that if a state can perform clock start events then it cannot perform clock termination events. Such a property derives from the assumption of *urgency of delays*: clock start events cannot be delayed but must be performed immediately, hence they prevent the execution of clock termination transitions. The third requirement checks that clock starting and termination transitions are consistent with the set of clocks that should be active in each state. This is done by defining a function S that maps each state onto the expected set of active clocks. In particular, such a set is empty in the initial state. The fourth requirement implements the following constraint: The unique role of clock start transitions in an IGSMP must be to lead to a time state where the started clocks are actually executed; therefore, the execution of such transitions cannot cause new behaviours to be performable by the IGSMP.

The semantics of IGSMPs has been defined in [4] in terms of the so called *interactive stochastic timed transition systems (ISTTS* following the spent lifetime model. Basically, ISTTSs are a particular form of PTTSs. We redefine IGSMPs semantics in terms of PTTS without altering the original definition[3].

Recall that IGSMP is defined under the assumption that clocks cannot terminate at the same time. That is, the probability that two different clocks take the same value is 0. Let $Term_k = \{(t_1,\ldots,t_k) \mid \exists i,j.1 \le i < j \le k \wedge t_i = t_j\}$ and $\mathcal{R}(f_1,\ldots,f_k) = (\mathbb{R}^k, \mathcal{F}, P)$. Define $\check{\mathcal{R}}(f_1,\ldots,f_k)$ to be the probability space $(\check{\mathbb{R}}^k, \check{\mathcal{F}}, \check{P})$ where,

---

[3] It is straightforward to see that the new semantics preserves bisimulation with respect to the original one in terms of ISTTS.

Table 1. Semantics of IGSMPs

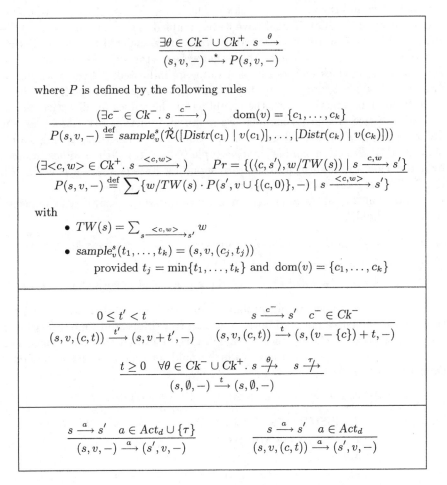

1. $\mathring{\mathbb{R}}^k = \mathbb{R}^k - Term_k$,
2. $\breve{\mathcal{F}} = \{E \subseteq \mathring{\mathbb{R}} \mid E \subseteq \mathcal{F}\}$, and
3. $\breve{P}(E) = P(E)$ for all $E \in \breve{\mathcal{F}}$.

Definition 8. *The semantics of* IGSMP *is defined by the PTTS* $[\![\text{IGSMP}]\!] = (\Sigma, Act \cup \mathbb{R}_{\geq 0}, \longrightarrow, \pi_0)$ *where:*

- $\Sigma \stackrel{\text{def}}{=} St \times Val \times (\{-\} \cup (Ck \times \mathbb{R}_{>0}))$,
- $Act = Act_d \cup Act_u$ *with* $Act_u = \{\tau, \star\}$,
- $\pi_0(s_0, \emptyset, -) = 1$, *and*
- $\longrightarrow$ *is defined according to the rules in Table 1.*

In addition to the control state s, a state $(s, v, (c, t))$ contains (i) the set of active clocks together with its spent lifetimes (represented by the partial valuation v), and (ii) a pair (c, t) containing the time value sampled by the winning clock and the name of this clock. The latter field is set to "$-$" whenever active clocks of the IGSMP still have to be sampled. The sampling $(s, v, -) \xrightarrow{\ast}$ $P(s, v, -)$ leads to states where starting clocks are associated to a spent lifetime 0, active clocks in $\mathrm{dom}(v)$ (except those that were re-started) preserve their value, and the winning clock and its sampled value are indicated. All the work lies on function P which aggregates probabilities of the preselection policy on clock start transitions together with the sampling of the values of all active clocks. To define P two auxiliary functions are required. Function $TW : St \to \mathbb{R}_{>0}$ in Table 1 computes the overall weight of the clock start transitions leaving a state of an IGSMP. Function $sample_v^s$ maps a tuple $(t_1, \ldots, t_k)$ of time values sampled by active clocks in $\mathrm{dom}(v)$ into the corresponding state $(s, v, (c_j, t_j))$ where j is the index of the clock which sampled with the least value.

Notice that, for all states (s, v, e) of any interpretation $[\![\mathsf{IGSMP}]\!]$ the following statements hold,

1. $(s, v, e) \xrightarrow{\tau}$ implies $(s, v, e) \xnrightarrow{\ast}$ and $(s, v, e) \xnrightarrow{t}$ for any $t \in \mathbb{R}_{\geq 0}$;
2. $(s, v, e) \xrightarrow{\ast}$ implies $(s, v, e) \xnrightarrow{t}$ for any $t \in \mathbb{R}_{\geq 0}$;
3. Either $(s, v, e) \xrightarrow{\tau}$, $(s, v, e) \xrightarrow{\ast}$, or there is a $t \in \mathbb{R}_{>0}$ such that $(s, v, e) \xrightarrow{t}$

ISTTS considers separately probabilistic transitions and non-deterministic transitions (the latter are used both for action transitions and time transitions). We tried to preserve this characteristic. Observe in Table 1 that only transitions $(s, v, -) \xrightarrow{\ast} P(s, v, -)$ are not trivial which would correspond to the probabilistic transition of ISTTS.

An IGSMP can be encoded in terms of a PSA as follows.

Definition 9. *Let* $\mathsf{IGSMP} = (St, Ck, Distr, Act_d \cup \{\tau\}, \longrightarrow, s_0)$. *Its interpretation in terms of PSA is given by* $\mathsf{PSA(IGSMP)} \stackrel{\text{def}}{=} (St, Ck, Distr, Act, \longrightarrow, s_0, \emptyset)$ *where* $Act = Act_d \cup Act_u$; $Act_u = \{\tau, -\} \cup \{\bar{w} | w \in \mathbb{R}_{>0}\}$ *with* $\prec$ *the least priority relation satisfying* $\alpha \prec \tau$ *for every* $\alpha \in Act_u - \{\tau\}$ *and* $- \prec \bar{w}$ *for every* $w \in \mathbb{R}_{>0}$; *and* $\longrightarrow$ *is defined by the following rules:*

$$\frac{s \xrightarrow{a} s' \quad a \in Act_d \cup \{\tau\}}{s \xrightarrow{\emptyset, a, \emptyset} s'} \qquad\qquad \frac{s \xrightarrow{c^-} s'}{s \xrightarrow{\{c\}, -, \emptyset} s'}$$

$$\frac{\exists <c, w> \in Ck^+. \; s \xrightarrow{<c,w>}}{\rho(C, s') = \text{if } (C = \{c\}) \text{ then } \sum \{w/TW(s) \mid s \xrightarrow{<c,w>} s'\} \text{ else } 0}{s \xrightarrow{\emptyset, \overline{TW(s)}} \rho}$$

where TW is as in Table 1.

The encoding is quite simple: each of the three type of transitions in IGSMP is encoded in a PSA control transition containing the only ingredient it represents. However, clock start transitions deserve particular attention. Since weights determine a probabilistic jumps, all clock start transitions emanating from one state should be encoded in a unique probabilistic transition. Thus, for instance, if $s_0 \xrightarrow{<c_1,1>} s_1$ and $s_0 \xrightarrow{<c_2,2>} s_2$ are the only two clock start transitions leaving s_0, then $s_0 \xrightarrow{\emptyset,\overline{3}} \{\langle c_1, s_1 \rangle \mapsto \frac{1}{3}, \langle c_2, s_2 \rangle \mapsto \frac{2}{3}\}$. Label $\overline{3}$ is kept for compositional matters. (Weights do not behave in the same way as probability values.) This will become more apparent in Section 7.1.

The following theorem states the adequacy of the translation of IGSMPs into PSAs. The notion of bisimulation on IGSMPs has been defined in [4].

Theorem 2. *Given* IGSMP$_1$ *and* IGSMP$_2$ *the following statements hold:*

1. IGSMP$_1 \sim$ IGSMP$_2$ *if and only if* PSA(IGSMP$_1$) $\sim$ PSA(IGSMP$_2$).
2. $[\![PSA(IGSMP_1)]\!]_s \sim [\![PSA(IGSMP_2)]\!]_s$ *implies* $[\![IGSMP_1]\!] \sim [\![IGSMP_2]\!]$.

Proof (Sketch).

1. It is routine to prove that the same relation R is a bisimulation on IGSMPs if and only if it is a bisimulation on their translation.
2. Let R be a bisimulation relation such that $[\![PSA(IGSMP_1)]\!]_s \sim [\![PSA(IGSMP_2)]\!]_s$ Define R' by

$$R' \stackrel{\text{def}}{=} \{\langle(s_1, v_1, -), (s_2, v_2, -)\rangle \mid \langle(s_1, v_1, e), (s_2, v_2, e')\rangle \in R\}$$
$$\cup \{\langle(s_1, v_1, (c_1, t_1)), (s_2, v_2, (c_2, t_2))\rangle \mid$$
$$\langle(s_1, v_1, \{c_1 \mapsto t_1\}), (s_2, v_2, \{c_2 \mapsto t_2\})\rangle \in R$$
$$\wedge \forall i \in \{1, 2\}. \forall c, c' \in \text{dom}(v_i). v_i(c) = v_i(c') \implies c = c'\}$$

It can be proved that R' is a bisimulation. $\qquad\square$

Notice that, although bisimulation is preserved forth and back at symbolic level, these is not the case at concrete level. The direct interpretation of IGSMP is weaker in the sense that more states are equated. In fact, the semantics of IGSMP makes an aggregation of the sampling transitions (those labelled with $\star$) while this is not present in the semantics of PSA.

6.2   Stochastic Automata and PSA

Another extension to generalized semi-Markov process which allows for compositional description of concurrent systems was introduced in [14, 11]. Rather than splitting actions, starting clocks, and terminating clocks in three different transitions, this model includes this three ingredients in only one symbolic transition just like PSA.

A *stochastic automata* (SA) [14, 11] is a PSA $(St, Ck, Distr, Act, \rightarrow, s_0, C_0)$ such that for all $s \xrightarrow{C,a} \rho$, ρ is a trivial distribution function. Following nomenclature in [14, 11], if $Act_u = \emptyset$, we say that the SA is *open*, that is, it represents a system that cooperates with the environment or is intended to be part of a larger system. If $Act_d = \emptyset$, the SA is *closed*, i.e., it represents a system that is complete by itself and no external interaction is required. In this last case there is a flat order in Act_u (i.e., actions in Act_u cannot be compared with each other).

As a matter of fact, in its original definition, a control transition in a SA had the form $s \xrightarrow{C,a} s'$ and there was a clock resetting function κ that took a control state and returned the clocks that should be started at the moment of reaching the state. The translation to this setting is straightforwardly given by $s \xrightarrow{C,a,\kappa(s')} s'$. We stick to the definition of SA given above since the difference does not give any sensible insight.

6.3   Stochastic Bundle Event Structures and PSA

Stochastic bundle event structures (SBES) were introduced in [9, 23]. They present a true concurrency framework rather than an interleaving one like PSA, IGSMP, or SA. We also present an encoding of this model in terms of PSA.

We briefly recall the definition of SBESs.

Definition 10. *A* bundle event structure (BES) *is a structure* $(E, \rightsquigarrow, \rightarrowtail, l, Act)$ *where E is a set of* events, $\rightsquigarrow \subseteq E \times E$ *is the* asymmetric conflict *relation,* $\rightarrowtail \subseteq 2^E \times E$ *is the* bundle *relation, and $l : E \rightarrow Act$ is the* action-labelling *function, such that $\rightsquigarrow$ is irreflexive, and for all $X \subseteq E$, $e \in E$, if $X \rightarrowtail e$ then for all $e', e'' \in X$, $e' \neq e''$ implies $e' \rightsquigarrow e''$.*

A stochastic bundle event structure (SBES) *is a triple* $\langle \mathcal{E}, \mathcal{F}, \mathcal{G} \rangle$ *where $\mathcal{E}$ is a bundle event structure* $(E, \rightsquigarrow, \rightarrowtail, l, Act)$*, and $\mathcal{F} : E \rightarrow PDF$ and $\mathcal{G} : \rightarrowtail \rightarrow PDF$ are two functions that associate distribution functions to events and bundles, respectively.*

Bundle event structures are variation of event structures [37], a well known causal based model and inherently different to the kind of model we have seen so far. Bundles indicate cause: if $X \rightarrowtail e$, one event of X *must* occur before e. Moreover, e can only occur if exactly one event of each of its bundles have already occurred. Besides, there are events whose occurrence prevent the occurrence of other events. This is indicated by the asymmetric conflict relation: if $e \rightsquigarrow e'$, then e cannot occur after e'.

A SBES is a BES decorated with stochastic information indicating the time in which actions are allowed to occur. Function $\mathcal{F}$ indicates that the occurrence time of an event e since the beginning of the execution is distributed according to $\mathcal{F}(e)$. $\mathcal{G}(X \rightarrowtail e)$ is the distribution of the time elapsed between the occurrence of X's only executed event and e.

In a causal based model such as event structures, executions are not represented by a total (linear) order such as a trace or a sequence of transitions (as it is the case in automata based models). In order to represent the independence of occurrence of concurrent events, the execution is defined in terms of partial orders. A partial execution is called a configuration and is defined as follows.

Definition 11. *Given a BES, a* configuration *of it is a set* Cf $\subseteq E$ *such that there is a strict total order* $\prec$ *in* Cf *where for all* $e \in$ Cf,

1. $X \rightarrowtail e$ *implies* $X \cap \{e' \in$ Cf $\mid e' \prec e\} \neq \emptyset$ *(intuitively: every bundle that causes e was already visited), and*
2. *for all* $e' \prec e$, $e \rightsquigarrow e'$ *does not hold, i.e. e is not in conflict with its predecessors.*

$\prec$ *is an order that* determines *that* Cf *is a configuration.*

We say that a configuration Cf *is* right before *a configuration* Cf$\cup\{e\}$ *(e $\notin$ Cf) if whenever $\prec$ is an order that determines that* Cf *is a configuration,* $\prec \cup($Cf $\times \{e\})$ *determines that* Cf $\cup \{e\}$ *is a configuration.*

The notion of "right before" makes it possible to obtain a notion of transition: Cf $\xrightarrow{l(e)}$ Cf $\cup \{e\}$ if Cf is right before Cf $\cup \{e\}$. This notion is central to the translation of SBES into PSA.

Definition 12. *Given* SBES $= \langle \mathcal{E}, \mathcal{F}, \mathcal{G} \rangle$ *with* $\mathcal{E} = (E, \rightsquigarrow, \rightarrowtail, l, Act)$, *its interpretation in terms of PSA is given by* PSA(SBES) $\overset{\text{def}}{=} (St, Ck, Distr, Act, \rightarrow, s_0, C_0)$ *where*

- *St is the set of all configurations in* $\mathcal{E}$, *with* $s_0 = \emptyset$, *the empty configuration;*
- *Ck $= E \cup \rightarrowtail$ with $C_0 = E$;*
- *Distr $= \mathcal{F} \cup \mathcal{G}$;*
- $\longrightarrow$ *is the least relation satisfying*

$$\text{Cf } is \ right \ before \text{ Cf} \cup \{e\}$$
$$C_g = \{e\} \cup \{(X, e') \in \rightarrowtail \mid e' = e\} \qquad C_s = \{(X, e') \in \rightarrowtail \mid e \in X\}$$
$$\rule{10cm}{0.4pt}$$
$$\text{Cf} \xrightarrow{C_g, l(e), C_s} \text{Cf} \cup \{e\}$$

If $Act_d = Act$ and $Act_u = \emptyset$ we say that the interpretation PSA(SBES) *is* open; *if $Act_d = \emptyset$ and $Act_u = Act$, we say it is* closed.

The translation is rather simple. Since in a SBES timing is associated to bundles (with $\mathcal{G}$) and to events (with $\mathcal{F}$), we take $Ck = E \cup \rightarrowtail$. $\mathcal{F}(e)$ associates the distribution of the time elapsed since the system starts. Therefore all "clocks $e \in E$" are started at initialization, that is, $C_0 = E$. "Clocks in $\rightarrowtail$" indicates time between two control transitions, then they are set on some control transition. The control transition is obtained from the same transition of the BES Cf $\xrightarrow{l(e)}$ Cf $\cup \{e\}$ which is decorated with the stochastic information obtained from the SBES. Therefore C_g contains the clocks that enable event e, that is all bundles that cause it. In addition it contains "clock e" which was set at initialization. On executing this transition, all bundles caused by e must start counting time. Then $(X, e') \in \rightarrowtail$ will start if and only if $e \in X$. Notice that PSA(SBES) is infinite if SBES is infinite.

Unfortunately, no equivalence relation is actually provided for SBES. We could imagine that equivalences on event structure can be lifted up to SBES. In this case, the translation PSA(SBES) would only be adequate for interleaving based equivalences such as bisimulation relations, since any causal information is lost in the translation.

7   Semantics of Stochastic Process Algebras

Stochastic process algebras are extensions of traditional process algebras [21, 31, 2] with some mean of representing stochastic time delay and occasionally probabilistic jumps. The syntax of a classic (non-stochastic) process algebra is defined by the following grammar:

$$P ::= \mathbf{0} \mid X \mid a.P \mid P + P \mid P/L \mid P[\phi] \mid P \parallel_s P \mid recX.P \qquad (6)$$

where $a \in Act$ is an *action name*, X is a process variable, $L, S \subseteq Act - \{\tau\}$, $\phi : (Act - \{\tau\}) \to (Act - \{\tau\})$.

Intuitively, their interpretation is as follows. $\mathbf{0}$ is a process that does not do anything. The *prefix* $a.P$ first performs the action a and then behaves as P. The *choice* $P + Q$ provides the possibility of executing one out of two possible behaviours P and Q. Though usually the choice is resolved non-deterministically, it can also be resolved depending on the stochastic information, which very much depends on the language choices of every stochastic process algebra. P/L behaves like P except that actions in L are hidden to the environment. $P[\phi]$ behaves like process P but actions are renamed according to ϕ. $P \parallel_s Q$ defines the *parallel composition*. It describes a process that executes P and Q in parallel forcing synchronization of actions in the set S. Other actions can be executed independently of the partner process. $recX.P$ defines the recursion on the variable X in the usual way.

In the remaining of this section we give semantics to several stochastic process algebras in terms of PSA and show that their semantics are equivalent to the originally given in terms of their original models.

7.1   The Calculus of IGSMP and PSA

The calculus for IGSMP [8, 4] extends traditional process algebra with a prefix operation that makes it possible to represent stochastic time delay and probabilistic jump. Its full syntax is given by adding the *delay prefix* $<f, w>.P$ to that of the classic calculus (6), where $w \in \mathbb{R}_{>0}$ is a weight, and $f \in PDF$ is a distribution function. Given $<f, w>.P$, w determines the probability of actually executing this process (this probability depends on the context), and f determines the probability of the waiting time before executing P in case this process has been selected to be executed. Therefore, in a process like $<f, 1>.P + <g, 2>.Q$, one third of the times the system waits a random time depending on f and then behaves like P, and the other $2/3$, it waits for a random time according to g and then behaves like Q.

IGSMP semantics has to make possible the distinction of, e.g., the time event on the left-hand side of $\parallel$ from the one in the right-hand side in process $<f, 1>.\mathbf{0} \parallel <f, 1>.\mathbf{0}$. The most problematic part is to keep the relation between start and termination events (i.e. the problem of expressing ST semantics, see Section 2.5). To do so IGSMP semantics uses a dynamic technique to name clocks (well-naming rule). When a new f-distributed time event appears, a fresh name $\langle f, i \rangle$ is generated. $i \in \mathbb{N}$ is the least index not yet used by other active delays with distribution f. Since start events and termination events are represented in different control transitions, IGSMP requires an additional operator $f_i^-.P$, which is associated to the termination of a clock meaning that clock $\langle f, i \rangle$ should terminate before executing P.

Since the problem of clock naming occurs because of parallel composition, to define IGSMP semantics, it needs an additional parameter: $P \|_{S,M} Q$ extends the parallel composition with a set $M \subseteq Ck \times (\{r_i \mid i \in \mathbb{N}\} \cup \{l_i \mid i \in \mathbb{N}\})$. M records the association between the name $\langle f, i \rangle$, generated according to the well naming rule for identifying f at the level of $P \|_{S,M} Q$, and the name $\langle f, j \rangle$, generated according to the well naming rule for identifying f at the level of P (or Q). In this way, when afterwards such a delay f terminates in P (or Q), the name $\langle f, j \rangle$ can be remapped to the correct name $\langle f, i \rangle$ at the level of $P \|_{S,M} Q$ by using the information recorded in M. More precisely, in a tuple $(\langle f, i \rangle, l_j)$, M records that the event $\langle f, i \rangle$ in $P \|_{S,M} Q$ is actually named $\langle f, j \rangle$ in P ("l" stands for *left*). In a tuple $(\langle f, i \rangle, r_j)$, M records that the event $\langle f, i \rangle$ comes from an event named $\langle f, j \rangle$ in Q ("r" is for *right*). In this context, $P \|_S Q$ is defined to be $P \|_{S,\emptyset} Q$.

Let $IGSMP_{sg}$ be the set of all strongly guarded processes defined with this new operations $f_1^-.P$ and $P \|_{S,M} Q$.

The semantics of this calculus in terms of IGSMP is given in [8, 4]. Its semantics in terms of PSA is as follows.

Definition 13. *Let* $\longrightarrow$ *be the least relation satisfying rules in Tables 2, 3, and 4. The* interpretation *of an IGSMP process P is given by* $\mathsf{PSA}(P) \overset{def}{=} (St_P, Ck, Distr, Act, \longrightarrow, P, \emptyset)$ *where*

- Ck, $Distr$, Act_d and Act_u are as in Definitions 7 and 9, and
- St_P is the subset of $IGSMP_{sg}$ such that (a) $P \in St_P$, and (b) if $Q \in St_P$, $Q \overset{C,a}{\longrightarrow} \rho$ and $\rho(C', Q') > 0$, then $Q' \in St_P$.

Rules in Table 2 are standard in process algebra. Rules in Table 3 define the clock start transitions. Notice that τ transitions are taken into account in

Table 2. Standard rules for the IGSMP calculus ($a \in Act \cup \{\tau\}$)

$$a.P \xrightarrow{\emptyset,a,\emptyset} P \qquad \frac{P \xrightarrow{\emptyset,a,\emptyset} P' \quad a \in L}{P/L \xrightarrow{\emptyset,\tau,\emptyset} P'/L} \qquad \frac{P \xrightarrow{\emptyset,a,\emptyset} P'}{P[\phi] \xrightarrow{\emptyset,\phi(a),\emptyset} P'[\phi]}$$

$$\frac{P \xrightarrow{\emptyset,a,\emptyset} P'}{\begin{array}{c} P+Q \xrightarrow{\emptyset,a,\emptyset} P' \\ Q+P \xrightarrow{\emptyset,a,\emptyset} P' \end{array}} \qquad \frac{P \xrightarrow{\emptyset,a,\emptyset} P' \quad a \notin L}{P/L \xrightarrow{\emptyset,a,\emptyset} P'/L} \qquad \frac{P\{rec\,X.P/X\} \xrightarrow{\emptyset,a,\emptyset} P'}{rec\,X.P \xrightarrow{\emptyset,a,\emptyset} P'}$$

$$\frac{P \xrightarrow{\emptyset,a,\emptyset} P' \quad a \notin S}{\begin{array}{c} P \|_{S,M} Q \xrightarrow{\emptyset,a,\emptyset} P' \|_{S,M} Q \\ Q \|_{S,M} P \xrightarrow{\emptyset,a,\emptyset} Q \|_{S,M} P' \end{array}} \qquad \frac{P \xrightarrow{\emptyset,a,\emptyset} P' \quad Q \xrightarrow{\emptyset,a,\emptyset} Q' \quad a \in S}{P \|_{S,M} Q \xrightarrow{\emptyset,a,\emptyset} P' \|_{S,M} Q'}$$

Table 3. Rules for start moves in the IGSMP calculus

$$<f,w>.P \xrightarrow{\emptyset,\bar{w},\{\langle f,1\rangle\}} f_1^-.P$$

$$\frac{P \xrightarrow{\emptyset,\bar{w}} \rho \quad Q \xrightarrow{\emptyset,\bar{w}'} \rho'}{P+Q \xrightarrow{\emptyset,\overline{w+w'}} (\frac{w}{w+w'}\cdot\rho + \frac{w'}{w+w'}\cdot\rho')} \qquad \frac{P \xrightarrow{\emptyset,\bar{w}} \rho \quad Q \xrightarrow{\emptyset,\bar{w}'}\!\!\!\!\!\not\to \quad Q \xrightarrow{\emptyset,\tau}\!\!\!\!\!\not\to}{\begin{array}{c} P+Q \xrightarrow{\emptyset,\bar{w}} \rho \\ Q+P \xrightarrow{\emptyset,\bar{w}} \rho \end{array}}$$

$$\frac{P \xrightarrow{\emptyset,\bar{w}} \rho \quad Q \xrightarrow{\emptyset,\bar{w}'} \rho'}{P \|_{S,M} Q \xrightarrow{\emptyset,\overline{w+w'}} (\frac{w}{w+w'}\cdot\rho \|_{S,M}^{P,Q} \frac{w'}{w+w'}\cdot\rho')} \qquad \frac{P \xrightarrow{\emptyset,\bar{w}} \rho \quad Q \xrightarrow{\emptyset,\bar{w}'}\!\!\!\!\!\not\to \quad Q \xrightarrow{\emptyset,\tau}\!\!\!\!\!\not\to}{\begin{array}{c} P \|_{S,M} Q \xrightarrow{\emptyset,\overline{w}} (\rho \|_{S,M}^{P,Q} null) \\ Q \|_{S,M} P \xrightarrow{\emptyset,\overline{w}} (null \|_{S,M}^{Q,P} \rho) \end{array}}$$

$$\frac{P \xrightarrow{\emptyset,\bar{w}} \rho \quad \forall a \in L.\ P \xrightarrow{\emptyset,a}\!\!\!\!\!\not\to}{P/L \xrightarrow{\emptyset,\bar{w}} \rho/L} \qquad \frac{P \xrightarrow{\emptyset,\bar{w}} \rho}{P[\phi] \xrightarrow{\emptyset,\bar{w}} \rho[\phi]} \qquad \frac{P\{rec\,X.P/X\} \xrightarrow{\emptyset,\bar{w}} \rho}{rec\,X.P \xrightarrow{\emptyset,\bar{w}} \rho}$$

where

$$(\rho \|_{S,M}^{P,Q} \rho')(R) \overset{\text{def}}{=} \begin{cases} \rho(\{\langle f,i\rangle\},P') & \text{if } R \equiv (\{\langle f,n(M_f)\rangle\},P' \|_{S,M\cup\{(\langle f,n(M_f)\rangle,l_i)\}} Q) \\ \rho'(\{\langle f,i\rangle\},Q') & \text{if } R \equiv (\{\langle f,n(M_f)\rangle\},P \|_{S,M\cup\{(\langle f,n(M_f)\rangle,r_i)\}} Q') \\ 0 & \text{otherwise} \end{cases}$$

$$null(P) \overset{\text{def}}{=} 0$$

$$(\rho/L)(C,Q) \overset{\text{def}}{=} \begin{cases} \rho(C,P) & \text{if } Q \equiv P/L \\ 0 & \text{otherwise} \end{cases} \qquad (\rho[\phi])(C,Q) \overset{\text{def}}{=} \begin{cases} \rho(C,P) & \text{if } Q \equiv P[\phi] \\ 0 & \text{otherwise} \end{cases}$$

and

$$M_f = \{i \in \mathbb{N} \mid \exists j \in \mathbb{N}.\ \exists d \in \{r_j,l_j\}.\ (\langle f,i\rangle,d) \in M\}$$

$$n(M_f) = \min\{j \in \mathbb{N} \mid j \notin M_f\}$$

rules for summation and parallel composition in order to ensure their priority over delays. A similar consideration is taken in the rule of hiding for actions in set L. Since weights define a probabilistic choice, clock start transitions need to be combined appropriately in a summation or a parallel composition. Suppose the left operand is willing to perform a clock start transition with weight w and the right operand is willing to perform a clock start transition with weight w'. Then, the left-hand side processes will be performed with probability $\frac{w}{w+w'}$ and the right ones, with probability $\frac{w'}{w+w'}$. These factors are henceforth used to construct the new distribution. Besides, the new transition carries the weight of both operands, that is $w + w'$. Finally, we address the attention to auxiliary function $(\rho \|_{S,M}^{P,Q} \rho')$. Apart from appropriately distributing measures, it has the duty to extend set M. Notice that M is extended with $(\langle f,n(M_f)\rangle,l_i)$ (or

Table 4. Rules for termination moves in the IGSMP calculus

$$f_i^- . P \xrightarrow{\{\langle f,i\rangle\},-,\emptyset} P$$

$$\dfrac{P \xrightarrow{\{\langle f,i\rangle\},-,\emptyset} P' \quad \forall a \in L.P \xrightarrow{\emptyset,a} }{P/L \xrightarrow{\{\langle f,i\rangle\},-,\emptyset} P'/L}$$

$$\dfrac{P \xrightarrow{\{\langle f,i\rangle\},-,\emptyset} P' \quad Q \xrightarrow{\emptyset,\bar{w}'} \quad Q \xrightarrow{\emptyset,\tau} }{P+Q \xrightarrow{\{\langle f,i\rangle\},-,\emptyset} P'}$$

$$Q + P \xrightarrow{\{\langle f,i\rangle\},-,\emptyset} P'$$

$$\dfrac{P \xrightarrow{\{\langle f,i\rangle\},-,\emptyset} P'}{P[\phi] \xrightarrow{\{\langle f,i\rangle\},-,\emptyset} P'[\phi]}$$

$$\dfrac{P\{rec\,X.P/X\} \xrightarrow{\{\langle f,i\rangle\},-,\emptyset} P'}{rec\,X.P \xrightarrow{\{\langle f,i\rangle\},-,\emptyset} P'}$$

$$\dfrac{P \xrightarrow{\{\langle f,i\rangle\},-,\emptyset} P' \quad (\langle f,j\rangle,l_i) \in M \quad Q \xrightarrow{\emptyset,\bar{w}'} \quad Q \xrightarrow{\emptyset,\tau} }{P \parallel_{S,M} Q \xrightarrow{\{\langle f,j\rangle\},-,\emptyset} P' \parallel_{S,M-\{(\langle f,j\rangle,l_i)\}} Q}$$

$$\dfrac{Q \xrightarrow{\{\langle f,i\rangle\},-,\emptyset} Q' \quad (\langle f,j\rangle,r_i) \in M \quad P \xrightarrow{\emptyset,\bar{w}'} \quad P \xrightarrow{\emptyset,\tau} }{P \parallel_{S,M} Q \xrightarrow{\{\langle f,j\rangle\},-,\emptyset} P \parallel_{S,M-\{(\langle f,j\rangle,r_i)\}} Q'}$$

$(\langle f, n(M_f)\rangle, r_i))$ if the left (or right) process performs $\langle f, i\rangle$; the term $n(M_f)$ is in charge of choosing the least $j \in \mathbb{N}$ such that $\langle f, j\rangle$ is not yet used in M.

Rules in Table 4 define the clock termination transitions. They also take into account the priority of τ transitions over delays but, in addition, they take into account the priority of clock start over clock terminations. In particular, notice the rules for $P \parallel_{S,M} Q$. When P terminates clock $\langle f, i\rangle$, clock $\langle f, j\rangle$ associated to l_i in M terminates at the level of the parallel composition (and hence eliminated from M). A similar mechanism takes place if Q terminates clock $\langle f, i\rangle$.

The following theorem states that the semantic of the IGSMP calculus in terms of PSA is equivalent to the original semantics in terms of IGSMPs.

Theorem 3. *For any IGSMP process P, $\mathsf{PSA}(P) \sim \mathsf{PSA}(\mathsf{IGSMP}(P))$, where $\mathsf{IGSMP}(P)$ is the IGSMP semantics of P as defined in [4].*

Proof (Sketch). More precisely, $\mathsf{PSA}(P)$ and $\mathsf{PSA}(\mathsf{IGSMP}(P))$ are identical. It can be proved using structural induction that the identity function is an isomorphism. □

7.2   ♤ and PSA

♤ (read "spades") extends traditional process algebras with two new operations: one that makes it possible to start a clock and the other that waits for its termination. ♤'s full syntax is given by adding the *clock setting* operation $\{|C|\}\,P$ and the *clock triggering* operation $C \mapsto P$ to that of the classic calculus (6), where $C \in Ck$. Process $\{|C|\}\,P$ behaves just like P except that initially it starts clocks in C and sample their termination value from their respective distribution given by function *Distr*. Process $C \mapsto P$ waits for all clocks in C to terminate and then executes P.

Unlike IGSMP, clock naming in $\mathbb{Q}$ is a syntactic issue. Semantic rules assume terms are already well named. Processes $P \parallel_A Q$ and $P + Q$ are well named if bounded clocks in P (i.e. those clocks in C of a subterm $\{C\} P'$ of P) do not occur in Q and those bounded in Q do not occur in P. Then $\{x\} \{x, z\} \mapsto a; 0 \parallel_a \{y\} \{y, z\} \mapsto a; 0$ is well named, while $\{x\} \{x\} \mapsto a; 0 \parallel_a \{x\} \{x\} \mapsto a; 0$ is not. All guarded terms are α-congruent to some well named processes [11].

The semantics of $\mathbb{Q}$ in terms of SA has been defined in [14, 11] and, up to some minor notational changes, it is the same as the one we give here.

Definition 14. *Let κ be the least function satisfying equations in Table 5. Let $\longrightarrow$ be the least relation satisfying rules in Table 6. The semantics of $P \in \mathbb{Q}$ is given by the SA $\mathsf{PSA}(P) \stackrel{\text{def}}{=} (St_P, Ck, Distr, Act, \longrightarrow, P, \kappa(P))$ where:*

- *St_P is the subset of $\mathbb{Q}$ such that (a) $P \in St_P$, and (b) if $Q \in St_P$, $Q \xrightarrow{C_g, a, C_r} Q'$, then $Q' \in St_P$;*
- *Ck and $Distr$ are just like in $\mathbb{Q}$; and*
- *if Act_d is the same set as the set of $\mathbb{Q}$ action names, and $Act_u = \emptyset$ we say that the semantics is open, if instead $Act_d = \emptyset$ and Act_u is the set of $\mathbb{Q}$ action names, we say that the semantics is closed*

Function $\kappa(P)$, given in Table 5, defines the set of clocks that needs to be started before executing P. For instance, in process $\{x\} \{x\} \mapsto a; Q$, clock x has to be started in order to wait for it to terminate and then enables the execution of a. Then, $\kappa(\{C\} P)$ has to include C and those clocks that are started in P (it could be that $P \equiv \{C'\} Q$ for some C' and Q). $P + Q$ needs to start all clocks started by both P and Q and similarly for $P \parallel_A Q$. Function κ is used to define the clocks to be set in a control transition and to define the clocks to be started at initialization.

Rules in Table 6 define the control transition. $a.P$ can perform action a at any moment; therefore it does not wait for any clock to terminate. When this transition is executed all clocks in $\kappa(P)$ are started. $C \mapsto P$ performs any activity P does but after all clocks in C are terminated. Notice that $\{C\} P$ proceeds exactly like P. The behavioural difference lies in the clocks to start at initialization. Rules for $P + Q$ and $P \parallel_A Q$ are quite standard except the rule for synchronization. In a synchronizing action $a \in A$ in $P \parallel_A Q$, both P and Q should be ready to perform it, so all clocks controlling a in P and all clocks controlling a in Q have to terminate. Moreover, when the transition is executed it starts all clocks that P and Q would have started independently.

Rules for the other operators follow the usual definitions. Notice, in particular, that the hiding operation is hiding to a silent action $\tilde{\tau} \in Act$ which is *not* the same action τ considered as the maximum of Act_u in Def. 1. $\mathbb{Q}$ does not impose maximal progress on the silent step. In this sense, it is not different from any other action. However, the silent step cannot synchronize and cannot be renamed.

Table 5. Clock resetting function in $\mathcal{Q}$

$$\kappa(a.P) = \kappa(\mathbf{0}) = \emptyset \qquad \kappa(P + Q) = \kappa(P \parallel_S Q) = \kappa(P) \cup \kappa(Q)$$
$$\kappa(\{\!\!\{C\}\!\!\} P) = \kappa(P) \cup C \qquad \kappa(C \mapsto P) = \kappa(P/L) = \kappa(P[\phi]) = \kappa(rec\, X.P) = \kappa(P)$$

Table 6. Rules for $\mathcal{Q}$

$$a.P \xrightarrow{\emptyset,a,\kappa(P)} P \qquad\qquad \dfrac{P \xrightarrow{C_g,a,C_r} P' \quad a \in L}{P/L \xrightarrow{C_g,\bar{\tau},C_r} P'/L}$$

$$\dfrac{P \xrightarrow{C_g,a,C_r} P'}{\{\!\!\{C\}\!\!\} P \xrightarrow{C_g,a,C_r} P'} \qquad \dfrac{P \xrightarrow{C_g,a,C_r} P'}{C \mapsto P \xrightarrow{C \cup C_g,a,C_r} P'} \qquad \dfrac{P \xrightarrow{C_g,a,C_r} P' \quad a \notin L}{P/L \xrightarrow{C_g,a,C_r} P'/L}$$

$$\dfrac{P \xrightarrow{C_g,a,C_r} P'}{\begin{array}{c} P + Q \xrightarrow{C_g,a,C_r} P' \\ Q + P \xrightarrow{C_g,a,C_r} P' \end{array}} \qquad \dfrac{P \xrightarrow{C_g,a,C_r} P' \quad a \notin S}{\begin{array}{c} P \parallel_S Q \xrightarrow{C_g,a,C_r} P' \parallel_S Q \\ Q \parallel_S P \xrightarrow{C_g,a,C_r} Q \parallel_S P' \end{array}} \qquad \dfrac{P \xrightarrow{C_g,a,C_r} P'}{P[\phi] \xrightarrow{C_g,\phi(a),C_r} P'[\phi]}$$

$$\dfrac{P \xrightarrow{C_g,a,C_r} P' \quad Q \xrightarrow{C'_g,a,C'_r} Q' \quad a \in S}{P \parallel_S Q \xrightarrow{C_g \cup C'_g,a,C_r \cup C'_r} P' \parallel_S Q'} \qquad \dfrac{P\{rec\,X.P/X\} \xrightarrow{C_g,a,C_r} P'}{rec\,X.P \xrightarrow{C_g,a,C_r} P'}$$

7.3   GSPA and PSA

Katoen et al. [9, 23] introduced a generalized stochastic process algebra (GSPA) and gave semantics to it in terms of SBES. GSPA introduces a stochastic timed action prefix $(f)a.P$ which replaces the action prefix in the classic calculus (6). $(f)a.P$, where $f \in PDF$, executes action a after waiting an amount of time sampled according to f, and then it behaves like P.

Notice that SBESs do not contain clocks. Instead, distributions are associated to each causal link (i.e. either bundles or the execution starting time). So clock naming was not a problem in GSPA's original semantics. However, since each execution of an action is considered a different event in this setting, event naming had to be considered with care [23].

In order to give semantics to GSPA in terms of PSA, we use a static clock naming technique explained in the following. First, define an auxiliary term $\langle f, i \rangle a.P$, with $\langle f, i \rangle \in PDF \times \mathbb{N}$. This term is a particular clock naming of the distribution f governing the delay of $(f)a.P$ and it is assigned according to function π defined in Table 7. More precisely, π is a function that looks for the set of clocks to be started on arriving to P, activates them and assigns them a name: given a GSPA process P and a set $\mathcal{C}$ of already active clock names, it returns an appropriately named process P' together with the new set of active clock names (which extends $\mathcal{C}$) and the set of clocks that it has activated. Then, if GSPA' is the set of all terms in this extended syntax, π : $(\text{GSPA}' \times 2^{Ck}) \rightarrow (2^{Ck} \times (\text{GSPA}' \times 2^{Ck}))$. In particular, $\pi((f)a.P,\mathcal{C})$ returns process $\langle f,i\rangle a.P$ provided $\langle f,i\rangle \notin \mathcal{C}$; then $\mathcal{C} \cup \{\langle f,i\rangle\}$ is the new set of active clock names and $\langle f,i\rangle$ is the only clock that needs to be set on arriving to $\langle f,i\rangle a.P$. Compare this with $\pi((\langle f,i\rangle a.P,\mathcal{C}) = (\emptyset,(\langle f,i\rangle a.P,\mathcal{C}))$. Since $\langle f,i\rangle$ is a clock name, no new clock has to be created. Moreover, no clock has to be set on arriving to $\langle f,i\rangle a.P$ in this case, since it was already set when $\langle f,i\rangle$ was created. Notice that π defines indeed a static naming: to name clocks for Q in $P\|_A Q$ or $P+Q$, π "looks" how clocks were named on P (first line in Table 7). In fact, $P\|_A Q$ and $Q\|_A P$ have associated different semantics objects (and similarly for $P+Q$ and $Q+P$). This does not happen under a dynamic technique (compare to IGSMP).

The interpretation of a GSPA process in terms of PSA is given in the following.

Definition 15. *Let π be the least function satisfying equations in Table 7. Let $\longrightarrow$ be the least relation satisfying rules in Table 8. The semantics of the GSPA term P in terms of PSA is given by* $\text{PSA}(P) \stackrel{\text{def}}{=} (St_P, Ck, Distr, Act, \longrightarrow, (P',\mathcal{C}), C)$ *where:*

- $St_P \subseteq GSPA' \times Ck$ *such that (a)* $(P',\mathcal{C}) \in St_P$, *and (b) if* $(Q,\mathcal{C}) \in St_P$, $(Q,\mathcal{C}) \xrightarrow{C_g,a,C_r} (Q',\mathcal{C}')$, *then* $(Q',\mathcal{C}') \in St_P$;
- $Ck = PDF \times I\!N$ *is a set of clocks, and Distr is defined by* $Distr(\langle f,i\rangle) = f$;
- P', $\mathcal{C}$, *and C are such that* $\pi(P,\emptyset) = (C,(P',\mathcal{C}))$; *and*
- *if Act_d is the same set as the set of GSPA action names, and $Act_u = \emptyset$ we say that the semantics is* open, *if instead $Act_d = \emptyset$ and Act_u is the set of GSPA action names, we say that the semantics is* closed

States in $\text{PSA}(P)$ are pairs $(Q,\mathcal{C})$ where Q is an extended GSPA process and $\mathcal{C}$ is the set of active clock names. GSPA does not have preselection policy; therefore it does not include probabilistic jump and transitions are trivial. Hence $\text{PSA}(P)$ is a stochastic automata.

Rules in Table 8 define the transition relation. There is no rule for $(f)a.P$ since clocks are not named. The rule for the auxiliary prefix states that whenever the system is in control state $(\langle f,i\rangle a.P,\mathcal{C})$ (for any $\mathcal{C} \in Ck$), it can perform action a provided clock $\langle f,i\rangle$ has terminated; afterwards, it sets clocks in C_P and moves to $(P',\mathcal{C}_P)$ where P' is the clock named version of P, $\mathcal{C}_P$ is the new set of active clocks (notice that $\langle f,i\rangle$ was removed from the set of active clocks), and C_P is the set of clocks that becomes active in P'. For the other operators, rules are very similar to $\diamondsuit$, except synchronization that needs special care on clock naming: notice that the set of active clocks for the source of Q transition is the set of active clocks for the target of P transition (namely, $\mathcal{C}_P$). Besides, like in $\diamondsuit$ the silent step $\tilde{\tau} \in Act$ is not action $\tau \in Act_u$. It is only special in the sense that it cannot synchronize neither be renamed.

Table 7. Defining the clocks to set and the next state in GSPA

Provided
$$\pi(P,\mathcal{C}) = (C_P,(P',\mathcal{C}_P)) \quad \text{and} \quad \pi(Q,\mathcal{C}_P) = (C_Q,(Q',\mathcal{C}_Q))$$
we define

$$\pi(\mathbf{0},\mathcal{C}) = (\emptyset,(\mathbf{0},\mathcal{C}))$$
$$\pi((f)a.P,\mathcal{C}) = (\{\langle f,i\rangle\},(\langle f,i\rangle a.P,\mathcal{C} \cup \{\langle f,i\rangle\}))$$
$$\text{with } i = \min\{j \in \mathbb{N} \mid \langle f,j\rangle \notin \mathcal{C}\}$$
$$\pi(\langle f,i\rangle a.P,\mathcal{C}) = (\emptyset,(\langle f,i\rangle a.P,\mathcal{C}))$$
$$\pi(P+Q,\mathcal{C}) = (C_P \cup C_Q,(P'+Q',\mathcal{C}_Q))$$
$$\pi(P \parallel_A Q,\mathcal{C}) = (C_P \cup C_Q,(P' \parallel_A Q',\mathcal{C}_Q))$$
$$\pi(P/L,\mathcal{C}) = (C_P,(P'/L,\mathcal{C}_P))$$
$$\pi(P[\phi],\mathcal{C}) = (C_P,(P'[\phi],\mathcal{C}_P))$$

Table 8. Rules for GSPA

$$\frac{\pi(P,\mathcal{C} - \{\langle f,i\rangle\}) = (C_P,(P',\mathcal{C}_P))}{(\langle f,i\rangle a.P,\mathcal{C}) \xrightarrow{\{\langle f,i\rangle\},a,C_P} (P',\mathcal{C}_P)}$$

$$\frac{(P,\mathcal{C}) \xrightarrow{C_g,a,C_r} (P',\mathcal{C}')}{\substack{(P+Q,\mathcal{C}) \xrightarrow{C_g,a,C_r} (P',\mathcal{C}') \\ (Q+P,\mathcal{C}) \xrightarrow{C_g,a,C_r} (P',\mathcal{C}')}} \qquad \frac{(P,\mathcal{C}) \xrightarrow{C_g,a,C_r} (P',\mathcal{C}') \quad a \notin S}{\substack{(P \parallel_S Q,\mathcal{C}) \xrightarrow{C_g,a,C_r} (P' \parallel_S Q,\mathcal{C}') \\ (Q \parallel_S P,\mathcal{C}) \xrightarrow{C_g,a,C_r} (Q \parallel_S P',\mathcal{C}')}}$$

$$\frac{(P,\mathcal{C}) \xrightarrow{C_g,a,C_r} (P',\mathcal{C}_P) \quad (Q,\mathcal{C}_P) \xrightarrow{C'_g,a,C'_r} (Q',\mathcal{C}_Q) \quad a \in S}{(P \parallel_S Q,\mathcal{C}) \xrightarrow{C_g \cup C'_g,a,C_r \cup C'_r} (P' \parallel_S Q',\mathcal{C}_Q)}$$

$$\frac{(P,\mathcal{C}) \xrightarrow{C_g,a,C_r} (P',\mathcal{C}') \quad a \in L}{(P/L,\mathcal{C}) \xrightarrow{C_g,\bar{\tau},C_r} (P'/L,\mathcal{C}')} \qquad \frac{(P,\mathcal{C}) \xrightarrow{C_g,a,C_r} (P',\mathcal{C}') \quad a \notin L}{(P/L,\mathcal{C}) \xrightarrow{C_g,a,C_r} (P'/L,\mathcal{C}')}$$

$$\frac{(P,\mathcal{C}) \xrightarrow{C_g,a,C_r} (P',\mathcal{C}')}{(P[\phi],\mathcal{C}) \xrightarrow{C_g,\phi(a),C_r} (P'[\phi],\mathcal{C}')}$$

In the following we state that the translation to PSA of the process, or the translation to PSA of its semantics in terms of SBES is the same up to bisimulation of its interpretations.

Theorem 4. *Let P be a GSPA process. Then $[\![PSA(P)]\!] \sim [\![PSA(\mathcal{E}_S[\![P]\!])]\!]$ both in the open and closed interpretation.*

The proof of this theorem makes use of the definition of symbolic bisimulation given in [11]. We omit it here as it requires some technical knowledge of both [23] and [11]. Nevertheless we observe that bisimulation (as defined in this paper) of $PSA(P)$ and $PSA(SBES(P))$ is not possible as it can be seen in the next example. Consider the GSPA process $P = (f)a.\mathbf{0} \|_a (g)a.\mathbf{0}$. Then

1. $PSA(P)$ initially sets clocks $\langle f, 1\rangle$ and $\langle g, 1\rangle$ and contains the only transition:

$$(\langle f, 1\rangle a.\mathbf{0} \|_a \langle g, 1\rangle a.\mathbf{0}, \{\langle f, 1\rangle, \langle g, 1\rangle\}) \xrightarrow{\{\langle f,1\rangle, \langle g,1\rangle\}, a, \emptyset} (\mathbf{0} \|_a \mathbf{0}, \emptyset)$$

2. According to [23], the interpretation of P in terms of SBES is given by $SBES(P)$ where $E = \{(e_l, e_r)\}$, $\leadsto = \emptyset$, $\rightarrowtail = \emptyset$, $l(e_l, e_r) = a$, $\mathcal{F}(e_l, e_r) = \max(f, g)$, and $\mathcal{G} = \emptyset$. That is, the synchronization of two events e_l and e_r is given by a new event (e_l, e_r) that couples them, and whose time delay is given by a random variable which is the maximum of the random variables associated to each synchronizing event. Hence, according to Def. 12, $PSA(SBES(P))$ initially sets the *only* clock (e_l, e_r) and has the only transition
$\emptyset \xrightarrow{\{(e_l, e_r)\}, a, \emptyset} \{(e_l, e_r)\}$.

The structures of $PSA(P)$ and $PSA(SBES(P))$ are clearly different and not bisimilar. Still, though so different in structure, it is not difficult to notice that their stochastic behavior is the same: both processes will have to wait some time which will be the maximum of two values, one sampled according to f, and the other sampled according to g.

8   Notes and Discussions

Recall that a control transition $s \xrightarrow{C, a} \rho$ in PSA performs four activities:

1. *waits for termination of timers,* that is, it waits for clock in C to terminate, hence enabling the transition,
2. *executes an action,* in this case action a, as part of the transition to a new state,
3. *start timers* as part of a probabilistic jump (recall that ρ is a distribution on $2^{Ck} \times St$), and
4. *probabilistically selects the next state* according to the distribution ρ.

Let us analyze how each of this issues are treated by different frameworks including those above discussed.

IGSMP and its calculus. The four ingredients are present in this framework. IGSMP provides three different kind of transitions: one to start a timer, another to wait for its termination, and the last one to show the execution of an action. The probabilistic jump is provided together with the start transition, and only at this level a probabilistic selection is possible. Action transitions are selected non-deterministically and termination transitions are selected in the moment their clocks terminate. In this sense, we could say that start transitions provide preselection policy, while termination transition provide a race selection policy.

Though the calculus provides the action prefix operation $a.P$ which relates to action execution, the other three ingredients provided by PSA are gathered in only one operation: the delay prefix $<f, w>.P$ which is given an ST-like semantics. Still the calculus provides auxiliary separated operations for clock start and clock termination. This auxiliary operations are used in order to achieve a sound and complete axiomatization of bisimulation, and to provide an expansion law.

SA and $\mathcal{Q}$. As transitions in SA are PSA transitions with a trivial probabilistic jump, the fourth ingredient is missing in this framework. The rest is treated as in PSA. Therefore SA and $\mathcal{Q}$ do not provide preselection policy.

$\mathcal{Q}$ provides the same three ingredients in three different basic operations: $\{C\} P$, $C \mapsto P$, and $a.P$. Again, the separations of concern makes it possible to achieve a sound and complete axiomatization of the algebra, and to obtain an expansion law.

SBES and GSPA. Rather than an interleaving model, Katoen et al. decided to pursue a true concurrency approach dropping, therefore, the idea of having an expansion law and any sensible axiomatization. Hence, parallel composition is considered a basic operation. From this viewpoint, there is no need to separate the different issues in different ingredients. Notice that in SBES clock start and termination are hidden and associated to bundles that enable events. Similarly, GSPA provides the clock start and termination, and action execution in only one prefix operation $(f)a.P$. Like SA and $\mathcal{Q}$, this framework does not provide preselection policy since there is no probabilistic selection.

Strulo's SPADES. SPADES [36, 18] is one of the earliest stochastic process algebras with generally distributed delays. Its design is such that the process algebra is directly mapped in the concrete model (similar to our PTTS) and no intermediate symbolic model is defined. Contrarily to the process algebras discussed above, SPADES provides a considerably larger variety of basic operations including four different prefix operations, a non-deterministic choice and a probabilistic choice (which allows for preselection policy).

Among the prefix operations, we encounter the action prefix $a.P$, a sampling prefix $\mathcal{R}[t \leftarrow f].P$, which samples a value according to distribution f, saves it in variable t and executes P, and a delay operation $(t).P$ which delays the amount of time indicated by variable t. This separation of concerns is different for the one proposed by $\mathcal{Q}$ and IGSMP. Notice that, $(t).P$ involves both start and termination of a timer, but its sampling is explicitly done in a separate operation $\mathcal{R}[t \leftarrow f].P$. As a consequence it cannot be encoded in PSA. Besides, the fact that start and termination of a timer is not separated in different operations prevents to decompose the parallel composition by means of an expansion law. So, strictly speaking, parallel composition should also be considered a basic operation as a general expansion law is not possible in this framework. As a consequence, though a large set of sound laws have been given for SPADES, finding a complete axiomatization is not easy.

NMSPA. NMSPA [28] has semantics in a particular kind of model that combines both symbolic and concrete views. Probability measures are not explicit. Rather, it labels the transition with a random variable. Besides, transitions are also labelled with a non-negative real value indicating passage of time. As a consequence, the semantic object associated to a process is uncountably large.

Some further remarks. For most of the algebras, a decision was taken to separate action execution, timer start, and timer termination in different operations. Such a design choice was made specially to achieve a solid algebraic framework, and therefore to facilitate the syntactic manipulation of the language. However, at the moment of modelling a system, the stochastic timed action prefix of GSPA seems to be enough.

A remark that holds in general is that race policy is present in all the previously discussed frameworks. In most of them, it can be simply present in a choice operation. The calculus for IGSMP requires a little of encoding but it is still present as the underlying model (the IGSMP) allows it.

Other process algebras that we have not explicitly discussed but is worth to mention are TIPP [17], stochastic π calculus [32], and GSMPA [6], among others. The first two were given semantics on infinite symbolic transition systems (i.e. based on distributions and not on real valued transitions) that required to carried conditional distribution function in a similar manner that we do in the spent lifetime semantics. Besides, expansion laws were not possible in these contexts. GSMPA can be regarded as a variant of IGSMPs which is based strictly on ST semantics as IGSMPs, but where names are generated statically (instead of dynamically) according to syntactic locations of delays w.r.t. to parallel composition operators in the algebraic term. Notably in [6] a definition of symbolic bisimulation equivalence based on clock associations is defined which accounts for probabilistic choices.

9 Conclusion

We thoroughly discussed the characteristics of stochastic process algebra with general distributions and presented a unifying setting, namely PSA. We could map several well known stochastic formal frameworks on PSA and see that their differences are not that significant. This says that, as far as the methodology for representing general distributions is concerned, the expressiveness of all the existing stochastic process algebras and their respective semantic models is basically the same. That is, stochastic action prefix (like in GSPA) non-deterministic choice, parallel composition, renaming (and hiding), and recursion constitute a "core" algebra that will suffice for representing general distributions in concurrent systems at, basically, the same expressive level as in all such algebras. The only one exception is probabilistic choice that turns to be a useful operator and is not present in all algebras. It appears in SPADES and in IGSMP when two weighted delayed prefixes are involved, e.g., in a choice.

Though residual and spent lifetime semantics were studied in different contexts, they are actually orthogonal to the original symbolic model. This is shown by the fact that PSA can be interpreted under both viewpoints.

Finally, it is worth to mention that the different frameworks were designed with some different objectives in mind. For instance, the true concurrent approach of SBES was originally pursued to propose an alternative semantic view of the cumbersome interleaving models for general distributions of its time. The design of GSMPA/IGSMP, instead, was strongly based on a conceptual study of the kind of semantics needed for a process algebra with general distributions, ending up with ST semantics (see [4]). On the other hand, the design of SA is related to that of timed automata with the idea that SA could be model checked at least in a stochastically abstracted setting (see [1, 12, 25]).

We finish with a brief note on open problems. An important development of the work on stochastic systems would be to understand how to develop an equivalence notion which equates different patterns of generally distributed delays as follows. In the definition of [31] the internal computations of processes are standard "τ" actions. In an algebra with generally distributed delays we can see also a delay as an internal computation (a *timed* "τ"). Therefore the idea is to extend the notion of bisimulation in such a way that it can equate, e.g., a sequence of timed τ with a single timed τ provided that distribution of durations are in the correct relationship. For example a sequence (or a more complex pattern) of exponential timed τ could be equated by a phase-type distributed timed τ. It is worth noting that the possibility of extending the notion of bisimulation in this way strictly depends on the fact that we can express delays with any duration distribution (in languages expressing exponential distributions only there is not such a possibility). This is desirable because it would lead to a significant state space reduction of semantic models. See [30] for a solution of this problem in the context of semi-Markov processes, i.e. in the absence of a parallel composition operator.

Acknowledgements

We thank the anonymous referees for their useful remarks. Moreover, we thank the organisers of the VOSS event to bring us two together on writing this article, specially Holger Hermanns who has been very helpful on suggestions and mediations.

References

1. R. Alur, C. Courcoubetis, and D. Dill. Model-checking for probabilistic real-time systems. In J. Leach Albert, B. Monien, and M. Rodríguez, editors, *Proceedings of the 18th International Colloquium Automata, Languages and Programming (ICALP'91)*, Madrid, volume 510 of *Lecture Notes in Computer Science*, pages 113–126. Springer-Verlag, 1991.

2. J.C.M. Baeten and W.P. Weijland. *Process Algebra*, volume 18 of *Cambridge Tracts in Theoretical Computer Science*. Cambridge University Press, 1990.
3. M. Bernardo. *Theory and Application of Extended Markovian Process Algebras*. PhD thesis, Dottorato di Ricerca in Informatica. Università di Bologna, Padova, Venezia, February 1999.
4. M. Bravetti. *Specification and Analysis of Stochastic Real-Time Systems*. PhD thesis, Dottorato di Ricerca in Informatica. Università di Bologna, Padova, Venezia, February 2002. Available at http://www.cs.unibo.it/~bravetti/.
5. M. Bravetti and M. Bernardo. Compositional asymmetric cooperations for process algebras with probabilities, priorities, and time. In *Proc. of the 1st Int. Workshop on Models for Time-Critical Systems, MTCS 2000*, State College (PA), volume 39(3) of *Electronic Notes in Theoretical Computer Science*. Elsevier, 2000.
6. M. Bravetti, M. Bernardo, and R. Gorrieri. Towards performance evaluation with general distributions in process algebra. In D. Sangiorgi and R. de Simone, editors, *Proceedings CONCUR 98*, Nice, France, volume 1466 of *Lecture Notes in Computer Science*, pages 405–422. Springer-Verlag, 1998.
7. M. Bravetti and R. Gorrieri. Deciding and axiomatizing weak st bisimulation for a process algebra with recursion and action refinement. *ACM Transactions on Computational Logic*, 3(4):465–520, 2002.
8. M. Bravetti and R. Gorrieri. The theory of interactive generalized semi-Markov processes. *Theoretical Computer Science*, 282:5–32, 2002.
9. E. Brinksma, J.-P. Katoen, R. Langerak, and D. Latella. A stochastic causality-based process algebra. *The Computer Journal*, 38(6):552–565, 1995.
10. D.R. Cox. The analysis of non-markovian stochastic processes by the inclusion of supplementary variables. In *Proc. of the Cambridge Philosophical Society*, volume 51, pages 433–440, 1955.
11. P.R. D'Argenio. *Algebras and Automata for Timed and Stochastic Systems*. PhD thesis, Department of Computer Science, University of Twente, 1999.
12. P.R. D'Argenio. A compositional translation of stochastic automata into timed automata. Technical Report CTIT 00-08, Department of Computer Science, University of Twente, 2000.
13. P.R. D'Argenio, H. Hermanns, J.-P. Katoen, and R. Klaren. MoDeST – a modelling and description language for stochastic timed systems. In L. de Alfaro and S. Gilmore, editors, *Proc. of PAPM/PROBMIV 2001*, Aachen, Germany, volume 2165 of *Lecture Notes in Computer Science*, pages 87–104. Springer-Verlag, September 2001.
14. P.R. D'Argenio, J.-P. Katoen, and E. Brinksma. An algebraic approach to the specification of stochastic systems (extended abstract). In D. Gries and W.-P. de Roever, editors, *Proceedings of the IFIP Working Conference on Programming Concepts and Methods, PROCOMET'98*, Shelter Island, New York, USA, IFIP Series, pages 126–147. Chapman & Hall, 1998.
15. R.J. van Glabbeek and F.W. Vaandrager. Petri net models for algebraic theories of concurrency. In J.W. de Bakker, A.J. Nijman, and P.C. Treleaven, editors, *Proceedings PARLE conference*, Eindhoven, *Vol. II (Parallel Languages)*, volume 259 of *Lecture Notes in Computer Science*, pages 224–242. Springer-Verlag, 1987.
16. P.W. Glynn. A GSMP formalism for discrete event simulation. *Proceedings of the IEEE*, 77(1):14–23, 1989.

17. N. Götz, U. Herzog, and M. Rettelbach. TIPP - Introduction and application to protocol performance analysis. In H. König, editor, *Formale Beschreibungstechniken für verteilte Systeme*, FOKUS series. Saur Publishers, 1993.

18. P.G. Harrison and B. Strulo. SPADES: Stochastic process algebra for discrete event simulation. *Journal of Logic and Computation*, 10(1):3–42, 2000.

19. H. Hermanns. *Interactive Markov Chains*, volume 2428 of *Lecture Notes in Computer Science*. Springer-Verlag, 2002.

20. J. Hillston. *A Compositional Approach to Performance Modelling*. Distinguished Dissertation in Computer Science. Cambridge University Press, 1996.

21. C.A.R. Hoare. *Communicating Sequential Processes*. Prentice-Hall International, 1985.

22. K. Kanani. *A Unified Framework for Systematic Quantitative and Qualitative Analysis of Communicating Systems*. PhD thesis, Imperial College (UK), 1998.

23. J.-P. Katoen. *Quantitative and Qualitative Extensions of Event Structures*. PhD thesis, Department of Computer Science, University of Twente, April 1996.

24. J.-P. Katoen, E. Brinksma, D. Latella, and R. Langerak. Stochastic simulation of event structures. In Ribaudo [33], pages 21–40.

25. M. Kwiatkowska, G. Norman, R. Segala, and J. Sproston. Verifying quantitative properties of continuous probabilistic timed automata. In C. Palamidessi, editor, *Proceedings CONCUR 2000*, State College, Pennsylvania, USA, volume 1877 of *Lecture Notes in Computer Science*, pages 123–137. Springer-Verlag, 2000.

26. K.G. Larsen and A. Skou. Bisimulation through probabilistic testing. *Information and Computation*, 94:1–28, 1991.

27. A.M. Law and W.D. Kelton. *Simulation Modelling and Analysis*. McGraw-Hill Inc., second edition, 1991.

28. N. López and M. Núñez. NMSPA: A non-Markovian model for stochastic processes. In *Proceedings of the International Workshop on Distributed System Validation and Verification (DSVV'2000)*, Taipei, Taiwan, ROC, 2000. Available at http://www.math.ntu.edu.tw/~eric/dsvv_proc/.

29. N. López and M. Núñez. A testing theory for generally distributed stochastic processes (extended abstract). In K. Larsen and M. Nielsen, editors, *Proceedings CONCUR 2001*, Aalborg, Denmark, volume 2154 of *Lecture Notes in Computer Science*, pages 321–335. Springer-Verlag, 2001.

30. N. López and M. Núñez. Weak stochastic bisimulation for non-markovian processes, 2003. Submitted for publication.

31. R. Milner. *Communication and Concurrency*. Prentice-Hall International, 1989.

32. C. Priami. Stochastic π-calculus with general distributions. In Ribaudo [33], pages 41–57.

33. M. Ribaudo, editor. *Proc. of the 4th Workshop on Process Algebras and Performance Modelling, PAPM'96*, Torino, Italy. Università di Torino, 1996.

34. Theo C. Ruys, Rom Langerak, Joost-Pieter Katoen, Diego Latella, and Mieke Massink. First passage time analysis of stochastic process algebra using partial orders. In Tiziana Margaria and Wang Yi, editors, *Proc. of the 7th International Conference on Tools and Algorithms for the Construction and Analysis of Systems, TACAS 2001*, Genova, Italy, volume 2031 of *Lecture Notes in Computer Science*, pages 220–235. Springer-Verlag, 2001.

35. R. Segala. *Modeling and Verification of Randomized Distributed Real-Time Systems*. PhD thesis, Department of Electrical Engineering and Computer Science, Massachusetts Institute of Technology, 1995.

36. B. Strulo. *Process Algebra for Discrete Event Simulation*. PhD thesis, Department of Computing, Imperial College, University of London, 1993.
37. G. Winskel. *Events in Computation*. PhD thesis, Department of Computer Science, University of Edinburgh, 1980.

An Overview of Probabilistic Process Algebras and Their Equivalences

Natalia López and Manuel Núñez

Dpt. Sistemas Informáticos y Programación
Universidad Complutense de Madrid
{natalia,mn}@sip.ucm.es

Abstract. In order to describe probabilistic processes by means of a formal model, some considerations have to be taken into account. In this paper we present some of the ideas appeared in the literature that could help to define appropriate formal frameworks for the specification of probabilistic processes. First, we will explain the different interpretations of the probabilistic information included in this kind of models. After that, the different choice operators used in the most common probabilistic languages are enumerated. Once we have an appropriate language, we have to give its semantics. Thus, we will review some of the theories based on bisimulation and testing semantics. We will conclude by studying the extensions of the chosen languages with other operators such as parallel composition and hiding.

1 Introduction

In order to specify the functional behavior of concurrent and distributed systems, process algebras [35, 32, 47, 11, 8] have been shown to be a powerful mechanism. In addition to several languages, there has been intensive research in the study of semantics that could appropriately capture equivalent behavior of syntactically different processes. Nevertheless, the original formulations were not able to accurately represent systems where quantitative information, as time or probabilities, play a fundamental role. For example, if a lossy channel is specified without using a probabilistic estimation of the faulty rate, all that can be known is that a message may arrive or not. On the contrary, if one specifies such a probability and the sending of the message is iterated, it can be proved that, with probability 1, the message will arrive.

In order to define a process algebra there exists two main decisions that have to be taken:

- The mechanism to model the *choice*[1] among a set of available actions. Usually, process algebraic languages consider either a (unique) CCS-like choice operator or a pair of choice operators as in CSP.

---

[1] Let us remark that in the majority of the semantic frameworks we have that other operations, such as parallel and hiding, can be *derived* from the choice operator (some notable exceptions are the π-calculus and true concurrency semantics). Thus, the *choice* of the choice operator is usually more relevant than other design decisions.

C. Baier et al. (Eds.): Validation of Stochastic Systems, LNCS 2925, pp. 89–123, 2004.
© Springer-Verlag Berlin Heidelberg 2004

– The *semantics* to assign meaning to processes. In this case, we can consider testing semantics, bisimulation semantics, trace semantics, etc.

In the case of process algebras with probabilistic information these two decisions are far from obvious. Specifically, in order to model the choice there are several possibilities. First, we may have probabilistic and non-probabilistic versions of the CCS and CSP choice operators as well as their possible combinations. In this paper, the symbol + will represent the usual CCS choice while $+_p$ will denote a probabilistic version of this operator. Besides, □ represents the usual external choice operator appearing in CSP-like languages while ⊕ denotes the internal choice operator. Their probabilistic versions are denoted by $□_p$ and $⊕_p$, respectively. In addition to the problem of the *choice of probabilistic choices*, we may have different probabilistic models depending on the way probabilities are treated.[2] For example, we may have a unique probability distribution to (probabilistically) resolve choices among actions, that is, either a generative or a stratified model. Another option is that probabilities are used only to resolve the non-determinism generated by different occurrences of the same action, that is, a reactive model. Besides, we will consider several approaches that are adaptations of one or several of these standard models. Finally, as it can be expected, the incorporation of probabilistic information makes harder the formal definition and study of semantic frameworks for the new languages.

The research on probabilistic models of computation is not new. Actually, the first work on probabilistic automata [57] originated back in the 1960's. During the 80's, there were several studies extending previous concepts, mainly logics, with probabilistic information (e.g. [25, 39, 36, 65, 19, 37]). However, it is in the last fifteen years that an explosion of models for probabilistic processes has happened, with different languages, different interpretations of the probabilistic information, etc. The first models related to labeled transition systems and probabilistic information appeared in the end of the 1980's. In [43], Larsen and Skou introduce a probabilistic extension of the classical notion of strong bisimulation. They characterize this relation by using a testing semantics where the discriminatory power of the tests is increased with respect to the tests in the classical framework [24, 32]. Specifically, they allow the use of multiple copies of the tested process so that the tester may experiment with one copy at a time. They show that if two processes are not probabilistically (strongly) bisimilar then there exists a test to distinguish them with a certain probability $1 - \epsilon$, where ϵ is arbitrarily small. This first notion of probabilistic strong bisimulation has become the standard definition and it will be shown with more detail in Section 4. Following this work, Bloom and Meyer [7] introduce a notion of probabilistic bisimulation in terms of *probabilizations* of non-probabilistic processes. They show that two labeled transition systems are bisimilar (without probabilistic information) iff weights can be assigned to obtain new systems that cannot

---

[2] Following the nomenclature introduced in [29], we consider the reactive, generative, and stratified models of probabilities. We will elaborate on the differences between these models in the next section.

be distinguished by a general notion of probabilistic testing. The first *proba-bilistic process algebra* appears one year later. In [26] a probabilistic version of SCCS [46], called PCCS, is presented. They replace the (binary) non-deterministic choice operator of SCCS by a (n-ary) probabilistic choice operator, so that they use a syntax as $\sum_{i \in I}[p_i]P_i$. In this case, the probabilities p_i are assigned to the actions offered by the corresponding process P_i. They present a system of equational rules for PCCS based on the strong bisimulation defined in [43].

The aim of this paper is to be useful as an *index* covering the main contributions related to models of probabilistic processes and their equivalences. In the first case, we will concentrate both on probabilistic extensions of process algebras and on those probabilizations of labeled transition systems such that there exists a straight translation into a process algebraic notation. In the latter case we will only consider probabilistic extensions of bisimulations and testing semantics. Let us clearly state that this paper does not try to be self-contained in the sense that all the models are completely presented. On the contrary, we have omitted most of the technicalities so that the reader can get an overview of probabilistic models without getting lost in technical details. That is, technical concepts will be given only in the case that they facilitate the understanding of the notions. Thus, it can be said that in the trade-off between *descriptiveness* and *formality* we have preferred to choose the first one. Nevertheless, the reader who desires to acquire a more thorough knowledge of some of the models is pointed to the original papers. Moreover, there are topics that would definitively fit in our study but that, in order to keep a reasonable length of the paper, we could not include. Most notably, we may remark the study of the probabilistic counterparts of simulation relations (e.g. [61, 62, 64]). Besides, we mainly focus on the different possibilities for choice operators while other operators are set aside (in Section 6, we review some of the proposals for parallel and hiding operators).

In order to get a first contact with probabilistic process algebras, next we present a simple language and its operational semantics. Despite its simplicity, most of the models presented in this paper are based on variations of this language. In particular, so-called fully probabilistic models can be described by languages similar to this one. As usual, *Act* denotes the set of visible actions that processes may perform. We also consider a special action $\tau \notin Act$ that represents internal behavior of processes. The set $Act \cup \{\tau\}$ is denoted by Act_τ. The set of probabilistic processes, denoted by *Proc*, is given by the following EBNF expression:

$$P ::= X \mid \text{stop} \mid \alpha\,;P \mid P +_p P \mid \text{rec}X.P$$

where X is a process variable, $\alpha \in Act_\tau$, and $p \in (0,1)$. The term stop represents a process that may perform no action, that is, a deadlocked process. The term $\alpha\,;P$ represents the process that performs the action α and then it behaves as P. $P +_p Q$ denotes the probabilistic choice between P and Q. Intuitively, p (resp. $1-p$) denotes the *probability* assigned to the actions that P (resp. Q) may perform. The idea is that the probabilities with which one of the components,

$$\frac{}{\alpha;P\overset{\alpha}{\longrightarrow}_1 P} \qquad \frac{P\overset{\alpha}{\longrightarrow}_q P'}{P+_p Q\overset{\alpha}{\longrightarrow}_{p\cdot q} P'} \qquad \frac{Q\overset{\alpha}{\longrightarrow}_q Q'}{P+_p Q\overset{\alpha}{\longrightarrow}_{(1-p)\cdot q} Q'} \qquad \frac{P\overset{\alpha}{\longrightarrow}_p P'}{recX.P\overset{\alpha}{\longrightarrow}_p P'[recX.P/X]}$$

Fig. 1. Operational semantics of a basic process algebra

for example P, performs actions are multiplied by a probability taken from the choice operator (in this case p). Let us note that we do not consider *extreme* values of probabilities, that is, the parameter associated with a probabilistic choice fulfills $0 < p < 1$. If we would allow, for example, $p = 1$ we would obtain a notion of priority. Such models are complex (even more if probabilities and priorities are mixed) and they are out of the scope of this paper (see [17] for an overview of the inclusion of priority in process algebras). Finally, the term $recX.P$ is used to define (possibly) recursive processes. During the rest of the paper we will omit trailing occurrences of stop. For example, we will write b instead of b ; stop.

In Figure 1 we present the operational semantics of our language. The intuitive meaning of a transition as $P \overset{\alpha}{\longrightarrow}_p P'$ is that the process P may perform the action α with probability p. After this action is performed, the process behaves as P'. Let us note that, despite the probabilities labeling transitions, the operational rules are similar to those for the corresponding subset of CCS. The first rule defines the behavior of the prefix operator: The term α ; P performs α with probability 1 and after that it behaves as P. The last rule in Figure 1 represents the usual definition for the behavior of recursive processes. The remaining two rules describe the behavior of the choice operator. If the left hand side of a choice $P +_p Q$ (that is, the process P) can perform an action with a certain probability q, the choice is resolved and the action is performed with probability $p\cdot q$. A similar situation appears for the right hand side of the choice. In this case the corresponding probabilities associated with actions of Q are multiplied by $1 - p$. This simple definition of the choice operator considers that the choice is resolved only by using its probabilistic parameter. In other words, the operator $+_p$ behaves as a pure (probabilistic) internal choice. For instance, we have that the processes a ; $P +_p$ stop and a ; P are not equivalent. In order to achieve such equivalence we need to include a *normalization* factor. Let us consider the function *live* defined as:

$$live(\text{stop}) = live(X) = 0 \qquad\qquad live(a ; P) = 1$$
$$live(P +_p Q) = \max(live(P), live(Q)) \qquad\qquad live(recX.P) = live(P)$$

It is easy to check that $live(P) = 1$ iff P is able to perform, at least, a transition. Then, the rules for the choice operators have to be modified accordingly:

$$\frac{P\overset{\alpha}{\longrightarrow}_q P'}{P+_p Q\overset{\alpha}{\longrightarrow}_{q\cdot\frac{p}{p+(1-p)\cdot live(Q)}} P'} \qquad \frac{Q\overset{\alpha}{\longrightarrow}_q Q'}{P+_p Q\overset{\alpha}{\longrightarrow}_{q\cdot\frac{1-p}{p\cdot live(P)+(1-p)}} Q'}$$

One of the classical problems when defining an operational semantics for probabilistic process algebras consists in taking into account different occurrences of the same transition. For example, consider the process $P = a +_{\frac{1}{2}} a$. If we were not careful, we would have the transition $P \xrightarrow{a} _{\frac{1}{2}}$stop only once, while we should have this transition twice. Some approaches to solve this problem are to index transitions (e.g. [28]), to increase the number of rules (e.g. [44]), to define a transition probability function (e.g. [63]), to add the probabilities associated with the same transition (e.g. [69]), or to consider that if a transition can be derived in several ways then each derivation generates a different instance (e.g. [52]). In the last case, multisets of transitions are considered instead of sets of transitions. The rest of the paper is structured as follows. Section 2 is devoted to show the different interpretations of probabilistic information in process algebras. Section 3 describes the different possibilities to add probabilities to the choice operator(s). In Section 4, the most significant examples of *strong* and *weak bisimulation* definitions will be given. Meanwhile, Section 5 is devoted to review the main contributions on probabilistic *testing semantics*.

In Section 6 we sketch some of the proposals for including parallel and hiding operators in probabilistic process algebras. A more detailed study of the parallel operator can be found in [59]. Finally, in Section 7 we present our conclusions.

2   Interpretation of Probabilities

As we have already commented, one of the design decisions when defining a probabilistic process algebra is the way probabilistic information is related to usual actions. In [29, 28] van Glabbeek et al. present three models of probabilistic processes based on the language PCCS. They are called *reactive, generative*, and *stratified*. While the first two models have been widely used when defining probabilistic processes, the stratified model has received less attention (maybe due to its inherent complexity). Actually, all the proposals analyzed in this paper take (somehow) either the reactive or the generative interpretation. We will introduce these models in terms of the actions that can be offered by the environment.

The *reactive model* was already introduced by Larsen and Skou in [43] for labeled transition systems. In [29], however, this interpretation of probabilities is considered in the context of probabilistic process algebras. A process *reacts* to the stimuli given by its environment, that is, only one action is offered by the environment. So, the process chooses among the actions of that type taking into account their associated probabilities. Thus, if we consider a process as

$$P = \left[\frac{1}{5}\right] a ; P_1 + \left[\frac{4}{5}\right] a ; P_2 + \left[\frac{1}{3}\right] b ; Q_1 + \left[\frac{2}{3}\right] b ; Q_2$$

and the environment offers the action a, then P will perform a and, after that, it will behave as P_1 with probability $\frac{1}{5}$ and as P_2 with probability $\frac{4}{5}$. Besides, if the environment offers b then P will perform it, and after that it will behave

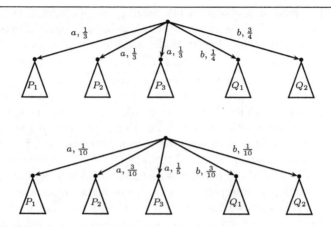

Fig. 2. Examples of a reactive process (top) and a generative process (bottom)

either as Q_1 or as Q_2, with probabilities $\frac{1}{3}$ and $\frac{2}{3}$, respectively. As this simple example shows, there is no probabilistic relation among different actions in the reactive model.

Intuitively, we say that a process as $P = \sum_{i \in I}[p_i]a_i \, ; P_i$ is a *reactive process* if the following condition holds:

$$\forall \alpha \in Act \text{ we have } \sum \{\, p_i \mid a_i = \alpha \,\wedge\, i \in I \,\} \in \{0,1\}$$

That is, the (sum of the) probability associated with each action of the alphabet is equal either to 0 or to 1. In Figure 2 a graphical example of a reactive process is given. Let us remark that the reactive model, as explained before, allows only a unique probabilistic distribution for each type of action, that is, a unique *bundle* for each type of action. However, there exist other reactive-based formalisms where this constraint is relaxed (e.g. in [61] the simple Segala's model allows several bundles for a given type of action, being the choice between these bundles non-deterministically resolved).

In the *generative model* the environment may simultaneously offer several actions and the process makes a choice among them, by considering the probability distribution assigned to these actions. In this model, there is a probabilistic relation among all the actions. For instance, an example of generative process is given by

$$P = \left[\frac{1}{6}\right] a \, ; P_1 + \left[\frac{1}{3}\right] a \, ; P_2 + \left[\frac{1}{2}\right] b \, ; P_3$$

Let us remark that in this case the sum of the probabilities associated with all the actions is equal to 1. However, there is a *redistribution* of probabilities among those actions that are offered by the environment. For instance, let us suppose that the environment offers only the action a.[3] Then, after performing a (with probability 1) the process P behaves as P_1 with probability $\frac{\frac{1}{6}}{\frac{1}{6}+\frac{1}{3}} = \frac{1}{3}$ and as P_2 with probability $\frac{\frac{1}{3}}{\frac{1}{6}+\frac{1}{3}} = \frac{2}{3}$. Let us suppose now that the environment offers both a and b. Then P performs a with a total probability equal to $\frac{1}{2}$ and b with the same probability. Afterwards, it behaves as P_1 with probability $\frac{1}{6}$, as P_2 with probability $\frac{1}{3}$, and as P_3 with probability $\frac{1}{2}$.

The syntax for *generative processes* can be given as $\sum_{i \in I}[p_i]a_i \; ; P_i$ where $\sum_{i \in I} p_i = 1$. An example of generative process is given in Figure 2 (bottom). The main difference with respect to the reactive model is that in the generative model there is a unique probability distribution among all the different action types. An example of this short of processes is the fully probabilistic basic process algebra given in the introduction of the paper.

In the *stratified* model the interpretation of probabilities is similar to the generative one but the probabilistic branching is kept, that is, the redistribution of probabilities is made *locally*. We will illustrate this difference by means of a simple example.

Example 1. Let us consider the process $P = [\frac{1}{2}]a + [\frac{1}{2}]([\frac{1}{2}]b + [\frac{1}{2}]c)$, and let us suppose that the occurrences of c are restricted in P, that is, let $P' = P\backslash\{c\}$. In the generative model, we have that P' will perform a with probability $\frac{\frac{1}{2}}{\frac{1}{4}+\frac{1}{2}} = \frac{2}{3}$. That is, the probability associated with c is proportionally distributed between the offered actions, a and b. In contrast, by using a stratified interpretation, we have that P' performs a with probability $\frac{1}{2}$ while b is also performed with probability $\frac{1}{2}$. This is so because the probability associated with c is *given* only to b. In Figure 3 we show a graphical representation of P in both the generative and stratified models.   □

Even though most of the models for probabilistic processes are based on one of the previous interpretations of probabilities, there are a few proposals combining more than one of these interpretations. This is the case of the probabilistic extension of I/O automata [45] presented in [66, 67] and recently formalized in a process algebraic style in [58] by using a subcalculus of the stochastic process algebra EMPA$_{gr}$ [4]. The idea is that input actions can be controlled by the environment, that is, the environment may decide which action is performed. Thus, it is more appropriate to give a probability distribution for each of the input actions. In other words, the model is reactive for input actions. On the contrary, output and internal actions cannot be controlled by the environment. So, a unique probability distribution is appropriate to resolve the choice among these actions. That is, the model is generative for output and internal actions. In addition, there is no probabilistic relation either among different input actions or between them and output/internal actions. The process will non-deterministically decide whether it performs either the input provided by the environment

---

[3] Actually, this is equivalent to consider that the action b has been *restricted*, that is, a restriction operator with parameter $\{b\}$ has been applied to P.

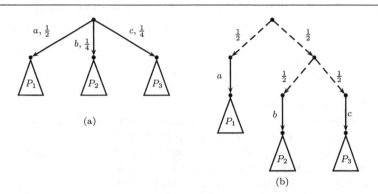

Fig. 3. Interpretation of $P = [\frac{1}{2}]a + [\frac{1}{2}]([\frac{1}{2}]b + [\frac{1}{2}]c)$ in the generative model (a) and in the stratified model (b)

or an output or internal action. It is worth to point out that an exponentially timed delay is associated with each state of the model in order to avoid the inclusion of non-determinism within input/output actions in the context of the parallel operator. A similar combination of probabilistic models appears in [2, 3] to define an algebraic model where processes with different advancing speeds are modeled. This advancing speed is achieved by using a probabilistic parallel operator where, in contrast with the previous model, time delays are not needed. In general, generative-reactive models present some very interesting properties. For example, if we consider a complete system where all the inputs are resolved, then a fully probabilistic system from the specification of the complete system is obtained. That is, non-determinism is used just to express a clear form of internal system control via input/output synchronization.

3   The Choice of Probabilistic Choices

In this section we briefly discuss how the choice among actions should be modeled. Let us remark that this question has not a standard answer even in the non-probabilistic setting. Most of the models have decided to choose either the CCS [47] or the CSP [35] approaches. In CSP there is an external choice, denoted in this paper by □, and there is also an internal choice, denoted by ⊕. The difference between these two operators is that while an external choice must be resolved by a decision taken from the environment (in other words, a visible action) internal choices are non-deterministically resolved. Thus, internal transitions do not resolve external choices. For instance, a process as $a \,\square\, (b \oplus c)$ can internally evolve either into $a \,\square\, b$ or into $a \,\square\, c$. The unique choice operator of CCS, denoted by +, is a mixture between external and internal choices. In this case, internal actions do resolve choices. For example, the process $(a\,;c) + (\tau\,;b)$ evolves after performing an internal transition into the process b.

It is worth to point out that, in the non-probabilistic setting, the choice operators of CSP can be (more or less) simulated with the ones from CCS, and vice versa. For example, the CSP process $a \oplus b$ can be expressed in CCS as the process $\tau \, ; a + \tau \, ; b$. Thus, the process $(a \oplus b) \, \Box \, c$ can be translated into CCS as $\tau \, ; (a+c) + \tau \, ; (b+c)$. Processes in CCS as $a + b$ and $\tau \, ; a + \tau \, ; b$ can be simulated by processes in CSP as $a \, \Box \, b$ and $a \oplus b$, respectively. *Mixed choices*, that is choices between visible and internal actions, can also be described in CSP, but in this case we need to use the hiding operator. Let us consider the CCS process $a + \tau \, ; b$. This process can be translated into CSP as $(a \, \Box \, c \, ; b) \backslash \{c\}$, where $P \backslash \{c\}$ means that all the occurrences of c in P are hidden, that is, they are transformed into internal actions.

Unfortunately, in a probabilistic setting the choice of choice operators is highly relevant because the previous simulations are not that easy. Most of the probabilistic extensions appeared in the literature are based on the ideas underlying CCS. That is, the most used notation is similar to the process algebra that we were describing in the introduction of the paper. However, there are other CCS-like models combining probabilistic choice operators and non-probabilistic ones. This is the case, for example, of the process algebra described in [69]. This model has two choice operators: The usual (non-probabilistic) CCS choice operator and a *purely* probabilistic choice operator. This last operator is the probabilization of the CSP internal choice operator, that is, the choice is resolved in a purely probabilistic manner.

Extensions based in CSP are more scarce. In [42, 52] both CSP choice operators are extended with probabilities. A probabilistic external choice operator behaves as the probabilistic CCS choice operator if internal actions are not involved in the choice. A probabilistic internal choice as $P \oplus_p Q$ is equivalent to the probabilistic CCS process $\tau \, ; a +_p \tau \, ; b$. In the case of CSP there are also proposals including both probabilistic and non-deterministic choices. In [60] the external choice operator is non-probabilistic but it is parameterized by a set of traces. This parameter must be considered as a *scheduler* to resolve non-deterministic choices. In [14] the authors introduce a language combining a (non-probabilistic) external choice and a probabilistic internal choice. Finally, in [13, 12] they considered both (non-probabilistic) CSP choice operators as well as a probabilistic internal choice.

Even more important than the choice operators that a language is using, it is to consider whether all the choices are (probabilistically) quantified. Actually, this is the main distinction that we will use during the rest of the paper. That is, probabilistic models can be roughly classified in the following two groups:

– *Fully Probabilistic Models.* In the case of work following these models we have that choices are always quantified. Intuitively, the underlying (probabilistic) transition systems have transitions labeled by both an action and a probability, while the interpretation of probabilities is generative. Examples of fully probabilistic models have appeared in [18, 52, 6, 15]. Finally, let us remark that the process algebra defined in the introduction of this paper is a basic language to describe this type of processes. We will use this fact

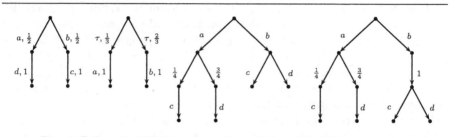

Fig. 4. Fully probabilistic and non-determinism probabilistic processes

during the rest of the paper when introducing semantic frameworks for fully probabilistic models.

- *Non-deterministic Probabilistic Models.* In these models some choices are quantified (usually by means of a purely probabilistic internal choice) while there are also non-deterministic choice operators. Thus, the reactive interpretation of probabilities as well as its variations fall into this category. Intuitively, the underlying (probabilistic) transition systems have transitions labeled either by an action or by a probability. In this case, we have a further distinction between *alternating models* (where there is a strict alternation of probabilistic and action transitions) and *non-alternating models.* Examples of studies where a non-deterministic probabilistic model has been used are [33, 69, 62, 56, 10]. A reactive probabilistic bundle of transitions with the same action, as described in the previous section, is interpreted as a single transition labeled by an action followed by a probabilistic bundle of pure probabilistic transitions.

In Figure 4 some examples of fully probabilistic processes (the first two ones) and non-deterministic probabilistic processes (the other two processes) are given. From now on we consider that $\xrightarrow{\alpha}_{p}$, with $\alpha \in Act_{\tau}$ and $p \in (0, 1]$, denotes a transition of a fully probabilistic process. While fully probabilistic models can be, more or less, represented by using the process algebra presented in the introduction of the paper, we need to introduce some notation to deal with non-deterministic probabilistic models. In the rest of this section we will go through the main differences between the alternating and the non-alternating models. We will also present definitions of basic process algebras for both models. In the following we will consider that in the context of non-deterministic probabilistic processes the transition $\xrightarrow{\alpha}$ denotes a non-deterministic transition performing action α while the transition $\longrightarrow_{p}$ denotes a probabilistic transition performed with probability p.

3.1   Alternating Model

Two types of processes can be distinguished in an alternating model: *Probabilistic* processes and *non-deterministic* processes. Besides, there usually exists

$$\frac{}{\bigoplus_{i \in I} [p_i] N_i \xrightarrow{\quad} _{p_j} N_j} \qquad \frac{}{\sum_{i \in I} \alpha_i; P_i \xrightarrow{\alpha_j} P_j}$$

$$\frac{P \xrightarrow{\quad} _p N}{\mathrm{rec} X.P \xrightarrow{\quad} _p N[\mathrm{rec} X.P/X]} \qquad \frac{N \xrightarrow{\alpha} P}{\mathrm{rec} X.N \xrightarrow{\alpha} P[\mathrm{rec} X.N/X]}$$

Fig. 5. Operational semantics for an alternating basic language

two types of transitions: *Probabilistic* transitions, outgoing from probabilistic states, labeled with probabilities, reaching non-deterministic processes, and *non-deterministic* transitions, outgoing from non-deterministic states, labeled with actions, reaching probabilistic processes.

Definition 1. Let $\mathcal{P}_N$ be the set of *non-deterministic* process expressions, ranging over $N, N', \ldots$, and let $\mathcal{P}_P$ be the set of *probabilistic* process expressions, ranging over $P, Q, \ldots$. We denote by *Proc* the set of processes in $\mathcal{P}_N \cup \mathcal{P}_P$, ranging over $G, G', \ldots$ The syntax for *probabilistic processes* is given by the EBNF expression:

$$P ::= \mathrm{stop} \mid X \mid \bigoplus_{i \in I} [p_i] N_i \mid \mathrm{rec} X.P$$

The syntax for *non-deterministic processes* is given by the EBNF expression:

$$N ::= \mathrm{stop} \mid X \mid \sum_{i \in I} \alpha_i ; P_i \mid \mathrm{rec} X.N$$

where $p_i \in [0, 1]$, $\sum_{i \in I} p_i = 1$, X is a process variable, and I is a set of indexes. The probabilistic operator $\bigoplus$ represents a pure probabilistic (internal) choice, while $\sum$ represents a non-deterministic choice. Finally, $\mathrm{rec} X.G$ is used to define recursive processes. □

The operational behavior of this process algebra is given in Figure 5. The left hand side rules describe probabilistic behaviors, while the right hand side rules describe non-deterministic behaviors. The first rule shows how a probabilistic choice is resolved by performing a probabilistic transition. In the second rule, non-deterministic choices are resolved by performing an action. Finally, the last two rules express the behavior of the recursive processes. Examples of studies using alternating models are given in [33, 10].

3.2 Non-alternating Model

As in the previous case, we have to consider the same two types of processes and transitions: *Probabilistic* and *non-deterministic*. The main difference among non-alternating models is the kind of states reached after a transition. For example, in the particular basic process algebra described below we consider that probabilistic transitions lead to non-deterministic states, while non-deterministic transitions may reach any kind of states. Thus, the syntax of our process algebra is defined as follows:

$$\frac{}{\bigoplus_{i \in I}[p_i]N_i \longrightarrow_{p_j} N_j} \qquad \frac{}{\sum_{i \in I}\alpha_i;G_i \xrightarrow{\alpha_j} G_j}$$

$$\frac{P \longrightarrow_p N}{recX.P \longrightarrow_p N[recX.P/X]} \qquad \frac{N \xrightarrow{\alpha} G}{recX.N \xrightarrow{\alpha} G[recX.N/X]}$$

Fig. 6. Operational semantics for a non-alternating basic language

Definition 2. Let $\mathcal{P}_N$ be the set of non-deterministic process expressions ranging over $N, N', \ldots$, and let $\mathcal{P}_P$ be the set of probabilistic process expressions, ranging over $P, Q, \ldots$. We denote by *Proc* the set of processes in $\mathcal{P}_N \cup \mathcal{P}_P$, ranging over $G, G', \ldots$ The syntax for *probabilistic processes* is given by the EBNF expression:

$$P ::= \text{stop} \mid X \mid \bigoplus_{i \in I}[p_i]N_i \mid recX.P$$

The syntax for *non-deterministic processes* is given by the EBNF expression:

$$N ::= \text{stop} \mid X \mid \sum_{i \in I}\alpha_i ; G_i \mid recX.N$$

And the syntax for the processes in *Proc* is given by the EBNF expression:

$$G ::= P \mid N$$

where $p_i \in [0,1]$, $\sum_{i \in I} p_i = 1$, X is a process variable, and I is a set of indexes. The probabilistic operator $\bigoplus$ represents a pure probabilistic (internal) choice, while $\sum$ represents a non-deterministic choice. As usually, $recX.G$ defines a recursive process. □

The operational semantics of this process algebra appears in Figure 6. Examples of languages following a non-alternating model are [69, 61, 56].

4   Bisimulation Semantics

In this section we will review some of the proposals for probabilistic extensions of the classical notions of strong and weak bisimulation. The definition of these two kinds of equivalence for fully probabilistic processes and for non-determinism probabilistic processes differ in several points. Thus, we will present them separately.

4.1   Probabilistic Strong Bisimulation

The definition of strong bisimulation for non-probabilistic processes is very intuitive because it represents the idea of a game of *imitations*. That is, if one process performs an action then the other one has to imitate it by performing the same action. Afterwards, the new processes have to be able to imitate each other, and so on.

Definition 3. Let *Proc* be a set of (non-probabilistic) processes. We say that an equivalence relation $\mathcal{R}$ is a *strong bisimulation* on *Proc* iff for any pair of processes $P, Q \in Proc$ we have that $P\mathcal{R}Q$ implies

$$\forall \alpha \in Act_\tau \text{ we have } P \xrightarrow{\alpha} P' \text{ implies } \exists Q' : Q \xrightarrow{\alpha} Q' \wedge P'\mathcal{R}Q'$$

We say that two processes $P, Q \in Proc$ are *strongly bisimilar*, denoted by $P \sim Q$, if there exists a strong bisimulation that contains the pair (P, Q).   □

In the probabilistic setting a new problem appears: The probabilities of reaching an equivalence class have to be computed. In the non-probabilistic setting we simply require $P \xrightarrow{a} P'$ implies $Q \xrightarrow{a} Q'$. That is, it is enough that the second *player* has a possibility to imitate the step of the first player. If we work with probabilistic processes, in addition to require that the second process is able to imitate the first one, we also need that it does it with the same probability. In other words, we have to consider all the possible ways to imitate the execution of the action and we have to add the probabilities associated with these possibilities. Moreover, we have to consider all the probabilities associated with evolutions into equivalent processes.

As we have already mentioned the first definition of (strong) probabilistic bisimulation was given in [43]. In that work, Larsen and Skou introduce a notion of strong bisimulation for probabilistic processes considering that two processes are bisimilar if they perform the same actions with the same *cumulative* probability. In [34] this notion is adapted to the alternating model while [61] also considers this equivalence for the case of the non-alternating model.

Strong Bisimulation for Fully Probabilistic Processes. The definition of probabilistic strong bisimulation has not varied since it was firstly introduced in [43, 44]. It is worth to note that this notion of bisimulation is based on the corresponding notion of *lumpability* initially introduced for Markov chains (see e.g. [40]).

Definition 4. Let *Proc* be the set of fully probabilistic processes defined by the EBNF given in the introduction of this paper. For any equivalence relation $\mathcal{R}$ on *Proc* we denote by $Proc/\mathcal{R}$ the set of equivalence classes induced by $\mathcal{R}$. We say that an equivalence relation $\mathcal{R}$ is a *strong bisimulation* on *Proc* iff for any pair of processes $P, Q \in Proc$ we have that $P\mathcal{R}Q$ implies

$$\forall \alpha \in Act_\tau, C \in Proc/\mathcal{R} \text{ we have } Prob(P, \alpha, C) = Prob(Q, \alpha, C)$$

where $Prob(P, \alpha, C) = \sum \{\, p \mid P \xrightarrow{\alpha}_p P' \wedge P' \in C \,\}$. We say that two probabilistic processes $P, Q \in Proc_p$ are *strongly bisimilar*, denoted by $P \sim Q$, if there exists a strong bisimulation that contains the pair (P, Q).   □

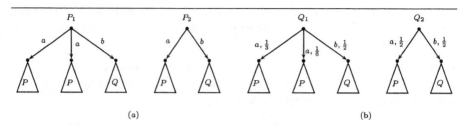

(a)   (b)

Fig. 7. Strongly bisimilar processes, in the non-probabilistic (a) and probabilistic (b) settings

Intuitively, $Prob(P, \alpha, C)$ computes the probability associated with those transitions leaving from P, labeled by α, and reaching a process belonging to the set C. For example, in the left hand side of Figure 7 we can see that if P_2 performs a then P_1 can perform any of its a-transitions to imitate it. However, in the probabilistic case (right hand side of Figure 7) the situation is slightly different. If Q_2 performs a with probability $\frac{1}{2}$ then the (sum of the) probabilities of both a-transitions in Q_1 have to be considered in order to imitate the movement performed by Q_2.

Strong Bisimulation for Non-deterministic Probabilistic Processes. As a representative case we consider the notion of strong bisimulation introduced in [34] for the alternating model. Their language is defined in two steps. First, probabilistic information is included. Then, this language is extended with temporal information. As we are only interested in probabilistic languages, we will limit our comments to the probabilistic features. Actually, the syntax and operational semantics of the corresponding probabilistic sub-language are already given in the previous section while describing a basic alternating process algebra (see Definition 1).

In order to define a notion of probabilistic bisimulation, the considerations about adding probabilities that were commented for the fully probabilistic case must also be taken into account.

Definition 5. Let $P \in Proc$ be a process. For any set of non-deterministic processes $C \subseteq \mathcal{P}_N$ the *cumulative distribution function* of P to reach C, denoted by $\mu(P, C)$, is defined as

$$\mu(P, C) = \sum \{ p \mid P \longrightarrow_p N \ \wedge \ N \in C \}$$

$\square$

Intuitively, $\mu(P, C)$ computes the probability of reaching a process in C from P after performing a probabilistic transition.

Definition 6. Let $\mathcal{R}$ be an equivalence relation on $\mathcal{P}_N$. We say that $\mathcal{R}$ is an *alternating bisimulation* iff for any pair of processes $N_1, N_2 \in \mathcal{P}_N$ we have that $N_1 \mathcal{R} N_2$ implies

$\forall \alpha \in Act_\tau$ we have $N_1 \xrightarrow{\alpha} P_1$ implies $\exists P_2 : N_2 \xrightarrow{\alpha} P_2 \wedge$
$$\forall C \in \mathcal{P}_N/\mathcal{R} : \ \mu(P_1, C) = \mu(P_2, C)$$

We say that two processes $N_1, N_2 \in \mathcal{P}_N$ are *alternating bisimulation equivalent*, denoted by $N_1 \sim N_2$, if there exists an alternating bisimulation that contains the pair (N_1, N_2). □

Intuitively, for non-deterministic processes their strong bisimulation behaves as the classical notion while for probabilistic processes the probabilities are computed following [43].

4.2   Probabilistic Weak Bisimulation

The imitation metaphor that we were using for strong bisimulation can be also applied to weak bisimulation. The main difference consists in how internal movements are imitated. In the case of strong bisimulation, an internal movement of the first player has to be imitated by exactly one internal movement of the second player. In the case of weak bisimulation, an internal movement can be imitated by any number (including zero) of internal movements. Moreover, in order to imitate the performance of a visible action, the second process may use as many internal transitions as needed.

Definition 7. Let *Proc* be a set of (non-probabilistic) processes. We say that an equivalence relation $\mathcal{R}$ is a *weak bisimulation* on *Proc* iff for any pair of processes $P, Q \in Proc$ we have that $P\mathcal{R}Q$ implies

$$\forall \alpha \in Act_\tau \text{ we have } P \xrightarrow{\alpha} P' \text{ implies } \exists Q' : Q \xRightarrow{\alpha} Q' \wedge P'\mathcal{R}Q'$$

We say that two processes $P, Q \in Proc$ are *weakly bisimilar*, denoted by $P \approx Q$, if there exists a weak bisimulation that contains the pair (P, Q). □

In the previous definition, if $\alpha \in Act$ then the transition $\xRightarrow{\alpha}$ represents the sequence of transitions $\xrightarrow{\tau}{}^* \xrightarrow{\alpha} \xrightarrow{\tau}{}^*$, while we have that $\xRightarrow{\tau}$ denotes the sequence $\xrightarrow{\tau}{}^*$; in turn $\xrightarrow{\tau}{}^*$ represents the reflexive and transitive closure of $\xrightarrow{\tau}$, that is, a (possibly empty) sequence of internal transitions.

    In contrast with the probabilistic extension of strong bisimulation, there is no common agreement about what a *good* definition of probabilistic weak bisimulation is. Actually, it is worth to point out that a similar situation appears in the non-probabilistic case where several alternative *weak* notions of bisimulation have appeared (see e.g. [48, 30, 21]). In the rest of this section we introduce some of the definitions that have appeared in the probabilistic setting. As we did for strong bisimulation, we split the presentation in two parts: Fully probabilistic processes and non-deterministic probabilistic processes.

Weak Bisimulation for Fully Probabilistic Processes. The first proposal for fully probabilistic processes of a probabilistic weak bisimulation does not appear until 1997 [6]. If we consider that probabilistic strong bisimulation was already introduced in 1989, we may see that it was not a trivial task to define an appropriate extension of weak bisimulation for probabilistic processes. In [6] processes are described as *fully probabilistic transition systems*, that is, labeled transition systems where transitions are labeled by an action and a probability. Thus, these transition systems are equivalent to the ones induced by the simple process algebra presented in the introduction of this paper when describing a fully probabilistic process algebra. First, we need to consider an auxiliary function to compute the probability of performing a given action from a given process reaching another process.

Definition 8. Let *Proc* be the set of fully probabilistic processes, $P, Q \in Proc$, and $\alpha \in Act_\tau$. We define the probability of P to reach Q by performing α, denoted by $\mathrm{prob}(P, \alpha, Q)$, as

$$\mathrm{prob}(P, \alpha, Q) = \sum \{ p \mid P \xrightarrow{\alpha}_p Q \}$$

□

Given that internal movements have to be (somehow) abstracted, it is also necessary to define the probability of reaching a set of processes after a sequence of actions is performed.

Definition 9. Let $P \in Proc$ be a process and $C \subseteq Proc$ be a set of processes. The *probability* to *reach* C from P by performing the action $\alpha \in Act_\tau$, denoted by $Prob(P, \alpha, C)$, is defined as $Prob(P, \alpha, C) = \sum_{Q \in C} \mathrm{prob}(P, \alpha, Q)$.

Let $P \in Proc$ be a process, $\gamma \subseteq Act_\tau^*$ be a set of sequences of actions, and $C \subseteq Proc$ be a set of processes. The *probability* of P to *reach* C *after* the sequences belonging to γ are performed, denoted by $Prob(P, \gamma, C)$, is defined as:

$$Prob(P, \gamma, C) = \begin{cases} 1 & \text{if } P \in C \wedge \epsilon \in \gamma \\ 0 & \text{if } P \notin C \wedge \epsilon \in \gamma \\ \sum_{(\alpha, Q) \in Act_\tau \times Proc} \mathrm{prob}(P, \alpha, Q) \cdot Prob(Q, \gamma', C) & \text{if } \gamma' = \gamma/\alpha \end{cases}$$

where ϵ denotes the empty trace and $\gamma/\alpha = \{\lambda \mid \alpha\lambda \in \gamma\}$.   □

In order to define this notion of probabilistic weak bisimulation, it is necessary to compute the probability with which different sets of processes (actually, different equivalence classes) are reached after performing sequences as $\tau^* a \tau^*$. That is, if a process performs an external action a, possibly preceded and succeeded by sequences of internal actions, then the other one has to simulate the action a in the same way. Let us remind that in order to simulate an internal movement a process may perform no action at all.

Definition 10. An equivalence relation $\mathcal{R}$ is a *probabilistic weak bisimulation* iff for any pair of processes $P, Q \in Proc$ we have that $P\mathcal{R}Q$ implies:

$\forall \alpha \in Act \cup \{\epsilon\}$, $C \in Proc/\mathcal{R}$ we have $Prob(P, \tau^* \alpha \tau^*, C) = Prob(Q, \tau^* \alpha \tau^*, C)$

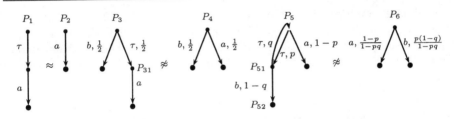

Fig. 8. Examples of weakly bisimilar and not bisimilar processes

where we consider $\tau^* \epsilon \tau^* = \tau^*$. We say that two processes $P, Q \in Proc$ are *probabilistically weakly bisimilar*, denoted by $P \approx Q$, if there exists a weak bisimulation that contains the pair (P, Q).   □

In Figure 8 we present some examples illustrating this notion of weak bisimulation. The first pair of processes P_1 and P_2 shows that the processes $\tau; a$ and a are weakly equivalent. However, the processes P_3 and P_4 are not equivalent. The idea is that abstracting internal transitions is not the same as deleting them. If P_3 performs its τ transition then the process P_{31} is reached. At that point, P_4 cannot simulate the movement performed by P_3. Actually, P_4 may try to *simulate* this movement by performing no transitions. However, P_4 is not equivalent to P_{31} because the former may perform, with a probability greater than zero, the action b while the latter cannot. Let us note that even if we delete probabilities, these two processes are not (non-probabilistic) weakly bisimilar.

Let us consider now the processes P_5 and P_6. They are not equivalent for the notion introduced in [6]. In order to show this non-equivalence, let us note that these two processes would be equivalent only if they were also equivalent to the process P_{51}. But this is not the case since, for example, the probabilities of reaching the (class of the) process P_{52} from P_5 and P_{51} are not equal. In the first case we obtain that this probability t_b is defined as $t_b = p \cdot (q \cdot t_b + 1 - q)$, that is, $t_b = \frac{p \cdot (1-q)}{1 - p \cdot q}$. In the case of P_{51} we obtain $t'_b = (1-q) + q \cdot p \cdot t'_b$, that is, $t'_b = \frac{1-q}{1-p \cdot q}$. So, these states are not equivalent. A similar situation appears if we consider the action a. In this case we obtain for P_5 and P_{51} the values $t_a = \frac{1-p}{1-p \cdot q}$ and $t'_a = \frac{q \cdot (1-p)}{1-p \cdot q}$, respectively.

It has been sometimes argued that this notion of bisimulation is too *fine*, that is, it distinguishes processes that should be equivalent. As an example, it may be questionable that P_5 and P_6 are not equivalent. Based on this result, [1] proposes an alternative notion of weak bisimulation without considering intermediate processes in sequences of τ transitions. In particular, these two processes are equivalent. Intuitively, they consider that it is not possible to decide at a given moment whether we are in the process P_5 or in the process P_{51} because each other can be reached by performing internal actions. So, there are some internal processes that are not considered in order to decide whether two processes are equivalent. In this case, by taking into account the previously computed probabilities, we have that the probability of performing both a and b, possibly preceded by a sequence of internal actions, is the same for P_5 and P_6, while the corresponding probabilities for P_{51} are not computed. Thus, the idea underlying [1] is that, in some cases, abstracting internal transitions is the same as (somehow) deleting them.

Weak Bisimulation for Non-deterministic Probabilistic Processes. The model described in [6] has also been used, as a starting point, for models considering both probabilistic and non-deterministic choices. In [56] a probabilistic weak bisimulation for a class of *labeled concurrent Markov chains* is introduced. Let us remark that this formalism can be easily translated into the language for the non-alternating model that we have introduced in Section 3.2. Thus, during the rest of this section we will consider that the set of processes *Proc* is the one presented in Definition 2.

In order to compute the probability with which a process performs a particular sequence of actions, it is necessary to resolve the non-determinism appearing in the process. For this reason they introduce the notion of *scheduler*. The idea is that for any partial computation (ending in a non-deterministic process) the scheduler chooses the next transition to be performed.

Before introducing their notion of weak bisimulation we will explain the role of two auxiliary functions. Given a process $G \in Proc$, $\gamma \in Act_\tau^*$, $C \subseteq Proc$, and a scheduler σ, the function $Prob(G, \gamma, C, \sigma)$ computes the cumulative probability to perform the sequence γ from G to reach any process in C with respect to σ. The main difference between this function and the one in [6] (Definition 9 in this survey) consists in the addition of a scheduler to resolve non-determinism. The formal definition of this function can be found in the original paper [56]. Besides, the function $\mu_{\mathcal{R}}(G, C)$ computes the cumulative probability of reaching any state of C by performing only one probabilistic transition, but considering only the probabilities of leaving the equivalence class of G.

Definition 11. Let $\mathcal{R}$ be an equivalence relation, $G \in Proc$, and $C \subseteq Proc$. The function $\mu_{\mathcal{R}}(G, C)$ is defined as:

$$\mu_{\mathcal{R}}(G, C) = \begin{cases} \frac{\mu(G,C)}{1-\mu(G,[G]_{\mathcal{R}})} & \text{if } \mu(G, [G]_{\mathcal{R}}) \neq 1 \\ \mu(G, C) & \text{otherwise} \end{cases}$$

where for any $G \in Proc$ and $C \subseteq Proc$ we have that $\mu(G, C) = \sum_{G' \in C} \mathrm{pr}(G, G')$. Additionally, the value $\mathrm{pr}(G, G')$ is defined as:

$$\mathrm{pr}(G, G') = \begin{cases} p & \text{if } G \longrightarrow_p G' \\ 1 & \text{if } G = G' \wedge G \in \mathcal{P}_N \\ 0 & \text{otherwise} \end{cases}$$

□

Definition 12. Let $Proc = \mathcal{P}_N \cup \mathcal{P}_P$ be the set of processes, where $\mathcal{P}_N$ denotes the set of non-deterministic processes and $\mathcal{P}_P$ denotes the set of probabilistic processes. An equivalence relation $\mathcal{R} \in Proc \times Proc$ is a *weak bisimulation* if for any pair of processes $G, G' \in Proc$ whenever $G\mathcal{R}G'$ we have

- $\forall \alpha \in Act_\tau$, if $G, G' \in \mathcal{P}_N$ and $G \xrightarrow{\alpha} G'$ then there exists a scheduler σ such that $Prob(G', \tau^* \overline{\alpha} \tau^*, [G']_\mathcal{R}, \sigma) = 1$, where $\overline{\alpha}$ denotes α if $\alpha \in Act$ and the empty string ϵ if $\alpha = \tau$.
- There exists a scheduler σ such that

$$\forall\, C \in (Proc/\mathcal{R}) - [G]_\mathcal{R} : \ \mu_\mathcal{R}(G, C) = Prob(G', \tau^*, C, \sigma)$$

We say that the processes G and G' are *weakly bisimilar*, denoted by $G \approx G'$, if there exists a weak bisimulation that contains the pair (G, G').   □

Intuitively, an equivalence relation is a weak bisimulation if two conditions hold. First, for any pair of related non-deterministic processes, if one of them can perform a visible action then the other one imitates it by performing several internal actions before and after this action; if the performed action is internal then the other process can imitate it by performing any number of internal actions, in particular none. Second, for any probabilistic process the cumulative probability to reach a set C through probabilistic transitions is equal to the probability needed by the other process to reach the same set by performing τ actions and probabilistic transitions.

Example 2. Consider the processes depicted in Figure 9. We have that P_1 and P_2 are weakly bisimilar since the probability with which both processes reach the non-deterministic processes that are able to perform a, that is, N_{11} and N_{21} respectively, is the same. In particular, we have that $\mu_\mathcal{R}(P_1, [N_{11}]_\mathcal{R}) = \frac{1}{2}$ while $Prob(P_2, \tau^*, [N_{11}]_\mathcal{R}, \sigma) = \sum_{i=0}^{\infty} \frac{1}{4} \cdot \frac{1}{2^i} = \frac{1}{2}$, for the appropriate scheduler σ.

The processes N_3 and P_4 are also weakly bisimilar since N_3 can perform a non-deterministic transition labeled by a, and for a particular scheduler σ we have $Prob(P_4, \tau^* a \tau^*, C, \sigma) = \frac{1}{2} + \frac{1}{2} = 1$, where C contains the processes associated with the leaves of the tree.   □

This model as well as the notion of weak bisimulation is also considered in [10]. They deal with both an alternating and a non-alternating model. Actually, the syntax of the languages for both models is the same. The difference appears in the definition of the operational semantics. In the alternating model a process as α ; P performs α and it reaches the probabilistic distribution of P after an internal step. On the contrary, in the non-alternating model the same process performs the action α and automatically reaches the probabilistic distribution of P. In the non-alternating model, their notions of strong and weak bisimulation turn to be equal to those introduced in [61]. Meanwhile, in the alternating model strong bisimulation is equivalent to the definition given in [33] while weak bisimulation coincides with [56]. Another study whose notion of weak bisimulation is also equivalent to the weak bisimulation defined in [56], but in the alternating case, is presented in [22].

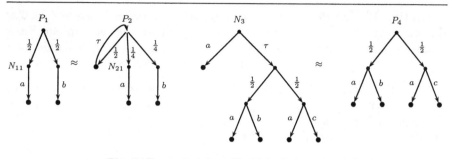

Fig. 9. Examples of weakly bisimilar processes

5   Testing Semantics

In the previous section we showed that bisimulation semantics can be interpreted in terms of a game of simulation. In the case of the classical theory of testing [24, 32] we can give an intuitive interpretation as well. The idea consists in an experimenter testing a process with a test. Given two processes, if there does not exist a test which returns different results for these two processes, then we say that they are equivalent. Depending on the way single tests are carried out, different testing semantics can be defined. Next we briefly sketch the different possibilities to define such a testing semantics. In order to test a process we consider the different interactions between the tested process and a set of tests. This interaction is usually modeled by the parallel composition of the process and the test. A test is (essentially) defined as a process but it may contain a special action, denoted by ω, indicating the successful termination of the testing procedure. If the interaction between a process and a test leads the test to a state where ω can be performed, we say that this computation is *successful*. Thus, if we consider all the possible interactions (i.e. computations) between a process P and a test T we may have three results:

- All of the computations are successful, and we say P *must* pass T.
- Some of the computations are successful, and we say P *may* pass T.
- There are no successful computations, and we say P does not pass T.

The previous possibilities induce, according to [24, 32] and depending on how two processes react to a set of tests, the following three testing equivalences.

- We say that two processes P and Q are *may* equivalent if for any test T we have P may pass T iff Q may pass T.
- We say that two processes P and Q are *must* equivalent if for any test T we have P must pass T iff Q must pass T.
- We say that two processes P and Q are *may-must* equivalent if for any test T we have both P may pass T iff Q may pass T and P must pass T iff Q must pass T.

It turns out that *may* equivalence identifies the same processes as trace equivalence does. Besides, if we consider non-divergent processes (that is, processes that cannot perform an infinite sequence of internal actions) we obtain that *must* equivalence is *equivalent* to the usual semantic model for CSP: Failure semantics. Finally, *may-must* equivalence is a little bit finer than *must* equivalence because it may partially *see* the behavior of a process in the case of a divergent behavior. For non-divergent processes, *must* and *may-must* equivalences identify the same processes. In addition to these equivalences, testing preorders can be defined in the usual way. For example, we may say that Q is *better* than P in the *may* sense if for any test T we have P may pass T implies Q may pass T. Finally, let us remark that several alternative notions of testing for non-probabilistic processes have appeared in the literature (e.g. [31, 55, 49, 9, 20]).

In order to define probabilistic testing semantics, the main question that has to be answered is: *With which probability does a process pass a test?* Depending on the underlying probabilistic model we have two possibilities. In the case of fully probabilistic models, there is a well established mechanism: The probability with which a process passes a test is given by the sum of the probabilities associated with successful computations. Thus, two processes are equivalent if they pass with the same probability any test. If we consider non-deterministic probabilistic models, we do not obtain a unique probability for passing a test, because of the presence of non-deterministic choices that are not quantified. In this case we have several possibilities. For example, we may consider that two processes are equivalent if the minimum/maximum probabilities with which these processes pass any test are the same. Even though probabilistic testing semantics have not received so much attention as the probabilizations of bisimulation, there are also numerous proposals dealing with the topic (e.g. [16, 69, 18, 51, 50, 38]). In this section we will review the main concepts underlying the definition of these semantic frameworks.

5.1   Testing Semantics for Fully Probabilistic Processes

As we have explained before, testing semantics for fully probabilistic processes share a common pattern: We compute the (unique) probability with which a process passes a test. Nevertheless, depending on the characteristics that we are interested to analyze there exist more than a unique possibility to define an equivalence.

In [16] three equivalences, as well as their corresponding denotational characterizations, are introduced. In that study, both processes and tests are defined in terms of labeled transitions systems. As usual, processes are identified with the initial state of the corresponding labeled transition system. However, processes include probabilistic information (by using a generative interpretation of probabilities) while tests are non-probabilistic and deterministic. In this approach, *testing systems* are defined as the parallel composition of a process and a test. In order to relate this syntax to the ones previously presented in this survey we have that his processes are similar to the probabilistic labeled transition systems induced by the basic process algebra given in the introduction. Moreover, tests can be generated in the same way but replacing the operator $+_p$ by a CCS choice operator without probabilities and where non-determinism is not allowed. Thus, the relevant choices are quantified because tests cannot introduce non-determinism.

Definition 13. Let $Proc$ be the set of fully probabilistic processes, $\mathcal{T}$ the set of tests, P be a process, T be a test, and $\gamma \in Act^*$ be a sequence of visible actions. The *total probability of performing* γ starting from (P,T) is given by

$$\mathrm{prob}((P,T),\gamma) = \sum_{(Q,T')\in Proc\times\mathcal{T}} \mathrm{prob}'((P,T),\gamma,(Q,T'))$$

where $\mathrm{prob}'((P,T),\gamma,(Q,T'))$ computes the probability of performing γ from (P,T) reaching (Q,T') and is defined as $Prob((P,T),\overline{\gamma},\{(Q,T')\})$ (see Definition 9). In addition, $\overline{\gamma}$ is defined as follows: $\overline{\epsilon} = \epsilon$, $\overline{a} = \tau^* a$, and $\overline{a\sigma} = \tau^* a\overline{\sigma}$. □

According to the different properties that the tester can be interested in, three partial orders are introduced.

– We write $P \leq_{tr} Q$ if for any sequential test T (that is, a trace of visible actions finishing with an acceptance action), and any trace of visible actions γ, we have that the probability of deadlock in the testing system conformed by Q and T after performing γ is less than or equal to the one for the testing system associated with P and T. Formally,

$$\forall T \in \mathcal{T}_{seq}, \gamma \in Act^* \text{ we have } \mathrm{prob}((P,T),\gamma) \leq \mathrm{prob}((Q,T),\gamma)$$

– We write $P \leq_{wte} Q$ if for any test T we have that after performing any sequence of visible actions the probability of deadlock, with respect to the visible actions that can be (weakly) performed in the next step, in the testing system conformed by Q and T is less than or equal to the one for the testing system associated with P and T. Formally,

$$\forall T \in \mathcal{T}, \gamma \in Act^* \text{ we have } \sum_{a\in Act} \frac{\mathrm{prob}((P,T),\gamma a)}{\mathrm{prob}((P,T),\gamma)} \leq \sum_{a\in Act} \frac{\mathrm{prob}((Q,T),\gamma a)}{\mathrm{prob}((Q,T),\gamma)}$$

– We write $P \leq_{ste} Q$ if for any test T we have that after performing any sequence of visible actions the probability of deadlock in the testing system conformed by Q and T is less than or equal to the one for the testing system associated with P and T. Formally,

$$\forall T \in \mathcal{T}, \gamma \in Act^* \text{ we have } \mathrm{prob}((P,T),\gamma) \leq \mathrm{prob}((Q,T),\gamma)$$

It is easy to show that the relation between these partial orders is as follows:

$$(P \leq_{tr} P') \Longleftarrow (P \leq_{wte} P') \Longleftarrow (P \leq_{ste} P')$$

The first *fully probabilistic* testing semantics, in the sense that both processes and tests are probabilistic, appears in [18]. This work was extended in [68,

15] with alternative characterizations of the corresponding equivalence. They consider probabilistic labeled transition systems to describe both processes and tests. However, in order to introduce this work, we will use the probabilistic process algebra defined in the introduction. The only difference between processes and tests is that the latter can perform a distinguished action ω that represents that the computation is successful. If the test can perform an acceptance action then the testing procedure will (successfully) finish. Let us remark that in addition to visible actions, that is those belonging to *Act*, internal actions may appear both in processes and tests. The interaction between processes and tests is given by composing them in parallel and assuming a total synchronization.

Definition 14. Let *Proc* be the set of fully probabilistic processes and $\mathcal{T}$ be the set of tests. Let us consider a process $P \in Proc$ and a test $T \in \mathcal{T}$. The *interaction system* for P and T, denoted by $P \parallel T$, is defined as a test where the new probability distribution function, denoted by prob_I, is defined as follows:

$$\text{prob}_I(P \parallel T, \alpha, Q \parallel T') = \begin{cases} 0 & \text{if } \nu(P,T) = 0 \\ \frac{\text{prob}(T,\tau,T') \cdot (1 - \text{prob}_P(P,\tau))}{\nu(P,T)} & \text{if } \nu(P,T) \neq 0 \wedge \alpha = \tau \wedge P = P' \\ \frac{\text{prob}(P,\tau,Q) \cdot (1 - \text{prob}_T(T,\tau))}{\nu(P,T)} & \text{if } \nu(P,T) \neq 0 \wedge \alpha = \tau \wedge T = T' \\ \frac{\text{prob}(P,\alpha,Q) \cdot \text{prob}(T,\alpha,T')}{\nu(P,T)} & \text{otherwise} \end{cases}$$

where $\alpha \in Act_\tau$, the cumulative probability transition functions prob, for P and T, follow Definition 8, and $\text{prob}_R(r,\alpha) = \sum_{r' \in R} \text{prob}_R(r,\alpha,r')$. We write $P \parallel T \xrightarrow{\alpha}_p Q \parallel T'$ if $\text{prob}_I(P \parallel T, \alpha, Q \parallel T') = p$.

In the definition of the function prob_I we have used a *normalization factor* $\nu : Proc \times \mathcal{T} \longrightarrow [0,1]$. The formal definition of this factor is given by

$$\nu(P,T) = \sum_{a \in Act} \text{prob}_P(P,a) \cdot \text{prob}_T(T,a)$$
$$+ \text{prob}_T(T,\tau) + \text{prob}_P(P,\tau) - \text{prob}_T(T,\tau) \cdot \text{prob}_P(P,\tau)$$

$\square$

Intuitively, the normalization factor computes the total probability with which $P \parallel T$ may perform actions. So, by dividing by this factor we automatically get that the sum of the probabilities associated with outgoing transitions of the interaction system equals 1. Let us justify why the normalization factor ν is defined in this way. First, an interaction system may perform an external action only if both process and test are able to. Thus, the probability of performing any visible action is given by

$$\sum_{a \in Act} \text{prob}_P(P,a) \cdot \text{prob}_T(T,a) \tag{1}$$

Regarding internal actions, the interaction system $P \parallel T$ can internally evolve if either the process or the test may autonomously perform an internal action. Nevertheless, if both of them decide to do it then these two transitions are merged into one (this is contemplated in the last case of the definition of prob_I). The probability with which P may perform τ while T does not is given by $\text{prob}_P(P, \tau) \cdot (1 - \text{prob}_T(T, \tau))$. Symmetrically, $P \parallel T$ may internally evolve because T performs a τ action and P does not with probability $\text{prob}_T(T, \tau) \cdot (1 - \text{prob}_P(P, \tau))$. Finally, the probability with which the process and the test simultaneously perform a τ action is given by $\text{prob}_P(P, \tau) \cdot \text{prob}_T(T, \tau)$. If we put together these three values and we unfold the products we obtain

$$\text{prob}_T(T, \tau) + \text{prob}_P(P, \tau) - \text{prob}_T(T, \tau) \cdot \text{prob}_P(P, \tau) \qquad (2)$$

If we add the values given in expressions (1) and (2) we finally get the definition of the normalization factor ν.

The probability with which a process P passes a test T is given by the sum of all the successful computations from $P \parallel T$. In order to compute this probability we need to define some auxiliary concepts.

Definition 15. Let P be a process, T be a test, and $P \parallel T$ be the associated interaction system. A *computation* is a sequence of transitions

$$C = P \parallel T \xrightarrow{\alpha_1} {}_{p_1} P_1 \parallel T_1 \xrightarrow{\alpha_2} {}_{p_2} \cdots P_{n-1} \parallel T_{n-1} \xrightarrow{\alpha_n} {}_{p_n} P_n \parallel T_n$$

such that for any $i < n$ we have $T_i \xrightarrow{\omega} \!\!\!\!\!/\,$ and there do not exist $p > 0$, α, Q, and T' such that $P_n \parallel T_n \xrightarrow{\alpha} {}_p Q \parallel T'$. If $T_n \xrightarrow{\omega}$ then we say that the computation is *successful*. We denote by $S_{P\parallel T}$ the set of *successful computations* from $P \parallel T$. The probability of a successful computation C, denoted by $Prob(C)$, is inductively defined as

$$Prob(P \parallel T) \qquad = 1$$
$$Prob(P \parallel T \xrightarrow{\alpha} {}_p C') = p \cdot Prob(C')$$

For any $\mathcal{C} \subseteq S_{P\parallel T}$ we define $Prob(\mathcal{C}) = \sum_{C \in \mathcal{C}} Prob(C)$. Finally, we say that p is the *probability* with which the process P *passes* the test T, denoted by $P\,pass_p\,T$, if $Prob(S_{P\parallel T}) = p$.

Let $T_0 \subseteq T$ be a set of probabilistic tests and P, Q two processes. We say that Q is *better* than P with respect to T_0, denoted by $P \sqsubseteq_{T_0} Q$, if for any test $T \in T_0$, we have $P\,pass_p\,T$ and $Q\,pass_q\,T$ implies $p \leq q$. Besides, we say that P and Q are *testing equivalent* with respect to T_0, denoted by $P \approx_{T_0} Q$, if for any test $T \in T_0$, if $P\,pass_p\,T$ and $Q\,pass_q\,T$ then we have $p = q$. □

Let us note that successful computations have finite length since, by definition, they reach in a finite amount of steps a deadlocked process (since no more actions can be performed). The testing equivalence previously defined is parameterized by a set of tests. In [68] they consider two of these sets: The whole

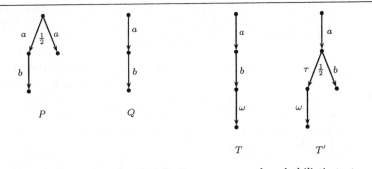

Fig. 10. Examples of probabilistic processes and probabilistic tests

family of tests and the set of tests that do not contain the internal action τ. The following example shows that the former set of tests has (strictly) more distinguishing power than the latter. We consider that $\sqsubseteq_0$ denotes the previous preorder with respect to τ-free tests while $\sqsubseteq$ denotes the preorder with respect to the whole family of tests.

Example 3. Let us consider the processes P and Q depicted in Figure 10. These processes are related if we consider only τ-free tests, that is, $P \sqsubseteq_0 Q$. However, this is not the case if tests are not restricted. For instance, let us consider the tests T and T' (see Figure 10). We obtain $Prob(S_{P\|T}) = \frac{1}{2}$ and $Prob(S_{Q\|T}) = 1$. Thus, we have $Q \not\sqsubseteq P$. Besides, $Prob(S_{P\|T'}) = \frac{3}{4}$ and $Prob(S_{Q\|T'}) = \frac{1}{2}$. Thus, we have $P \not\sqsubseteq Q$. □

Actually, the whole set of test is too discriminative since the preorder coincides in fact with the induced equivalence relation.

In [51] testing semantics for a probabilistic extension of a subset of the specification language LOTOS [41] are studied. The underlying model, that is the induced probabilistic labeled transition systems, is equivalent to the one in [18] since probability is introduced by replacing the LOTOS choice operator by a probabilistic one. In this work the reactive and generative interpretations of probabilities are characterized in terms of tests. So, two processes are *reactive* testing equivalent if they pass with the same probability all the tests consisting of sequences of actions. The intuition is that only one action may be offered simultaneously, being that the reason why tests are traces. In the *generative* testing equivalence all the possible tests are allowed, that is, more than one action may be offered at the same time. In a third testing semantics, called *limited generative*, tests may offer more than an action simultaneously but all of them must be offered with the same probability.

Another approach for probabilistic testing, based also on the notion introduced in [18], is presented in [53]. In this work, fair testing [49, 9] is characterized in terms of a probabilistic testing semantics. Intuitively, it is shown that a (non-probabilistic) process P *fairly* passes a test T iff for any probabilizations of P

and T, that is P_p and T_p, we have that P_p passes T_p with probability equal to 1. Let us remark that in the probabilistic setting the following two properties are not equal: all the computations of the system $P \| T$ are successful and $P\,pass_1\,T$.

Example 4. Let $P = recX.\ (a\,;stop + \tau\,;X)$ be a (non-probabilistic) process and $T = a\,;\omega$ be a the test. We have that P must pass T does not hold because of the infinite sequence of τ actions. However, this computation can be considered *unfair* because the left hand side of the choice is never taken. On the contrary, if we consider the probabilistic process $P_p = recX.\ (a\,;stop +_p \tau\,;X)$ then for any $0 < p < 1$ we have $P_p\,pass_1\,T$. □

The models described above have in common that they consider probabilistic versions of basic CCS (or, equivalently, probabilistic labeled transition systems). Even though these languages are usually simpler, the corresponding testing semantics present a drawback: They do not (sufficiently) abstract internal actions.

Example 5. Let us consider the processes $P_1 = a$ and $P_2 = \tau\,;a$. If we take the notion of composition defined in [18] or in [51] then these two processes are not testing equivalent. For example, for the test $T = a\,;\omega +_{\frac{1}{2}}\tau\,;stop$ we have that, on the one hand, P_1 passes T with probability $\frac{1}{2}$ while, on the other hand, P_2 passes T with probability $\frac{1}{4}$. □

Thus, it is worth to define formalisms where the previous shortcomings are corrected. Actually, this problem appears because of the possibility of *mixed* choices, that is, choices where both internal and external actions are offered simultaneously. We may either restrict the language (as they do in [68], where a testing semantics excluding τ actions in tests is studied) or to consider a new language. In [52], a CSP like language is considered where both choice operators are extended with probabilities. Tests are defined as processes but possibly containing the ω action. In addition to the definition of a testing equivalence, in [52] an alternative characterization and a fully abstract denotational semantics are also provided. Both of them are based on the corresponding ones for (non-probabilistic) must testing. In [54] the framework is extended with a sound and complete axiomatization. This language is also used in [27] in order to define a probabilistic extension of refusal testing [55].

5.2   Testing Semantics for Non-deterministic Probabilistic Processes

In [69] a language featuring both a (non-probabilistic) CCS choice operator and a probabilistic internal choice operator is presented. In this model, tests are also considered as processes that can perform an acceptance action to express the success of a computation. They introduced a notion of probabilistic testing by considering the (set of) probabilities with which a process may pass a test. Actually, given the fact that not all the choices are quantified, a unique probability cannot be (in general) determined.

Example 6. Let us consider the process $Q = \tau \,;\, a + \tau \,;\, (a \oplus_{\frac{1}{2}} b) + \tau \,;\, b$ and the test $T = b \,;\, \omega$. The process Q has three possible choices for resolving its non-determinism: a, $a \oplus_{\frac{1}{2}} b$, and b. Considering the first choice, Q fails the test T, that is, Q passes the test T with probability 0; with the second choice, Q passes the test T with probability $\frac{1}{2}$, meanwhile for the last choice, the test is passed with probability 1 by Q. Thus, Q passes the test T with the set of probabilities $\{0, \frac{1}{2}, 1\}$. □

In [69] the authors define probabilistic interpretations of the classical may and must semantics as follows:

- A process P *must* pass a test T if the minimum of the set of probabilities with which P passes T is equal to 1.
- A process P *may* pass a test T if the maximum of the set of probabilities with which P passes T is greater than 0.

Another proposal for testing semantics in the case of non-deterministic probabilistic processes is presented in [13, 12]. They consider a language featuring the two non-probabilistic choice operators from CSP as well as a probabilistic internal choice. Even though their notions of testing are similar to those of [69], in order to introduce alternative characterizations they follow [52, 14] by providing adequate probabilizations of acceptance sets and acceptance trees. In addition, they also present sound and complete axiomatizations for their testing equivalences.

6   Other Operators

In the previous sections we have concentrated on the possible considerations for defining probabilistic choice operators as well as on the semantic models for probabilistic processes. Given the fact that the considered semantic frameworks, either bisimulations or testing semantics, are defined from the operational behavior of processes, the language that it is used to compose processes can be set (partially) apart. For instance, even though in testing semantics the application of tests to processes is usually defined by means of a parallel operator, this operator can be *hidden* in the presentation of the models. Actually, the only relevant feature that we have considered so far was whether the language had only quantified choices or whether non-deterministic choices were also allowed. In this section we will study how other operators are included in probabilistic process algebras. Specifically, we will consider the parallel composition and hiding operators. More additional comments about the parallel operator in the probabilistic setting are given in [59].

6.1   Parallel Composition Operator

The inclusion of parallel operators in probabilistic process algebras presents some additional problems, with respect to the non-probabilistic setting, depending on

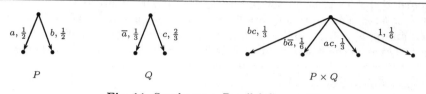

Fig. 11. Synchronous Parallel Composition

how probabilities associated with the different components are combined (see [23] for a discussion and classification of different possibilities). In this section we restrict the study to fully probabilistic models, because such kind of models is the most difficult one to be dealt with in the presence of a parallel operator, hence it needs extensive discussion. If the simple Segala's model [61] is considered for standard non-deterministic parallel operator, or the generative-reactive model [2, 3] for probabilistic parallel, and [66, 67] for exponentially timed parallel, then parallel composition becomes straightforward.

The first probabilistic process algebraic notations included a *synchronous parallel composition* because the definition was simpler in the context of generative models. Intuitively, we suppose that there is an operation $*$ on the set of actions *Act* such that $(Act, *)$ is a monoid. Thus, if we have the parallel composition of P and Q, P may perform a, and Q may perform b, then the parallel composition will perform the action belonging to *Act* given by $a * b$. For example, this is the case of the probabilistic version of SCCS defined in [26]. In that language, the parallel composition of two processes P and Q, denoted by $P \times Q$, behaves as the following inference rule describes

$$\frac{P \xrightarrow{\alpha}_p P', \; Q \xrightarrow{\beta}_q Q'}{P \times Q \xrightarrow{\alpha*\beta}_{p \cdot q} P' \times Q'}$$

The probabilities of the transitions of $P \times Q$ are simply determined by the product of the corresponding probabilities. The action performed by the composition, that is $\alpha * \beta$, is considered as the simultaneous (unordered) occurrence of both actions.

Example 7. Let us consider the processes $P = a +_{\frac{1}{2}} b$ and $Q = \bar{a} +_{\frac{1}{3}} c$ depicted in Figure 11. The parallel composition of both processes, $P \times Q$, will have four possible *atomic* actions to be performed. The first action will be 1 (i.e. the result of $a * \bar{a}$) with probability $\frac{1}{6}$, another action will be $a * c$ with probability $\frac{1}{3}$, the third one will be $b * \bar{a}$ with probability $\frac{1}{6}$, and, finally, $b * c$ with probability $\frac{1}{3}$ (see Figure 11). □

The definition of an asynchronous parallel composition is not so simple.[4] First, we consider the case of a CCS-like operator. Let us remind that the parallel composition of two non-probabilistic processes P and Q, denoted by $P \mid Q$, is defined by the following operational rules:

$$\frac{P \xrightarrow{a} P'}{P \mid Q \xrightarrow{a} P' \mid Q} \qquad \frac{Q \xrightarrow{a} Q'}{P \mid Q \xrightarrow{a} P \mid Q'} \qquad \frac{P \xrightarrow{a} P', \ Q \xrightarrow{\bar{a}} Q'}{P \mid Q \xrightarrow{\tau} P' \mid Q'}$$

The first two rules express the fact that processes are allowed to asynchronously perform actions. The third rule defines the behavior in the presence of synchronization. If one of the processes may perform an action a and the other one can perform the complementary one, that is $\bar{a}$, then a synchronization may take place. However, it is worth to point out that a simple adaptation of the previous rules to the case of fully probabilistic processes may add non-determinism to the model. The following example illustrates this problem.

Example 8. Let us consider again the processes $P = a +_{\frac{1}{2}} b$ and $Q = \bar{a} +_{\frac{1}{3}} c$. If we try to make an analogy with the synchronous case then we have that there exist four different possibilities: P and Q synchronize by performing a and $\bar{a}$ respectively, P performs a and Q performs c, P performs b and Q performs $\bar{a}$, or P performs b and Q performs c. The problem appears because the performance of any two actions is not *atomic*. Thus, there are two ways of performing b and c by P and Q, respectively: First b and after that c, or vice versa. However, we only know the total probability of performing b and c, that is, $\frac{1}{2} \cdot \frac{2}{3} = \frac{1}{3}$. A similar situation appears for the other cases.   □

In [5] a way to solve the problems described before in a generative framework is presented. They propose an operator $\|^{p,q}$ where $p, q \in (0,1)$. The term $P\|^{p,q}Q$ represents a process that asynchronously performs actions from P with probability p, provided that P and Q are not going to synchronize; the probability $1 - p$ plays a similar role for Q. The value q denotes the probability of either P or Q to perform an action asynchronously when a synchronization is possible. Thus, synchronization is performed with probability $1 - q$.

In the non-probabilistic setting, a CSP-like asynchronous parallel composition operator has a parameter representing the set of synchronization actions. All the actions included in it have to be performed synchronously by the components of the parallel composition, while the rest of the actions are performed autonomously. The operational rules for the parallel composition of

---

[4] Let us remark that a first approach has been already explained in Definition 14 (see also [15]). There, the composition of process and test is based on the general idea of considering all possible pairs of actions arising from an independent choice in the two processes (product of probabilities) and then restricting to the acceptable pairs. However, in this composition interleaving actions (actually, the main source of problems when introducing parallel operators in a probabilistic setting) are not considered.

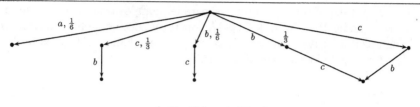

$$(a \,\square_{\frac{1}{2}}\, b) \,\|_{\{a\}}\, (a \,\square_{\frac{1}{3}}\, c)$$

Fig. 12. Asynchronous Parallel Composition for CSP

processes P and Q synchronizing in A, denoted by $P \,\|_A\, Q$, is defined as follows:

$$\frac{P \xrightarrow{a} P', \ a \notin A}{P \,\|_A\, Q \xrightarrow{a} P' \,\|_A\, Q} \qquad \frac{Q \xrightarrow{a} Q', \ a \notin A}{P \,\|_A\, Q \xrightarrow{a} P \,\|_A\, Q'} \qquad \frac{P \xrightarrow{a} P', \ Q \xrightarrow{a} Q', \ a \in A}{P \,\|_A\, Q \xrightarrow{a} P' \,\|_A\, Q'}$$

In contrast with the CCS operator previously presented, synchronization actions cannot be performed autonomously, as it was the case for a and $\bar{a}$ in Example 8. However, non-determinism still arises as the following example shows.

Example 9. Let us consider the processes $P = a \,\square_{\frac{1}{2}}\, b$ and $Q = a \,\square_{\frac{1}{3}}\, c$, and their parallel composition $R = P\|_{\{a\}}Q$. Intuitively, R may perform a with probability $\frac{1}{6}$. In addition, R can also asynchronously perform either b or c. However, there are several ways to perform them. If P performs b then Q has two options: It can try to perform either a or c. Thus, if Q decides to synchronize with P by performing a, then P should perform b with probability $\frac{1}{6}$, that is, $\frac{1}{2} \cdot \frac{1}{3}$. For the same reason, c can be performed with probability $\frac{1}{3} = \frac{1}{2} \cdot \frac{2}{3}$ when P desires to perform a. Non-determinism appears in the other possible option: P decides to perform b while Q chooses to perform c. As in the CCS case, there are again two possibilities: Either b or c is the first action to be performed. In this case, all we know is the combined probability of these two options, that is, $\frac{1}{3}$. This example is graphically presented in Figure 12. □

In the case of this parallel operator we have an additional problem: We need a *normalization factor*. Let us explain this point by using the previous example.

Example 10. Let us consider again the processes defined in Example 9. If the process $R = P \,\|_{\{a\}}\, Q$ performs c with probability $\frac{1}{3}$ then we obtain the process $P\|_{\{a\}}$stop. At that moment, P can perform a with probability $\frac{1}{2}$ but there exists no possibility of synchronization. So, all the probability should be assigned to b, that is, it seems logical that $P \,\|_{\{a\}}\,$ stop performs b with probability 1. □

The problem may be solved by introducing a parallel operator with two parameters, denoted by $\|_A^p$, where $p \in (0,1)$ and A is the synchronization set (see e.g. [51]). The process $P\|_A^p Q$ can evolve either by performing non-synchronizing actions from P or Q (multiplying its associated probabilities by p and $1 - p$ respectively) or by synchronizing in an action belonging to A.

6.2   Hiding Operators

There exist some proposals to include either a hiding operator or a restriction operator into probabilistic process algebras. For example in [12], where a process algebra featuring both probabilistic and non-deterministic behaviors is presented, the operational definition of hiding is given by the following rules:

$$\frac{P \longrightarrow_p P'}{P \backslash A \longrightarrow_p P' \backslash A} \quad \frac{P \longrightarrow P'}{P \backslash A \longrightarrow P' \backslash A} \quad \frac{P \xrightarrow{a} P' \wedge a \notin A}{P \backslash A \xrightarrow{a} P' \backslash A} \quad \frac{P \xrightarrow{a} P' \wedge a \in A}{P \backslash A \longrightarrow P' \backslash A}$$

where $\longrightarrow_p$ represents the resolution of a probabilistic choice, $\longrightarrow$ the internal evolution of a process, and $\xrightarrow{a}$ represents the evolution of the process by performing the visible action a. That is, the definition is essentially the same as in the non-probabilistic case. A similar operator is given in [1].

7   Conclusions

In this paper we have presented the most relevant adaptations to the probabilistic setting of the classical notions of bisimulation and testing semantics. We have distinguished between fully probabilistic models, that is where all the choices are quantified, and non-deterministic probabilistic processes, that is where some choices are quantified and others are non-deterministically resolved. We expect that this paper can serve both to those researchers who are only interested in getting a brief overview of the field as well as to those who desire to get a more thorough knowledge by following the references that we provide.

Acknowledgments. The authors would like to thank the anonymous referees of this paper for the careful reading. Their very valuable comments have undoubtedly improved the quality of this survey.

References

1. S. Andova and J.C.M. Baeten. Abstraction in probabilistic process algebra. In *TACAS 2001, LNCS 2031*, pages 204–219. Springer, 2001.
2. M. Bravetti and A. Aldini. Expressing processes with different action durations through probabilities. In *PAPM-PROBMIV 2001, LNCS 2165*, pages 168–183. Springer, 2001.
3. M. Bravetti and A. Aldini. Discrete time generative-reactive probabilistic processes with different advancing speeds. *Theoretical Computer Science*, 290(1):355–406, 2003.

4. M. Bravetti and M. Bernardo. Compositional asymmetric cooperations for process algebras with probabilities, priorities, and time. In *MTCS 2000, Electronics Notes in Theoretical Computer Science 39(3)*. Elsevier, 2000.

5. J.C.M. Baeten, J.A. Bergstra, and S.A. Smolka. Axiomatizing probabilistic processes: ACP with generative probabilities. *Information and Computation*, 121(2):234–255, 1995.

6. C. Baier and H. Hermanns. Weak bisimulation for fully probabilistic processes. In *Computer Aided Verification'97, LNCS 1254*, pages 119–130. Springer, 1997.

7. B. Bloom and A.R. Meyer. A remark on bisimulation between probabilistic processes. In *Logic at Botik'89, LNCS 363*, pages 26–40. Springer, 1989.

8. J.A. Bergstra, A. Ponse, and S.A. Smolka, editors. *Handbook of Process Algebra*. North Holland, 2001.

9. E. Brinksma, A. Rensink, and W. Vogler. Fair testing. In *CONCUR'95, LNCS 962*, pages 313–327. Springer, 1995.

10. E. Bandini and R. Segala. Axiomatizations for probabilistic bisimulation. In *ICALP 2001, LNCS 2076*, pages 370–381. Springer, 2001.

11. J.C.M. Baeten and W.P. Weijland. *Process Algebra*. Cambridge Tracts in Computer Science 18. Cambridge University Press, 1990.

12. D. Cazorla, F. Cuartero, V. Valero, F.L. Pelayo, and J.J. Pardo. Algebraic theory of probabilistic and non-deterministic processes. *Journal of Logic and Algebraic Programming*, 55(1–2):57–103, 2003.

13. D. Cazorla, F. Cuartero, V. Valero, and F.L. Pelayo. A process algebra for probabilistic and nondeterministic processes. *Information Processing Letters*, 80:15–23, 2001.

14. F. Cuartero, D. de Frutos, and V. Valero. A sound and complete proof system for probabilistic processes. In *4th International AMAST Workshop on Real-Time Systems, Concurrent and Distributed Software, LNCS 1231*, pages 340–352. Springer, 1997.

15. R. Cleaveland, Z. Dayar, S.A. Smolka, and S. Yuen. Testing preorders for probabilistic processes. *Information and Computation*, 154(2):93–148, 1999.

16. I. Christoff. Testing equivalences and fully abstract models for probabilistic processes. In *CONCUR'90, LNCS 458*, pages 126–140. Springer, 1990.

17. R. Cleaveland, G. Lüttgen, and V. Natarajan. Priority in process algebra. In J.A. Bergstra, A. Ponse, and S.A. Smolka, editors, *Handbook of process algebra*, chapter 12. North Holland, 2001.

18. R. Cleaveland, S.A. Smolka, and A.E. Zwarico. Testing preorders for probabilistic processes. In *19th ICALP, LNCS 623*, pages 708–719. Springer, 1992.

19. C. Courcoubetis and M. Yannakakis. Verifying temporal properties of finite-state probabilistic programs. In *29th IEEE Symposium on Foundations of Computer Science*, pages 338–345. IEEE Computer Society Press, 1988.

20. D. de Frutos-Escrig, L.F. Llana-Díaz, and M. Núñez. Friendly testing as a conformance relation. In *Formal Description Techniques for Distributed Systems and Communication Protocols (X), and Protocol Specification, Testing, and Verification (XVII)*, pages 283–298. Chapman & Hall, 1997.

21. D. de Frutos-Escrig, N. López, and M. Núñez. Global timed bisimulation: An introduction. In *Formal Description Techniques for Distributed Systems and Communication Protocols (XII), and Protocol Specification, Testing, and Verification (XIX)*, pages 401–416. Kluwer Academic Publishers, 1999.

22. J. Desharnais, V. Gupta, R. Jagadeesan, and P. Panangaden. Weak bisimulation is sound and complete for PCTL*. In *CONCUR 2002, LNCS 2421*, pages 355–370. Springer, 2002.

23. P.R. D'Argenio, H. Hermanns, and J.-P. Katoen. On generative parallel composition. In *PROBMIV'98, Electronics Notes in Theoretical Computer Science 22.* Elsevier, 1999.
24. R. de Nicola and M.C.B. Hennessy. Testing equivalences for processes. *Theoretical Computer Science,* 34:83–133, 1984.
25. Y.A. Feldman and D. Harel. A probabilistic dinamic logic. In *14th ACM Symposium on Theory of Computing,* pages 181–195. ACM Press, 1982.
26. A. Giacalone, C.-C. Jou, and S.A. Smolka. Algebraic reasoning for probabilistic concurrent systems. In *Proceedings of Working Conference on Programming Concepts and Methods, IFIP TC 2.* North Holland, 1990.
27. C. Gregorio and M. Núñez. Denotational semantics for probabilistic refusal testing. In *PROBMIV'98, Electronic Notes in Theoretical Computer Science 22.* Elsevier, 1999.
28. R. van Glabbeek, S.A. Smolka, and B. Steffen. Reactive, generative and stratified models of probabilistic processes. *Information and Computation,* 121(1):59–80, 1995.
29. R. van Glabbeek, S.A. Smolka, B. Steffen, and C.M.N. Tofts. Reactive, generative, and stratified models of probabilistic processes. In *5th IEEE Symposium on Logic In Computer Science,* pages 130–141. IEEE Computer Society Press, 1990.
30. R. van Glabbeek and W.P. Weijland. Branching time and abstraction in bisimulation semantics. *Journal of the ACM,* 43(3):555–600, 1996.
31. M. Hennessy. An algebraic theory of fair asynchronous communicating processes. *Theoretical Computer Science,* 49:121–143, 1987.
32. M. Hennessy. *Algebraic Theory of Processes.* MIT Press, 1988.
33. H. Hansson and B. Jonsson. A framework for reasoning about time and realibility. In *10th IEEE Real-Time Systems Symposium.* IEEE Computer Society Press, 1989.
34. H. Hansson and B. Jonsson. A calculus for communicating systems with time and probabilities. In *11th IEEE Real-Time Systems Symposium,* pages 278–287. IEEE Computer Society Press, 1990.
35. C.A.R. Hoare. *Communicating Sequential Processes.* Prentice Hall, 1985.
36. S. Hart and M. Sharir. Probabilistic temporal logics for finite and bounded models. In *16th ACM Symposium on Theory of Computing,* pages 1–13. ACM Press, 1984.
37. C. Jones and G.D. Plotkin. A probabilistic powerdomain of evaluations. In *4th IEEE Symposium on Logic In Computer Science,* pages 186–195. IEEE Computer Society Press, 1989.
38. M. Kwiatkowska and G.J. Norman. A testing equivalence for reactive probabilistic processes. In *EXPRESS'98, Electronic Notes in Theoretical Computer Science 16.* Elsevier, 1998.
39. D. Kozen. A probabilistic PDL. In *15th ACM Symposium on Theory of Computing,* pages 291–297. ACM Press, 1983.
40. J.G. Kemeny and J.L. Snell. *Finite Markov Chains.* Springer, 1976.
41. LOTOS. A formal description technique based on the temporal ordering of observational behaviour. IS 8807, TC97/SC21, 1988.
42. G. Lowe. Probabilistic and prioritized models of timed CSP. *Theoretical Computer Science,* 138:315–352, 1995.
43. K. Larsen and A. Skou. Bisimulation through probabilistic testing. In *16th ACM Simposium on Principles of Programming Languages,* pages 344–352. ACM Press, 1989.
44. K. Larsen and A. Skou. Bisimulation through probabilistic testing. *Information and Computation,* 94(1):1–28, 1991.

45. N.A. Lynch and M.R. Tuttle. Hierarchical correctness proofs for distributed algorithms. In *6th ACM Symp. on Principles of Distributed Computing*, pages 137–151. ACM Press, 1987.
46. R. Milner. Calculi for synchrony and asynchrony. *Theoretical Computer Science*, 253:267–310, 1983.
47. R. Milner. *Communication and Concurrency*. Prentice Hall, 1989.
48. R. Milner and D. Sangiorgi. Barbed bisimulation. In *19th ICALP, LNCS 623*, pages 685–695. Springer, 1992.
49. V. Natarajan and R. Cleaveland. Divergence and fair testing. In *22nd ICALP, LNCS 944*, pages 648–659. Springer, 1995.
50. K. Narayan Kumar, R. Cleaveland, and S.A. Smolka. Infinite probabilistic and nonprobabilistic testing. In *18th Conference on Foundations of Software Technology and Theoretical Computer Science, LNCS 1530*, pages 209–220. Springer, 1998.
51. M. Núñez and D. de Frutos. Testing semantics for probabilistic LOTOS. In *Formal Description Techniques VIII*, pages 365–380. Chapman & Hall, 1995.
52. M. Núñez, D. de Frutos, and L. Llana. Acceptance trees for probabilistic processes. In *CONCUR'95, LNCS 962*, pages 249–263. Springer, 1995.
53. M. Núñez and D. Rupérez. Fair testing through probabilistic testing. In *Formal Description Techniques for Distributed Systems and Communication Protocols (XII), and Protocol Specification, Testing, and Verification (XIX)*, pages 135–150. Kluwer Academic Publishers, 1999.
54. M. Núñez. Algebraic theory of probabilistic processes. *Journal of Logic and Algebraic Programming*, 56(1–2):117–177, 2003.
55. I. Phillips. Refusal testing. *Theoretical Computer Science*, 50(3):241–284, 1987.
56. A. Philippou, I. Lee, and O. Sokolsky. Weak bisimulation for probabilistic systems. In *CONCUR 2000, LNCS 1877*, pages 334–349. Springer, 2000.
57. M.O. Rabin. Probabilistic automata. *Information and Control*, 6:230–245, 1963.
58. E.W. Stark, R. Cleaveland, and S.A. Smolka. A process-algebraic language for probabilistic I/O automata. In *CONCUR'03, LNCS 2761*. Springer, 2003.
59. A. Sokolova and E.P. de Vink. Probabilistic automata: System types, parallel composition and comparation. In *VOSS GI-Dagstuhl seminar, LNCS*. Springer, 2003. This volume.
60. K. Seidel. Probabilistic communicating processes. *Theoretical Computer Science*, 152:219–249, 1995.
61. R. Segala and N. Lynch. Probabilistic simulations for probabilistic processes. In *CONCUR'94, LNCS 836*, pages 481–496. Springer, 1994.
62. R. Segala and N. Lynch. Probabilistic simulations for probabilistic processes. *Nordic Journal of Computing*, 2(2):250–273, 1995.
63. E.W. Stark and S.A. Smolka. A complete axiom system for finite-state probabilistic processes. In *Proof, Language and Interaction: Essays in Honour of Robin Milner*. MIT Press, 2000.
64. M. Stoelinga and F.W. Vaandrager. Root contention in IEEE 1394. In *5th AMAST Workshop on Real-Time and Probabilistic Systems, LNCS 1601*, pages 53–74. Springer, 1999.
65. M.Y. Vardi. Automatic verification of probabilistic concurrent finite-state programs. In *26th IEEE Symposium on Foundations of Computer Science*, pages 327–338. IEEE Computer Society Press, 1985.
66. S.-H. Wu, S.A. Smolka, and E.W. Stark. Composition and behaviors of probabilistic I/O automata. In *CONCUR'94, LNCS 836*, pages 513–528. Springer, 1994.

67. S.-H. Wu, S.A. Smolka, and E.W. Stark. Composition and behaviors of probabilistic I/O automata. *Theoretical Computer Science*, 176(1-2):1–37, 1997.
68. S. Yuen, R. Cleaveland, Z. Dayar, and S.A. Smolka. Fully abstract characterizations of testing preorders for probabilistic processes. In *CONCUR'94, LNCS 836*, pages 497–512. Springer, 1994.
69. W. Yi and K.G. Larsen. Testing probabilistic and nondeterministic processes. In *Protocol Specification, Testing and Verification XII*, pages 47–61. North Holland, 1992.

Verifying Qualitative Properties of Probabilistic Programs

Benedikt Bollig[1]* and Martin Leucker[2]**

[1] Lehrstuhl für Informatik II, RWTH Aachen, Germany
bollig@informatik.rwth-aachen.de
[2] IT department, Uppsala University, Sweden
Martin.Leucker@it.uu.se

Abstract. In this chapter, we present procedures for checking linear temporal logic and automata specifications of sequential and concurrent probabilistic programs. We follow two different approaches: For LTL and sequential probabilistic programs, our method proceeds in a tableau style fashion, while the remaining procedures are based on automata theory.

1 Introduction

Randomization techniques have been employed to solve numerous problems of computing both sequentially and in parallel. Examples of probabilistic algorithms that are asymptotically better than their deterministic counterparts in solving various fundamental problems abound. It has also been shown that they allow solutions of problems which cannot be solved deterministically [7]. They have the advantages of simplicity and better performance both in theory and often in practice. An overview of the domain of randomized algorithms is given in [12].

As for any kind of hardware or software system, it is important to develop formal methods and tools for verifying their correctness, thus also for probabilistic programs. *Model checking*, introduced independently in [2] and [14], turned out to be one fruitful approach to automatically verify systems (see [3] for an overview of model checking in practice). In the model-checking approach, usually a finite system $\mathcal{M}$, often an abstraction of a real system, and a property, usually expressed as a temporal-logic formula φ or as an automaton describing the computations that adhere to the property, are given. The *model-checking procedure* decides whether the set or tree formed of all *computations* of $\mathcal{M}$ satisfy φ, or, in other words, whether the given system satisfies the required property. Temporal logics used for expressing requirements are usually linear temporal logic (LTL) (as initially proposed by Pnueli in [10]) or branching time logics like the computation-tree logics CTL and CTL*.

---

* Part of this work was done during the author's stay at the IT department, Uppsala University, Sweden. He is grateful for the hospitality and the overall support.
** This author is supported by the European Research Training Network "Games".

C. Baier et al. (Eds.): Validation of Stochastic Systems, LNCS 2925, pp. 124–146, 2004.
© Springer-Verlag Berlin Heidelberg 2004

In the probabilistic setting, one is no longer interested in *all* but only *almost all* computations. Thus, (sets or pathes of trees of) computations with zero probability are ignored when deciding whether a given property is satisfied. More general, one is also interested whether a given property is satisfied with some given probability.

In this chapter, we describe methods for checking whether almost all runs of a finite sequential or concurrent probabilistic program in the sense of [5] and [8] satisfy a given LTL formula or all accepting runs of an automaton describing the underlying property. Thus, we are concerned about quality aspects of a given probabilistic program rather than in estimating quantity measures for given properties or in temporal logics allowing to express quantitative aspects of properties. These issues are addressed in one of the subsequent chapters.

We present a procedure for checking LTL specifications of sequential probabilistic programs, which requires running time that is exponential in the size of the formula and linear in the size of the program. It can be shown that this is optimal.

Furthermore, we describe a procedure for checking automata specifications of sequential and concurrent probabilistic programs. It can be shown that this problem can be solved in time polynomial in the size of the program and exponential in the size of the formula. However, to simplify our presentation, we present a less technical construction based on the same idea, however, with a slightly inferior complexity than the one with optimal bounds. As LTL specifications can easily be transformed into automata specifications involving an exponential blow-up [17], the procedure can also be used to check LTL specifications (in time double exponential in the size of the formula).

This chapter is mainly based on [4], in which additionally methods for an extension of LTL—known as ETL [18]—are shown and proofs showing that all constructions are optimal are given. Since we only want to give an introduction to the field of verification of qualitative properties of probabilistic programs, we refer to this paper for a more detailed study of these extensions and proof ideas.

This chapter is organized as follows. In the next section, we introduce the necessary concepts and notation of automata, linear temporal logic, and probabilistic programs. Furthermore, we recall some mathematical background in the domain of graph and probability theory. In Section 3, we describe a model-checking procedure for LTL formulas and sequential probabilistic programs. Section 4 studies this question for automata specifications and sequential as well as concurrent probabilistic programs. We draw our conclusions in Section 5. This chapter ends giving suggestions for further reading in Section 6.

2   Preliminaries

2.1   Words and Graphs

Given an alphabet Σ, Σ^* denotes the set of *finite* and Σ^ω the set of *infinite words* over Σ. For a word $w = a_0 a_1 \ldots \in \Sigma^\omega$ and a natural i, let $w(i)$ denote

$a_i a_{i+1} \ldots$, the ith suffix of w. Furthermore, take $inf(w)$ as the *infinity set* $\{a_i \mid |\{j \mid a_j = a_i\}| = \infty\}$ of letters that occur infinitely often in w.

Given a directed graph G with nodes V and edges E, we call a node $v \in V$ (a set $D \subseteq V$ of nodes) *reachable* from $v' \in V$ if there is a path from v' to v (to a node contained in D). A *strongly connected component* (SCC) of G is a maximal set D of nodes such that every two nodes contained in D are reachable from each other. A SCC is called *bottom* if there is no edge leading out of it. We define G to be *strongly connected* if V forms a SCC. Furthermore, G is called *nontrivial* if it contains at least one edge. A set $D \subseteq V$ is said to be *nontrivial* if $G[D]$, the subgraph of G *induced* by D, is nontrivial. The *size* of G, denoted by $|G|$, is defined to be $|V| + |E|$.

2.2   Automata

Let Σ be an alphabet. An ω-*automaton* over Σ is a tuple $\mathcal{A} = (S, S_0, \delta, Acc)$ where S is its nonempty finite set of *states*, $S_0 \subseteq S$ is the set of *initial states*, $\delta : S \times \Sigma \to 2^S$ is its *transition function*, and Acc is an *acceptance component*. A *run* of $\mathcal{A}$ on a word $w = a_1 a_2 \ldots \in \Sigma^\omega$ is a sequence of states $\rho = s_0 s_1 s_2 \ldots \in S^\omega$ such that $s_0 \in S_0$ and, for $i = 0, 1, \ldots$, $s_{i+1} \in \delta(s_i, a_{i+1})$.

An ω-automaton $\mathcal{A} = (S, S_0, \delta, Acc)$ is called *Büchi automaton* if Acc is a set $F \subseteq S$. We then call a run ρ of $\mathcal{A}$ *accepting* if $inf(\rho) \cap F \neq \emptyset$.

A *Streett automaton* (*Rabin automaton*) over Σ differs from a Büchi automaton only in the acceptance component. More precisely, it is a structure $\mathcal{A} = (S, S_0, \delta, \mathcal{F})$ where $\mathcal{F}$ is a subset of $(2^S)^2$. A run ρ of $\mathcal{A}$ is called *accepting* if, for all pairs $(U, V) \in \mathcal{F}$, $inf(\rho) \cap U \neq \emptyset$ implies $inf(\rho) \cap V \neq \emptyset$ (there is a pair $(U, V) \in \mathcal{F}$ such that $inf(\rho) \cap U \neq \emptyset$ and $inf(\rho) \cap V = \emptyset$).

The *language* of an ω-automaton $\mathcal{A} = (S, S_0, \delta, Acc)$, denoted by $L(\mathcal{A})$, is defined to be the set $\{w \in \Sigma^\omega \mid$ there is an accepting run of $\mathcal{A}$ on $w\}$. $\mathcal{A}$ is called *deterministic (quasi-deterministic)* if both $|S_0| = 1$ and $|\delta(s, a)| = 1$ ($|\delta(s, a)| \leq 1$) for all $s \in S$, $a \in \Sigma$. Furthermore, the *size* $|\mathcal{A}|$ of $\mathcal{A}$ is defined to be the size of its (transition) graph.

2.3   Propositional Linear Temporal Logic

In the following, let $\mathcal{P}$ be a nonempty finite set of *propositions*. The set LTL($\mathcal{P}$) of (Propositional) Linear Temporal Logic (LTL) formulas over $\mathcal{P}$ is inductively defined as follows: $\mathcal{P}$ is a subset of LTL($\mathcal{P}$), and, for LTL($\mathcal{P}$) formulas φ and ψ, $\mathcal{X}\varphi$, $\varphi\mathcal{U}\psi$, $\neg\varphi$, and $\varphi \vee \psi$ are LTL($\mathcal{P}$) formulas as well. An LTL($\mathcal{P}$) formula is inductively interpreted over $\pi = \pi_0 \pi_1 \ldots \in (2^\mathcal{P})^\omega$ as follows:

- $\pi \models p \in \mathcal{P}$ if $p \in \pi_0$
- $\pi \models \neg\varphi$ if $\pi \not\models \varphi$
- $\pi \models \varphi \vee \psi$ if $\pi \models \varphi$ or $\pi \models \psi$
- $\pi \models \mathcal{X}\varphi$ if $\pi(1) \models \varphi$
- $\pi \models \varphi\mathcal{U}\psi$ if there is an $i \geq 0$ such that $\pi(i) \models \psi$ and, for each $j \in \{0, \ldots, i-1\}$, $\pi(j) \models \varphi$

The *language* of a formula $\varphi \in$ LTL($\mathcal{P}$), denoted by $L(\varphi)$, is defined to be the set $\{\pi \in (2^\mathcal{P})^\omega \mid \pi \models \varphi\}$.

2.4   Probability Spaces and Probabilistic Programs

A nonempty set of possible outcomes of an experiment of chance is called *sample space*. Let Ω be a sample space. A set $\mathfrak{B} \subseteq 2^{\Omega}$ is called *Borel field* over Ω if it contains Ω, $\Omega \setminus E$ for each $E \in \mathfrak{B}$, and the union of any countable sequence of sets from $\mathfrak{B}$. A Borel field $\mathfrak{B}$ is *generated* by an at most countable set $\mathcal{E}$, denoted by $\mathfrak{B} = \langle \mathcal{E} \rangle$, if $\mathfrak{B}$ is the closure of $\mathcal{E}$'s elements under complement and countable union.

A *probability space* is a triple $\mathcal{PS} = (\Omega, \mathfrak{B}, \mu)$ where Ω is a sample space, $\mathfrak{B}$ is a Borel field over Ω, and μ is a mapping $\mathfrak{B} \to [0,1]$ such that $\mu(\Omega) = 1$ and $\mu(\bigcup_{i=1}^{\infty} E_i) = \sum_{i=1}^{\infty} \mu(E_i)$ for any sequence $E_1, E_2, \ldots$ of pairwise disjoint sets from $\mathfrak{B}$. We call μ a *probability measure*. An event $E \in \mathfrak{B}$ is said to occur *almost surely* if $\mu(E) = 1$.

A *concurrent probabilistic program* over $\mathcal{P}$, the set of propositions, is a tuple $\mathcal{M} = (Q, N, R, \Delta, P_0, P, \mathcal{V})$ where

- Q is its nonempty finite set of *states*, which is partitioned into sets N and R of *nondeterministic* and *randomizing states*, respectively,
- $\Delta \subseteq Q \times Q$ is the set of transitions with $\Delta(q) := \{q' \mid (q,q') \in \Delta\} \neq \emptyset$ for each $q \in Q$,
- $P_0 : Q \to [0,1]$ and $P : \{(q,q') \in \Delta \mid q \in R\} \to (0,1]$ are the *initial* and *transition probability distributions*[1], respectively, where $\sum_{q \in Q} P_0(q) = 1$ and, for each $q \in R$, $\sum_{q' \in \Delta(q)} P_{q,q'} = 1$, and
- $\mathcal{V} : Q \to 2^{\mathcal{P}}$ is the *valuation function*, which is extended to words over Q as expected.

A concurrent probabilistic program over $\{p\}$ is shown in Figure 1. Hereby, the nondeterministic states are represented by circles, whereas the randomizing ones are given by rectangles.

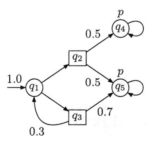

Fig. 1. A concurrent probabilistic program

In case that a concurrent probabilistic program $\mathcal{M}$ has no nondeterministic states, i.e., N is the empty set, we call it a *sequential probabilistic program* over $\mathcal{P}$ and identify $\mathcal{M}$ with $(Q, \Delta, P_0, P, \mathcal{V})$. $\mathcal{M}$ induces a probability space $\mathcal{PS}_{\mathcal{M}} = (\Omega_{\mathcal{M}}, \mathfrak{B}_{\mathcal{M}}, \mu_{\mathcal{M}})$ as follows: $\Omega_{\mathcal{M}} = \{q_1 q_2 \ldots \in Q^{\omega} \mid P_0(q_1) > 0$ and, for

---

[1] We usually write $P_{q,q'}$ instead of $P((q,q'))$.

all $i \geq 1$, $(q_i, q_{i+1}) \in \Delta\}$ is the set of *trajectories* of $\mathcal{M}$, $\mathfrak{B}_{\mathcal{M}}$ is generated by $\{\mathcal{C}_{\mathcal{M}}(x) \mid x \in Q^*\}$ where $\mathcal{C}_{\mathcal{M}}(x) = \{x' \in \Omega_{\mathcal{M}} \mid x \text{ is a prefix of } x'\}$ is the *basic cylinder set* of x wrt. $\mathcal{M}$. The measure $\mu_{\mathcal{M}}$ is uniquely given by $\mu_{\mathcal{M}}(\Omega_{\mathcal{M}}) = 1$ and, for $n \geq 1$, $\mu_{\mathcal{M}}(\mathcal{C}_{\mathcal{M}}(q_1 \ldots q_n)) = P_0(q_1) \cdot P_{q_1,q_2} \cdot \ldots \cdot P_{q_{n-1},q_n}$ (where we assume $P_{q_i,q_{i+1}}$ to be 0 if $(q_i, q_{i+1}) \notin \Delta$).

A *scheduler* of a concurrent probabilistic program $\mathcal{M} = (Q, N, R, \Delta, P_0, P, \mathcal{V})$ over $\mathcal{P}$ is a mapping $u : Q^*N \to Q$ such that $u(xq) = q'$ implies $(q, q') \in \Delta$.

Similarly to the sequential case, a scheduler u of a concurrent program $\mathcal{M} = (Q, N, R, \Delta, P_0, P, \mathcal{V})$ induces a probability space $\mathcal{PS}_{\mathcal{M},u} = (\Omega_{\mathcal{M},u}, \mathfrak{B}_{\mathcal{M},u}, \mu_{\mathcal{M},u})$ as follows: $\Omega_{\mathcal{M},u} = \{q_1 q_2 \ldots \in Q^\omega \mid P_0(q_1) > 0 \text{ and, for all } i \geq 1, (q_i \in R \text{ and } (q_i, q_{i+1}) \in \Delta) \text{ or } (q_i \in N \text{ and } u(q_1 \ldots q_i) = q_{i+1})\}$ is the set of trajectories of $\mathcal{M}$ wrt. u, $\mathfrak{B}_{\mathcal{M},u} = \langle\{\mathcal{C}_{\mathcal{M},u}(x) \mid x \in Q^*\}\rangle$ where $\mathcal{C}_{\mathcal{M},u}(x) = \{x' \in \Omega_{\mathcal{M},u} \mid x \text{ is a prefix of } x'\}$ is the basic cylinder set of x wrt. $\mathcal{M}$ and u. The measure $\mu_{\mathcal{M},u}$ is uniquely given by $\mu_{\mathcal{M},u}(\Omega_{\mathcal{M},u}) = 1$ and, for $n \geq 1$, $\mu_{\mathcal{M},u}(\mathcal{C}_{\mathcal{M},u}(q_1 \ldots q_n)) = P_0(q_1) \cdot P'_{q_1,q_2} \cdot \ldots \cdot P'_{q_{n-1},q_n}$ where

$$
P'_{q_i,q_{i+1}} = \begin{cases} 0 & \text{if } (q_i, q_{i+1}) \notin \Delta \\ P_{q_i,q_{i+1}} & \text{if } (q_i, q_{i+1}) \in \Delta \text{ and } q_i \in R \\ 1 & \text{if } (q_i, q_{i+1}) \in \Delta \text{ and } q_i \in N \end{cases}.
$$

When we concentrate on topological aspects of probabilistic programs, we consider a program to be a graph where the states become (nondeterministic or randomizing) nodes and the transitions become edges. In the following, unless the context requires, a program will also denote its graph. From this point of view, we call a state *absorbing* if it forms a (nontrivial) bottom SCC. Otherwise, it is called *transient*.

3   Checking LTL Specifications of Sequential Programs

For a concurrent probabilistic program $\mathcal{M}$ over $\mathcal{P}$ with valuation function $\mathcal{V}$, a scheduler u of $\mathcal{M}$, and a formula $\varphi \in \text{LTL}(\mathcal{P})$, let $L_{\mathcal{M},u}(\varphi) := \{x \in \Omega_{\mathcal{M},u} \mid \mathcal{V}(x) \in L(\varphi)\}$. Similarly, for a sequential program $\mathcal{M}$ with valuation function $\mathcal{V}$, we define $L_{\mathcal{M}}(\varphi) := \{x \in \Omega_{\mathcal{M}} \mid \mathcal{V}(x) \in L(\varphi)\}$. It can be shown, using structural induction, that these sets are measurable:

Proposition 1 ([16]). *Given a concurrent (sequential) probabilistic program $\mathcal{M}$ over $\mathcal{P}$ and a formula $\varphi \in \text{LTL}(\mathcal{P})$, we have $L_{\mathcal{M},u}(\varphi) \in \mathfrak{B}_{\mathcal{M},u}$ for each scheduler u of $\mathcal{M}$ ($L_{\mathcal{M}}(\varphi) \in \mathfrak{B}_{\mathcal{M}}$).*

Definition 1. *Given a formula $\varphi \in \text{LTL}(\mathcal{P})$, a concurrent probabilistic program $\mathcal{M}$ over $\mathcal{P}$ is said to satisfy φ if, for all schedulers u of $\mathcal{M}$, $\mu_{\mathcal{M},u}(L_{\mathcal{M},u}(\varphi)) = 1$ (we say that u satisfies φ). A sequential probabilistic program is said to satisfy φ if $\mu_{\mathcal{M}}(L_{\mathcal{M}}(\varphi)) = 1$.*

For example, the concurrent probabilistic program $\mathcal{M}$ shown in Figure 1 and, in particular, the scheduler of $\mathcal{M}$ from Figure 2, satisfy the $\text{LTL}(\{p\})$ formula $\Diamond \Box p$. As usual, $\Diamond \varphi$ is an abbreviation for $True \mathcal{U} \varphi$ and $\Box \varphi$ for $\neg \Diamond \neg \varphi$.

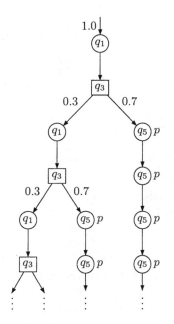

Fig. 2. A scheduler

Theorem 1. *We can test whether a sequential probabilistic program $\mathcal{M}$ satisfies a formula φ in time $O(|\mathcal{M}|2^{|\varphi|})$, or in space polynomial in φ and polylogarithmic in $\mathcal{M}$. We can compute the probability of satisfaction $\mu_{\mathcal{M}}(L_{\mathcal{M}}(\varphi))$ in time exponential in φ and polynomial in $\mathcal{M}$.*

In the remainder of this section, we proof the previous theorem by formulating a procedure checking whether a sequential probabilistic program $\mathcal{M}$ satisfies an LTL formula φ. The procedure processes φ according to its inductive structure in a bottom-up manner, eliminating its temporal connectives one-by-one and transforms $\mathcal{M}$ in each step to preserve the probability of satisfiability.

To simplify our presentation, we sometimes say that a state satisfies an LTL formula instead of saying that all trajectories starting from this state satisfy this formula.

There are two transformations $\mathcal{T}_{\mathcal{U}}$ and $\mathcal{T}_{\mathcal{X}}$ corresponding to the two temporal connectives $\mathcal{U}$ and $\mathcal{X}$. We describe $\mathcal{T}_{\mathcal{U}}$ in detail and list the modifications for $\mathcal{T}_{\mathcal{X}}$. We illustrate the procedure with the program given in Figure 3.

Transformation $\mathcal{T}_{\mathcal{U}}$ Let $p\mathcal{U}p'$ be the innermost subformula of φ.[2] Let W^+ contain all states satisfying p' and W_1^- contain all states satisfying $\neg p$ and $\neg p'$. Let W_2^- comprise all states of bottom strongly connected components (of $\mathcal{M}$ considered as a graph) which have no state in W^+. Thus, states of W_2^- possibly satisfy p but no state satisfying p' is reachable. Let $W^- = W_1^- \cup W_2^-$. Clearly, all tra-

---

[2] As Boolean combinations of propositions can be evaluated in each state, we simplify our life and think of p and p' as propositions.

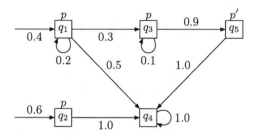

Fig. 3. A sequential probabilistic program

jectories starting in W^+ satisfy $p\mathcal{U}p'$ and all trajectories starting in W^- satisfy $\neg(p\mathcal{U}p')$. For our example (Figure 3), we get $W^+ = \{q_5\}$, $W_1^- = W_2^- = \{q_4\}$.

Let Q^+ (Q^-) be the states q such that for all trajectories starting at q the first state in $W^+ \cup W^-$ is in W^+ (W^-). Note that $W^+ \subseteq Q^+$ and $W^- \subseteq Q^-$. For $\mathcal{M}$ of Figure 3, we have $Q^+ = \{q_3, q_5\}$ and $Q^- = \{q_2, q_4\}$.

Almost surely, every trajectory reaches eventually a bottom strongly connected component. Trajectories starting in a state of Q^+ satisfy $p\mathcal{U}p'$ and those starting in a state of Q^- never reach a state satisfying p' before reaching a state not satisfying p or never reach a state satisfying p' at all. Using P_q as a shorthand for the measure of trajectories starting in q satisfying $p\mathcal{U}p'$, we conclude that $P_q = 1$ for all $q \in Q^+$ and $P_q = 0$ for $q \in Q^-$.

Let $Q^?$ comprise the remaining states, so $Q^? = \{q_1\}$ in our example. As $q \in Q^?$ satisfies p but not p', a trajectory π starting in q satisfies $p\mathcal{U}p'$ iff $\pi(1)$ satisfies $p\mathcal{U}p'$. Thus, $P_q = \sum_{q'} P_{q,q'} P_{q'}$. We summarize:

Lemma 1.

$$P_q = \sum_{q'} P_{q,q'} P_{q'} \quad \textit{if } q \in Q^?$$
$$P_q = 1 \quad\quad\quad\quad \textit{if } q \in Q^+$$
$$P_q = 0 \quad\quad\quad\quad \textit{if } q \in Q^-$$

This set of equations has a unique solution.

Proof. It remains to show that the set of equations has a unique solution. Clearly, any two solutions can only differ on $Q^?$. Let us form a new sequential probabilistic program $\mathcal{M}^?$ whose state space consists of $Q^?$ and one additional absorbing state $\hat{q}$ with a transition to itself with probability 1. For pairs of states of $Q^?$, the transitions and their probabilities agree with those of $\mathcal{M}$. Furthermore, for each $q \in Q^?$ with transitions to states $q' \notin Q^?$ we add a single transition q to $\hat{q}$ with probability $\sum_{q' \notin Q^?} P_{q,q'}$. The state $\hat{q}$ forms the only bottom s.c.c.: Since all states of $Q^?$ satisfy p and $\neg p'$ any further bottom s.c.c. would be in W_2^-, thus not in $Q^?$. Hence, every trajectory will reach state $\hat{q}$ with probability one.

Let T be the matrix of transition probabilities $P_{q,q'}$ restricted to states of $Q^?$. Furthermore, for T^k, the kth power of the matrix T, the entry $T_{q,q'}^k$ is equal to the probability that a path starting in q reaches q' in k steps. Since every trajectory goes eventually to $\hat{q}$ (and stays there) with probability one, we get

$\lim_{k \to \infty} T^k = 0$. For any two solutions $a = (a_q)_{q \in Q^?}$ and $b = (b_q)_{q \in Q^?}$ of the equation system, we have $a - b = T(a - b)$, thus $a - b = T^k(a - b)$, and taking the limit as k tends to infinity we conclude that $a = b$.   □

For the example shown in Figure 3, we get $P_{q_3} = P_{q_5} = 1$ and $P_{q_2} = P_{q_4} = 0$. Furthermore, $P_{q_1} = P_{q_1,q_1} P_{q_1} + P_{q_1,q_3} P_{q_3} + P_{q_1,q_4} P_{q_4} = 0.2 \cdot P_{q_1} + 0.3 \cdot 1 + 0.5 \cdot 0 = \frac{3}{8}$.

We are ready for the construction of the new program $\mathcal{M}'$ and the new formula φ'. $\mathcal{M}'$ has a larger state space Q' and is defined over the set of atomic propositions $\mathcal{P}$ enriched with ξ which serves in φ' as a substitute for $p\mathcal{U}p'$. Let

- $Q' = \{(q, \xi) \mid q \in Q^+ \cup Q^?\} \cup \{(q, \neg\xi) \mid q \in Q^- \cup Q^?\}$.
- $\mathcal{V}'((q, \xi)) = \mathcal{V}(q) \cup \{\xi\}$ and $\mathcal{V}'((q, \neg\xi)) = \mathcal{V}(q)$.
- we define the transitions of $\mathcal{M}'$ implicitly by giving non-zero transition probabilities. ξ_1 and ξ_2 stand for ξ or $\neg\xi$ and $\overline{P}$ for $1 - P$. We distinguish the following cases based on the states of $\mathcal{M}$:
  - $q, q' \in Q^+ \cup Q^-$. There is exactly one state (q, ξ_1) and (q', ξ_2) in Q'. We set $P'_{(q,\xi_1),(q',\xi_2)} = P_{q,q'}$.
  - $q \in Q^+ \cup Q^-$, $q' \in Q^?$. There is exactly one state with first component q and two states with first component q'. We set $P'_{(q,\xi_1),(q',\xi)} = P_{q,q'} P_{q'}$ and $P'_{(q,\xi_1),(q',\neg\xi)} = P_{q,q'} \overline{P_{q'}}$.
  - $q \in Q^?$. If $q' \in Q^?$, we define two transitions by $P'_{(q,\xi),(q',\xi)} = P_{q,q'} P_{q'}/P_q$ and $P'_{(q,\neg\xi),(q',\neg\xi)} = P_{q,q'} \overline{P_{q'}}/\overline{P_q}$. If $q' \in Q^+$ ($q' \in Q^-$), we define a single transition by $P'_{(q,\xi),(q',\xi)} = P_{q,q'}/P_q$ (resp. $P'_{(q,\neg\xi),(q',\neg\xi)} = P_{q,q'}/\overline{P_q}$).
- P'_0 is defined by $P'_0((q, \xi)) = P_0(q)P_q$ and $P'_0((q, \neg\xi)) = P_0(q)\overline{P_q}$, for all $(q, \xi), (q, \neg\xi) \in Q'$.

The new sequential probabilistic program is $\mathcal{M}' = (Q', \Delta', P', P'_0, \mathcal{V}')$ and the formula φ' is obtained by substituting $p\mathcal{U}p'$ by ξ in φ.

For the sequential probabilistic program $\mathcal{M}$ of Figure 3 and $\varphi = p\mathcal{U}p'$, we obtain $\mathcal{M}'$ as given in Figure 4 and $\varphi' = \xi$.

Let us show that the probability of $\mathcal{M}$ satisfying φ coincides with the one of $\mathcal{M}'$ satisfying φ'. Therefore, we analyze our construction in more detail.

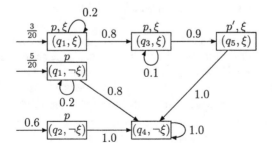

Fig. 4. A sequential probabilistic program

Let $\mathcal{G}'^+$ be the subgraph of $\mathcal{M}'$ induced by states (q, ξ) with $q \in Q^+$, $\mathcal{G}'^-$ be the one induced by all states $(q, \neg\xi)$ with $q \in Q^-$. Furthermore, let $\mathcal{G}'^{?,+}$ be the subgraph induced by states (q, ξ) and $\mathcal{G}'^{?,-}$ the one induced by states $(q, \neg \xi)$, for $q \in Q^?$. By construction $\mathcal{G}'^+$ $(\mathcal{G}'^-)$ is isomorphic to $\mathcal{G}^+$ $(\mathcal{G}^-)$, the subgraph of $\mathcal{M}$ induced by states in Q^+ (Q^-). Furthermore, $\mathcal{G}'^{?,+}$ is isomorphic to $\mathcal{G}'^{?,-}$ and $\mathcal{G}^?$, the subgraph of $\mathcal{M}$ induced by the states of $Q^?$.

Additionally, every transition from a state of $\mathcal{G}'^{?,+}$ $(\mathcal{G}'^{?,-})$ leading to a state not in $\mathcal{G}'^{?,+}$ $(\mathcal{G}'^{?,-})$, leads to a state in $\mathcal{G}'^+$ $(\mathcal{G}'^-)$. Thus, if ξ holds in a state of Q', almost surely every trajectory will eventually reach a state satisfying p' and every state in between satisfies p, and, if $\neg \xi$, almost all trajectories satisfy $\neg(p \mathcal{U} p')$. In other words, $\xi \equiv p \mathcal{U} p'$ holds almost surely in every state of $\mathcal{M}'$.

Let us see how the transition probabilities of $\mathcal{M}$ and $\mathcal{M}'$ relate: For every finite path $q'_1 \ldots q'_m$ in $\mathcal{M}'$, we have by construction that $\mu_{\mathcal{M}'}(q'_1 \ldots q'_m) = P_{q_1,q_2} \cdot \ldots \cdot P_{q_{m-1},q_m} \cdot P_{q_m}$ or $\mu_{\mathcal{M}'}(q'_1 \ldots q'_m) = P_{q_1,q_2} \cdot \ldots \cdot P_{q_{m-1},q_m} \cdot \overline{P_{q_m}}$, depending on whether $q'_m = (q, \xi)$ or $q'_m = (q_m, \neg \xi)$.

Let g be the mapping that projects a trajectory of $\mathcal{M}'$ to $\mathcal{M}$ by projecting to the first component. If $g(q'_m) \in Q^+ \cup Q^-$ then $q'_1 \ldots q'_m$ is the only trajectory with projection to $q_1 \ldots q_m$. If $g(q_m) \in Q^?$, there are exactly two trajectories with projection $q_1 \ldots q_m$: $(q_1, \xi) \ldots (q_m, \xi)$ and $(q_1, \neg \xi) \ldots (q_m, \neg \xi)$.

We conclude that for every measurable set E of trajectories of $\mathcal{M}$ the set $\{w' \in Q'^\omega \mid g(w) \in E\}$ is measurable and both sets have the same measure. This implies that $\mu_{\mathcal{M}'}(L_{\mathcal{M}'}(\varphi)) = \mu_{\mathcal{M}}(L_{\mathcal{M}}(\varphi))$. Since $p \mathcal{U} p' \equiv \xi$ for almost all trajectories in $\mathcal{M}'$, we get

Proposition 2.

$$\mu_{\mathcal{M}'}(L_{\mathcal{M}'}(\varphi')) = \mu_{\mathcal{M}}(L_{\mathcal{M}}(\varphi))$$

We conclude the presentation of $\mathcal{T}_\mathcal{U}$ analyzing its complexity. Let us first consider its time complexity. The construction of the new formula can be carried out in linear time with respect to $|\varphi|$. The information whether a state of $\mathcal{M}$ belongs to Q^+, Q^-, or $Q^?$ can be evaluated in linear time, as we show. We assume that $\mathcal{M}$ is given as a graph. Note that a Boolean combination of atomic propositions can be evaluated in time linear in its size in every state, so that our treatment of p and p' as propositions does not change the complexity.

1. Mark every state of $\mathcal{M}$ by Q^+ in which p' holds and by Q^- in which neither p nor p' holds.
2. Partition the remaining graph into strongly connected components. Process those in a bottom-up manner, i.e., when a s.c.c. D is processed all s.c.c. that are reachable from D have already been processed. Assign the nodes of D in the following manner:
   (a) if there is no edge leading to a state in Q^+ then add the nodes of D to Q^-.
   (b) if all edges leaving D lead to states in Q^+ then add the nodes of D to Q^+.
   (c) otherwise add the nodes of D to $Q^?$.

It is easy to see that the algorithm is correct. (1) corresponds to identifying W^+ and W_1^-. It can be done in linear time using a standard depth-first search.

Partitioning a graph into strongly connected components can also be done in linear time. In (2), the bottom strongly connected components are processed first, which also finds the states of W_2^+. The further processing is obviously correct. The classification of every s.c.c. can be carried out in time linear in the size of the component.

For constructing $\mathcal{M}'$ we have to solve the linear equation system of Lemma 1, which can be done in polynomial time in the size of $\mathcal{M}$. For checking whether the formula is satisfied, however, the exact transition probabilities are not important, but it has to be decided whether they are 0 or > 0. Thus, we need either linear or polynomial time for construction $\mathcal{M}'$ depending on whether we are interested in the question of satisfaction or in the exact measure.

Let us consider the space complexity of $\mathcal{T}_\mathcal{U}$. First note that checking whether there is path from a node u to a node v can be tested in space $\log^2 |\mathcal{M}|$. For every state q of $\mathcal{M}$, we have to check whether (q, ξ) or $(q, \neg\xi)$ are states of $\mathcal{M}'$. $(q, \xi) \in Q'$ iff $q \in Q^+ \cup Q^?$. This is the case iff q satisfies p' or there is a node q' reachable satisfying p' by a path of which every node satisfies p. This can be checked in space $\log^2 |\mathcal{M}|$. $(q, \neg\xi) \in Q'$ iff $q \in Q^- \cup Q^?$. This is the case if q satisfies $\neg p$ and $\neg p'$, or satisfies p but not p' and there is a path by states satisfying p to a state $q' \notin Q^+ \cup Q^?$ (tested as described above). Again, this can be tested in space $\log^2 |\mathcal{M}|$. The edges of $\mathcal{M}'$ can be constructed using similar ideas.

Transformation $\mathcal{T}_\mathcal{X}$ Transformation $\mathcal{T}_\mathcal{X}$ for an innermost formula of the form $\mathcal{X}p$ is patterned after the transformation $\mathcal{T}_\mathcal{U}$. We first partition the states of $\mathcal{M}$ into three disjoint subsets Q^+, Q^-, and $Q^?$, defined as follows: Q^+ comprises all states q for which all transitions are into states satisfying p, Q^- contains the ones for which all transitions yield states not satisfying p, and the remaining states having transitions to states satisfying p as well as transitions to states not satisfying p are assigned to $Q^?$. Again, we will see that for a trajectory starting in a state of Q^+ (Q^-), it will satisfy $\mathcal{X}p$ with probability 1 (0, respectively). For trajectories starting in states of $Q^?$ both events have a non-zero probability.

As before, let P_q denote the probability that $\mathcal{X}p$ is satisfied starting from state q, which is given as

Lemma 2.

$$
\begin{aligned}
P_q &= \sum_{p \in \mathcal{V}(q')} P_{q,q'} &&\text{if } q \in Q^? \\
P_q &= 1 &&\text{if } q \in Q^+ \\
P_q &= 0 &&\text{if } q \in Q^-
\end{aligned}
$$

As there is no recursion involved, the previous set of equations trivially has a unique solution.

The new sequential probabilistic program $\mathcal{M}'$ has a larger state space Q' and is defined over the set of atomic propositions $\mathcal{P}$ enriched with ξ which serves in φ' as a substitute for $\mathcal{X}p$. The state space and the propositions valid in each state are similarly defined as in the case of $\mathcal{T}_\mathcal{U}$. We define the transitions of $\mathcal{M}'$ implicitly by giving non-zero transition probabilities, as follows:

Let ξ_1 and ξ_2 stand for ξ or $\neg\xi$ and $\overline{P}$ for $1 - P$. We distinguish the following cases based on the states of $\mathcal{M}$:

- $q, q' \in Q^+ \cup Q^-$. There is exactly one state (q, ξ_1) and (q', ξ_2) in Q'. We set $P'_{(q,\xi_1),(q',\xi_2)} = P_{q,q'}$.
- $q \in Q^+ \cup Q^-$, $q' \in Q^?$. There is exactly one state with first component q and two states with first component q'. We set $P'_{(q,\xi_1),(q',\xi)} = P_{q,q'}P_{q'}$ and $P'_{(q,\xi_1),(q',\neg\xi)} = P_{q,q'}\overline{P_{q'}}$.
- $q \in Q^?$.
  - If $q' \in Q^?$ and q' satisfies p, we define two transitions by $P'_{(q,\xi),(q',\xi)} = P_{q,q'}P_{q'}/P_q$ and $P'_{(q,\xi),(q',\neg\xi)} = P_{q,q'}\overline{P_{q'}}/P_q$.
  - If $q' \in Q^?$ and q' satisfies $\neg p$, we define two transitions by $P'_{(q,\neg\xi),(q',\xi)} = P_{q,q'}P_{q'}/\overline{P_q}$ and $P'_{(q,\neg\xi),(q',\neg\xi)} = P_{q,q'}\overline{P_{q'}}/\overline{P_q}$.
  - If $q' \in Q^+ \cup Q^-$ and q' satisfies p, we define a single transition by $P'_{(q,\xi),(q',\xi_2)} = P_{q,q'}/P_q$ where (q',ξ_2) is the unique state of $\mathcal{M}'$ with first component q'.
  - If $q' \in Q^+ \cup Q^-$ and q' satisfies $\neg p$, we define a single transition by $P'_{(q,\neg\xi),(q',\xi_2)} = P_{q,q'}/\overline{P_q}$ where (q',ξ_2) is the unique state of $\mathcal{M}'$ with first component q'.

The initial distribution P'_0 is defined by $P'_0((q,\xi)) = P_0(q)P_q$ and $P'_0((q,\neg\xi)) = P_0(q)\overline{P_q}$, for all $(q,\xi), (q,\neg\xi) \in Q'$.

The new sequential probabilistic program is $\mathcal{M}' = (Q', \Delta', P', P'_0, \mathcal{V}')$ and the formula φ' is obtained by substituting $\mathcal{X}p$ by ξ in φ. Similar as in case of $\mathcal{T}_{\mathcal{U}}$, we get

Proposition 3.
$$\mu_{\mathcal{M}'}(L_{\mathcal{M}'}(\varphi')) = \mu_{\mathcal{M}}(L_{\mathcal{M}}(\varphi))$$

Using similar arguments as in the case of $\mathcal{T}_{\mathcal{U}}$, we learn that $\mathcal{T}_{\mathcal{X}}$ can be carried out within the same time complexity of $\mathcal{T}_{\mathcal{U}}$.

The overall procedure If φ has k temporal operators, we can compute the measure $\mu_{\mathcal{M}}(L_{\mathcal{M}}(\varphi))$ as follows: We apply k times the appropriate transformations $\mathcal{T}_{\mathcal{U}}$ or $\mathcal{T}_{\mathcal{X}}$ and obtain the sequence $\varphi^1, \mathcal{M}^1, \dots, \varphi^k, \mathcal{M}^k$, where φ^k is a simple propositional formula. Then $\mu_{\mathcal{M}}(L_{\mathcal{M}}(\varphi)) = \mu_{\mathcal{M}^k}(L_{\mathcal{M}^k}(\varphi^k))$, which is simply the sum of the initial probabilities in $\mathcal{M}^k$ over all states satisfying φ^k.

To conclude the proof of Theorem 1, it remains to show that our procedure can be carried out within the given time and space boundaries. As both transformations can be carried out in time linear in the size of the program (or polynomial if we are interested in the exact transition probabilities) and in every step the state space is at most doubled, we get $O(2^{|\varphi|}|\mathcal{M}|)$ as an upper bound for checking whether $\mathcal{M}$ satisfies φ. When we are interested in the exact measure, we need time polynomially in the size of $\mathcal{M}$ and exponentially in the size of φ. Note that it was shown in [16] that this problem is PSPACE-hard (wrt. the size of the formula). Thus, it is PSPACE-complete (as in the non-probabilistic case) and the given procedure is optimal.

It can be shown that the space required to carry out this computation is proportional to $|\varphi|(|\varphi| + \log^2(2^{|\varphi|}|\mathcal{M}|))$. Thus, the space complexity is in $O(|\varphi|^3 + |\varphi| \log^2 |\mathcal{M}|)$.

Note that the presented procedure can easily be extended to deal with *past tense connectives* defined in [9]. Using the same ideas, one can achieve also the same time and space boundaries.

4 Model Checking Probabilistic Programs Against Automata Specifications

Let $\mathcal{A}$ be an ω-automaton over $2^{\mathcal{P}}$. Given a concurrent probabilistic program $\mathcal{M}$ over $\mathcal{P}$ with valuation function $\mathcal{V}$ and a scheduler u of $\mathcal{M}$, we denote by $L_{\mathcal{M},u}(\mathcal{A})$, similarly to the case of LTL, the set $\{x \in \Omega_{\mathcal{M},u} \mid \mathcal{V}(x) \in L(\mathcal{A})\}$. In the same manner, for a sequential program $\mathcal{M}$ over $\mathcal{P}$ with valuation function $\mathcal{V}$, we define $L_{\mathcal{M}}(\mathcal{A})$ to be $\{x \in \Omega_{\mathcal{M}} \mid \mathcal{V}(x) \in L(\mathcal{A})\}$.

In the following, probabilistic programs will be defined over $\mathcal{P}$ while ω-automata will be defined over $2^{\mathcal{P}}$. It is a precondition for model checking programs against automata specifications that languages defined by automata are measurable. This is the case:

Proposition 4 ([16]). *For a concurrent probabilistic program $\mathcal{M}$, a scheduler u of $\mathcal{M}$, and a Büchi automaton $\mathcal{A}$, $L_{\mathcal{M},u}(\mathcal{A}) \in \mathfrak{B}_{\mathcal{M},u}$.*

Of course, the corresponding holds for sequential probabilistic programs.

Given a concurrent probabilistic program $\mathcal{M}$ and a Büchi automaton $\mathcal{A}$, the *concurrent emptiness problem (for automata)* is to decide whether it holds $\mu_{\mathcal{M},u}(L_{\mathcal{M},u}(\mathcal{A})) = 0$ for all schedulers u of $\mathcal{M}$. In contrast, the *concurrent universality problem* is to decide whether $\mu_{\mathcal{M},u}(L_{\mathcal{M},u}(\mathcal{A})) = 1$ for all schedulers u. *Sequential emptiness* and *sequential universality for automata* are defined analogously.

From the computational point of view, the above problems become easier to handle in case the automata specification at hand is a deterministic one. So, the transformation of a Büchi automaton into an equivalent ω-automaton that behaves deterministically forms the basis of forthcoming algorithms.

Theorem 2 (Safra [15]). *For a Büchi automaton $\mathcal{B}$ over Σ with n states, there is a deterministic Streett (Rabin) automaton $\mathcal{A}$ over Σ with $2^{O(n \log n)}$ states and $O(n)$ pairs in the acceptance component such that $L(\mathcal{A}) = L(\mathcal{B})$.*

4.1 Checking Concurrent Probabilistic Programs

We assume an automata specification to represent undesired behaviour. Thus, we are interested in solving the emptiness problem for automata.

Let us make some assumptions which make life easier in the following. Without loss of generality, we can assume that, in a concurrent probabilistic program $\mathcal{M} = (Q, N, R, \Delta, P_0, P, \mathcal{V})$, it holds $Q \subseteq 2^{\mathcal{P}}$ and, for each $q \in Q$, $\mathcal{V}(q) = q$. This can be achieved by sufficiently adding propositions to the program (to distinguish states) and accordingly adding transitions to the automaton. Then $\mathcal{M}$ can easily be transformed into an equivalent (in a sense described below) concurrent probabilistic program $\mathcal{M}'$

- that has no longer an initial probability distribution but a (randomizing) initial state,
- whose transitions are additionally labelled with elements of $2^{\mathcal{P}}$, and
- that can be viewed as a quasi-deterministic ω-automaton over $2^{\mathcal{P}}$.

The transformation proceeds as follows:

1. Add a new randomizing (initial) state $q_{\mathcal{M}}$ to $\mathcal{M}$ and, for each state q of $\mathcal{M}$ with $P_0(q) > 0$, add a transition $q_{\mathcal{M}} \to q$, for which we set $P'_{q_{\mathcal{M}}, q} = P_0(q)$.
2. Including the new ones, label each transition $q \to q'$ with $\mathcal{V}(q')$.

For an example how to transform a probabilistic program, look at Figure 5 where, for the sake of clarity, we assume a and b to be (different) sets of propositions.

A trajectory of $\mathcal{M}'$ is no longer a sequence of states but a sequence of transition labellings $\mathcal{V}(q)$ that $\mathcal{M}'$ finds along a path starting at $q_{\mathcal{M}}$. We will freely use the notation introduced for probabilistic programs also in the setting of such transformed programs giving them the obvious meaning. Regardless of the probabilities some transitions are equipped with, $\mathcal{M}'$ can be viewed as a quasi-deterministic Büchi automaton with single initial state $q_{\mathcal{M}}$ and acceptance component Q. The product of $\mathcal{M}'$ and a deterministic ω-automaton $\mathcal{A}$, denoted by $\mathcal{M}' \times \mathcal{A}$, is defined as usual and at least quasi-deterministic. On the other hand, preserving the probabilities, we can view $\mathcal{M}' \times \mathcal{A}$ also as a concurrent probabilistic program where a state is defined to be nondeterministic or randomizing according to its first component and trajectories are sequences of transition labellings. More precisely, $(q, s) \to (q', s')$ becomes a transition of $\mathcal{M}' \times \mathcal{A}$ if $q \to q'$

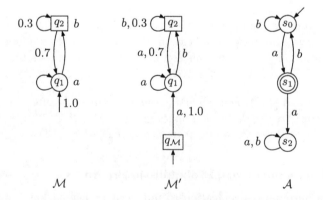

Fig. 5. A program, its transformation, and a deterministic Streett automaton

and $s \rightarrow s'$ are $\mathcal{V}(q')$-labelled transitions of $\mathcal{M}'$ and $\mathcal{A}$, respectively. In that case, $(q, s) \rightarrow (q', s')$ is also labelled $\mathcal{V}(q')$ and, if q is randomizing, equipped with probability $P'_{q,q'}$. The product of the concurrent probabilistic program $\mathcal{M}$ and the deterministic Streett automaton $\mathcal{A}$ from Figure 5 with acceptance condition $\{(\{s_0, s_2\}, \{s_1\})\}$ is illustrated in Figure 6 (note that, for the sake of clarity, the probabilities of transitions that go out from randomizing states are omitted).

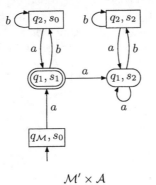

$$\mathcal{M}' \times \mathcal{A}$$

Fig. 6. The product of a program and an automaton

Given a probabilistic program $\mathcal{M}$ and a deterministic Streett automaton $\mathcal{A}$ with acceptance component $\mathcal{F}$, we want to mark some SCCs of $\mathcal{M}' \times \mathcal{A}$ to be good in some sense. We call a set D of its states *accepting* if, for all pairs $(U, V) \in \mathcal{F}$, the following holds:

$$\{s \mid (q, s) \in D \text{ for some } q\} \cap U \neq \emptyset \quad \text{implies} \quad \{s \mid (q, s) \in D \text{ for some } q\} \cap V \neq \emptyset$$

Otherwise, D is called *rejecting*. We say that a state (r, f) of a rejecting set D is *rejecting* if there is a pair $(U, V) \in \mathcal{F}$ such that $f \in U$ and D contains no state (q, s) with $s \in V$.

The following proposition is crucial in this subsection and immediately leads to an algorithm that solves the concurrent emptiness problem for automata.

Proposition 5. *For a concurrent program $\mathcal{M} = (Q, N, R, \Delta, P_0, P, \mathcal{V})$ and a deterministic Streett automaton $\mathcal{A} = (S, \{s_0\}, \delta, \mathcal{F})$, there exists a scheduler u of $\mathcal{M}$ with $\mu_{\mathcal{M},u}(L_{\mathcal{M},u}(\mathcal{A})) > 0$ iff there is a set D of states of $\mathcal{M}' \times \mathcal{A}$ satisfying the following:*

(1) $(\mathcal{M}' \times \mathcal{A})[D]$ is nontrivial and strongly connected,
(2) D is accepting and reachable from $(q_{\mathcal{M}}, s_0)$, and
(3) for all transitions $(q, s) \rightarrow (q', s')$ of $\mathcal{M}' \times \mathcal{A}$ with $(q, s) \in D$ and $(q', s') \notin D$, (q, s) is nondeterministic, i.e., it holds $q \in N$.

Proof. It is easy to see that $\mu_{\mathcal{M},u}(L_{\mathcal{M},u}(\mathcal{A})) > 0$ for a scheduler u of $\mathcal{M}$ iff there is a scheduler u' of $\mathcal{M}'$ such that $\mu_{\mathcal{M}',u'}(L_{\mathcal{M}',u'}(\mathcal{A})) > 0$.

($\Leftarrow$) We fix a path $\beta \in ((Q \cup \{q_{\mathcal{M}}\}) \times S)^*$ through $\mathcal{M}' \times \mathcal{A}$ from $(q_{\mathcal{M}}, s_0)$ to a state of D. Let β' be the projection of β onto the first component. The scheduler u' of $\mathcal{M}'$ satisfying $\mu_{\mathcal{M}',u'}(L_{\mathcal{M}',u'}(\mathcal{A})) > 0$ follows β' taking $\mathcal{M}' \times \mathcal{A}$ from the initial state to D and, henceforth, forces the trajectory both to stay within D and to almost surely visit each state of D infinitely often. This can be accomplished by, for a given nondeterministic state (q, s), alternately choosing the transitions $(q, s) \to (q', s')$ of $\mathcal{M}' \times \mathcal{A}$ with $(q', s') \in D$ (recall that the history of a trajectory is at the scheduler's disposal.) Clearly, $\mu_{\mathcal{M}',u'}(\mathcal{C}_{\mathcal{M}',u'}(\beta'))$ is nonzero. Given $\mathcal{C}_{\mathcal{M}',u'}(\beta')$, the conditional probability that $\mathcal{M}'$, wrt. u', follows a trajectory that visits exactly the states of D infinitely often is one. As such a trajectory is contained in $L_{\mathcal{M}',u'}(\mathcal{A})$, we conclude $\mu_{\mathcal{M}',u'}(L_{\mathcal{M}',u'}(\mathcal{A})) > 0$.

($\Rightarrow$) Note that a trajectory x of $\mathcal{M}'$ wrt. u' unambiguously defines a path $\tilde{x}$ through $\mathcal{M}' \times \mathcal{A}$ starting from $(q_{\mathcal{M}}, s_0)$. This is due to the fact that $\mathcal{A}$ is deterministic and that a transition of $\mathcal{M}'$ has a unique label. Let $\mathcal{D}$ contain the subsets D of states of $\mathcal{M}' \times \mathcal{A}$ such that $(\mathcal{M}' \times \mathcal{A})[D]$ is strongly connected. Furthermore, for $D \in \mathcal{D}$, let $E(D) := \{x \in \Omega_{\mathcal{M}',u'} \mid inf(\tilde{x}) = D\}$. Now suppose that $\mu_{\mathcal{M}',u'}(L_{\mathcal{M}',u'}(\mathcal{A})) > 0$ for a scheduler u' of $\mathcal{M}'$. As

$$L_{\mathcal{M}',u'}(\mathcal{A}) = \bigcup_{D \in \mathcal{D} \text{ is accepting}} E(D),$$

we can find an accepting set $D \in \mathcal{D}$ that satisfies $\mu_{\mathcal{M}',u'}(E(D)) > 0$. (Otherwise, the probability of the countable union $L_{\mathcal{M}',u'}(\mathcal{A})$ of events would be zero.) As D is the infinity set of at least one infinite path through $\mathcal{M}' \times \mathcal{A}$ starting from $(q_{\mathcal{M}}, s_0)$, it is reachable from the initial state, thus satisfies condition (2), and forms a nontrivial (strongly connected) subgraph of $\mathcal{M}' \times \mathcal{A}$, satisfying condition (1). Now suppose there is a transition $(q, s) \to (q', s')$ of $\mathcal{M}' \times \mathcal{A}$ with $(q, s) \in D$, $(q', s') \notin D$, and $q \in R$. As, for every trajectory $x \in E(D)$, $\tilde{x}$ visits (q, s) infinitely often (and each time the probability to exit D is nonzero), it will almost surely leave D infinitely often so that we have $\mu_{\mathcal{M}',u'}(E(D)) = 0$ contradicting our assumption. It follows that D also satisfies condition (3) from Proposition 5, which concludes our proof.   □

Note that, in the above proof, we explicitly make use of the fact that a trajectory of $\mathcal{M}'$ determines exactly one corresponding run of $\mathcal{M}' \times \mathcal{A}$ starting from $(q_{\mathcal{M}}, s_0)$ (recall that $\mathcal{A}$ is deterministic).

Let us fall back on the example contributed by Figures 5 and 6, respectively. There is an accepting set $D = \{(q_1, s_1), (q_2, s_0)\}$ of states of $\mathcal{M}' \times \mathcal{A}$ that is reachable from $(q_{\mathcal{M}}, s_0)$ via $\beta = (q_{\mathcal{M}}, s_0)(q_1, s_1)$ and forms a nontrivial and strongly connected subgraph of $\mathcal{M}' \times \mathcal{A}$. The only transition leading from a state of D to a state outside of D is $(q_1, s_1) \to (q_1, s_2)$, thus, arises from a nondeterministic state. In fact, for the scheduler u' of $\mathcal{M}'$ with $u'(x) = q_2$ for every $x \in (Q \cup \{q_{\mathcal{M}}\})^*$ that ends in a nondeterministic state (i.e. in q_1), it holds $\mu_{\mathcal{M}',u'}(L_{\mathcal{M}',u'}(\mathcal{A})) > 0$.

Based on Proposition 5, we now provide an algorithm that solves the concurrent emptiness problem, i.e., decides, given a concurrent probabilistic program $\mathcal{M}$ and a Büchi automaton $\mathcal{B}$, whether there is a scheduler u of $\mathcal{M}$ such that $\mu_{\mathcal{M},u}(L_{\mathcal{M},u}(\mathcal{B})) > 0$:

1. From $\mathcal{B}$, build a deterministic Streett automaton $\mathcal{A}$ with $L(\mathcal{A}) = L(\mathcal{B})$.
2. Compute (the graph of) $\mathcal{M}' \times \mathcal{A}$, remove those states that are not reachable from $(q_{\mathcal{M}}, s_0)$, and let G denote the resulting graph.
3. Repeat
   (a) Determine the sets $\mathcal{AC}$ of nontrivial and accepting and $\mathcal{RC}$ of nontrivial and rejecting SCCs of G, respectively.
   (b) For each $C \in \mathcal{RC}$, remove the transitions going out from rejecting states.
   (c) For each $C \in \mathcal{AC}$, do the following:
      i. Find the set H of states $(q, s) \in C$ with randomizing q from where there is a transition leaving C.
      ii. If H is the empty set, then return "Yes" (thus, halt with success). Otherwise, remove the transitions going out from the states of H.
   until $\mathcal{AC} \cup \mathcal{RC} = \emptyset$.
4. Return "No".

Obviously, the algorithm terminates. Furthermore, it returns the answer "Yes" iff there is a scheduler u of $\mathcal{M}$ such that $\mu_{\mathcal{M},u}(L_{\mathcal{M},u}(\mathcal{B})) > 0$. Let us discuss its complexity in the following. Starting from a Büchi automaton with n states, construct an equivalent deterministic Streett automaton with $2^{O(n \log n)}$ states and $O(n)$ pairs in the acceptance component. Assume that $\mathcal{M}'$ has m states. The number of states of $\mathcal{M}' \times \mathcal{A}$ is not greater than $m \cdot 2^{O(n \log n)}$. Thus, steps (a), (b), and (c) are repeated at most $m \cdot 2^{O(n \log n)}$-times, respectively. Determining the SCCs of G can be done in time linear in the size of $\mathcal{M}' \times \mathcal{A}$. Overall, the algorithm runs in time $O(m^3 \cdot 2^{O(n \log n)})$, i.e., it is quadratic in $|\mathcal{M}|$ and exponential in $|\mathcal{A}|$.

Theorem 3. *Given a concurrent probabilistic program $\mathcal{M}$ and a Büchi automaton $\mathcal{A}$, the concurrent emptiness problem can be solved in time $O(|\mathcal{M}|^2 \cdot 2^{O(|\mathcal{A}|)})$.*

The complexity of the above algorithm can be improved by constructing, instead of a (fully) deterministic ω-automaton, a Büchi automaton that is *deterministic in the limit* [4], i.e., that behaves deterministically as soon as it reaches a final state. Such an automaton only requires $2^{O(n)}$ states and finally leads to an $O(m^3 \cdot 2^{O(n)})$-complexity of the algorithm. But as it does not provide further insights in the principles of probabilistic model checking, we omit the corresponding construction. Courcoubetis and Yannakakis furthermore show that to decide the concurrent emptiness problem requires exponential time in the total input size $|\mathcal{M}| + |\mathcal{A}|$ [4].

The concurrent universality problem for automata can be solved analogously, but, from the computational point of view, it seems to be more appropriate to construct a deterministic Rabin automaton from a Büchi automaton. Accordingly, we call a set D of states of $\mathcal{M}' \times \mathcal{A}$ *accepting* if there is a pair (U, V) from the acceptance component of $\mathcal{A}$ and a state $(r, f) \in D$ such that $f \in U$ and D contains no state (q, s) with $s \in V$. Otherwise, we call D *rejecting*.

Proposition 6. *Given a concurrent program $\mathcal{M} = (Q, N, R, \Delta, P_0, P, \mathcal{V})$ and a deterministic Rabin automaton $\mathcal{A} = (S, \{s_0\}, \delta, \mathcal{F})$, there is a scheduler u of $\mathcal{M}$*

with $\mu_{\mathcal{M},u}(L_{\mathcal{M},u}(\mathcal{A})) < 1$ iff there is a set D of states of $\mathcal{M}' \times \mathcal{A}$ satisfying the following:

(1) $(\mathcal{M}' \times \mathcal{A})[D]$ is nontrivial and strongly connected,
(2) D is rejecting and reachable from $(q_{\mathcal{M}}, s_0)$, and
(3) for all transitions $(q, s) \to (q', s')$ of $\mathcal{M}' \times \mathcal{A}$ with $(q, s) \in D$ and $(q', s') \notin D$, it holds $q \in N$.

Proving Proposition 6 mainly applies techniques used to prove Proposition 5. The algorithm for the concurrent universality problem as suggested by Proposition 6 runs in time $O(|\mathcal{M}|^2 \cdot 2^{O(|\mathcal{A}|)})$.

However, to solve the concurrent universality problem, our algorithm cannot fall back on using ω-automata that are deterministic in the limit for the following reason: We then cannot be sure that a path leading to D does not have an "accepting" counterpart elsewhere. In [4], it is mentioned that the concurrent universality problem for Büchi automata that are deterministic in the limit requires exponential time.

4.2   Checking Sequential Probabilistic Programs

Sequential probabilistic programs were defined as a special case of concurrent ones. Thus, the above algorithms carry over to the case the probabilistic program has no nondeterministic state. Let us formulate an algorithm explicitly for this case. Two basic properties wrt. a sequential probabilistic program $\mathcal{M}$ and a deterministic ω-automaton $\mathcal{A}$ will play a crucial role in both qualitative and quantitative analysis:

- Almost surely, a trajectory of $\mathcal{M}'$ will take $\mathcal{M}' \times \mathcal{A}$ to a bottom SCC.
- Almost surely, a trajectory of $\mathcal{M}'$ that leads $\mathcal{M}' \times \mathcal{A}$ into a bottom SCC D will visit each state of D infinitely often.

Wrt. the emptiness problem, a sequential probabilistic program does not satisfy its automata specification (and thus contains undesired behaviour), iff, within the product of the transformed program and the specification, one can find an accepting bottom SCC that is reachable from the initial state. The following characterization immediately follows from Proposition 5.

Proposition 7. Given a sequential probabilistic program $\mathcal{M}$ and a deterministic Streett automaton $\mathcal{A}$, it holds $\mu_{\mathcal{M}}(L_{\mathcal{M}}(\mathcal{A})) > 0$ iff there is a reachable (from the initial state) and accepting bottom SCC of $\mathcal{M}' \times \mathcal{A}$.

Thus, after having computed a deterministic ω-automaton and a product automaton, the algorithm for solving the sequential emptiness problem only needs to determine whether there is an accepting bottom SCC that is reachable from the initial state. This can be done in time linear in the size of the product automaton. The algorithm works in time $O(m^2 \cdot 2^{O(n \log n)})$ where m is assumed to be the number of states of the program and n the number of states of the Büchi automaton. Again, the usage of ω-automata that are deterministic in the limit reduces the time complexity to $O(m^2 \cdot 2^{O(n)})$.

Checking universality wrt. sequential programs is closely related to checking emptiness.

Proposition 8. *For a sequential probabilistic program $\mathcal{M}$ and a deterministic Streett automaton $\mathcal{A}$, we have $\mu_{\mathcal{M}}(L_{\mathcal{M}}(\mathcal{A})) = 1$ iff all the reachable bottom SCCs of $\mathcal{M}' \times \mathcal{A}$ are accepting.*

Proof. Clearly, $\mu_{\mathcal{M}}(L_{\mathcal{M}}(\mathcal{A})) = 1$ iff $\mu_{\mathcal{M}'}(L_{\mathcal{M}'}(\mathcal{A})) = 1$.

($\Leftarrow$) Almost surely, a trajectory of $\mathcal{M}'$ will lead $\mathcal{M}' \times \mathcal{A}$ into a bottom SCC. Let x be a finite sequence of states of $\mathcal{M}'$ that takes $\mathcal{M}' \times \mathcal{A}$ from the initial state to a bottom SCC D. As D is nontrivial and accepting, $\mu_{\mathcal{M}'}(L_{\mathcal{M}'}(\mathcal{A}) \mid \mathcal{C}_{\mathcal{M}'}(x)) = 1$. Overall, we get $\mu_{\mathcal{M}'}(L_{\mathcal{M}'}(\mathcal{A})) = 1$.

($\Rightarrow$) Assume that $\mu_{\mathcal{M}'}(L_{\mathcal{M}'}(\mathcal{A})) = 1$ and suppose there is a reachable bottom SCC D of $\mathcal{M}' \times \mathcal{A}$ that is rejecting. We can find a finite sequence x of states of $\mathcal{M}'$ with $\mu_{\mathcal{M}'}(\mathcal{C}_{\mathcal{M}'}(x)) > 0$ that takes $\mathcal{M}' \times \mathcal{A}$ from the initial state to D. As D is rejecting, $\mu_{\mathcal{M}'}(L_{\mathcal{M}'}(\mathcal{A}) \mid \mathcal{C}_{\mathcal{M}'}(x)) = 0$. Thus, with nonzero probability, $\mathcal{M}'$ will produce a trajectory that is not contained in $L_{\mathcal{M}'}(\mathcal{A})$, contradicting the assumption. □

For example, the product of a (transformed) sequential probabilistic program $\mathcal{M}'$ and a deterministic Streett automaton $\mathcal{A}$ may have the structure as shown in Figure 7 where the bottom SCCs $\{\bar{q}_4\}$ and $\{\bar{q}_5\}$ are both assumed to be accepting. As $\{\bar{q}_4\}$ and $\{\bar{q}_5\}$ are the only bottom SCCs, it follows from Proposition 8 that $\mu_{\mathcal{M}}(L_{\mathcal{M}}(\mathcal{A})) = 1$.

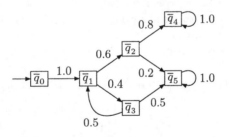

Fig. 7. A sequential probabilistic program

4.3   Quantitative Analysis

So far, we provided algorithms for checking probabilistic programs against qualitative properties, i.e., that do not compute the exact probability of satisfaction. We have seen that the algorithms amount to simple graph analyses as satisfaction of qualitative properties does not depend on exact probabilities but only on the topology of the respective program and its specification at hand. In the following, we generalize the above results.

Proposition 9. *Given a sequential probabilistic program $\mathcal{M}$ and a deterministic Streett automaton $\mathcal{A}$, $\mu_{\mathcal{M}}(L_{\mathcal{M}}(\mathcal{A}))$ is equal to the probability that a trajectory of $\mathcal{M}'$ takes $\mathcal{M}' \times \mathcal{A}$ from the initial state to an accepting bottom SCC.*

To determine $\mu_{\mathcal{M}}(L_{\mathcal{M}}(\mathcal{A}))$, we compute, starting from $\mathcal{M}' \times \mathcal{A}$, a new sequential probabilistic program $abs(\mathcal{M}' \times \mathcal{A})$ by

- removing all states that lie in a bottom SCC D and do not have an immediate predecessor outside of D,
- removing all transitions inside a bottom SCC, and
- adding, for each remaining state (q, s) of a bottom SCC, a transition from (q, s) to itself, which is equipped with probability 1, respectively.

The resulting program $abs(\mathcal{M}' \times \mathcal{A})$ is *absorbing*, i.e., each of its bottom SCCs consists of a single absorbing state. Employing the *fundamental matrix* N of $abs(\mathcal{M}' \times \mathcal{A})$ [6], we can easily compute the probability that a trajectory of $abs(\mathcal{M}' \times \mathcal{A})$ ends in an absorbing state. Suppose $\{\bar{q}_0, \ldots, \bar{q}_{n-1}\}$ is the set of states of $abs(\mathcal{M}' \times \mathcal{A})$ with only initial state $\bar{q}_0$. Without loss of generality, $\{\bar{q}_0, \ldots, \bar{q}_{m-1}\}$ is the set of transient states for a natural $m < n$. Let $T = (t_{ij})_{i,j \in \{0,\ldots,n-1\}}$ be the transition matrix of $abs(\mathcal{M}' \times \mathcal{A})$, i.e., t_{ij} is the transition probability of going from $\bar{q}_i$ to $\bar{q}_j$. It is of the form

$$T = \begin{pmatrix} Q & R \\ 0 & I_{n-m} \end{pmatrix}$$

where I_{n-m} is the $(n-m) \times (n-m)$ matrix that has 1's on the main diagonal and 0's elsewhere, 0 is the matrix of dimension $(n-m) \times m$ that has all components 0, the $m \times (n-m)$ matrix R gives information about the transition probabilities of going from a transient to an absorbing state, and Q is of dimension $m \times m$ and contains the transition probabilities exclusively respecting the set of transient states. For example, assume $abs(\mathcal{M}' \times \mathcal{A})$ to be partly shown in Figure 7 (note that, in fact, $abs(\mathcal{M}' \times \mathcal{A})$ is absorbing) and suppose the only state that emanates from an accepting bottom SCC is $\bar{q}_4$. Its transition matrix is given by

$$T = \begin{pmatrix} 0 & 1 & 0 & 0 & 0 & 0 \\ 0 & 0 & 0.6 & 0.4 & 0 & 0 \\ 0 & 0 & 0 & 0 & 0.8 & 0.2 \\ 0 & 0.5 & 0 & 0 & 0 & 0.5 \\ 0 & 0 & 0 & 0 & 1 & 0 \\ 0 & 0 & 0 & 0 & 0 & 1 \end{pmatrix}.$$

The fundamental matrix of $abs(\mathcal{M}' \times \mathcal{A})$ now arises as follows:

$$N = (n_{ij})_{i,j \in \{0,\ldots,m-1\}} = (I_m - Q)^{-1}$$

Hereby, n_{ij} tells us how often a trajectory of $abs(\mathcal{M}' \times \mathcal{A})$ starting at $\bar{q}_i$ is expected to be in $\bar{q}_j$ before absorption. In our example,

$$Q = \begin{pmatrix} 0 & 1 & 0 & 0 \\ 0 & 0 & 0.6 & 0.4 \\ 0 & 0 & 0 & 0 \\ 0 & 0.5 & 0 & 0 \end{pmatrix}, \quad I_4 - Q = \begin{pmatrix} 1 & -1 & 0 & 0 \\ 0 & 1 & -0.6 & -0.4 \\ 0 & 0 & 1 & 0 \\ 0 & -0.5 & 0 & 1 \end{pmatrix}, \text{ and}$$

$$N = (I_4 - Q)^{-1} = \begin{pmatrix} 1 & 1.25 & 0.75 & 0.5 \\ 0 & 1.25 & 0.75 & 0.5 \\ 0 & 0 & 1 & 0 \\ 0 & 0.625 & 0.375 & 1.25 \end{pmatrix}.$$

Finally,

$$B = (b_{ij})_{i \in \{0,\dots,m-1\}, j \in \{0,\dots,n-m-1\}} = NR$$

contains the probabilities b_{ij} that $abs(\mathcal{M}' \times \mathcal{A})$, starting at $\bar{q}_i$, reaches the absorbing state $\bar{q}_{j+m}$. Let A be the subset of $\{m, \dots, n-1\}$ such that $\bar{q}_j$ is an absorbing state of $abs(\mathcal{M}' \times \mathcal{A})$ that emanates from an accepting bottom SCC of $\mathcal{M}' \times \mathcal{A}$ iff $j \in A$. If A is empty, $\mu_{\mathcal{M}}(L_{\mathcal{M}}(\mathcal{A})) = 0$. Otherwise,

$$\mu_{\mathcal{M}}(L_{\mathcal{M}}(\mathcal{A})) = \sum_{j \in A} b_{0(j-m)}.$$

To complete our example,

$$B = NR = \begin{pmatrix} 1 & 1.25 & 0.75 & 0.5 \\ 0 & 1.25 & 0.75 & 0.5 \\ 0 & 0 & 1 & 0 \\ 0 & 0.625 & 0.375 & 1.25 \end{pmatrix} \begin{pmatrix} 0 & 0 \\ 0 & 0 \\ 0.8 & 0.2 \\ 0 & 0.5 \end{pmatrix} = \begin{pmatrix} 0.6 & 0.4 \\ 0.6 & 0.4 \\ 0.8 & 0.2 \\ 0.3 & 0.7 \end{pmatrix}$$

lets us come to the conclusion that the probability the underlying sequential probabilistic program will produce a trajectory accepted by the automata specification is 0.6.

Under consideration of the matrix operations, we get an algorithm that determines the exact probability of satisfaction in time single exponential in the size of the Büchi automaton $\mathcal{B}$ and polynomial in the size of the sequential probabilistic program $\mathcal{M}$:

1. Construct a deterministic Streett automaton $\mathcal{A}$ with $L(\mathcal{A}) = L(\mathcal{B})$.
2. Build $\mathcal{M}' \times \mathcal{A}$ and determine its accepting bottom SCCs.
3. Compute $abs(\mathcal{M}' \times \mathcal{A})$ as well as its fundamental matrix N.
4. Compute $B = NR$ and, within the first of its rows, sum up the probabilities that belong to states respectively emanating from an accepting bottom SCC of $\mathcal{M}' \times \mathcal{A}$.

We can summarize the above observations as follows:

Theorem 4. *For a sequential probabilistic program $\mathcal{M}$ and a Büchi automaton $\mathcal{A}$, sequential emptiness and sequential universality can be solved in time $O(|\mathcal{M}| \cdot 2^{O(|\mathcal{A}|)})$, respectively. Furthermore, we are able to determine the exact probability $\mu_{\mathcal{M}}(L_{\mathcal{M}}(\mathcal{A}))$ in time single exponential in the size of $\mathcal{A}$ and polynomial in $|\mathcal{M}|$.*

Note that both sequential emptiness and sequential universality for automata are PSPACE-complete in the size of the automaton ([16], [4]).

5   Conclusion

In this chapter, we presented procedures for checking linear temporal logic specifications and automata specifications of sequential and concurrent probabilistic programs. The complexities of these problems are summarized in Table 1. We followed two different approaches: For LTL and sequential probabilistic programs, our method proceeded in a tableau style manner, while for the remaining problems, we followed the automata theoretic approach.

Table 1. Complexity of model checking problems

	LTL formulas	ω-automata
sequential programs	$O(|\mathcal{M}| \cdot 2^{O(|\varphi|)})$	$O(|\mathcal{M}| \cdot 2^{O(|\mathcal{A}|)})$
concurrent programs	$O(|\mathcal{M}|^2 \cdot 2^{2^{O(|\varphi|)}})$	$O(|\mathcal{M}|^2 \cdot 2^{O(|\mathcal{A}|)})$

To gain further insight of probabilistic programs, one might be interested in specifying boundaries for the probabilities of properties. These extension to temporal logics are studied in the next chapters.

6   Bibliographic Remarks and Further Reading

Initial papers studying temporal logics for probabilistic programs are [8] and [5]. Concurrent probabilistic programs were first studied in [16].

Our chapter is mainly based on [4]. The model checking procedure for LTL and sequential probabilistic programs (Theorem 1) is presented in [4]. PSPACE-hardness (wrt. length of formula) of this problem was first shown in [16]. Our algorithms for checking automata specifications are presented or at least suggested in [4]. Note that for solving the sequential emptiness problem, the authors make use of a slightly different kind of product construction. First complexity results were achieved by Vardi [16] and extended by [4].

A different approach to study probabilistic systems is to understand randomness as a kind of *fairness*. Let us reconsider the example shown in Figure 1, and recall, that nondeterministic states are denoted by circles while probabilistic states are drawn as squares. Intuitively, nondeterminism abstracts that it is not known which possible transition the system under consideration takes. For example, it is possible that the given system chooses to move to state q_3 whenever it is in state q_1. For random states, however, we know that the system chooses a transition with a fixed probability. For example, with probability 0.3, the transition from q_3 to q_1 is taken. This implies that it is unlikely (with probability 0) that the system will choose transition q_3 to q_1 infinitely often. In almost all runs, the system will be *fair* and eventually take the other possible outcome in state q_3 and move to state q_5 (if it moves to state q_3 at all).

The idea of using concepts of fairness to study probabilistic systems first appeared in [11] where the notion of *extreme fairness* was introduced. In [13], α-*fairness* was defined to deal with checking a restricted version of linear temporal logic (but extended with past-tense modalities) specifications of concurrent probabilistic programs.[3] A notion of γ-fairness is introduced in [1] to analyze parameterized probabilistic systems.

References

1. T. Arons, L. Zuck, and A. Pnueli. Automatic verification by probabilistic abstraction. In *submitted to Fossacs'03*, 2003.
2. E. M. Clarke, E. A. Emerson, and A. P. Sistla. Automatic verification of finite state concurrent systems using temporal logic specifications: A practical approach. In *Conference Record of the Tenth Annual ACM Symposium on Principles of Programming Languages*, pages 117–126, Austin, Texas, January 24–26, 1983. ACM SIGACT-SIGPLAN.
3. Edmund M. Clarke and Jeanette M. Wing. Formal methods: State of the art and future directions. *ACM Computing Surveys*, 28(4):626–643, December 1996.
4. Costas Courcoubetis and Mihalis Yannakakis. The complexity of probabilistic verification. *Journal of the ACM*, 42(4):857–907, July 1995.
5. S. Hart and M. Sharir. Probabilistic temporal logics for finite and bounded models. In *ACM Symposium on Theory of Computing (STOC '84)*, pages 1–13, Baltimore, USA, April 1984. ACM Press.
6. J. G. Kemeny and J. L. Snell. *Finite Markov Chains*. Van Nostrand Reinhold, New York, 1960.
7. D. Lehman and M. O. Rabin. On the advantage of free choice: A fully symmetric and fully distributed solution to the dining philosophers problem. In *Proceedings of 10th ACM Symposium of Principles of Programming Languages*, pages 133–138, Williamsburg, 1981.
8. Daniel Lehmann and Saharon Shelah. Reasoning with time and chance. *Information and Control*, 53(3):165–198, June 1982.
9. O. Lichtenstein and A. Pnueli. Checking that finite state concurrent programs satisfy their linear specification. In *Proceedings of the Twelfth Annual ACM Symposium on Principles of Programming Languages*, pages 97–107, New York, January 1985. ACM.
10. Amir Pnueli. The temporal logic of programs. In *Proceedings of the 18th IEEE Symposium on the Foundations of Computer Science (FOCS-77)*, pages 46–57, Providence, Rhode Island, October 31–November 2 1977. IEEE Computer Society Press.
11. Amir Pnueli. On the extremely fair treatment of probabilistic algorithms. In *Proceedings of the Fifteenth Annual ACM Symposium on Theory of Computing*, pages 278–290, Boston, Massachusetts, 25–27 April 1983.

---

[3] Note that Pnueli follows a different model of sequential and concurrent probabilistic programs. Thus, [13] speaks of sequential programs which correspond to concurrent programs in our terminology.

12. P. Pardalos, S. Rajasekaran, J. Reif, and J. Rolim, editors. *Handbook on Randomized Computing*. Kluwer Academic Publishers, Dordrecht, The Netherlands, June 2001.
13. Amir Pnueli and Lenore D. Zuck. Probabilistic verification. *Information and Computation*, 103(1):1–29, March 1993.
14. J.P. Queille and J. Sifakis. Specification and verification of concurrent systems in CESAR. In *Proceedings of the Fifth International Symposium in Programming*, volume 137 of *Lecture Notes in Computer Science*, pages 337–351, New York, 1982. Springer.
15. Shmuel Safra. On the complexity of omega-automata. In *Proceedings of the 29th Annual Symposium on Foundations of Computer Science, FoCS'88*, pages 319–327, Los Alamitos, California, October 1988. IEEE Computer Society Press.
16. Moshe Y. Vardi. Automatic verification of probabilistic concurrent finite-state programs. In *26th Annual Symposium on Foundations of Computer Science*, pages 327–338, Portland, Oregon, 21–23 October 1985. IEEE.
17. Moshe Y. Vardi. *An Automata-Theoretic Approach to Linear Temporal Logic*, volume 1043 of *Lecture Notes in Computer Science*, pages 238–266. Springer, New York, NY, USA, 1996.
18. Pierre Wolper. Temporal logic can be more expressive. *Information and Control*, 56(1/2):72–99, January/February 1983.

On Probabilistic Computation Tree Logic

Frank Ciesinski* and Marcus Größer**

Rheinische Friedrich Wilhelms Universität zu Bonn,
Institut für Informatik - Abt. I, Römerstraße 164, 53117 Bonn
{ciesinsk,groesser}@cs.uni-bonn.de

Abstract. In this survey we motivate, define and explain model checking of probabilistic deterministic and nondeterministic systems using the probabilistic computation tree logics $PCTL$ and $PCTL^*$. Juxtapositions to non-deterministic computation tree logic are made and algorithms are presented.

Keywords: $PCTL$, $PCTL^*$, discrete time Markov chains, Markov decision processes, scheduler, fairness, probabilistic deterministic systems, probabilistic nondeterministic systems, quantitative model checking.

1   Introduction and Overview

Model Checking is a *fully automatic* verification method for both hard- and software systems that has seen a very promising development for now almost 25 years. Numerous efficient methods have been proposed very successfully to attack the discouraging general complexity of Model Checking (see e.g. [12]). Various temporal logics were proposed to emphasize certain different aspects of models to be checked. Among them are logics to cover the behaviour of probabilistic systems. Taking probabilities into account in addition to nondeterministic behaviour expands the possibilities of modeling certain aspects of the system under consideration. While nondeterministic systems are considered in connection to underspecification, interleaving of several processes and interaction with the specified system from the outside, the probabilities can be exploited to model a certain probability of error or other stochastic behaviour both occurring in various real world applications, e.g. certain randomized algorithms or communication protocols over faulty media.

The development of probabilistic Model Checking can roughly be divided into two periods. With *qualitative* Model Checking, whichs development merely took place in the 1980s, it is possible to decide whether a system satisfies its specification with probability 1. Many aspects of qualitative probabilistic Model Checking were worked out, such as the termination of probabilistic programs [27], the omega-automaton-based approach for Model Checking probabilistic (as well as non-probabilistic) systems [49,51,13], the tableaux based method [40]

---

* Supported by the DFG-NWO-Project "VOSS".
** Supported by the DFG-Project "VERIAM" and the DFG-NWO-Project "VOSS".

C. Baier et al. (Eds.): Validation of Stochastic Systems, LNCS 2925, pp. 147–188, 2004.

and Model Checking of probabilistic real-time systems [2]. In the 1990s the idea of qualitative Model Checking was extended to *quantitative* Model Checking where it becomes possible not only to check whether the specification is satisfied with probability 1 but to specify arbitrary probability values $p \in [0,1]$. Model Checking algorithms were developed for the deterministic case (Discrete Time Markov Chains, $DTMCs$) [25, 3] as well as for the nondeterministic case (Markov decision processes, $MDPs$) [14, 15, 17, 18, 9, 7], including fairness [10].

In this survey we summarize the main results presented in the papers mentioned above and use the logic $PCTL$ (Probabilistic Computation Tree Logic), which is commonly used corresponding to Discrete Markov Chains and Markov decision processes as the basic temporal logic [26, 9]. It extends the temporal logic $RTCTL$ (Real Time CTL, [19]). While $RTCTL$, as an extension to the common known temporal logic CTL, additionally provides discrete time properties such as *A property holds within a certain amount of (discrete) time* , $PCTL$ allows to state properties such as *A property holds within a certain amount of (discrete) time with a probability of at least 99%*. Thus, $PCTL$ supplies a formalism which adds probabilities in a certain way. Intuitively $RTCTL$ allows it to state "hard deadlines" while $PCTL$ is capable of stating so called "soft deadlines".

After making some basic definitions concerning probabilistic processes, the syntax and semantics of $PCTL$ and $PCTL^*$ are defined, similarities and differences between $PCTL$ and CTL (resp. $PCTL^*$ and CTL^*) are outlined. Then for both the deterministic and the nondeterministic case model checking procedures are described and fairness is taken into account as well.

The reader is supposed to be familiar with temporal logics and basic notions of probability theory. See e.g. [22, 21, 37, 12].

2   Validation of Probabilistic Deterministic Systems

$PCTL$ - like it is presented in this chapter - was introduced by *Hansson* and *Jonsson* [25] and concentrates on probabilistic model checking with respect to probabilistic deterministic systems (PDS). In essence a PDS is a state labelled Markov chain and $PCTL$ allows to check properties of the kind "starting from a state s, a certain $PCTL$ path formula holds with a probability of at least p" using a probabilistic operator. This probabilistic operator may be seen as a counterpart to the $\forall$ and $\exists$ operator in CTL (i.e. it can be applied to path-formulae only). Similar to CTL, $PCTL$ can as well be extended by relaxing the $PCTL$-syntax to $PCTL^*$, which will be dealt with in Section 2.5. We will firstly define essential structures for modeling probabilistic processes and the syntax and semantics of $PCTL$. As descriptions go along, certain similarities and differences to CTL are emphasized. Finally a model checking algorithm for $PCTL$ is outlined.

2.1   Probabilistic Deterministic Systems

We define *Probabilistic Deterministic Systems* (PDS) which essentially are state labelled *Finite Markov chains* (we restrict ourselves to finite systems).

Definition 1. [Probability Distribution]
Given a finite set S, a probability distribution on S is a function

$$\mu : S \to [0,1] \quad such\ that \quad \sum_{s \in S} \mu(s) = 1.$$

Given a probability distribution on S, $supp(\mu)$ denotes the *support*, i.e. the states s of S with $\mu(s) > 0$. For $s \in S$, μ_s^1 denotes the unique distribution on S that satisfies $\mu_s^1(s) = 1$. With $Distr(S)$ we denote the set of all probability distributions on S.

Definition 2. [Finite Markov Chain]
A finite Markov chain is a tuple

$$M = (S, T, p),$$

where

- *S is a finite set,*
- *$T \subseteq S \times S$ is a set of transitions and*
- *$p : S \times S \to [0,1]$ is a function such that for all $s \in S$*
  - *$p(s,.)$ is a probability distribution on S*
  - *$(s,t) \in T$ iff $t \in supp(p(s,.))$.*

  Throughout this survey we will use the notation p_{uv} instead of $p(u,v)$ for $u, v \in S$.

We also refer to (S,T) as the *underlying graph* of M. Note that T is total, so that (S,T) has no terminal nodes.

Normally a Markov chain is equipped with a starting probability distribution. A Markov chain with a starting distribution induces a stochastic process on the set S of its states in a natural way. The probability that the process starts in a certain state with step 0 is determined by the starting distribution. Moreover, being in state u in the $(n-1)$th step, the probability that the process is in state v in the nth step is equal to p_{uv}. The fact, that those probabilities do not depend on the previous steps (*history-independent* or *memoryless*) is called the *Markov property*.

Definition 3. [Ergodic Set]
Let $G = (V, E)$ be a finite directed graph. We call $C \subseteq V$ an ergodic set of G, iff C is a terminal strongly connected component of G, i.e.

- *$\forall\ (u,v) \in E\ :\ u \in C\ \Rightarrow v \in C$*
- *$\forall\ u, v \in C\ :\ \exists$ a path from u to v in G*

An ergodic set is therefore a strongly connected component that cannot be left once the execution sequence reached one of its states. Given a Markov chain $M = (S, T, p)$, we call $C \subseteq S$ an ergodic set of M, if C is an ergodic set of the underlying graph of M.

Definition 4. [Probabilistic Deterministic System (PDS)]
A probabilistic deterministic system (PDS) is a tuple

$$T_{PDS} = (M, AP, L, \bar{s}),$$

where

- $M = (S, T, p)$ *is a finite Markov chain,*
- AP *is a finite set of atomic propositions,*
- L *is a labelling function* $L : S \to 2^{AP}$ *that labels any state* $s \in S$ *with those atomic propositions that are supposed to hold in s and*
- $\bar{s} \in S$ *is the starting state.*

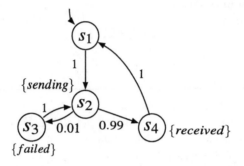

Fig. 1. PDS example: A very simple communication protocol

In the sequel we just write T instead of T_{PDS}. We now define the essential executional issues of a probabilistic deterministic system. Given a *PDS* $T = (M, AP, L, \bar{s})$ with $M = (S, T, p)$ we define the following:

Definition 5. [Path]
A path of T is a finite or infinite sequence $\pi = s_0, s_1, \ldots$ *of states, such that* $(s_i, s_{i+1}) \in T$ *for all i under consideration.*
    Given a finite path $\rho = s_0, s_1, \ldots, s_n$, *we denote s_0 by $first(\rho)$ and s_n by* $last(\rho)$. *The length $|\rho|$ of ρ is equal to n. For an infinite path π, the length is equal to ∞. Given a path* $\pi = s_0, s_1, \ldots$ *(finite or infinite) and $i \le |\pi|$, we denote the ith state of π by π^i (i.e. $\pi^i = s_i$) and the i-th prefix by $\pi\uparrow^i = s_0, s_1, \ldots, s_i$.*
    Given an infinite path $\pi = s_0, s_1, \ldots$, *we denote the suffix starting at π^i by* $\pi\uparrow_i$, *i.e.* $\pi\uparrow_i = s_i, s_{i+1}, \ldots$. *Furthermore we denote by $Paths_{fin}$ (resp. $Paths_{inf}$) the set of finite (resp. infinite) paths of a given PDS and by $Paths_{fin}(s)$ (resp. $Paths_{inf}(s)$) the set of finite (resp. infinite) paths of a given PDS starting at the state s.*

Definition 6. [Trace]
We define the trace of a (finite) path $\pi = s_0, s_1, \ldots$ *to be the (finite) word over the alphabet 2^{AP} which we get from the following:*

$$trace(\pi) = L(\pi^0), L(\pi^1), \ldots = L(s_0), L(s_1), \ldots.$$

Definition 7. [Basic Cylinder]
For $\rho \in Paths_{fin}$ the basic cylinder of ρ is defined as

$$\Delta(\rho) = \{\pi \in Paths_{inf} \quad : \quad \pi{\uparrow}^{|\rho|} = \rho\}.$$

Let $s \in S$. Following Markov chain theory and measure theory [32] we now define a probability space on the set of paths starting in s.

Definition 8. [Probability Space of a Markov Chain]
Given a Finite Markov chain $M = (S, T, p)$ and $s \in S$, we define a probability space

$$\Psi^s = (\Delta(s), \Delta^s, prob_s),$$

such that

- Δ^s is the σ-algebra generated by the empty set and the basic cylinders over S that are contained in $\Delta(s)$.
- $prob_s$ is the uniquely induced probability measure which satisfies the following:
  $prob_s(\Delta(s)) = 1$ and for all basic cylinders $\Delta(s, s_1, \ldots, s_n)$ over S :

$$prob_s(\Delta(s, s_1, \ldots, s_n)) = p_{ss_1} \cdots p_{s_{n-1}s_n}$$

We now state a fact about ergodic sets that we will need later.

Lemma 1. [Reaching an Ergodic Set]
Given a Markov chain $M = (S, T, p)$, the following holds for all $s \in S$:

$$prob_s(\{\pi \in Paths_{inf}(s) \ : \ \exists C \text{ ergodic set of } M \text{ } s.t. \forall c \in C : \pi^i = c$$
$$\text{for infinitely many } i\text{'s}\}) = 1.$$

Proof. see [32]

This means that given a Markov chain $M = (S, T, p)$ and an arbitrary starting state $\bar{s}$ it holds that the stochastic process described by M and starting in $\bar{s}$ will reach an ergodic set C of M and visit each state of C infinitely often with probability 1.

This concludes the preliminary definitions needed for $PCTL$-model checking. We now define the syntax of $PCTL$.

2.2   The Syntax of $PCTL$

We define the $PCTL$ syntax along with some intuitive explanations about the corresponding semantics when meaningful. The precise semantics of $PCTL$ will be defined in a formal way in Section 2.3 afterwards.

Definition 9. [*PCTL*-Syntax]
The syntax of PCTL is defined by the following grammar:[1]

PCTL-state formulae

$$\Phi ::= \text{true} \quad | \quad a \quad | \quad \Phi_1 \wedge \Phi_2 \quad | \quad \neg\Phi \quad | \quad [\phi]_{\bowtie p}$$

PCTL-path formulae:

$$\phi ::= \Phi_1 \mathcal{U}^{\leq t}\Phi_2 \quad | \quad \Phi_1\mathcal{U}\Phi_2 \quad | \quad \mathcal{X}\Phi$$

where $t \in I\!N$, $p \in [0,1] \subset I\!R$, and $a \in AP, \bowtie \in \{>,\geq,\leq,<\}$

The set of all *PCTL*-formulae is $form_{PCTL} = \{f : f$ is a $PCTL$ formula$\}$ where we assume a fixed set of atomic propositions. The symbol $\bowtie$ is used for ">, $\geq$", "<" or "$\leq$" respectively.

Definition 10. [**Length of a** *PCTL*-**Formula**] *We define the length of a PCTL-formula Φ as the number of atomic propositions, temporal, Boolean and probabilistic operators that are contained inside the formula and write $|\Phi|$* [2].

Other Boolean operators (i.e. $\vee, \rightarrow, \leftrightarrow$) are not defined explicitly but can be derived from $\wedge$ and $\neg$ as usual. State formulae represent properties which can be any atomic proposition (i.e. $a \in AP$) or Boolean combinations of them as well as $[\ ]_{\bowtie p}$ properties that require a certain amount (w.r.t. a measure, see definition 8) of paths to be existing starting in the current state and satisfying the enclosed path formula. Path formulae involve the strong until (be it the bounded $\mathcal{U}^{\leq t}$ or the unbounded $\mathcal{U}$) or the nextstep operator $\mathcal{X}$. The nextstep can neither be derived from $\mathcal{U}$ nor from $\mathcal{U}^{\leq t}$ and neither can the bounded or unbounded until be derived from $\mathcal{X}$ [3]. Thus, we define $\mathcal{U}^{\leq t}$, $\mathcal{U}$ and $\mathcal{X}$ explicitly as three independent operators (and we will later give three different algorithms to cover these operators). Intuitively the path formula $\Phi_1\mathcal{U}^{\leq t}\Phi_2$ states that Φ_1 holds continuously

---

[1] Indexed terminal symbols (here: Φ_1 and Φ_2) are identified with terminal symbols Φ in order to simplify the *PCTL*-grammar. The symbol a represents arbitrary atomic propositions. It is assumed that $p \in [0,1]$ and $t \in I\!N$. Hence this is an abstract grammar.

[2] Please note that here $|\Phi|$ does not depend on the step values contained by bounded until operators (i.e. on the t in $\mathcal{U}^{\leq t}$). These steps *do* affect the complexity of the algorithms presented later in this survey but are dealt with separately.

[3] Informally speaking $[\Phi_1\mathcal{U}^{\leq t}\Phi_2]_{\bowtie p, p\neq 0}$ cannot be expressed as nested $\mathcal{X}$ expressions because every $\mathcal{X}$ would have to be enclosed by another $[\ ]_{\bowtie p'}$, which cannot be done because each p, p', etc. must be constant. Furthermore it is true that $\neg\Phi \wedge [true\mathcal{U}^{\leq 1}\Phi]_{\bowtie p} \not\equiv [\mathcal{X}\Phi]_{\bowtie p}$. Nextstep states only about the future, until about the present *and* the future.

from now on until within at most t time units Φ_2 becomes true. The unbounded operator $\Phi_1 \mathcal{U} \Phi_2$ does not require a certain bound but nevertheless requires Φ_2 to become true *eventually*.

However, the weak until $(\widetilde{\mathcal{U}})$ operator can be derived by using $\mathcal{U}$ via

$$[\Phi_1 \widetilde{\mathcal{U}}^{\leq t} \Phi_2]_{\geq p} := \neg[(\neg \Phi_2) \mathcal{U}^{\leq t}(\neg \Phi_1 \wedge \neg \Phi_2)]_{>(1-p)},$$
$$[\Phi_1 \widetilde{\mathcal{U}}^{\leq t} \Phi_2]_{> p} := \neg[(\neg \Phi_2) \mathcal{U}^{\leq t}(\neg \Phi_1 \wedge \neg \Phi_2)]_{\geq(1-p)},$$
$$[\Phi_1 \widetilde{\mathcal{U}}^{\leq t} \Phi_2]_{\leq p} := \neg[(\neg \Phi_2) \mathcal{U}^{\leq t}(\neg \Phi_1 \wedge \neg \Phi_2)]_{<(1-p)},$$
$$[\Phi_1 \widetilde{\mathcal{U}}^{\leq t} \Phi_2]_{< p} := \neg[(\neg \Phi_2) \mathcal{U}^{\leq t}(\neg \Phi_1 \wedge \neg \Phi_2)]_{\leq(1-p)}.$$

The weak until $\widetilde{\mathcal{U}}^{\leq t}$ does not require Φ_2 to become *true*. Either Φ_1 holds for at least the next t time units or Φ_2 becomes *true* in the next t time units and Φ_1 holds continuously until this happens. The unbounded weak until operator can be defined similarly.

Similar to "always $\square$" and "eventually $\lozenge$" operators in CTL there are pendants in $PCTL$.

$$[\lozenge^{\leq t} \Phi]_{\bowtie p} := [true \mathcal{U}^{\leq t} \Phi]_{\bowtie p},$$
$$[\square^{\leq t} \Phi]_{\bowtie p} := \neg[true \mathcal{U}^{\leq t} \neg \Phi]_{\bowtie(1-p)}.$$
$$= \neg[\lozenge^{\leq t} \neg \Phi]_{\bowtie(1-p)}.$$

Again, the definitions are the same for the unbounded versions of these operators. The main difference between CTL and $PCTL$ lies in the extended ability to quantify over paths. While in CTL $\forall \phi$ indicates that the path formula ϕ must hold for all paths a $PCTL$ formula like $[\phi]_{>0.99}$ requires the set of paths starting in the current state satisfying ϕ to have the probability measure of more than 0.99. The formula $[\Phi_1 \mathcal{U}^{\leq 10} \Phi_2]_{>0.5}$ states that the set of paths in which Φ_1 will hold continuously until in at most 10 time units Φ_2 will eventually become *true* has the probability measure of more than 0.5. Some authors write statements like $\Phi_1 \mathcal{U}^{\leq t}_{>p} \Phi_2$ instead of $[\Phi_1 \mathcal{U}^{\leq t} \Phi_2]_{>p}$ for the sake of convenience. We omit this for the sake of a clear notation since the $[\ ]_{\bowtie p}$ quantifier can be applied to every path formula and not only to $\mathcal{U}$ or $\widetilde{\mathcal{U}}$.

Remark 1. PCTL versus pCTL
The syntax and semantics of $PCTL$ as defined above follow the work of *Hansson* and *Jonsson* [25]. In some papers (for instance [9]) the slightly different language *pCTL* is used instead of $PCTL$. In *pCTL* there is no bounded until-operator. To keep this survey consistent all results for the bounded until operator $\mathcal{U}^{\leq t}$ are presented as well.

Further-going relations between the CTL-$\forall$ and -$\exists$ quantifier and the $PCTL$-$[\ ]_{\bowtie p}$ operator will be given after the formal definition of $PCTL$ semantics.

2.3   The Semantics of $PCTL$

While specifications written in CTL or $RTCTL$ syntax are interpreted over transitions systems (also known as Kripke structures), $PCTL$ formulae (and later $PCTL^*$ formulae) are considered in the context of probabilistic systems, be they *deterministic (PDS)* or *nondeterministic (PNS)*. First of all we attend the deterministic case. For a *PDS* T we define the satisfaction relation $\models_{PDS}$, that is $\models_{PDS} \subseteq (S_T \cup Paths_{inf}) \times form_{PCTL}$, where S_T denotes the state space of T.

Definition 11. [*PCTL*-**Semantics for** *PDS*]
Let $\models_{PDS}$ be the smallest relation s.t. the following constraints are satisfied. We simply write $s \models \Phi$ instead of $(s, \Phi) \in \models_{PDS}$ and $\pi \models \phi$ instead of $(\pi, \phi) \in \models_{PDS}$.

$$s \models true$$
$$s \models a \qquad \Leftrightarrow a \in L(s)$$
$$s \models \Phi_1 \wedge \Phi_2 \Leftrightarrow s \models \Phi_1 \text{ and } s \models \Phi_2$$
$$s \models \neg \Phi \qquad \Leftrightarrow s \not\models \Phi$$
$$s \models [\phi]_{\bowtie p} \qquad \Leftrightarrow prob_s(\pi \in Paths_{inf}(s) : \pi \models \phi) \bowtie p$$
$$\pi \models \Phi_1 \mathcal{U}^{\leq t} \Phi_2 \Leftrightarrow \exists i \leq t : \pi^i \models \Phi_2 \text{ and } \forall j : 0 \leq j < i : \pi^j \models \Phi_1$$
$$\pi \models \Phi_1 \mathcal{U} \Phi_2 \quad \Leftrightarrow \exists i \in I\!\!N : \pi^i \models \Phi_2 \text{ and } \forall j : 0 \leq j < i : \pi^j \models \Phi_1$$
$$\pi \models \mathcal{X} \Phi \qquad \Leftrightarrow \pi^1 \models \Phi$$

We define the equivalence relation $\equiv \subset form_{PCTL} \times form_{PCTL}$ of two PCTL-formulae Φ_1 and Φ_2 as

$$(\Phi_1, \Phi_2) \in \equiv \quad iff \; s \models \Phi_1 \Leftrightarrow s \models \Phi_2 \text{ for all PDS } T \text{ and for all } s \in S_T,$$

and write $\Phi_1 \equiv \Phi_2$ instead of $(\Phi_1, \Phi_2) \in \equiv$.

Remark 2. In order to define the satisfaction relation $\models$ it is necessary that the set $\{\pi \in Paths_{inf}(s) : \pi \models \Phi\}$ is measurable w.r.t. $prob_s$ (see definition 8). This can be shown by structural induction [49].

We now show certain connections between *PCTL* and *CTL*.

$\forall$ *and* $\exists$ *in CTL versus* $[\;]_{\bowtie}$ *in PCTL* :
Contrary to *CTL* the syntactical definition of *PCTL* does neither contain an universal quantification $\forall$ nor an existential $\exists$. Nevertheless the following equivalences for *CTL* formulae on the left and *PCTL* formulae on the right do hold (where a, b are atomic propositions):

(1) $\forall(\mathcal{X}a) \equiv [\mathcal{X}a]_{\geq 1}$
(2) $\exists(\mathcal{X}a) \equiv [\mathcal{X}a]_{> 0}$
(3) $\exists(a\mathcal{U}b) \equiv [a\mathcal{U}b]_{> 0}$, but
(4) $\forall(a\mathcal{U}b) \not\equiv [a\mathcal{U}b]_{\geq 1}$.
(5) $\forall(\square a) \equiv [\square a]_{\geq 1}$ or more general $\forall(a\widetilde{\mathcal{U}}b) \equiv [a\widetilde{\mathcal{U}}b]_{\geq 1}$ and
(6) $\exists(\square a) \not\equiv [\square a]_{> 0}$ and hence $\exists(a\widetilde{\mathcal{U}}b) \not\equiv [a\widetilde{\mathcal{U}}b]_{> 0}$.

Proofs for (3),(4) and (6)

(3) If at least one path π exists that fulfills the until formula, the finite prefix $\pi \uparrow^i$ of π fulfills $a\mathcal{U}b$ (for some $i \in I\!\!N$). It is clear that the basic cylinder of $\pi \uparrow^i$ has a measure > 0 and all paths of $\Delta(\pi \uparrow^i)$ satisfy $a\mathcal{U}b$ as well.
(4) Figure 2 shows a *PDS* that satisfies the *PCTL* formula but does not satisfy the *CTL* formula.
(6) See also figure 2.

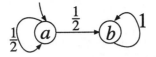

Fig. 2. PDS counterexample for (4) and (6)

In $PCTL$ it would be sufficient to only include $>$ and $\geq$ in the syntax, because $<$ and $\leq$ can be derived using syntactical transformations. More precisely:

(7) $[\phi]_{>p} \equiv \neg[\phi]_{\leq p}$

(8) $[\phi]_{\geq p} \equiv \neg[\phi]_{<p}$

The reason for including $<$ and $\leq$ in the syntax is that the transformation rules (7) and (8) become invalid for the nondeterministic case. Without too much anticipation it should be mentioned here that in order to be able to use the same $PCTL$-syntax for the deterministic case (this Section) and the nondeterministic case (Section 3) we include $<$ and $\leq$ but also stress the existence of (7) and (8).

2.4   Model Checking of $PCTL$ Against a PDS

The basic idea of model checking a $PCTL$ (state) formula Φ against a probabilistic deterministic system (PDS) $\mathcal{T}$ follows the idea of CTL Model Checking à la Clarke, Emerson and Sistla [11] and involves the calculation of *satisfaction sets* $Sat(\Phi)$, where $Sat(\Phi) = \{s \in S_{\mathcal{T}} : s \models \Phi\}$. In order to calculate these sets the syntax tree of Φ is constructed and the subformulae are processed in a *bottom-up* manner, more precisely the leafs are labelled with atomic propositions or *true*, while the inner nodes are labelled with $\neg$ _ , _∧ _, $[\mathcal{U}^{\leq t}\text{_}]_{\bowtie p}, [\mathcal{U}\text{_}]_{\bowtie p}$ and $[\mathcal{X}\text{_}]_{\bowtie p}$. Nodes labelled with $\neg$ _ or $[\mathcal{X}\text{_}]_{\bowtie p}$ have exactly one son. Other inner nodes have two sons. The algorithm traverses the syntax tree in this (*postfix*) order and calculates the satisfaction sets recursively (i.e. the syntax tree is not constructed explicitly, see also figure 4).

We now present schemes for calculating solutions for the bounded and unbounded until operator followed by an algorithm that sketches a complete model checking process for $PCTL$. The calculation of a solution for the nextstep operator $(\mathcal{X})$ is presented only in the algorithm and can be understood without any further comment. See also [25].

The Bounded Until

The case $[\Phi_1\mathcal{U}^{\leq t}\Phi_2]_{\bowtie p}, p$ *arbitrary* : $P^t(s)$, as defined in the following recursive equation, is exactly $prob_s(\Phi_1\mathcal{U}^{\leq t}\Phi_2)^4$ that must be calculated in order to decide whether the $PCTL$-state formula involving a certain probability is satisfied by a certain state $s \in S$.

---

[4] Which here is used as an abbreviation for $prob_s(\{\pi \in Paths_{inf}(s) : \pi \models \Phi_1\mathcal{U}^{\leq t}\Phi_2\})$.

(recursion 1)

$$P^i(s) = \begin{cases} 1, & \text{if } s \models \Phi_2 \\ 0, & \text{if } s \not\models \Phi_1 \text{ and } s \not\models \Phi_2, \\ 0, & \text{if } i = 0 \text{ and } s \not\models \Phi_2, \\ \sum_{s' \in S_T} p_{ss'} \cdot P^{i-1}(s') & \text{otherwise.} \end{cases}$$

This recursion follows the semantical definition given in definition 11 and states a solution for the bounded until operator. Algorithm 2 performs the calculation of this recursion.

The Unbounded Until (Special Case)

The case $[\Phi_1 \mathcal{U} \Phi_2]_{>0}$: We will now see that also a special class of $PCTL$-formulae with *infinite* until bounds can be reduced to this scheme easily (we refer to the case $[\Phi_1 \mathcal{U} \Phi_2]_{>0}$). Later a recursion is defined whichs solutions cover arbitrary unbounded until formulae (i.e. $[\Phi_1 \mathcal{U} \Phi_2]_{\bowtie p}$). However because of equivalence (3) (see Section 2.3) it is clear that also a conventional graph analysis as performed by an ordinary CTL-model checker could be used here (if Φ_1 and Φ_2 are already checked or happen to be atomic propositions). Though the use of a CTL model checker would be the preferable method in this case, we give a proof outline.

Lemma 2. $s \models [\Phi_1 \mathcal{U} \Phi_2]_{>0} \Leftrightarrow s \models [\Phi_1 \mathcal{U}^{\leq |S_T|} \Phi_2]_{>0}$ $(\Leftrightarrow s \models \exists(\Phi_1 \mathcal{U} \Phi_2))$.

Proof. "$\Rightarrow$" If $s \models [\Phi_1 \mathcal{U} \Phi_2]_{>0}$ then $\{\pi \in paths(s) : \pi \models \Phi_1 \mathcal{U} \Phi_2\} \neq \emptyset$. It is clear that because of the finiteness of S these paths can be shortened by removing cycles such that the bounded until formula holds.
"$\Leftarrow$" obvious .

The Unbounded Until (General Form)

The case $[\Phi_1 \mathcal{U} \Phi_2]_{\bowtie p}, p$ *arbitrary:* For the solution of this case the calculation schemes as presented above cannot be used. Due to the absence of a finite bound the algorithm could run infinitely and the "trick" with $\mathcal{U}^{\leq |S_T|}$ cannot be applied since the exact calculation of the probability mass function is necessary to decide whether a certain sharp bound $p > 0$ is reached or not.

A recursion is presented [25] that is based on a partitioning of S_T into three subsets U_0, U_1 and $U_?$.

$S_s = Sat(\Phi_2),$
$S_f = \{s \in S_T : s \notin Sat(\Phi_1) \wedge s \notin Sat(\Phi_2)\},$
$S_i = \{s \in S_T : s \in Sat(\Phi_1) \wedge s \notin Sat(\Phi_2)\},$
$U_0 = S_f \cup \{s \in S_i : \text{ there exists no path in } S_i \text{ from } s \text{ to any } s' \in S_s\},$
$U_1 = S_s \cup \{s \in S_i : \text{ the measure of reaching a state in } S_s$
$\qquad\qquad\qquad \text{through } S_i \text{ starting in } s \text{ is } 1\}$,
$U_? = S \backslash (U_1 \cup U_0),$

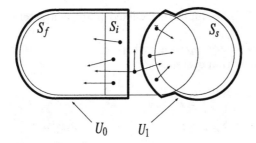

Fig. 3. The sets S_f, S_i, S_f, U_0, U_1 and $U_?$

These sets can be calculated in the following way:

S_s, S_f, S_i: These sets can be calculated by investigating the labelling, respectively the satisfaction sets of the maximal proper subformulae for each state $s \in S_T$.

Preliminary Step: Remove every edge $(s, s') \in T$ for each $s \in S_f \cup S_s$.

U_0: Perform a backward search starting in S_s. Obviously every state in $S \backslash U_0$ will be visited, which yields U_0.

U_1: Perform (with U_0 at hand) a backward search starting in U_0. Now every state in $S \backslash U_1$ is visited, which yields U_1.

$U_?$: With U_1 and U_0, calculate $U_? = S \backslash (U_1 \cup U_0)$.

With these sets the following recursion describes the measure for the unbounded until operator.

(recursion 2)

$$P^\infty(s) = \begin{cases} 1, & \text{if } s \in U_1 \\ 0, & \text{if } s \in U_0, \\ \sum_{s' \in S_T} p_{ss'} \cdot P^\infty(s') & \text{otherwise} \end{cases}$$

This recursion is a linear equation system, which has an unique solution [25, 15].

$$x_s = \sum_{s' \in U_?} p_{ss'} \cdot x_{s'} + \sum_{s'' \in U_1} p_{ss''}, s \in U_?$$

This linear equation system can be solved in polynomial time for instance using Gaussian-elimination. However, for large probability matrices, iterative methods like the Jacobi- or the Gauss-Seidel-method perform much better.

Remark 3.

The until operator of CTL has a least fixpoint semantic, i.e. $Sat(\exists(\Phi_1 \mathcal{U} \Phi_2))$ and $Sat(\forall(\Phi_1 \mathcal{U} \Phi_2))$ are the least fixed points of certain monotonehigher-order operators. For $PCTL$, a similar least fixed point characterization of $\mathcal{U}$ can be established (see e.g. [4]):

The probability vector $(q_s)_{s \in S_T}$ where $q_s = P^\infty(s)$ is the *least solution* in $([0,1]^{|S_T|})$ of the equation system

$$x_s = 0 \text{ if } s \in U_0'$$
$$x_s = 1 \text{ if } s \in U_1'$$
$$x_s = \sum_{s' \in U_?'} p_{ss'} \cdot x_{s'} + \sum_{s'' \in U_1'} p_{ss''} \text{ if } s \in U_?'$$

where $U_0', U_1', U_?'$ are arbitrary sets with $U_0' \subseteq U_0$, $U_1' \subseteq U_1$ and $U_?' \subseteq S_T \backslash (U_0' \cup U_1')$. In order to obtain a *unique solution* the condition $U_0' = U_0$ is needed. The reason is that otherwise, e.g. for states $s \in S_T \backslash (Sat(\Phi_1) \cup Sat(\Phi_2))$ with $p_{ss'} = 1$, we have the equation $x_s = x_s$ which, of course, has no unique solution.

The Algorithm

Pseudocode for model checking $PCTL$ against a PDS is presented (algorithm 1 and algorithm 2) that uses the connections stated in the paragraphs before. In algorithm 2 we are not interested in particular if the starting state $\bar{s} \in S_T$ satisfies the $PCTL$ specification. Instead the satisfaction set $Sat(\Phi)$ is calculated that contains all states $s \in S_T$ s.t. $s \models \Phi$. In addition to that we need a stub algorithm (algorithm 1) which is called with the starting state and whichs only task is to check whether it is true or not that the starting state is contained in $Sat(\Phi)$ (which was calculated by algorithm 2).

In algorithm 2 it is stated (implicitly due to the recursive execution) that a syntax tree of Φ is built and processed in bottom up manner. This assures an efficient complexity for the processing of the subformulae and is illustrated in figure 4. The syntax tree is built according to the $PCTL$-grammar. Atomic propositions will always be leafs of this tree, the Sat sets are calculated for these leafs. The next inner nodes to be processed will either be Boolean combinations (i.e. $\wedge, \vee$ or $\neg$) or nextstep- or until-expressions. Sat sets for these are calculated according to the cases in the algorithm.

Algorithm 1 PCTL_MODELCHECK($\Phi, T, \bar{s}$)

Input: PDS $T, PCTL$ formula $\Phi, \bar{s} \in S_T$
Output: truth value of $\bar{s} \models_{PDS} \Phi$

---

calculate $Sat(\Phi)$ with algorithm 2.
IF $s \in Sat(\Phi)$ **THEN**
        return *true*
ELSE
        return *false*

Algorithm 2 $Sat(\Phi)$

Input: $PCTL$ state formula,
Output: set of all $s \in S$ that satisfy Φ

<u>**CASE**</u> Φ is

true	: return S_T;
a	: return $\{s \in S_T : a \in L(s)\}$;
$\Phi_1 \wedge \Phi_2$	: return $Sat(\Phi_1) \cap Sat(\Phi_2)$;
$\neg\Phi'$	: return $S_T \setminus Sat(\Phi')$;
$[\mathcal{X}\Phi']_{\bowtie p}$	: calculate $Sat(\Phi')$ and return $\{s \in S_T \; : \; \Sigma_{s' \in Sat(\Phi')} p_{ss'} \bowtie p\}$;

$[\Phi_1 \mathcal{U}^{\leq t}\Phi_2]_{\bowtie p}$: calculate $Sat(\Phi_1)$ and $Sat(\Phi_2)$;

$$\forall s \in S_T : P^0(s) = \begin{cases} 1 & \text{if } s \in Sat(\Phi_2), \\ 0 & \text{otherwise} \end{cases}$$

<u>**FOR**</u> $i = 1$ to t <u>**DO**</u>

$$\forall s \in S_T : P^i(s) = \begin{cases} 0, \text{ if } \; s \notin Sat(\Phi_1) \text{ and } s \notin Sat(\Phi_2), \\ 1, \text{ if } \; s \in Sat(\Phi_2), \\ \sum_{s' \in S_T} p_{ss'} \cdot P^{i-1}(s') \text{ otherwise .} \end{cases}$$

<u>**OD**</u>;
return $\{s \in S_T : P^t(s)\bowtie p\}$;

$[\Phi_1 \mathcal{U}\Phi_2]_{>0}$: calculate $Sat(\Phi_1)$ and $Sat(\Phi_2)$ and do conventional graph analysis;
return $\{s \in S_T : s \models \exists(\Phi_1 \mathcal{U}\Phi_2)\}$;

$[\Phi_1 \mathcal{U}\Phi_2]_{\bowtie p}$: calculate $Sat(\Phi_1)$ and $Sat(\Phi_2)$ and solve linear equation system;
return $\{s \in S_T : P^\infty(s)_{\bowtie p}\}$;

<u>**END CASE**</u>

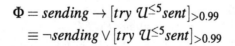

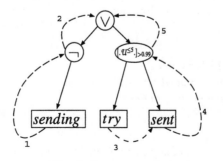

Fig. 4. Syntax tree example

The Complexity of $PCTL$ Model Checking Against a PDS

We now outline the complexity of the model checking algorithm for checking a $PCTL$ formula Φ against a PDS. Given a PDS $T = (M, AP, L, \bar{s})$, $M = (S, T, p)$, we give asymptotic complexity results. Obviously, the number of subformulae to be checked is $\leq length(\Phi)$.

- For the case that Φ' is an atomic proposition, its negation or a Boolean combination of them, the satisfaction set $Sat(\Phi')$ can be computed in time $O(|S|)$.
- If Φ' is a nextstep expression $[\mathcal{X}\Phi'']_{\bowtie p}$ it can be checked in time $O(|S| + |T|)$ (where we assume a sparse matrix representation).
- The case $\Phi' = [\Phi_1 \mathcal{U}^{\leq t}\Phi_2]_{\bowtie p}$ can be checked in time $O(|S_T| \cdot |T| \cdot t)$.
- The case $\Phi' = [\Phi_1 \mathcal{U}\Phi_2]_{\bowtie p}$ can be checked by computing the sets U_0, U_1 and $U_?$ in time $O(|S| + |T|)$ each and then solving a linear equation system of size $(|S| \times |S|)$ which is possible to do in time $O(poly(|S|))$.
- The recursion tree of algorithm 2 (which can be seen as building and processing the syntaxtree of Φ) requires $O(length(\Phi))$ steps. We sum up our exposition:

Theorem 1. *Let T be a PNS and Φ a $PCTL$ state formula over the set of atomic propositions of T. Then $Sat(\Phi)$ can be computed in time $O(poly(|S|) \cdot t_\Phi \cdot length(\Phi))$.*[5]

Proof. The correctness of theorem 1 follows from the statements about the complexities made before.

2.5   The Syntax and Semantics of $PCTL^*$

We now introduce the logic $PCTL^*$ and explain how to model check a PDS against a $PCTL^*$-formula. Throughout this Section let $T = (M, AP, L, \bar{s})$ be a PDS with $M = (S, T, p)$.

Like standard CTL we can extend $PCTL$ to a richer logic $PCTL^*$ which allows the negation and conjunction of path formulae and also the combination of temporal modalities. In addition, every state formula is as well a path formula.

Definition 12. [PCTL*-Syntax]
Given the set of atomic propositions AP, the syntax of $PCTL^*$-formulae over AP is defined by the following grammar:

---

[5] Where t_Φ stands for the maximal t in subformulae of the form $\Phi_1\mathcal{U}^{\leq t}\Phi_2$ (and is at least 1).

$PCTL^*$-state formulae

$$\Phi ::= true \mid a \mid \Phi_1 \wedge \Phi_2 \mid \neg\Phi \mid [\phi]_{\bowtie p}$$

$PCTL^*$-path formulae:

$$\phi ::= \neg\phi \mid \phi_1 \wedge \phi_2 \mid \Phi \mid \phi_1 \mathcal{U}^{\le t} \phi_2 \mid \phi_1 \mathcal{U} \phi_2 \mid \mathcal{X}\phi$$

where $a \in AP$, $t \in I\!N, p \in [0,1] \subset I\!R$ and $\bowtie \in \{\ge, >, \le, <\}$.

With $State_{PCTL^*} = \{\Phi : \Phi$ is a $PCTL^*$ state formula$\}$ we denote the set of all $PCTL^*$ state formulae and with $Path_{PCTL^*} = \{\phi : \phi$ is a $PCTL^*$ path formula$\}$ we denote the set of all $PCTL^*$ path formulae for a fixed set AP of atomic propositions. Observe that $State_{PCTL^*} \subset Path_{PCTL^*}$ holds. The length of a $PCTL^*$ formula ϕ is defined similarly to the length of a $PCTL$ formula (see definition 10, section 2.2) and is denoted by $|\phi|$.

For the PDS T we define the satisfaction relation $\models^*_{PDS}$, which is a relation on $(S \cup Paths_{inf}) \times Path_{PCTL^*}$.

Definition 13. [PCTL*-Semantics for Probabilistic Deterministic Systems]
Let $\models^*_{PDS}$ be the smallest relation on $(S \cup Paths_{inf}) \times Path_{PCTL^*}$ s.t. for $a \in AP$, $\Phi, \Phi_1, \Phi_2 \in State_{PCTL^*}$, $\phi, \phi_1, \phi_2 \in Path_{PCTL^*}$ and $s \in S$:[6]

$$s \models true$$
$$s \models a \qquad \Leftrightarrow a \in L(s)$$
$$s \models \Phi_1 \wedge \Phi_2 \Leftrightarrow s \models \Phi_1 \text{ and } s \models \Phi_2$$
$$s \models \neg\Phi \qquad \Leftrightarrow s \not\models \Phi$$
$$s \models [\phi]_{\bowtie p} \qquad \Leftrightarrow prob_s(\{\pi \in Paths_{inf}(s) : \pi \models \phi\}) \bowtie p$$
$$\pi \models \phi_1 \wedge \phi_2 \Leftrightarrow \pi \models \phi_1 \text{ and } \pi \models \phi_2$$
$$\pi \models \neg\phi \qquad \Leftrightarrow \pi \not\models \phi$$
$$\pi \models \Phi \qquad \Leftrightarrow \pi^0 \models \Phi$$
$$\pi \models \phi_1 \mathcal{U}^{\le t} \phi_2 \Leftrightarrow \exists \ i \le t \text{ such that } \pi{\uparrow}_i \models \phi_2 \text{ and } \forall j : 0 \le j < i : \pi{\uparrow}_j \models \phi_1$$
$$\pi \models \phi_1 \mathcal{U} \phi_2 \qquad \Leftrightarrow \exists \ i \in I\!N \text{ such that } \pi{\uparrow}_i \models \phi_2 \text{ and } \forall j : 0 \le j < i : \pi{\uparrow}_j \models \phi_1$$
$$\pi \models \mathcal{X}\phi \qquad \Leftrightarrow \pi{\uparrow}_1 \models \phi$$

Note, that the definition of the semantics requires that the set $\{\pi \in Paths_{inf}(s) : \pi \models \phi\}$ is measurable for a $PCTL^*$ formula ϕ. Like for $PCTL$ formulae, this can be shown by structural induction (see remark 2).

Remark 4.
Note, that for a $PCTL^*$-path formula the following equivalences hold, since $Paths_{inf}(s)$ is the disjoint union of $\{\pi \in Paths_{inf}(s) : \pi \models \phi\}$ and $\{\pi \in Paths_{inf}(s) : \pi \models \neg\phi\}$.

---

[6] Observe that we use the notation $s \models \Phi$ (resp. $\pi \models \phi$) instead of $(s, \Phi) \in \models^*_{PDS}$ (resp. $(\pi, \phi) \in \models^*_{PDS}$).

$$- s \models [\phi]_{<p} \quad \Leftrightarrow \quad s \models [\neg\phi]_{>(1-p)}$$
$$- s \models [\phi]_{\leq p} \quad \Leftrightarrow \quad s \models [\neg\phi]_{\geq(1-p)}$$

Due to this fact, it is sufficient to consider formulae which do not make use of $[.]_{<p}$ and $[.]_{\leq p}$. Note however, that the above equivalences can not be applied for $PCTL$ formulae, since it is not allowed to negate $PCTL$-path formulae.

As in the non-probabilistic case, one could also define the logic $PCTL^+$, which extends $PCTL$ by negation and conjunction of path formulae, but does not allow the combination of temporal modalities (e.g. $\phi_1 \mathcal{U} \phi_2$, where ϕ_1 and ϕ_2 are path formulae). In contrast to the non-probabilistic case, where CTL^+ is not more expressive than CTL, we believe that $PCTL^+$ is strictly more expressive than $PCTL$. We only give an informal argument and show where the transformation method from CTL^+ to CTL fails in the probabilistic case. Consider the CTL^+ formula $\Phi = \exists(\Diamond a \wedge \Diamond b)$. Then the following equivalence holds:

$$\exists(\Diamond a \wedge \Diamond b) \quad \equiv \quad \exists\Diamond(a \wedge \exists\Diamond b) \vee \exists\Diamond(b \wedge \exists\Diamond a),$$

where the formula on the right hand side is a CTL formula. Considering only one path (due to the existential quantification), the above equivalence is easy to accomplish. But in the probabilistic setting, we have the following problem:

$$[\Diamond a \wedge \Diamond b]_{\geq p} \quad \not\equiv \quad [\Diamond(a \wedge [\Diamond b]_?)]_? \vee [\Diamond(b \wedge [\Diamond a]_?)]_?,$$

whatever might be in place for the question marks. Even the following idea with an infinite disjunction of $PCTL$ formulae does not work:

$$[\Diamond a \wedge \Diamond b]_{\geq p} \quad \not\equiv \quad \bigvee_{p_1 \cdot p_2 + p_3 \cdot p_4 \geq p} [\neg b\, \mathcal{U}(a \wedge [\Diamond b]_{\geq p_2})]_{\geq p_1} \wedge [\neg a\, \mathcal{U}(b \wedge [\Diamond a]_{\geq p_3})]_{\geq p_4}.$$

Note that the infinite disjunction is not a $PCTL$ formula, since $PCTL$ formulae are always finite (so the $\not\equiv$ is not really appropriate in this case). Moreover do we believe that there is no alternative way to express $[\Diamond a \wedge \Diamond b]_{\geq p}$ in an equivalent $PCTL$ formula.

For the sake of completeness we shortly describe the logic LTL since we will need it later.

Definition 14. [LTL-Syntax]
Given the set of atomic propositions AP, LTL-formulae over AP are $PCTL^*$-path formulae given by the following grammar:

LTL formulae:

$$\phi ::= true \mid a \mid \neg\phi \mid \phi_1 \wedge \phi_2 \mid \phi_1 \mathcal{U} \phi_2 \mid \mathcal{X}\phi$$

where $a \in AP$.

We define the language $\mathcal{L}_\omega(\phi)$ of an LTL formula ϕ over AP to be the set of infinite words w over 2^{AP} such that $w \models \phi$, assuming $L(x) = x \;\forall x \in 2^{AP}$ (recall definition 4 of the labelling function $L(.)$). Note that for a path π the following holds :

$$\pi \models \phi \quad \text{iff} \quad trace(\pi) \in \mathcal{L}_\omega(\phi).$$

Remark 5.
Note that there is no bounded Until operator in the LTL-syntax. One can ommit the $\mathcal{U}^{\leq t}$, since the following equivalences hold (where ϕ_1 and ϕ_2 are LTL-formulae) :

$$\phi_1 \mathcal{U}^{\leq 0} \phi_2 \equiv \phi_2$$

$$\phi_1 \mathcal{U}^{\leq t} \phi_2 \equiv \left(\phi_2 \vee (\phi_1 \wedge \mathcal{X}(\phi_1 \mathcal{U}^{\leq t-1} \phi_2)) \right)$$

So the formula $\phi_1 \mathcal{U}^{\leq t} \phi_2$ can be rewritten as a nested Next Step expression of length $O(t \cdot (|\phi_1| + |\phi_2|))$.

2.6   Model Checking of $PCTL^*$ Against a PDS

There are different approaches to $PCTL^*$ model checking against deterministic systems (see e.g. [3, 13, 50]). We follow the automata theoretic approach since we will do the same in Section 3.3 where we discuss nondeterministic systems. Given a state formula $\Phi \in State_{PCTL^*}$ we want to compute the set $Sat(\Phi)$. To do so, we proceed in the same way as with CTL^* formulae (see [20]).

As with $PCTL$ formulae, we first create the parse tree of our given formula and then work from the bottom to the top using satisfaction sets for all subformulae. Having a look at the above defined semantics tells us that conjunction (resp. negation) are just the intersection (resp. the complement) of satisfaction sets of subformulae[7]. The terminal cases are also easy to handle. That leaves us with the case $\Phi = [\phi]_{\bowtie p}$. We assume that we have already computed the satisfaction sets for all maximal state $PCTL^*$-subformulae $\gamma_1, \ldots, \gamma_n$ of ϕ and have labelled the states of S appropriately with new atomic propositions $r_1, \ldots, r_n$[8]. Let

$$\phi' = \phi\{^{\gamma_1}/_{r_1}, \ldots, ^{\gamma_n}/_{r_n}\}$$

be the formula we get from ϕ by replacing each occurrence of γ_i by r_i, $1 \leq i \leq n$. Let ϕ'' be the formula we get from ϕ' by rewriting all subformulae involving $\mathcal{U}^{\leq t}$ as an expression of nested Next Step operators as explained in remark 5 in section 2.5. Then ϕ'' is an LTL formula over the atomic propositions $r_1, \ldots, r_n$ and for all infinite paths π of the given system it holds that $\pi \models \phi$ if and only if $\pi \models \phi''$. Observe, that if ϕ has m occurences of the Unbounded Until operator and $t_1, \ldots, t_m$ are the corresponding bounds, we get $|\phi''| = O(\Pi_{i=1}^m t_i \cdot |\phi|)$.
So in order to decide whether $s \models_{PDS}^* [\phi]_{\bowtie p}$, we have to compute

$$prob_s(\{\pi \in Paths_{inf}(s) : \pi \models \phi''\}).$$

---

[7] $Sat(\Phi_1 \wedge \Phi_2) = Sat(\Phi_1) \cap Sat(\Phi_2)$ and $Sat(\neg\Phi) = S \setminus Sat(\Phi)$.
[8] $r_i \in L(s)$ iff $s \in Sat(\gamma_i)$, where AP is extended by $\{r_1, \ldots, r_n\}$.

This can be done by using the standard ω-automaton approach (see e.g. [47, 51]). An ω-automaton accepts infinite words over a given alphabet. In the probabilistic setting a kind of determinism is needed. Here we use deterministic Rabin automata.

Definition 15. [Deterministic Rabin Automaton]
A deterministic Rabin automaton is a tuple

$$\mathcal{A} = (\Sigma, Q, \bar{q}, \rho, \alpha),$$

where

- Σ *is a finite alphabet,*
- Q *is a finite set of states,*
- $\bar{q} \in Q$ *is an initial state,*
- $\rho : Q \times \Sigma \longrightarrow Q$ *is a transition function and*
- $\alpha \subseteq 2^Q \times 2^Q$ *is an acceptance condition.*

Definition 16. [Run, Limit and α-Satisfaction]
Given a deterministic Rabin automaton $\mathcal{A} = (\Sigma, Q, \bar{q}, \rho, \alpha)$ and an infinite word $w = a_1, a_2, \ldots \in \Sigma^\omega$ over Σ we call the unique sequence $r = \bar{q}, q_1, q_2, \ldots$ with

$$q_i = \rho(q_{i-1}, a_i) \qquad (q_0 = \bar{q})$$

the run of $\mathcal{A}$ over w.
We define the limit of a run $r = \bar{q}, q_1, q_2, \ldots$ as

$$lim(r) = \{q \ : \ q = q_i \ \text{for infinitely many } i's\}.$$

We say a subset $X \subseteq Q$ satisfies α if and only if

$$\exists\, (L, U) \in \alpha \ : L \cap X \neq \emptyset \ \wedge \ U \cap X = \emptyset.$$

We say a run r satisfies α if and only if $lim(r)$ satisfies α.
Observe that Rabin acceptance can be expressed with an LTL formula, i.e. r satisfies α iff r satisfies the LTL formula

$$\bigvee_{(L,U)\in\alpha} (\Box\Diamond L \wedge \Diamond\Box\neg U).$$

We now define the language of a deterministic Rabin automaton.

Definition 17. [$\mathcal{L}_\omega(\mathcal{A})$]
Let $\mathcal{A} = (\Sigma, Q, \bar{q}, \rho, \alpha)$ be a deterministic Rabin automaton and $w \in \Sigma^\omega$. Let r be the run of $\mathcal{A}$ over w. We say $\mathcal{A}$ accepts w if and only if r satisfies α. We call the set

$$\mathcal{L}_\omega(\mathcal{A}) = \{w \in \Sigma^\omega : \mathcal{A} \text{ accepts } w\}$$

the language of $\mathcal{A}$.

We now state a useful lemma (see [47, 51, 50, 43]).

Lemma 3. [**Rabin Automaton for a** *LTL* **formula**]
Given a LTL formula ϕ over AP, a deterministic Rabin automaton $\mathcal{A}$ with the alphabet 2^{AP} can be constructed such that

$$\mathcal{L}_\omega(\phi) = \mathcal{L}_\omega(\mathcal{A}).$$

Remember that we want to compute the value

$$prob_s(\{\pi \in Paths_{inf}(s) \ : \ \pi \models \phi''\}),$$

where ϕ'' is a *LTL* formula over the extended set *AP* of atomic propositions. We first construct a deterministic Rabin automaton $\mathcal{A} = (\Sigma, Q, \bar{q}, \rho, \alpha)$ with $\mathcal{L}_\omega(\phi'') = \mathcal{L}_\omega(\mathcal{A})$.

We then build a product Markov chain $\mathcal{T} \times \mathcal{A}$ for $\mathcal{T}$ and $\mathcal{A}$ by the following: $\mathcal{T} \times \mathcal{A} = (S \times Q, \ T', \ p')$ with

- $p'((s,q),(s',q')) = p_{ss'}$ if $q' = \rho(q, L(s))$ and 0 otherwise
- $T' = \{((s,q),(s',q')) \ : \ p'((s,q),(s',q')) \neq 0\}$.

It is easy to see that there is a one-to-one correspondence between the paths of $\mathcal{T}$ and those of $\mathcal{T} \times \mathcal{A}$ by path-lifting. In addition, taking a path π of $\mathcal{T}$, lifting it to a path π' of $\mathcal{T} \times \mathcal{A}$ and then reducing it to its second component, gives the run of $\mathcal{A}$ over $trace(\pi)$. So $\mathcal{T} \times \mathcal{A}$ also keeps track whether $trace(\pi)$ will be accepted by $\mathcal{A}$ or not, i.e. whether π satisfies ϕ'' or not. Taking this into account it is easy to show that

$$prob_s^{\mathcal{T}}(\{\pi \in Paths_{inf}(s) \ : \ \pi \models \phi''\}) =$$

$$prob_{(s,\bar{q})}^{\mathcal{T} \times \mathcal{A}}(\{\pi' \text{ path in } \mathcal{T} \times \mathcal{A} \ : \ \pi' \text{ satisfies } \alpha'\}),$$

where α' is the lifted accepted condition from $\mathcal{A}$, i.e.

$$\alpha' = \{(S \times L, \ S \times U) \ : \ (L, \ U) \in \alpha\}.$$

Let *Acc* be the union of the ergodic sets of $\mathcal{T} \times \mathcal{A}$ that satisfy α'. Using lemma 1 about ergodic sets, it follows that

$$prob_{(s,\bar{q})}^{\mathcal{T} \times \mathcal{A}}(\{\pi' \text{ path in } \mathcal{T} \times \mathcal{A} \ : \ \pi' \text{ satisfies } \alpha'\}) =$$

$$prob_{(s,\bar{q})}^{\mathcal{T} \times \mathcal{A}}(\{\pi' \text{ path in } \mathcal{T} \times \mathcal{A} \ : \ \pi' \text{ reaches } Acc\}).$$

The right side of this equation equals

$$prob_{(s,\bar{q})}^{\mathcal{T} \times \mathcal{A}}(\{\pi' \text{ path in } \mathcal{T} \times \mathcal{A} \ : \ \pi' \models \Diamond acc\}),$$

when exactly the states $\in Acc$ are labelled with *acc* and can therefore be computed by the algorithms introduced in the *PCTL* Section 2.4. The following flow chart (figure 5) sketches the main ideas:

As a closing remark we want to mention that the size of the constructed automaton $\mathcal{A}$ is doubly exponential in the size of the formula ϕ''. And the time needed to compute $prob_s(\{\pi \in Paths_{inf}(s) \ : \ \pi \models \phi\})$ is doubly exponential in

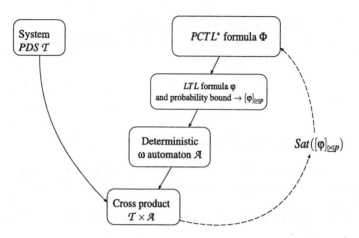

Fig. 5. Model Checking Scheme for $PCTL^*$

$O(\Pi_{i=1}^{m} t_i \cdot |\phi|)$, where m is the number of occurences of $\mathcal{U}^{\leq t}$ in the formula ϕ and $t_1, \ldots, t_m$ are the corresponding bounds. However, in [15] a different approach is discussed, which is only single exponential in $O(\Pi_{i=1}^{m} t_i \cdot |\phi|)$. In addition, both methods are polynomial in the size of the system. (See the article by B. Bollig and M. Leuker in this volume.)

3 Validation of Nondeterministic Probabilistic Systems

In section 2 methods were established to show how to model check $PCTL$ and $PCTL^*$ against a PDS. We now move on to *nondeterministic probabilistic systems* that involve a set of probability distributions per transition rather than only one as in the deterministic case. In non-probabilistic systems nondeterminism can be used to describe input behaviour from outside the system (e.g. user input) or certain fault possibilities. Also the interleaving behaviour of asynchronous parallel processes can be expressed using nondeterminism. This carries over to the probabilistic setting, in which also underspecification can be expressed with nondeterminism. In the nondeterministic case *schedulers* are introduced that pick the distributions as the execution of a nondeterministic system goes along. These probabilistic nondeterministic systems are essentially state-labelled *Markov decision processes (MDPs)*.

3.1 Nondeterministic Probabilistic Systems

We now define nondeterminism in the probabilistic context. Several types of nondeterministic systems were introduced in [49, 24, 26, 45] and [9]. We follow here the approach of [9].

Definition 18. [Probabilistic Nondeterministic System (PNS)]
A probabilistic nondeterministic system is a tuple

$$\mathcal{T}_{PNS} = (S, Steps, AP, L, \bar{s}),$$

where

- S *is a finite set of states,*
- *Steps is a function that assigns to each state $s \in S$ a finite, non-empty set of probability distributions on S, i.e.*

$$Steps : S \rightarrow 2^{Distr(S)},$$

such that $1 \leq |Steps(s)| < \infty$ $\forall s \in S$.
- AP *is a finite set of atomic propositions,*
- L *is a labelling function $L : S \rightarrow 2^{AP}$ that labels any state $s \in S$ with those atomic propositions that are supposed to hold in s and*
- $\bar{s} \in S$ *is the starting state.*

Similar to section 2.1 we simply write $\mathcal{T}$ instead of $\mathcal{T}_{PNS}$ in the rest of this chapter.

Thus the main difference between definition 4 of a PDS and the definition of a PNS is, that there is not only one transition distribution per state but a set of distributions and a function $Steps$ that yields them for each state. Intuitively, the set $Steps(s)$ represents the non-deterministic alternatives in state s. Having resolved the nondeterminism in a state s, i.e. having chosen a $\mu \in Steps(s)$, the process itself resolves the probabilistic choice by selecting some state s' with the probability $\mu(s')$. We refer to the elements of $Steps(s)$ as the transitions of s.

We now explain how a path is defined in a PNS.

Definition 19. [Path in a PNS]
Let $\mathcal{T} = (S, Steps, AP, L, \bar{s})$ be a probabilistic nondeterministic system. We define a path of $\mathcal{T}$ to be a (finite or infinite) sequence

$$\pi = s_0 \xrightarrow{\mu_1} s_1 \xrightarrow{\mu_2} s_2 \ \ldots$$

s.t. for all i under consideration : $s_i \in S$, $\mu_{i+1} \in Steps(s_i)$ and $\mu_{i+1}(s_{i+1}) > 0$. The definitions of the length of a path, of $first(\pi), last(\pi), \pi^k, \pi\uparrow^k$ and $\pi\uparrow_k$ and of $Paths_{fin}, Paths_{inf}$ (resp. $Paths_{fin}(s), Paths_{inf}(s)$) carry over from the deterministic case (see definition 5). If $k < |\pi|$, we put $step(\pi, k) = \mu_{k+1}$. As for deterministic Rabin automata, given an infinite path π, we denote by $lim(\pi)$ the set of states that occur infinitely often in π.

But how is the nondeterminism resolved? We give the definition of a scheduler (sometimes called strategy, adversary or policy) that decides, based on the past history of the system, which possible step to perform next.

Definition 20. [Scheduler]
Let $\mathcal{T} = (S, Steps, AP, L, \bar{s})$ be a probabilistic nondeterministic system. A scheduler of $\mathcal{T}$ is a function B mapping every finite path ρ of $\mathcal{T}$ to a distribution $B(\rho) \in Steps(last(\rho))$.

That means, after the process described by the PNS performed the first $|\rho|$ steps, the scheduler B chooses the next transition to be $\mu = B(\rho)$, thus resolving the nondeterminism. So if a system behaves according to a scheduler B and has followed the path $\rho = s_0 \xrightarrow{\mu_1} s_1 \xrightarrow{\mu_2} \ldots \xrightarrow{\mu_n} s_n$ so far, then it will be in state s in the next step with probability

$$B(\rho)(s).$$

We denote the set of infinite paths that comply with a given scheduler B with $Paths_{inf}^B$, i.e.

$$Paths_{inf}^B = \{\pi \in Paths_{inf} : B(\pi\uparrow^i) = step(\pi, i) \quad i = 0, 1, 2, \ldots\}.$$

By $Paths_{inf}^B(s)$ we denote $Paths_{inf}^B \cap Paths_{inf}(s)$.

We call a scheduler B

$$simple, \quad \text{iff} \quad B(\rho) = B(last(\rho)) \ \forall \text{ finite path } \rho,$$

i.e. B is history independent.

For a state s to satisfy a $PCTL^*$ state formula of type $[\phi]_{\bowtie p}$[9], we need to ensure that starting in s, for all schedulers under consideration the given system produces a path satisfying ϕ with a probability $\bowtie p$.

We therefore define for each state $s \in S$ and for each scheduler B a probability space on the set of infinite paths that start in s and comply with the scheduler B. Given a scheduler B and a state $s \in S$, we define the basic cylinders of $Paths_{inf}^B(s)$ as in the case for $PDSs$.

Definition 21. [Basic Cylinder (for a PNS)]
Given a scheduler B, we define for $\rho \in Paths_{fin}^B$ the basic cylinder of ρ w.r.t B as

$$\Delta^B(\rho) = \{\pi \in Paths_{inf}^B : \pi\uparrow^{|\rho|} = \rho\}.$$

Definition 22. [Probability Space of a PNS]
Given a PNS $T = (S, Steps, AP, L, \bar{s})$, a scheduler B and $s \in S$, we define a probability space

$$\Psi_s^B = (\Delta^B(s), \Delta_s^B, prob_s^B),$$

such that

- Δ_s^B *is the σ-algebra generated by the empty set and the basic cylinders that are contained in $\Delta^B(s) = Paths_{inf}^B(s)$.*
- $prob_s^B$ *is the uniquely induced probability measure which satisfies the following : $prob_s^B(\Delta^B(s)) = 1$ and for all basic cylinders $\Delta^B(\rho) \subseteq \Delta^B(s)$ with $\rho = s_0 \xrightarrow{\mu_1} s_1 \xrightarrow{\mu_2} \ldots \xrightarrow{\mu_n} s_n$[10]:*

$$prob_s^B(\Delta^B(\rho)) = \prod_{i=0}^{|\rho|-1} B(\rho\uparrow^i)(\rho^{i+1}) = \prod_{i=0}^{|\rho|-1} B(s_0 \xrightarrow{\mu_1} \ldots \xrightarrow{\mu_i} s_i)(s_{i+1}).$$

---

[9] Remember, that $\bowtie \in \{\geq, >, \leq, <\}$.
[10] Note that $s_0 = s$.

Remark 6.

To get the above probability space Ψ_s^B for some $s \in S$ and some scheduler B one can also consider the Markov chain

$$M_s^B = (S_s^B, T_s^B, p_s^B),\ ^{11}$$

where

- $S_s^B = Paths_{fin}^B(s)$ and

for all $\rho, \rho' \in S_s^B \times S_s^B$

- $p_s^B(\rho, \rho') = B(\rho)(last(\rho'))$,   if $\rho = \rho'\uparrow^{|\rho'|-1}$ and 0 otherwise.
- $(\rho, \rho') \in T_s^B \quad \Leftrightarrow \quad p_s^B(\rho, \rho') > 0$.

If we now identify $\rho \in Paths_{fin}^B(s)$ with ρ_s^B, where $(\rho_s^B)^i = \rho\uparrow^i$, $i = 0, 1, \ldots, |\rho|$ and $|\rho_s^B| = |\rho|$, we get that $prob_s^B(\Delta^B(\rho))$ equals the measure of $\Delta(\rho_s^B)$ in the Markov chain M_s^B.

Now we have everything at hand to define the semantics of $PCTL^*$ formulae over a PNS with respect to a given set of schedulers. We follow the work of [9]. For the rest of this section let $\mathcal{T} = (S, Steps, AP, L, \bar{s})$ be a PNS and let $Sched$ be a nonempty set of schedulers of $\mathcal{T}$.

Definition 23. [PCTL*-Semantics for Probabilistic Nondeterministic Systems]

Let $\models_{Sched}$ be the smallest relation on $(S \cup Paths_{inf}) \times Path_{PCTL^*}$ s.t. for $a \in AP$, $\Phi, \Phi_1, \Phi_2 \in State_{PCTL^*}$, $\phi, \phi_1, \phi_2 \in Path_{PCTL^*}$ and $s \in S$:

$s \models_{Sched} true$

$s \models_{Sched} a \qquad\qquad \Leftrightarrow a \in L(s)$

$s \models_{Sched} \Phi_1 \wedge \Phi_2 \quad \Leftrightarrow s \models_{Sched} \Phi_1$ and $s \models_{Sched} \Phi_2$

$s \models_{Sched} \neg\Phi \qquad\quad \Leftrightarrow s \not\models_{Sched} \Phi$

$s \models_{Sched} [\phi]_{\bowtie p} \quad \Leftrightarrow prob_s^B(\{\pi \in Paths_{inf}^B(s) : \pi \models_{Sched} \phi\}) \bowtie p \ \ \forall B \in Sched$

$\pi \models_{Sched} \phi_1 \wedge \phi_2 \ \Leftrightarrow \pi \models_{Sched} \phi_1$ and $\pi \models_{Sched} \phi_2$

$\pi \models_{Sched} \neg\phi \qquad\quad \Leftrightarrow \pi \not\models_{Sched} \phi$

$\pi \models_{Sched} \Phi \qquad\quad \Leftrightarrow \pi^0 \models_{Sched} \Phi$

$\pi \models_{Sched} \phi_1 \mathcal{U}^{\leq t}\phi_2 \Leftrightarrow \exists\ i \leq t$ such that $\pi\uparrow_i \models_{Sched} \phi_2$ and

$\qquad\qquad\qquad\qquad \forall j : 0 \leq j < i : \pi\uparrow_j \models_{Sched} \phi_1$

$\pi \models_{Sched} \phi_1 \mathcal{U}\phi_2 \quad \Leftrightarrow \exists\ i \in I\!N$ such that $\pi\uparrow_i \models_{Sched} \phi_2$ and

$\qquad\qquad\qquad\qquad \forall j : 0 \leq j < i : \pi\uparrow_j \models_{Sched} \phi_1$

$\pi \models_{Sched} \mathcal{X}\phi \qquad\quad \Leftrightarrow \pi\uparrow_1 \models_{Sched} \phi$

Again it is required, that the set $\{\pi \in Paths_{inf}^B : \pi \models_{Sched} \phi\}$ is measurable for a $PCTL^*$ formula ϕ and a scheduler B, which follows from the analogous

---

[11] The state set of M_s^B is countable. Nevertheless do the definitions given in section 2.1 for Finite Markov chains also work here.

fact in the deterministic setting and remark 6 (or can be shown by structural induction).

So in order to model check a formula of the kind $[\phi]_{\bowtie p}$ for a state s, we have to verify that for all schedulers $B \in Sched$ under consideration the value $prob_s^B(\{\pi \in Paths_{inf}^B(s) : \pi \models_{Sched} \phi\})$ is $\bowtie p$.

Remark 7.
Note, that for a *PNS* and a *PCTL* formula ϕ, we cannot decide whether $s \models_{Sched} [\phi]_{\trianglelefteq p}$ with ($\trianglelefteq \in \{\leq, <\}$) by just considering *PCTL* formulae of the type $s \models_{Sched} [\phi']_{\trianglerighteq p}$ with ($\trianglerighteq \in \{\geq, >\}$). This was possible in the case of *PDS* (see equivalences 7 and 8 in section 2.3), but becomes impossible here, because the semantics quantifies over all schedulers.

Contrary to that, it suffices for *PCTL** formulae to choose $\bowtie$ from $\{\geq, >\}$ because of remark 4.

What kind of sets *Sched* are we interested in? There are two different sets of schedulers we will focus on, the set $Sched_{all}$ of all schedulers and the set $Sched_{fair}$ of fair schedulers, which will be defined in the following. First we give a definition of a fair path. For simplicity, we restrict our attention to strong fairness for all transitions. Other notions of fairness are considered in e.g. [49, 39, 41].

Definition 24. [Fairness of an Infinite Path]
Let $T = (S, Steps, AP, L, \bar{s})$ be a *PNS* and π an infinite path of T. π is called fair, iff the following holds:

$$\forall s \in lim(\pi) : \forall \mu \in Steps(s) : |\{j : \pi^j = s \land step(\pi, j) = \mu\}| = \infty$$

We denote the set of fair paths by $Paths_{fair}$.[12]

That means a path π is fair, if and only if for each state s occurring infinitely often in π, each nondeterministic alternative in $Steps(s)$ is taken infinitely often in π (at s). For a scheduler B to be fair we require the set of fair paths to have measure 1.

Definition 25. [Fair Scheduler]
Let $T = (S, Steps, AP, L, \bar{s})$ be a *PNS* and B be a scheduler for T. We call B fair, if and only if $prob_s^B(Path_{inf}^B(s) \cap Path_{fair}) = 1$.

Note, that in the following we will write $\models_{all}$ instead of $\models_{Sched_{all}}$ and $\models_{fair}$, instead of $\models_{Sched_{fair}}$.

We now give an example of a *PNS*. We depict a *PNS* as follows. We use ellipses for the states. Thick arrows stand for the outgoing transitions from a state. A transition $\mu \in Steps(s) \setminus \{\mu_t^1 : t \in S\}$ is represented by a thick arrow that ends in a small circle, representing the probabilistic choice. We use directed thin arrows leading from the circle of a probabilistic choice to the possible

---

[12] Note, that $Paths_{fair} \subseteq Paths_{inf}$.

successor states (i.e. all states t where $\mu(t) > 0$). A distribution $\mu_t^1 \in Steps(s)$ is represented by a thick arrow leading from s to t. The starting state has an extra incoming arrow and the state labelling is written in brackets.

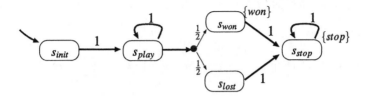

Fig. 6. The roulette player

Figure 6 shows a simplified roulette player who keeps placing bets until his wife comes to the casino. That explains the nondeterministic choice in the state s_{play} (i.e. $Steps(s_{play}) = \{\mu_{s_{play}}^1, \mu'\}$, with $\mu'(s_{won}) = \mu'(s_{lost}) = \frac{1}{2}$). We have $L(s_{won}) = won$, $L(s_{stop}) = stop$ and $L(s') = \emptyset$ for the other states.

Observe that starting from state s_{init}, $\pi = s_{init}, s_{play}, s_{play}, s_{play}, \ldots$ is the only path that is not fair. So if the system starts in s_{init} and follows a fair scheduler, it will eventually reach the state s_{stop}. Considering the formula

$$\Phi = [\Diamond(won \wedge \mathcal{X} stop)]_{\geq 0.5}$$

it is easy to see that

$$s_{init} \not\models_{all} \Phi \qquad \text{but} \qquad s_{init} \models_{fair} \Phi.$$

3.2   Model Checking of *PCTL* Against a *PNS*

Given a *PNS* $T = (S, Steps, AP, L, \bar{s})$, a set *Sched* of schedulers for T and a *PCTL* state formula Φ, we want to decide, which states in S satisfy the formula Φ with respect to the schedulers under consideration[13]. In this section we present a model checking algorithm for this problem which is very similar to the one for *PDS* (see section 2.4). As in section 2.4, we first build the parse tree of the given formula Φ. We then process the parse tree bottom up, computing the satisfaction sets of the subformulae of Φ (i.e. $Sat(\Phi') = \{s \in S : s \models_{Sched} \Phi'\}$, where Φ' is a subformula of Φ).

As before we have $Sat(true) = S$, $Sat(a) = \{s \in S : L(s) = a\}$, $Sat(\neg\Phi') = S \setminus Sat(\Phi')$ and $Sat(\Phi_1 \wedge \Phi_2) = Sat(\Phi_1) \cap Sat(\Phi_2)$ (independent of the chosen set *Sched* of schedulers). So the only case left is $\Phi' = [\phi]_{\bowtie p}$, where ϕ is a *PCTL* path formula, i.e. $\phi = \mathcal{X}\Phi'$ or $\phi = \Phi_1 \mathcal{U}^{\leq t}\Phi_2$ or $\phi = \Phi_1 \mathcal{U}\Phi_2$, with $t \in \mathbb{N}$. The cases concerning the next-step and bounded-until operator will be explained in the following section and how to deal with the unbounded-until operator will be explained afterwards.

---

[13] Remember, that we are only interested in $Sched \in \{Sched_{all}, Sched_{fair}\}$.

Next-Step and Bounded-Until Operator

We first describe the procedure for $s \models_{all} [\mathcal{X}\Phi']_{\bowtie p}$ in detail. Following the semantics we have

$$s \models_{all} [\mathcal{X}\Phi']_{\bowtie p} \Leftrightarrow prob_s^B(\{\pi \in Paths_{inf}^B(s) : \pi^1 \in Sat(\Phi')\}) \bowtie p \; \forall B \in Sched_{all}.$$

For $B \in Sched_{all}$ we have

$$prob_s^B(\{\pi \in Paths_{inf}^B(s) : \pi^1 \in Sat(\Phi')\}) = \sum_{s' \in Sat(\Phi')} B(\pi^0)(s').$$

Since every possible distribution in $Steps(s)$ can be chosen by a scheduler and since we have to quantify over all schedulers, we get

$$s \models_{all} [\mathcal{X}\Phi']_{\trianglerighteq p} \quad \Leftrightarrow \quad \min_{\mu \in Steps(s)} \sum_{s' \in Sat(\Phi')} \mu(s') \trianglerighteq p,$$

where the latter can easily be checked algorithmically. Observe that the case $s \models_{all} [\mathcal{X}\Phi']_{\trianglelefteq p}$ is dealt with in the same way by replacing min by max.

Due to the fact that each mapping

$$B : \{\rho \in Paths_{fin} : |\rho| \le t\} \longrightarrow \cup_{s \in S} Steps(s) \text{ with } B(\rho) \in Steps(last(\rho))$$

can be extended in a fair way, we get the same results for $s \models_{Sched} [\mathcal{X}\Phi']_{\bowtie p}$ and $s \models_{Sched} [\Phi_1 \mathcal{U}^{\le t}\Phi_2]_{\bowtie p}$ independent of what $Sched$ was chosen to be. That gives

Lemma 4.

$$s \models_{Sched} [\mathcal{X}\Phi']_{\trianglerighteq p} \quad \Leftrightarrow \quad min_{\mu \in Steps(s)} \sum_{s' \in Sat(\Phi')} \mu(s') \trianglerighteq p$$

$$s \models_{Sched} [\mathcal{X}\Phi']_{\trianglelefteq p} \quad \Leftrightarrow \quad max_{\mu \in Steps(s)} \sum_{s' \in Sat(\Phi')} \mu(s') \trianglelefteq p$$

for $Sched \in \{Sched_{all}, Sched_{fair}\}$.

Having taken care of the case $[\mathcal{X}\Phi']_{\bowtie p}$, we now turn our attention to the case $[\Phi_1 \mathcal{U}^{\le t}\Phi_2]_{\bowtie p}$ for which we give a recursive computation similar to the one in section 2.4 about PDS. The difference is, that this time we have to ensure to make the "worst" choice (of scheduler) in each step. The following lemma shows precisely what this means.

Lemma 5.
Let $Sched \in \{Sched_{all}, Sched_{fair}\}$. For all $s \in S$ and $j \in \mathbb{N}_0$ we define $p_{s,j}^{max}$ and $p_{s,j}^{min}$ as follows:

- *if $s \in Sat(\Phi_2)$, then $p_{s,j}^{max} = p_{s,j}^{min} = 1 \; \forall j \in \mathbb{N}_0$*
- *if $s \notin Sat(\Phi_1) \cup Sat(\Phi_2)$, then $p_{s,j}^{max} = p_{s,j}^{min} = 0 \; \forall j \in \mathbb{N}_0$*

- *if $s \in Sat(\Phi_1) \setminus Sat(\Phi_2)$, then $p_{s,0}^{max} = p_{s,0}^{min} = 0$,*

$$p_{s,j}^{min} = min_{\mu \in Steps(s)} \sum_{t \in S} \mu(t) \cdot p_{t,(j-1)}^{min},$$

$$p_{s,j}^{max} = max_{\mu \in Steps(s)} \sum_{t \in S} \mu(t) \cdot p_{t,(j-1)}^{max}.$$

Then, for all $s \in S$:

$$s \models_{Sched} [\Phi_1 \mathcal{U}^{\leq t} \Phi_2]_{\unrhd p} \quad iff \quad p_{s,t}^{min} \unrhd p \quad and \quad s \models_{Sched} [\Phi_1 \mathcal{U}^{\leq t} \Phi_2]_{\unlhd p} \quad iff \quad p_{s,t}^{max} \unlhd p$$

Proof:
It can be shown by induction on j, that

$$p_{s,j}^{extr} = extr_{B \in Sched} prob_s^B \{\pi \in Paths_{inf}^B(s) \ : \ \pi \models_{Sched} \Phi_1 \mathcal{U}^j \Phi_2\},$$

where $extr \in \{min, max\}$, thus yielding the claim.

The Unbounded-Until Operator. In contrary to the Next-Step and Bounded-Until operators, we have to distinguish different cases for the satisfaction relation of the Unbounded-Until operator according to the chosen set of schedulers.

So we first consider the case where the chosen set of schedulers is $Sched_{all}$. Given a state s we want to decide whether s satisfies $[\Phi_1 \mathcal{U} \Phi_2]_{\bowtie p}$, which is equivalent to

$$p_s^B(\Phi_1 \mathcal{U} \Phi_2) := prob_s^B(\{\pi \in Paths_{inf}^B(s) : \pi \models_{all} \Phi_1 \mathcal{U} \Phi_2\}) \bowtie p \quad \forall B \in Sched_{all}.$$

By the results of [9] we know that

$$sup\{p_s^B(\Phi_1 \mathcal{U} \Phi_2) : B \in Sched_{all}\} = max\{p_s^B(\Phi_1 \mathcal{U} \Phi_2) : B \in Sched_{simple}\} \text{ and}$$

$$inf\{p_s^B(\Phi_1 \mathcal{U} \Phi_2) : B \in Sched_{all}\} = min\{p_s^B(\Phi_1 \mathcal{U} \Phi_2) : B \in Sched_{simple}\}.[14]$$

From now on we denote $max\{p_s^B(\Phi_1 \mathcal{U} \Phi_2) : B \in Sched_{simple}\}$ by $p_s^{max}(\Phi_1 \mathcal{U} \Phi_2)$ and $min\{p_s^B(\Phi_1 \mathcal{U} \Phi_2) : B \in Sched_{simple}\}$ by $p_s^{min}(\Phi_1 \mathcal{U} \Phi_2)$.

Since $Sched_{simple} \subseteq Sched_{all}$, we know that the supremum and infimum are really maximum and minimum. This fact gives rise to the following lemma.

Lemma 6.

$$s \models_{all} [\Phi_1 \mathcal{U} \Phi_2]_{\unlhd p} \quad iff \quad p_s^{max}(\Phi_1 \mathcal{U} \Phi_2) \unlhd p \quad iff \quad s \models_{simple} [\Phi_1 \mathcal{U} \Phi_2]_{\unlhd p}$$
$$s \models_{all} [\Phi_1 \mathcal{U} \Phi_2]_{\unrhd p} \quad iff \quad p_s^{min}(\Phi_1 \mathcal{U} \Phi_2) \unrhd p \quad iff \quad s \models_{simple} [\Phi_1 \mathcal{U} \Phi_2]_{\unrhd p} \quad [15]$$

---

[14] Note, that the set $Sched_{simple}$ is finite.
[15] This does not hold for $PCTL^*$ formulae. For a counterexample see section 4.

We will describe later (see section 3.2) how to compute the values $p_s^{max}(\Phi_1\mathcal{U}\Phi_2)$ and $p_s^{min}(\Phi_1\mathcal{U}\Phi_2)$.

Now we consider the case where the chosen set of schedulers is $Sched_{fair}$. Given a state s, we want to decide, whether

$$prob_s^B(\{\pi \in Paths_{inf}^B(s) : \pi \models_{fair} \Phi_1\mathcal{U}\Phi_2\}) \bowtie p \quad \forall B \in Sched_{fair}.$$

Here we have to deal with $\bowtie \in \{\geq, >\}$ and $\bowtie \in \{\leq, <\}$ in two different ways.

– $\bowtie \in \{\leq, <\}$: Since $p_s^{max}(\Phi_1\mathcal{U}\Phi_2) = max\{p_s^B(\Phi_1\mathcal{U}\Phi_2) : B \in Sched_{all}\}$ (see above), it obviously holds that

$$p_s^{max}(\Phi_1\mathcal{U}\Phi_2) \geq p_s^F(\Phi_1\mathcal{U}\Phi_2) \quad \forall F \in Sched_{fair}. \tag{1}$$

Let now B be a scheduler such that $p_s^B(\Phi_1\mathcal{U}\Phi_2) = p_s^{max}(\Phi_1\mathcal{U}\Phi_2)$. It can be shown (see [10]) that there is a fair scheduler F_B with

$$p_s^{F_B}(\Phi_1\mathcal{U}\Phi_2) \geq p_s^B(\Phi_1\mathcal{U}\Phi_2) = p_s^{max}(\Phi_1\mathcal{U}\Phi_2). \tag{2}$$

From (1) and (2) it follows, that

$$\sup\{p_s^B(\Phi_1\mathcal{U}\Phi_2) : B \in Sched_{fair}\} = \max\{p_s^B(\Phi_1\mathcal{U}\Phi_2) : B \in Sched_{fair}\} =$$

$$p_s^{max}(\Phi_1\mathcal{U}\Phi_2),$$

which yields

Lemma 7.

$$s \models_{fair} [\Phi_1\mathcal{U}\Phi_2]_{\trianglelefteq p} \quad iff \quad p_s^{max}(\Phi_1\mathcal{U}\Phi_2)\trianglelefteq p.$$

– $\bowtie \in \{\geq, >\}$: This case differs a little from the others. First, we give an example that shows that this case cannot be handled like the other cases. Consider the following $PNS\ \mathcal{T} = (\{s,t\}, Steps, \{a,b\}, L, s)$ with $Steps(s) = \{\mu_s^1, \mu_t^1\}$, $Steps(t) = \{\mu_t^1\}$, $L(s) = \{a\}$ and $L(t) = \{b\}$.

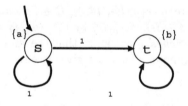

For the simple scheduler B with $B(s) = \mu_s^1$ (and $B(t) = \mu_t^1$) we get $p_s^B(a\mathcal{U}b) = 0$, thus $p_s^{min}(a\mathcal{U}b) = 0$. On the other hand we have $p_s^F(a\mathcal{U}b) = 1$ for each fair scheduler F and therefore $s \models_{fair} [a\mathcal{U}b]_{\geq 1}$. The problem in this example is, that the simple scheduler B with $B(s) = \mu_s^1$ forces the process to stay forever in the "non-successful" state s from which the "successful" state t can be reached. In contrary to this it can be shown for fair schedulers, that with probability 1 all states that are reachable from a state that is visited infinitely often, are also visited infinitely often. This explains, why $p_s^{min}(a\mathcal{U}b)$ cannot be "approximated" by fair schedulers. We now introduce some notation we will need to explain how to decide whether $s \models_{fair} [\Phi_1\mathcal{U}\Phi_2]_{\trianglerighteq p}$.

Definition 26. [**The Sets $U_{>0}$ and $U_?$ and the Corresponding Atomic Propositions**]
Let $U_{>0}$ be the set of states for which there is a scheduler B, such that $p_s^B(\Phi_1\mathcal{U}\Phi_2) > 0$, i.e.

$$U_{>0} = \{s \in S : p_s^{max}(\Phi_1\mathcal{U}\Phi_2) > 0\}.$$

Let $U_? = U_{>0} \setminus Sat(\Phi_2)$.
Extend the set AP by two new atomic propositions $a_{>0}$ and $a_?$ and the labelling by

$$a_{>0} \in L(s) \quad \Leftrightarrow \quad s \in U_{>0} \quad and \quad a_? \in L(s) \quad \Leftrightarrow \quad s \in U_?.$$

So $s \in U_{>0}$ if and only if starting from s it is possible to satisfy the formula $\Phi_1\mathcal{U}\Phi_2$.
For each fair scheduler F and each state s, it can be shown (see [10]), that

$$prob_s^F(\{\pi \in Paths_{inf}^F(s) : \pi \models_{fair} \Phi_1\mathcal{U}\Phi_2 \text{ or } \pi \models_{fair} true\mathcal{U}\neg a_{>0}\}) = 1.$$

That means, if the process follows the fair scheduler F, then with probability 1 the process will either satisfy the formula $\Phi_1\mathcal{U}\Phi_2$ or will eventually reach a state that is not contained in $U_{>0}$. In other words, the set of infinite paths complying with the scheduler F and staying in $U_?$ forever, is a nullset (with respect to the measure defined by F). Since the set $\{\pi \in Paths_{inf}^F(s) : \pi \models_{fair} \Phi_1\mathcal{U}\Phi_2 \text{ or } \pi \models_{fair} true\mathcal{U}\neg a_{>0}\}$ is the disjoint union of the two sets $\{\pi \in Paths_{inf}^F(s) : \pi \models_{fair} \Phi_1\mathcal{U}\Phi_2\}$ and $\{\pi \in Paths_{inf}^F(s) : \pi \models_{fair} a_?\mathcal{U}\neg a_{>0}\}$, it follows from the above, that

$$prob_s^F(\{\pi \in Paths_{inf}^F(s) : \pi \models_{fair} \Phi_1\mathcal{U}\Phi_2\}) =$$

$$1 - prob_s^F(\{\pi \in Paths_{inf}^F(s) : \pi \models_{fair} a_?\mathcal{U}\neg a_{>0}\}).$$

So we get

$$s \models_{fair} [\Phi_1\mathcal{U}\Phi_2]_{\trianglerighteq p} \quad \Leftrightarrow \quad p_s^F(\Phi_1\mathcal{U}\Phi_2) \trianglerighteq p \;\; \forall F \in Sched_{fair} \quad \Leftrightarrow$$

$$p_s^F(a_?\mathcal{U}\neg a_{>0}) \trianglelefteq 1 - p \;\; \forall F \in Sched_{fair} \quad \Leftrightarrow \quad p_s^{max}(a_?\mathcal{U}\neg a_{>0}) \trianglelefteq 1 - p,$$

which leads to the following lemma

Lemma 8.

$$s \models_{fair} [\Phi_1\mathcal{U}\Phi_2]_{\trianglerighteq p} \quad iff \quad p_s^{max}(a_?\mathcal{U}\neg a_{>0}) \trianglelefteq 1 - p.$$

The only remaining question is how to efficiently compute the set $U_{>0}$, $U_?$ and the values $p_s^{max}(\Phi_1\mathcal{U}\Phi_2)$ and $p_s^{min}(\Phi_1\mathcal{U}\Phi_2)$, if the sets $Sat(\Phi_1)$ and $Sat(\Phi_2)$ have already been computed.

The Computation of $p_s^{max}(\Phi_1 \mathcal{U} \Phi_2)$, $p_s^{min}(\Phi_1 \mathcal{U} \Phi_2)$ and the set $U_{>0}$

Recall, that

$$U_{>0} = \{s \in S : p_s^{max}(\Phi_1 \mathcal{U} \Phi_2) > 0\}.$$

So we could compute $U_{>0}$, by computing the values $p_s^{max}(\Phi_1 \mathcal{U} \Phi_2)$ for all $s \in S$. But it is also easy to see, that $s \in U_{>0}$ if and only if starting from s, one can reach a state in $Sat(\Phi_2)$ via a path in $Sat(\Phi_1)$, i.e. $\exists \rho \in Paths_{fin}(s)$ such that $last(\rho) \in Sat(\Phi_2) \wedge \rho^i \in Sat(\Phi_1)$, $i = 0, 1, \ldots, last(\rho) - 1$.

So a more efficient way for computing $U_{>0}$ is to perform a backwards reachability analysis from the states in $Sat(\Phi_2)$ within the states of $Sat(\Phi_1)$. Making this idea precise, we define the directed graph $G_{>0}(\Phi_1 \mathcal{U} \Phi_2) = (S, E_{>0})$, where

$$E_{>0} = \{(s,t) \in S \times S : t \in Sat(\Phi_1) \setminus Sat(\Phi_2) \wedge \exists \mu \in Steps(t) : \mu(s) > 0\}.$$

Then

$$U_{>0} = \{s \in S : s \text{ can be reached in } G_{>0}(\Phi_1 \mathcal{U} \Phi_2) \text{ from a state } s \in Sat(\Phi_2)\},$$

so $U_{>0}$ can be computed by a depth first search in $G_{>0}(\Phi_1 \mathcal{U} \Phi_2)$.

To compute $p_s^{max}(\Phi_1 \mathcal{U} \Phi_2), s \in S$, one can use linear programming techniques. It can be shown (see [9, 14]), that the values $p_s^{max}(\Phi_1 \mathcal{U} \Phi_2), s \in S$ are the unique solution of the linear minimization problem

$$x_s = 1 \quad \text{if } s \in Sat(\Phi_2)$$
$$x_s = 0 \quad \text{if } s \notin U_{>0}$$
$$x_s \geq \sum_{t \in S} \mu(t) \cdot x_t \quad \text{if } s \in Sat(\Phi_1) \setminus Sat(\Phi_2), \quad \mu \in Steps(s),$$

where $\sum_{s \in S} x_s$ is minimized.

This problem can be solved by well known methods like the ellipsoid method or the simplex method. For $p_s^{min}(\Phi_1 \mathcal{U} \Phi_2)$ one gets an analogous linear maximization problem.

For reasons of completeness we also want to mention that one can use the characterization of the function $s \mapsto p_s^{max}(\Phi_1 \mathcal{U} \Phi_2)$ as the least fixed points of certain operators on certain function spaces and compute (resp. approximate) those values with iterative methods (see [4]).

The Algorithm

Now we have everything on hand to present a model checking algorithm for *PCTL* formulae against a *PNS*. As mentioned, one can build the parse tree of the given formula (or the parse graph to avoid multiple computations) and compute the satisfaction sets for the subformulae in a bottom up manner. In algorithm 3 and 4 we present a method that calculates the satisfaction sets recursively, so in an actual implementation one had to ensure that there were no multiple calls for syntactically identical subformulae.

The Complexity of *PCTL* Model Checking Aainst a *PNS*

In this paragraph we briefly estimate the time and space complexity of our model checking algorithm. Given a *PNS* T and a *PCTL* state formula Φ we denote by n the number of states of T, i.e. $n = |S|$ and by m the number of transitions of T, i.e. $m = \sum_{s \in S} |Steps(s)|$. The number of subformulae of Φ is linear in $|\Phi|$.

Algorithm 3 PCTL_MODELCHECK

Input: PNS $\mathcal{T} = (S, Steps, AP, L, \bar{s})$, $PCTL$ state formula Φ over AP, $sched \in \{all, fair\}$

Output: truth value of $\bar{s} \models_{sched} \Phi$

IF $\bar{s} \in Sat(\Phi, sched)$ **THEN**
    return $true$
ELSE
    return $false$

- For a subformula Φ', where Φ' is $true$, an atomic proposition a or of the form $\neg\Phi''$ or $\Phi_1 \wedge \Phi_2$, the set $Sat(\Phi')$ can be computed in time $\mathcal{O}(n)$.
- If the subformula Φ' is of the form $[\mathcal{X}\Phi'']_{\bowtie p}$, the set $Sat(\Phi')$ can be computed in time $\mathcal{O}(m \cdot n)$, since for every $s \in S$ we have to compute the sum $\sum_{t \in Sat(\Phi'')} \mu(t)$ for every $\mu \in Steps(s)$.
- If the subformula Φ' is of the form $[\Phi_1 \mathcal{U}^{\leq t}\Phi_2]_{\bowtie p}$, the set $Sat(\Phi')$ can be computed in time $\mathcal{O}(t \cdot m \cdot n)$ (see algorithm).
- If the subformula Φ' is of the form $[\Phi_1 \mathcal{U}\Phi_2]_{\bowtie p}$, the set $Sat(\Phi')$ can be computed in time $\mathcal{O}(poly(m, n))$. As mentioned above, $p_s^{max}(\Phi_1 \mathcal{U}\Phi_2)$, resp. $p_s^{min}(\Phi_1 \mathcal{U}\Phi_2)$ can be computed with well known methods (e.g. Ellipsoid-, Karmakar-method), which need polynomial time in the size of the problem. The needed graph analyses have time complexity $\mathcal{O}(m \cdot n)$.

Summing up over all subformulae of Φ, we get, that the model checking algorithms needs time

$$\mathcal{O}(|\Phi| \cdot (t_\Phi \cdot m \cdot n + p(m, n))),$$

where t_Φ stands for the maximal t in subformulae of the form $\Phi_1 \mathcal{U}^{\leq t}\Phi_2$[16] and $p(n, m)$ is a polynomial that represents the costs for solving the linear optimization problems.

For the estimation of the space complexity we assume w.l.o.g. that $n \leq m$. To represent the system $\mathcal{T}$ itself, we need space $\mathcal{O}(m \cdot n)$ (neglecting the representation for the labelling function L). For each subformula Φ', the set $Sat(\Phi')$ requires space $\mathcal{O}(n)$. The needed graph analyses require space $\mathcal{O}(m \cdot n)$ and for the computation of $p_s^{max}(\Phi_1 \mathcal{U}\Phi_2)$, resp. $p_s^{min}(\Phi_1 \mathcal{U}\Phi_2)$ space $\mathcal{O}(n^2)$ is needed.

Altogether this leads us to the following theorem:

Theorem 2.
Let $\mathcal{T}$ be a PNS and Φ a PCTL state formula over the set of atomic propositions of $\mathcal{T}$. Then $Sat(\Phi)$ can be computed in time and space linear in the size of Φ and t_ϕ[17] and polynomial in the size of $\mathcal{T}$ if the set of schedulers under consideration is $Sched_{all}$ or $Sched_{fair}$.

---

[16] If there is no such subformula in Φ, then t_Φ is 1.

[17] Where t_Φ stands for the maximal t in subformulae of the form $\Phi_1 \mathcal{U}^{\leq t}\Phi_2$ (and is at least 1).

Algorithm 4 $Sat(\Phi, sched)$

Input: Y
Output: Set of all $s \in S$ that satisfy Φ with respect to the chosen set of schedulers

---

__CASE__ Φ is

$true$: return S

a : return $\{s \in S : a \in L(s)\}$

$\Phi_1 \wedge \Phi_2$: return $Sat(\Phi_1, sched) \cap Sat(\Phi_2, sched)$

$\neg\Phi'$: return $S \setminus Sat(\Phi', sched)$

$[\mathcal{X}\Phi']_{\unrhd p}$: compute $Sat(\Phi', sched)$;
$\quad \forall s \in S:$ compute $p_{\mathcal{X}}^{min}(s) = \min_{\mu \in Steps(s)} \sum_{t \in Sat(\Phi', sched)} \mu(t)$;
$\quad$ return $\{s \in S : p_{\mathcal{X}}^{min}(s) \unrhd p\}$

$[\mathcal{X}\Phi']_{\unlhd p}$: compute $Sat(\Phi', sched)$;
$\quad \forall s \in S:$ compute $p_{\mathcal{X}}^{max}(s) = \max_{\mu \in Steps(s)} \sum_{t \in Sat(\Phi', sched)} \mu(t)$;
$\quad$ return $\{s \in S : p_{\mathcal{X}}^{max}(s) \unlhd p\}$

$[\Phi_1 \mathcal{U}^{\leq t} \Phi_2]_{\unrhd p}$: compute $Sat(\Phi_1, sched)$ and $Sat(\Phi_2, sched)$;
$$\forall s \in S : p_0^{min}(s) = \begin{cases} 1 & \text{if } s \in Sat(\Phi_2, sched) \\ 0 & \text{otherwise} \end{cases}$$
$\quad$ __FOR__ $i = 1$ to t __DO__
$$\forall s \in S : p_i^{min}(s) = \begin{cases} 0, \text{ if } s \notin Sat(\Phi_1, sched) \cup Sat(\Phi_2, sched), \\ 1, \text{ if } s \in Sat(\Phi_2, sched), \\ \min_{\mu \in Steps(s)} \sum_{t \in S} \mu(t) \cdot p_{i-1}^{min}(t) \text{ otherwise} \end{cases}$$
$\quad$ __OD__
$\quad$ return $\{s \in S : p_t^{min}(s) \unrhd p\}$

$[\Phi_1 \mathcal{U}^{\leq t} \Phi_2]_{\unlhd p}$: as above, replace "$\unrhd$" by "$\unlhd$" and "min" by "max"

$[\Phi_1 \mathcal{U} \Phi_2]_{\unlhd p}$: compute $Sat(\Phi_1, sched)$ and $Sat(\Phi_2, sched)$;
$\quad$ compute $p_s^{max}(\Phi_1 \mathcal{U} \Phi_2)$ by solving a linear minimization problem;
$\quad$ return $\{s \in S : p_s^{max}(\Phi_1 \mathcal{U} \Phi_2) \unlhd p\}$

$[\Phi_1 \mathcal{U} \Phi_2]_{\unrhd p}$: compute $Sat(\Phi_1, sched)$ and $Sat(\Phi_2, sched)$;
$\quad$ __IF__ $sched = $ "all" __THEN__
$\quad\quad$ compute $p_s^{min}(\Phi_1 \mathcal{U} \Phi_2)$;
$\quad\quad$ return $\{s \in S : p_s^{min}(\Phi_1 \mathcal{U} \Phi_2) \unrhd p\}$
$\quad$ __ELSE__ $\quad$ (*$sched = $ "fair"*)
$\quad\quad$ compute $U_{>0}$, $U_?$ and the labelling for $a_{>0}$, $a_?$;
$\quad\quad$ compute $p_s^{max}(a_? \mathcal{U} \neg a_{>0})$;
$\quad\quad$ return $\{s \in S : p_s^{max}(a_? \mathcal{U} \neg a_{>0}) \unlhd 1 - p\}$
$\quad$ __ENDIF__

__END CASE__

Remark 8.

In [10] another classification of schedulers is considered, namely strictly fair schedulers. We call a scheduler B strictly fair if and only if all infinite paths that comply with B are fair. By $Sched_{sfair}$ we denote the set of strictly fair schedulers. For $\models_{Sched_{sfair}}$ some of the above calculations (the Unbounded-Until cases) are a little bit more complicated than for the fair schedulers (for details see [10]). However the results of theorem 2 still hold, if the set of schedulers under consideration is $Sched_{sfair}$.

3.3   Model Checking of $PCTL^*$ Against a PNS

As in the deterministic case (see section 2.5) the model checking of PNS is extended to $PCTL^*$-formulae as well. In this section we explain how to model check a PNS against a $PCTL^*$-formula. Throughout this section let $T = (S, Steps, AP, L, \bar{s})$ be a PNS. It follows from remark 4 that it is sufficient to consider formulae which do not make use of $[.]_{>p}$ and $[.]_{\geq p}$. The basic procedure is similar to the one in section 2.5. Let Φ be a $PCTL^*$ state formula. We only consider Φ to be of the form $\Phi = [\phi]_{\trianglelefteq p}$ since the other cases are either trivial or can be reduced to this one (see above). As in the deterministic case we assume, that we have already computed (recursively) the satisfaction sets of all maximal state $PCTL^*$-subformulae $\gamma_1, \ldots, \gamma_n$ of ϕ, so we can view them as atomic propositions. Let ϕ' be the formula we get from ϕ by rewriting all subformulae involving $\mathcal{U}^{\leq t}$ as an expression of nested Next Step operators as explained in remark 5 in section 2.5. Then ϕ' is a LTL formula over $\{\gamma_1, \ldots, \gamma_n\}$. [9] presents a model checking algorithm for $pCTL^*$ (which essentially coincides with our $PCTL^*$) that uses a normal form for the LTL formula ϕ'. In [18, 17], an alternative method is considered, which uses the representation of LTL formulae by ω-automata and runs in time polynomial in the size of the system and doubly-exponential in the size of the formula (thus, this method meets the lower bound presented in [15]). In [9] and [18, 17] only the relation $\models_{all}$ is considered. We will first present briefly the model checking algorithm of [18, 17] and then explain, how it can be modified to handle fairness (see [10]).

   The first step is the same for both cases ($\models_{all}$ and $\models_{fair}$). As in section 2.6, we construct a deterministic Rabin automaton $\mathcal{A} = (\Sigma, Q, \bar{q}, \rho, \alpha)$ such that $\mathcal{L}_\omega(\mathcal{A}) = \mathcal{L}_\omega(\phi')$. Recall that the size of the automaton $\mathcal{A}$ is doubly exponential in $|\phi'| = O(\Pi_{i=1}^m t_i \cdot |\phi|)^{18}$ and that a run r of $\mathcal{A}$ is accepting if and only if there exists $(L, U) \in \alpha$, such that $lim(r) \cap L \neq \emptyset$ and $lim(r) \cap U = \emptyset$. After constructing the automaton $\mathcal{A}$, we build the product PNS $T' = T \times \mathcal{A} = (S \times Q, Steps', AP, L', (\bar{s}, \bar{q}))$, where

---

[18] where m is the number of occurences of $\mathcal{U}^{\leq t}$ in the formula ϕ and $t_1, \ldots, t_m$ are the corresponding bounds

- $\mu' \in Steps'((s,q))$ $\iff$
  $\exists\, \mu \in Steps(s)$, s.t. $\mu'((t,p)) = \mu(t)$, if $p = \rho(q, L(s))$ and 0 otherwise.
- $L'((s,q)) = L(s)$.

Let α' be the lifted acceptance condition from $\mathcal{A}$, i.e.

$$\alpha' = \{(S \times L, S \times U) \;:\; (L, U) \in \alpha\}.$$

As in the deterministic case we get a one-to-one correspondence between the infinite paths of $\mathcal{T}$ and $\mathcal{T}'$ by path lifting which works as follows:

- First we embed S into $S \times Q$ by $s \mapsto (s, q_s)$, where $q_s = \rho(\bar{q}, L(s))$.
- Let $\pi = s_0 \xrightarrow{\mu_1} s_1 \xrightarrow{\mu_2} s_2 \ldots$ be an infinite path of $\mathcal{T}$. The induced infinite paths π' of $\mathcal{T}'$ is given by

$$\pi' = (s_0, p_0) \xrightarrow{\mu_1'} (s_1, p_1) \xrightarrow{\mu_2'} (s_2, p_2) \ldots,$$

where $p_0 = q_s$, $p_{i+1} = \rho(p_i, L(s_i))$ and

$$\mu_i'((t,p)) = \begin{cases} \mu_i(t), & \text{if } p = \rho(p_{i-1}, L(s_{i-1})) \\ 0, & \text{otherwise.} \end{cases}$$

The Case $\models_{all}$

This case is very similar to the deterministic case, where accepting ergodic sets played an important role. Since now we have to quantify over all schedulers, we give the definition of a *maximal end component*, which is the pendant to an ergodic set in a Markov chain.

Definition 27. [Sub-PNS]
Given a PNS $\mathcal{T} = (S, Steps, AP, L, \bar{s})$, a sub-PNS of $\mathcal{T}$ is a tuple $\mathcal{T}' = (S', Steps')$ such that

- $S' \subseteq S$ and
- $Steps'(s') \subseteq Steps(s')$ $\forall s' \in S'$.

Note, that a *sub-PNS* $\mathcal{T}'$ of a *PNS* $\mathcal{T}$ (even when equipped with a state labelling and a starting state) is not a *PNS*, because $Steps'$ maps a state s' from $\mathcal{T}'$ to a probability distribution on the states of $\mathcal{T}$, so it can happen, that $supp(\mu) \supset S'$ for some $\mu \in Steps'(s')$ for some $s' \in S'$.

Definition 28. [End Component]
We call a sub-PNS $\mathcal{T}' = (S', Steps')$ an end component if :

- $\forall s' \in S' \;:\; supp(\mu) \subseteq S'$ $\forall \mu \in Steps'(s')$.
- *The graph $(S', E_{\mathcal{T}'})$ is strongly connected,*
  where $E_{\mathcal{T}'} = \{(s', t') : s', t' \in S' \text{ and } \exists \mu \in Steps'(s') : \mu(t') > 0\}$.

Note that because of the first requirement, an end component is (accept for the lacking state labelling and the missing starting state) a *PNS*. The second requirement ensures, that for each pair (s', t') of states, there is always a scheduler, such that there is a finite path from s' to t'. It also holds, that given any starting state and a scheduler of a *PNS*, the induced process will end up with probability 1 in an end component (thus the name "end component")(see [18]).

We say that an end component $(S', Steps')$ is contained in another end component $(S'', Steps'')$ if

- $S' \subseteq S''$ and
- $Steps'(s') \subseteq Steps''(s')$ $\forall s' \in S'$.

Definition 29. [Maximal End Component]
An end component of a PNS T is called maximal if it is not contained in any other end component of T.

As said before, maximal end components are the pendant to ergodic sets in Markov chains. So the rest of the procedure is very similar to the deterministic case.

We call a maximal end component $(S', Steps')$ of T' accepting, if S' satisfies the acceptance condition α'. Let

$$Acc' = \{S' \;:\; (S', Steps') \text{ is an accepting maximal end component of } T'\}.$$

Let acc be a new atomic proposition such that $acc \in L'(s')$ if and only if $s' \in Acc'$. One can then show (see [18]), that

$$sup_{B \in Sched_{all}} \; prob_s^B(\{\pi \;:\; \pi \in Paths_{inf}^B(s) : \pi \models_{all} \phi'\}) \; =$$

$$sup_{B' \in Sched_{all}} \; prob_{(s,s_q)}^{B'}(\{\pi' \;:\; \pi' \in Paths_{inf}^{B'}((s, q_s)) : \pi' \models_{all} \Diamond acc\}),$$

where the latter quantity can be computed by means of section 3.2. This leads to the following lemma.

Lemma 9.

$$s \models_{all} [\phi']_{\unlhd p} \quad \textit{iff} \quad (s, q_s) \models_{all} [\Diamond acc]_{\unlhd p}.$$

So we get

$$Sat(\Phi) \;=\; \{s \in S \;:\; p_{(s,q_s)}^{max}(\Diamond acc) \unlhd p\}.$$

The Case $\models_{fair}$
Not only do we have a one-to-one correspondence between the infinite paths of T and T' but it holds also, that the infinite path π of T is fair if and only if the corresponding path π' of T' is fair. We even get that each fair scheduler B of T induces a fair scheduler B' of T', such that $Paths_{inf}^{B'} = \{\pi' \;:\; \pi \in Paths_{inf}^B\}$. In

addition, the associated probability spaces on $Paths_{inf}^B(s)$ and $Paths_{inf}^{B'}((s,q_s))$ are isomorphic.

For each $(L',U') \in \alpha'$, let $Acc'_{(L',U')}$ be the union of all subsets T' of $S \times Q$ such that $T' \cap U' = \emptyset$ and for all $t' \in T'$ the following hold:

- $supp(\mu') \subseteq T'$ for all $\mu' \in Steps(t')$
- $reach(t') \cap L' \neq \emptyset$.[19]

Let $Acc' = \cup_{(L',U') \in \alpha'} Acc'_{(L',U')}$ and let acc be an atomic proposition such that $acc \in L'(s')$ if and only if $s' \in Acc'$.

It then follows (see [10]) for all fair schedulers F' of T', that

$$prob_{s'}^{F'}(\Diamond acc) = prob_{s'}^{F'}(\{\pi' \in Paths_{inf}^{F'}(s') \; : \; L'(\pi'^0), L'(\pi'^1), \ldots \in \mathcal{L}_\omega(\mathcal{A})\}).$$

Therefore we get the following lemma.

Lemma 10.

$$s \models_{fair} [\phi']_{\unlhd p} \quad \Longleftrightarrow \quad (s, q_s) \models_{fair} [\Diamond acc]_{\unlhd p}.$$

So in order to compute the set $Sat(\Phi) = Sat([\phi]_{\unlhd p} = Sat([\phi']_{\unlhd p})$, we have to compute the values $p_{(s,q_s)}^{max}(\Diamond acc)$ for T' by means of section 3.2, i.e.

$$Sat(\Phi) = \{s \in S \; : \; p_{(s,q_s)}^{max}(\Diamond acc) \unlhd p\}.$$

4   Concluding Remarks

In this paper we gave an overview of the logics $PCTL$ and $PCTL^*$ and model checking algorithms for deterministic and nondeterministic (finite-state) probabilistic systems. We followed the state-labelled approach. Action-labelled variants of $PCTL/PCTL^*$ have been investigated as well, see e.g. [46, 44, 23, 8].

In the remainder of this Section we want to give a short overview on furthergoing topics.

In Section 3.1 we defined schedulers for a PNS. There is a more general definition of schedulers, the so called stochastic schedulers. Given a probabilistic nondeterministic system T, a stochastic scheduler of T is a function B mapping every finite path ρ of T to a distribution $B(\rho) \in Distr(Steps(last(\rho)))$.

That means, after the process described by the PNS performed the first $|\rho|$ steps, the scheduler B chooses with the probability $B(\rho)(\mu)$ the next transition to be μ, for $\mu \in Steps(last(\rho))$. So if a system behaves according to a scheduler

---

[19] Where $reach(t')$ denotes the set of states that are reachable from t', i.e. $reach(t') = \{s \in S \; : \; \exists \, \rho \in Paths_{fin} \; : \; t' = first(\rho)$ and $s = last(\rho)\}$.

B and has followed the path ρ so far, then it will be in state t in the next step with probability

$$\sum_{\mu \in Steps(last(\rho))} B(\rho)(\mu) \cdot \mu(t).$$

Giving a stochastic scheduler B we can define a probability space in the straightforward way.

Furthermore, we can classify the stochastic schedulers into different classes (see [42]). A scheduler B is said to be of the kind

- (HS), if it is history-dependent and stochastic.
- (MS), if it is stochastic and markovian, that means $B(\rho) = B(last(\rho))$ for all finite path ρ, i.e. B is history independent.
- (HD), if it is history-dependent and deterministic, that means for all finite path $\rho : B(\rho)(\mu) \in \{0,1\}$ $\forall \mu \in Steps(last(\rho))$, i.e. B schedules a unique next transition[20].
- (MD), if it is markovian and deterministic (also called simple).

Given a PNS, we denote the set of all its stochastic schedulers by $Sched_{stoch}$. From now on we denote the set of schedulers of the kind (HD) by $Sched_{det}$, i.e. with the notations of Section 3 $Sched_{det} = Sched_{all}$.

Given a PNS and a $PCTL$ path formula ϕ, we have already seen in Section 3.2, that

$$s \models_{Sched_{det}} [\phi]_{\bowtie p} \quad \Leftrightarrow \quad s \models_{Sched_{simple}} [\phi]_{\bowtie p}$$

Moreover the following holds (see [9, 27, 45]):

- Given a PNS and a $PCTL$ path formula ϕ, then for all states s

$$s \models_{Sched_{stoch}} [\phi]_{\bowtie p} \quad \Leftrightarrow \quad s \models_{Sched_{simple}} [\phi]_{\bowtie p}.$$

- Given a PNS and a $PCTL^*$ path formula ϕ, then for all states s

$$s \models_{Sched_{stoch}} [\phi]_{\bowtie p} \quad \Leftrightarrow \quad s \models_{Sched_{det}} [\phi]_{\bowtie p}.$$

An informal explanation for these results is the following: It can be shown (see [27, 45]), that the measure of a measurable set Ω w.r.t. a stochastic scheduler is a convex combination of measures of Ω w.r.t. deterministic schedulers. Since we are only interested in maximal and minimal measures, we get the above results. Hence, all methods presented in Section 3 apply to stochastic schedulers as well.

---

[20] This coincides with definition 20.

Note, that for a $PCTL^*$ formula ϕ, it is not sufficient to consider only the simple schedulers as the following example demonstrates:

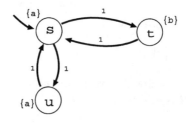

Consider the $PCTL^*$ formula $\phi = \Box a \vee \Box(a \to \mathcal{X}b)$.

Observe that due to the absence of probabilistic choice, starting in s, each scheduler will schedule a unique path. There are only two simple schedulers, the one that schedules the path $\pi = s, u, s, u, s, u, \ldots$. And the one that schedules the path $\pi' = s, t, s, t, s, t, \ldots$. Both paths satisfy the formula ϕ, so we get $s \models_{Sched_{simple}} [\phi]_{\geq 1}$. On the other hand, there is a scheduler B in $Sched_{det}$, that schedules the path $\pi'' = s, t, s, u, s, t, s, u, \ldots$. This path does not satisfy the given formula ϕ and we get $\{\pi \in Paths^B_{inf}(s) : \pi \models \phi\} = \emptyset$. So $prob^B_s(\{\pi \in Paths^B_{inf}(s) : \pi \models \phi\}) = 0$ and $s \not\models_{Sched_{det}} [\phi]_{\geq 1}$. Therefore $s \not\models_{Sched_{stoch}} [\phi]_{\geq 1}$.

As in the non probabilistic setting various simulation and bisimulation relations have been defined and studied in connection with probabilistic systems [31, 29, 45, 30]. It is possible to give logical characterizations for some of these relations, e.g. for $DTMCs$ bisimulation equivalence is the same as $PCTL^*$-equivalence (see e.g. [3]). Similar results for certain types of nondeterministic probabilistic systems and $PCTL^*$ have been established in [46, 28].

The state space explosion problem arises in the probabilistic setting to the same extent as in the non-probabilistic setting. In the last 10 years several methods have been developed to attack the state space explosion problem for nonprobabilistic systems. Some of these approaches have been extended to the probabilistic case. In [16] a technique was introduced that involves an iterative usage of a simulation relation by starting with a small abstract model and successively refining it, until the properties under consideration are safely checked. Another approach are symbolic methods that use $MTBDDs$ (or other Decision Diagram variants) as an underlying data structure. Some of the model checking algorithms presented in this article were implemented (with several modifications) in the tool $PRISM$ ([36]) which uses standard methods of symbolic model checking (see e.g. [38, 6] and the article by D. Parker and A. Miner in this volume) and in the tool $ProbVerus$ [48].

A probabilistic variant of timed automata ([1]) and continuous real time extensions of $PCTL$ were introduced and studied in [33–35]. (See the article by Jeremy Sproston in this volume.)

CSL, an extension of $PCTL$ for reasoning about performance measures and quantitative temporal properties of Continuous Time Markov chains was introduced and studied in [3, 5]. (See the article by Lucia Cloth in this volume.)

_3_3_3_3_3_3_3_3_3_3_3_3_3_3_3_3_3_3

References

1. Alur and Dill. A theory of timed automata. *Theoretical Computer Science*, pages 183–235, 1994.
2. Rajeev Alur, Costas Courcoubetis, and David L. Dill. Model-checking for probabilistic real-time systems (extended abstract). In Javier Leach Albert, Burkhard Monien, and Mario Rodríguez-Artalejo, editors, *Automata, Languages and Programming, 18th International Colloquium*, volume 510 of *Lecture Notes in Computer Science*, pages 115–126, Madrid, Spain, 8–12 July 1991. Springer-Verlag.
3. A. Aziz, V. Singhal, F. Balarin, and R. K. Brayton. It usually works: The temporal logic of stochastic systems. *Lecture Notes in Computer Science*, 939:155–165, 1995.
4. C. Baier. On algorithmic verification methods for probabilistic systems (habilitation thesis). November 1998.
5. C. Baier, J.-P. Katoen, and H. Hermanns. Approximate symbolic model checking of continuous-time markov chains. *Lecture Notes in Computer Science, 10th International Conference on Concurrency Theory, CONCUR'99*, 1664:146–161, 1999.
6. Christel Baier, Edmund M. Clarke, Vassili Hartonas-Garmhausen, Marta Z. Kwiatkowska, and Mark Ryan. Symbolic model checking for probabilistic processes. In Pierpaolo Degano, Robert Gorrieri, and Alberto Marchetti-Spaccamela, editors, *Automata, Languages and Programming, 24th International Colloquium*, volume 1256 of *Lecture Notes in Computer Science*, pages 430–440, Bologna, Italy, 7–11 July 1997. Springer-Verlag.
7. Beauquier and Slissenko. Polytime model checking for timed probabilistic computation tree logic. *Acta Informatica 35*, pages 645–664, 1998.
8. Danièle Beauquier and Anatol Slissenko. Polytime model checking for timed probabilistic computation tree logic. *Acta Informatica*, 35(8):645–664, 1998.
9. A. Bianco and L. De Alfaro. Model checking of probabilistic and nondeterministic systems. *Lecture Notes in Computer Science*, 1026:499–513, 1995.
10. C.Baier and M.Kwiatkowska. Model checking for a probabilistic branching time logic with fairness. *Distributed Computing*, 11, 1998.
11. E. M. Clarke, E. Allen Emerson, and A. P. Sistla. Automatic verification of finite state concurrent systems using temporal logic specifications: A practical approach. *ACM Transactions on Programming Languages and Systems*, 8(2):244–263, 1986.
12. Edmund M. Clarke, Orna Grumberg, and Doron A. Peled. *Model Checking*. The MIT Press, 1999.
13. Costas Courcoubetis and Mihalis Yannakakis. Verifying temporal properties of finite-state probabilistic programs. In *29th Annual Symposium on Foundations of Computer Science*, pages 338–345, White Plains, New York, 24–26 October 1988. IEEE.
14. Costas Courcoubetis and Mihalis Yannakakis. Markov decision processes and regular events (extended abstract). In Michael S. Paterson, editor, *Automata, Languages and Programming, 17th International Colloquium*, volume 443 of *Lecture Notes in Computer Science*, pages 336–349, Warwick University, England, 16–20 July 1990. Springer-Verlag.
15. Costas Courcoubetis and Mihalis Yannakakis. The complexity of probabilistic verification. *Journal of the ACM*, 42(4):857–907, July 1995.

16. P.R. D'Argenio, B. Jeannet, H.E. Jensen, and K.G. Larsen. Reduction and refinement strategies for probabilistic analysis. In H. Hermanns and R. Segala, editors, *Proceedings of Process Algebra and Probabilistic Methods. Performance Modeling and Verification. Joint International Workshop, PAPM-PROBMIV 2001*, Copenhagen, Denmark, Lecture Notes in Computer Science. Springer-Verlag, 2001.

17. L. De Alfaro. Temporal logics for the specification of performance and reliability. *Lecture Notes in Computer Science*, 1200:165–??, 1997.

18. Luca de Alfaro. Formal verification of probabilistic systems. Thesis CS-TR-98-1601, Stanford University, Department of Computer Science, June 1998.

19. Emerson, Mok, Sistla, and Srinivasan. Quantitative temporal reasoning. In *Journal of Real Time System*, pages 331–352, 1992.

20. E. A. Emerson and J. Y. Halpern. 'Sometimes' and 'not never' revisited: on branching time versus linear time temporal logic. *Journal of the ACM*, 33(1):151–178, 1986.

21. E. Allen Emerson. Temporal and modal logic. In Jan van Leeuwen, editor, *Handbook of Theoretical Computer Science, Volume B: Formal Models and Semantics*, pages 995–1072. Elsevier Science Publishers, Amsterdam, The Netherlands, 1990.

22. W. Feller. *An Intro. to Probability Theory and its application*. John Wiley and Sons, New York, 1968.

23. H. Hansson. *Time and Probability in Formal Design of Distributed Systems*. Series in Real-Time Safety Critical Systems. Elsevier, 1994.

24. H. Hansson and B. Jonsson. A calculus for communicating systems with time and probabitilies. In IEEE Computer Society Press, editor, *Proceedings of the Real-Time Systems Symposium - 1990*, pages 278–287, Lake Buena Vista, Florida, USA, December 1990. IEEE Computer Society Press.

25. Hans Hansson and Bengt Jonsson. A logic for reasoning about time and reliability. *Formal Aspects of Computing*, 6(5):512–535, 1994.

26. Hans A. Hansson. Time and probability in formal design of distributed systems. SICS Dissertation Series 05, Swedish Institute of Computer Science, Box 1263, S-164 28 Kista, Sweden, 1994.

27. Sergiu Hart, Micha Sharir, and Amir Pnueli. Termination of probabilistic concurrent programs. *ACM Transactions on Programming Languages and Systems (TOPLAS)*, 5(3):356–380, July 1983.

28. R. Jagadeesan J. Desharnais, V. Gupta and P.Panangaden. Weak bisimulation is sound and complete for pctl*. *Lecture Notes in Computer Science*, 2421:355–370, 2002.

29. Bengt Jonsson and Kim Guldstrand Larsen. Specification and refinement of probabilistic processes. In *Proceedings, Sixth Annual IEEE Symposium on Logic in Computer Science*, pages 266–277, Amsterdam, The Netherlands, 15–18 July 1991. IEEE Computer Society Press.

30. Chi-Chang Jou and Scott A. Smolka. Equivalences, congruences, and complete axiomatizations for probabilistic processes. In J. C. M. Baeten and J. W. Klop, editors, *CONCUR '90: Theories of Concurrency: Unification and Extension*, volume 458 of *Lecture Notes in Computer Science*, pages 367–383, Amsterdam, The Netherlands, 27–30August 1990. Springer-Verlag.

31. K. Larsen K and A. Skou. Bisimulation through probabilistic testing. *Inf Comput 94*, pages 1–28, 1991.

32. J.G. Kemeny and J.L. Snell. *Finite Markov Chains*. Princeton University Press, 1960.

33. Marta Kwiatkowska, Gethin Norman, Roberto Segala, and Jeremy Sproston. Automatic verification of real-time systems with discrete probability distributions. *Lecture Notes in Computer Science*, 1601:75–95, 1999.
34. Marta Kwiatkowska, Gethin Norman, Roberto Segala, and Jeremy Sproston. Verifying quantitative properties of continuous probabilistic timed automata. *Lecture Notes in Computer Science*, 1877, 2000.
35. Marta Kwiatkowska, Gethin Norman, and Jeremy Sproston. Probabilistic model checking of the ieee 802.11 wireless local area network protocol. *Process Algebra and Probabilistic Methods, Performance Modeling and Verification, Second Joint International Workshop PAPM-PROBMIV 2002, Copenhagen, Denmark, July 25-26, 2002, Proceedings*, LNCS 2399:169–187, 2002.
36. Marta Z. Kwiatkowska, Gethin Norman, and David Parker. PRISM: Probabilistic symbolic model checker. In *Computer Performance Evaluation / TOOLS*, pages 200–204, 2002.
37. Zohar Manna and Amir Pnueli. *The Temporal Logic of Reactive and Concurrent Systems*. Springer, New York, 1992.
38. K. L. McMillan. *Symbolic Model Checking*. PhD thesis, Carnegie Mellon University, Pittsburgh, 1993.
39. A. Pnueli and L. Zuck. Verification of multiprocess probabilistic protocols. *Distributed Computing*, 1:53–72, 1986.
40. Amir Pnueli and Lenore Zuck. Probabilistic verification by tableaux. In *Proceedings, Symposium on Logic in Computer Science*, pages 322–331, Cambridge, Massachusetts, 16–18 June 1986. IEEE Computer Society.
41. Amir Pnueli and Lenore D. Zuck. Probabilistic verification. *Information and Computation*, 103(1):1–29, March 1993.
42. Martin L. Puterman. *Markov Decision Processes—Discrete Stochastic Dynamic Programming*. John Wiley & Sons, Inc., New York, NY, 1994.
43. Shmuel Safra. On the complexity of ω-automata. In *29th Annual Symposium on Foundations of Computer Science*, pages 319–327, White Plains, New York, 24–26 October 1988. IEEE.
44. R. Segala. *Modeling and Verification of Randomized Distributed Real-Time Systems*. PhD thesis, Department of Electrical Engineering and Computer Science, Massachusetts Institute of Technology, June 1995. Available as Technical Report MIT/LCS/TR-676.
45. R. Segala and N. Lynch. Probabilistic simulations for probabilistic processes. *Lecture Notes in Computer Science*, 836:481–496, 1994.
46. Roberto Segala and Nancy Lynch. Probabilistic simulations for probabilistic processes. *Nordic Journal of Computing*, 2(2):250–273, Summer 1995.
47. Wolfgang Thomas. Automata on infinite objects. In J. van Leeuwen, editor, *Handbook of Theoretical Computer Science*, chapter 4, pages 133–191. Elsevier Science Publishers B. V., 1990.
48. S. Campos V. Hartonas-Garmhausen and E. Clarke. Probverus: Probabilistic symbolic model checking. *Lecture Notes in Computer Science*, 1601:96–110, 1999.
49. M. Y. Vardi. Automatic verification of probabilistic concurrent finite-state programs. In *26th Annual Symposium on Foundations of Computer Science*, pages 327–338, Los Angeles, Ca., USA, October 1985. IEEE Computer Society Press.

50. Moshe Y. Vardi. Probabilistic linear-time model checking: An overview of the automata-theoretic approach. *Lecture Notes in Computer Science*, 1601:265–276, 1999.
51. Moshe Y. Vardi and Pierre Wolper. An automata-theoretic approach to automatic program verification (preliminary report). In *Proceedings, Symposium on Logic in Computer Science*, pages 332–344, Cambridge, Massachusetts, 16–18 June 1986. IEEE Computer Society.

Model Checking for Probabilistic Timed Systems*

Jeremy Sproston

Dipartimento di Informatica, Università di Torino, 10149 Torino, Italy

Abstract. Application areas such as multimedia equipment, communication protocols and networks often feature systems which exhibit both
probabilistic and timed behaviour. In this paper, we consider analysis of
such probabilistic timed systems using the technique of model checking,
in which it is verified automatically whether a system satisfies a certain
desired property. In order to describe formally probabilistic timed systems, we consider probabilistic extensions of timed automata, such as
real-time probabilistic processes, probabilistic timed automata and continuous probabilistic timed automata, the underlying semantics of each
of which is an infinite-state structure. For each formalism, we consider
how the well-known region equivalence relation can be used to reduce
the infinite state-space model into a finite-state system, which can then
be used for model checking.

1 Introduction

The increasing reliance on complex computer systems in diverse fields such as
business, transport and medicine has led to increased interest in obtaining formal
guarantees of system correctness. In particular, *model checking* is an automatic method for guaranteeing that a mathematical model of a system satisfies a
formula representing a desired property [16]. Many real-life systems, such as
multimedia equipment, communication protocols, networks and fault-tolerant
systems, exhibit *probabilistic* behaviour, leading to the study of *probabilistic
model checking* of probabilistic and stochastic models [48, 25, 18, 11, 9]. Similarly, it is common to observe complex *real-time* behaviour in such systems. Model
checking of timed systems has been studied extensively in, for example, [4, 26,
50, 39, 12, 53], with much attention given to the formalism of *timed automata* [5].
In this paper, we present an overview of model checking techniques for probabilistic extensions of timed automata. Note that our aim is not to explore all
methods for model checking probabilistic timed models; model checking has also
been developed for stochastic processes such as Discrete-Time Markov Chains
[25], Continuous-Time Markov Chains [8], and Semi-Markov Chains [41].

A timed automaton consists of a finite graph equipped with a finite set of
real-valued variables called *clocks* which increase at the same rate as real-time.

---

* Supported in part by the EU within the DepAuDE project IST-2001-25434.

Changes in the values of the clocks can enable or disable edges of the graph; furthermore, an edge traversal may result in a clock being reset to 0. In general, we consider the verification of a timed automaton model against a timed property, for instance "a request will always be followed by a response within 5 milliseconds". Such timed properties can be expressed in timed extensions of temporal logic such as TCTL [4, 26]. A problem is that the underlying semantic model of a timed automaton is infinite-state, as clocks, which characterize partially each state, are real-valued. Recall that a fundamental requirement of traditional model checking, as exemplified by the seminal work of [15, 44], is that the state space of the labelled transition system is of *finite* cardinality. This restriction implies that the fixpoint computations of the model-checking algorithm, which compute progressively sets of states which satisfy given temporal logic formulae, will terminate, because there is a finite number of state sets that can be computed, implying in turn that the entire model-checking algorithm will terminate. A key notion for obtaining decidability and complexity results for timed automata model checking is *clock equivalence* [5, 4]. This equivalence relation guarantees that the future behaviour of the system from two clock-equivalent states will be sufficiently similar from the point of view of timed properties. More precisely, the same TCTL properties are satisfied by clock equivalent states. Furthermore, the fact that this equivalence has a finite number of classes for any timed automaton means that we are able to obtain a finite state space on which TCTL model checking can be performed. After a section of preliminaries, we review timed automata and their clock-equivalence-based model-checking methods in Section 3.

This paper surveys the following probabilistic extensions of timed automata. In Section 4, we consider *real-time probabilistic processes* [2, 3]. This formalism, which is similar to generalized semi-Markov processes [49, 24], resets clock values not to 0, as in the case of timed automata, but to negative values drawn from continuous density functions. Clocks then trigger edge traversals in the graph of the model when their value reaches 0. We focus on *qualitative* ("almost-sure") probabilistic timed properties, which refer to the satisfaction of a property with probability strictly greater than 0 or with probability 1 only. For example, we can express properties such as "a request will be followed by a response within 5 milliseconds with probability 1". Clock equivalence can be used to obtain a finite-state system which represents the behaviour of a real-time probabilistic process, which can then be subject to traditional graph-theoretic model-checking methods.

The formalism of *probabilistic timed automata* [30, 35] is considered in Section 5. Probabilistic timed automata extend both timed automata and *Markov decision processes* [23, 43], which are stochastic processes in which both nondeterministic behaviour and probabilistic behaviour coexist. Therefore, probabilistic timed automata are timed automata for which probabilistic choice may be made over certain edges of the graph. This formalism has proved particularly useful for the study of timed randomized algorithms, such as those used to obtain the election of a leader in the IEEE1394 FireWire standard [38] and those used in "backoff" rules in communication protocols [37]. Probabilistic timed automata may be verified against *quantitative* probabilistic timed properties, which refer to the satisfaction of a property with a certain probability. An example of such a property is "a request will be followed by a response within 5 milliseconds with probability 0.99 or greater". Model-checking methods for probabilistic timed automata can use clock equivalence to obtain a finite-state Markov decision process, which can subsequently be subject to probabilistic-model-checking methods [11, 9], such as those implemented in the tool PRISM [32].

Finally, *continuous probabilistic timed automata* [34] feature a combination of aspects of real-time probabilistic processes and probabilistic timed automata, and are considered in Section 6. This formalism exhibits both nondeterministic choice and probabilistic choice, like probabilistic timed automata, and has the ability to reset clocks according to continuous density functions, like real-time probabilistic processes. We consider model checking quantitative probabilistic timed properties of continuous probabilistic timed automata. The application of clock equivalence yields a finite-state Markov decision process which *conservatively approximates* the probability with which a property is satisfied, in the sense that we can obtain an interval of probabilities within which the true probability is contained.

For surveys considering timed automata and real-time probabilistic processes in more depth, we refer the reader to [53, 1] and [17], respectively.

2   Clocks

The models studied in this paper are defined as operating in *continuous real-time*. That is, system transitions can occur at any point within the continuum defined by the set of reals. In order to analyze the behaviour of the models, we use the notion of multidimensional real-valued space, the points in which characterize partially the particular system configurations at a certain point in time. Let $\mathcal{X}$ be a finite set of real-valued variables called *clocks*, the values of which are interpreted as increasing at the same rate as real-time. Using $\mathbb{R}$ to denote the set of real numbers, we refer to a point $val \in \mathbb{R}^{|\mathcal{X}|}$ within $|\mathcal{X}|$-dimensional, real-valued space as a *clock valuation*.[1] For a clock $x \in \mathcal{X}$ and a clock valuation val, we use val_x to denote the value of x with regard to val (formally, the value val_x is the projection of val on the x-axis). Let $\mathbf{0} \in \mathbb{R}^{|\mathcal{X}|}$ be the clock valuation which assigns 0 to all clocks in $\mathcal{X}$. Denoting by $\mathbb{R}_{\geq 0}$ the set of non-negative real numbers, then for any clock valuation $val \in \mathbb{R}^{|\mathcal{X}|}$ and value $\delta \in \mathbb{R}_{\geq 0}$, we use $val + \delta$ to denote the clock valuation obtained from the values $(val + \delta)_x = val_x + \delta$ for all clocks $x \in \mathcal{X}$. Finally, for a clock set $X \subseteq \mathcal{X}$, we let $val[X := 0]$ be the clock valuation for which all clocks within X are set to 0; formally, we let

---

[1] Our definition of clock valuations allows both positive and negative reals to be assigned to clocks. In later sections, we will consider usually clock valuations which assign *only* non-negative or *only* non-positive values to clocks.

$val[X := 0]$ be the clock valuation val' obtained from letting $val'_x = 0$ for all $x \in X$ and letting $val'_x = val_x$ for all other clocks $x \in \mathcal{X} \setminus X$. Operations such as $val + \delta$ are called *clock increments* and those such as $val[X := 0]$ are called *clock assignments*.

For a real number $q \in \mathbb{R}$, let $\lfloor q \rfloor$ be the integer part of q. Let max $\in \mathbb{N}$ be a natural number. Given the clock set $\mathcal{X}$, we define *clock equivalence (with respect to* max*)* to be the relation $\equiv_{\max} \subseteq \mathbb{R}_{\geq 0}^{|\mathcal{X}|} \times \mathbb{R}_{\geq 0}^{|\mathcal{X}|}$ over valuations within non-negative real-valued space, where $val \equiv_{\max} val'$ if and only if:

- for each clock $x \in \mathcal{X}$, either both $val_x >$ max and $val'_x >$ max or:
  - $\lfloor val_x \rfloor = \lfloor val'_x \rfloor$ (clock values have the same integer part), and
  - $val_x - \lfloor val_x \rfloor = 0$ if and only if $val'_x - \lfloor val'_x \rfloor = 0$ (the fractional parts of the clocks' values are either all zero or all positive); and
- for each pair of clocks $x, y \in \mathcal{X}$, either $\lfloor val_x - val_y \rfloor = \lfloor val'_x - val'_y \rfloor$ (the integer part of the difference between clocks values is the same) or both $val_x - val_y >$ max and $val'_x - val'_y >$ max.

Note that the second point implies that the ordering on the fractional parts of the clock values of x and y is the same, unless the difference between the clocks is above max. For an example of two clock valuations val, val' which have the same ordering on the fractional parts on clocks in the set $\mathcal{X} = \{x_1, x_2, x_3\}$, consider $val_{x_1} = 0.1$, $val_{x_2} = 0.25$ and $val_{x_3} = 0.8$ and $val'_{x_1} = 0.5$, $val'_{x_2} = 0.9$ and $val'_{x_3} = 0.95$. Instead, the clock valuation $val''_{x_1} = 0.3$, $val''_{x_2} = 0.75$ and $val''_{x_3} = 0.6$ does *not* have the same ordering on the fractional parts as val and val'. An example of the partition that clock equivalence induces on a space of valuations of two clocks x_1 and x_2 is shown in Figure 1, where the natural number max is equal to 2; each vertical, horizontal or diagonal line segment, open area, and point of intersection of lines is a distinct clock equivalence class.

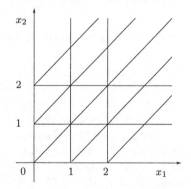

Fig. 1. The partition induced by clock equivalence with max $= 2$

We define the *time successor* of a clock equivalence class α to be the first distinct clock equivalence class reached from α by letting time pass. Formally, the time successor β of α is the clock equivalence class for which $\alpha \neq \beta$ and, for all $val \in \alpha$, there exists $\delta \in \mathbb{R}_{\geq 0}$ such that $val + \delta \in \beta$ and $val + \delta' \in \alpha \cup \beta$ for all $0 \leq \delta' \leq \delta$. A clock equivalence class α is *unbounded* if, for all $val \in \alpha$ and $\delta \in \mathbb{R}_{\geq 0}$, we have $val + \delta \in \alpha$; in the case of an unbounded clock equivalence class α, we let the time successor of α be α itself. Finally, we define the effect of clock assignment on clock equivalence classes. For a clock equivalence class α and clock set $X \subseteq \mathcal{X}$, we use $\alpha[X := 0]$ to denote the clock equivalence class defined by the set $\{val[X := 0] \mid val \in \alpha\}$ (that this set defines a clock equivalence class can be verified easily). For a clock equivalence class α and a subset $X \subseteq \mathcal{X}$ of clocks, let $\alpha \downarrow_X$ be the restriction of the valuations contained within α to the clocks in X.

The set $\Psi_{\mathcal{X}}$ of *clock constraints* over $\mathcal{X}$ is defined by the following syntax:

$$\psi ::= x \sim c \mid x - y \sim c \mid \psi \wedge \psi \mid \neg \psi$$

where $x, y \in \mathcal{X}$ are clocks, $\sim \in \{<, \leq\}$ is a comparison operator, and $c \in \mathbb{N}$ is a natural. We abbreviate the clock constraint $\neg(x < c)$ by $x \geq c$, the clock constraint $\neg(x \leq c)$ by $x > c$, and the clock constraint $x \leq c \wedge x \geq c$ by $x = c$. Furthermore, we denote a trivially-satisfied clock constraint such as $\neg(x < c \wedge x > c)$ by `true`. The clock valuation val *satisfies* the clock constraint $\psi \in \Psi_{\mathcal{X}}$, written $val \models \psi$, if and only if ψ resolves to true after substituting each clock $x \in \mathcal{X}$ with the corresponding clock value val_x from val. Similarly, a clock equivalence class α satisfies ψ, written $\alpha \models \psi$, if and only if $val \models \psi$ for all $val \in \alpha$; in this case, we say that α is ψ-*satisfying*. Observe that every clock equivalence class is a set of valuations definable by a clock constraint.

3   Timed Automata

3.1   Modelling with Timed Automata

We consider the task of modelling systems using the various formalisms featured in this paper by using the running example of a light switch. In this section, we model the action of a light which, after its switch is pressed, stays illuminated for between 2 and 3 minutes. A timed automaton modelling this simple system is shown in Figure 2. The model comprises of two nodes, called *locations*, which are connected by *edges*. The annotations on the nodes and edges denote clock constraints and clock assignments, both of which refer to the single clock x of the model. The behaviour of the timed automaton takes the following form. Time elapses while the timed automaton remains within a certain location, increasing the value of the clock x. It is possible to use a particular edge to switch between locations only if the condition on the values of clocks which annotate the edge is satisfied by the current values of the clocks. Edge traversals are instantaneous. On traversal of an edge, the clocks within the clock set which forms the second annotation to the edge (within { }) are reset to the value 0; the values of all other clocks remain the same. We also note that conditions on the values of clocks which label the locations of the timed automaton can prevent time elapsing beyond a certain bound. More precisely, the timed automaton can re-

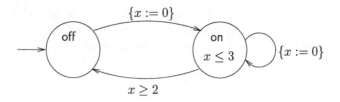

Fig. 2. A timed automaton modelling a switch

main within a location, letting time pass, only as long as the current values of the clocks satisfy the condition which annotates that location (for example, the clock constraint $x \leq 3$ which labels the location on in Figure 2). The absence of a clock constraint annotation on locations or edges refers to true, whereas the absence of a clock assignment annotation refers to the empty assignment $\{\emptyset := 0\}$.

The timed automaton in Figure 2 commences its behaviour in the location off. The edge to location on can be taken at any point, corresponding to the switch being pressed, and results in the value of x being reset to 0. In location on, the timed automaton can return to off after 2 minutes have elapsed, and not after 3 time units have elapsed, provided that the switch is not pressed again. The self-loop edge of on represents the switch being pressed while the light is on, in which case the timed automaton may leave (must leave) the location after 2 minutes (before more than 3 minutes, respectively).

3.2   Syntax of Timed Automata

Let AP be a set of atomic propositions, which we henceforth assume to be fixed. A *timed automaton* TA $= (L, \bar{l}, \mathcal{X}, inv, E, \mathcal{L})$ comprises:

- a finite set L of *locations*, of which one $\bar{l} \in L$ is designated as being the *initial location*,
- a finite set $\mathcal{X}$ of *clocks*,
- a function $inv : L \to \Psi_{\mathcal{X}}$ associating an *invariant condition* with each location,
- a set $E \subseteq L \times \Psi_{\mathcal{X}} \times 2^{\mathcal{X}} \times L$ of *edges*,
- and a *labelling function* $\mathcal{L} : L \to 2^{AP}$ associating a set of atomic propositions with each location.

The invariant condition inv takes the form of an assignment of a clock constraint to each location, which gives the set of clock valuations that are admissible in the location. Each edge $(l, g, X, l') \in E$ is determined by (1) its source location l and target location l', (2) its *guard* g, and (3) the set X of clocks called its *clock reset*. The role of a guard is to define a condition on the values of the clocks which enable the edge for choice, while the role of the clock reset is to specify which clocks should be reset to 0 on traversal of the edge.

The behaviour of a timed automaton starts in the initial location with all clocks set to 0. Time may then elapse, making the clocks increase in value, while the model remains in the initial location. Such passage of time is limited by the invariant condition of the location, in the sense that time can elapse only while the clock values satisfy the location's invariant condition. Control of the timed automaton can leave the location via one of its outgoing edges at any point when the current values of the clocks satisfy the guard of the relevant edge. When an edge is traversed, the clocks within the clock reset are set to 0; then the behaviour of the timed automaton continues anew in the manner described above in the new location. Observe that the choice of when an edge is taken, and of which edge is selected, is purely nondeterministic; there is no probabilistic choice within the model. The time spent in any given location may be 0, given that a guard is satisfied upon entry of the location.

Finally, note that the use of timed automata for modelling complex, real-life systems is aided by a naturally-defined parallel composition operator [5], which we omit from this paper for simplicity. In order to achieve parallel composition, the definition of timed automata is extended with a set of *actions*, where every edge is labelled by an action. Intuitively, two timed automata which are composed in parallel synchronize edge traversals with shared actions, and are able to traverse edges labelled with other, "non-shared" actions independently, following the precedent of process-algebraic theories of parallel composition.

3.3   Semantics of Timed Automata

A *transition system* $\mathsf{TS} = (S, \bar{s}, T, lab)$ comprises a set S of states, of which $\bar{s} \in S$ is declared as being the initial state, a transition relation $T \subseteq S \times S$, and a labelling function $lab : S \to 2^{AP}$ associating a set of atomic propositions with each state. We often denote a transition $(s, s') \in T$ as $s \to s'$, and choose occasionally to label the arrow with an extra symbol, for example $s \xrightarrow{symbol} s'$. A *path* of TS is an infinite sequence of transitions $s_0 \to s_1 \to \cdots$. We denote by $Path(s)$ the set of paths of TS starting in the state $s \in S$.

The semantics of a timed automaton $\mathsf{TA} = (L, \bar{l}, \mathcal{X}, inv, E, \mathcal{L})$ is expressed formally in terms of an infinite-state transition system $\mathsf{TS} = (S, \bar{s}, T, lab)$. The state set S of the transition system comprises pairs of the form (l, val), the first element of which is a location, and the second element of which is a clock valuation which satisfies the location's invariant condition. Formally, we let $S = \{(l, val) \in L \times \mathbb{R}_{\geq 0}^{|\mathcal{X}|} \mid val \models inv(l)\}$. The initial state $\bar{s}$ of the transition system is defined as $(\bar{l}, \mathbf{0})$. The labelling function lab of TS is derived from the labelling function of TA by letting $lab(l, val) = \mathcal{L}(l)$. Transitions between states are derived from the time-elapse and edge-traversal behaviour of the timed automaton. Formally, we define $T \subseteq S \times S$ to be the smallest relation such that $((l, val), (l', val')) \in T$ if both of the following conditions hold:

Time Elapse: there exists $\delta \in \mathbb{R}_{\geq 0}$ such that $val + \delta' \models inv(l)$ for all $0 \leq \delta' \leq \delta$, and

Edge Traversal: there exists an edge $(l, g, X, l') \in E$ such that $val + \delta \models g$, $val' \models inv(l')$ and $val' = val[X := 0]$.

We use $s \xrightarrow{\delta} s'$ to denote that $(s, s') \in T$ and that $\delta \in \mathbb{R}_{\geq 0}$ can be used in the time-elapse part of this transition. Intuitively, the duration of a transition $s \xrightarrow{\delta} s'$ is the non-negative real δ. Note that the fact that clocks are real-valued means that the transition system is generally infinite-state, and that the real-valued durations of time that may elapse from a state means that the transition system is generally infinitely branching.

A behaviour of the timed automaton corresponds to an infinite path comprising of transitions of its transition system. Not all behaviours of timed automata are of interest: certain behaviours correspond to an infinite number of edge traversals made within a bounded amount of time, which contradicts our intuition that the operation of a real-time system cannot be "infinitely fast". Therefore, we would like to disregard such unrealisable behaviours during model checking analyses. Formally, a path $s_0 \xrightarrow{\delta_0} s_1 \xrightarrow{\delta_1} \cdots$ of the transition system of a timed automaton is a *divergent* path if the infinite sum $\sum_{i \geq 0} \delta_i$ diverges. We use $Path_{div}(s)$ to denote the set of divergent paths starting in state $s \in S$. As will be seen in the next section, we consider model checking over divergent paths only. Finally, we say that a timed automaton is *non-Zeno* if there exists a divergent path commencing in each state.

3.4   Model Checking Timed Automata

We now proceed to describe how model checking of timed automata can be made possible using clock equivalence, following [4, 5]. First, we consider the property description languages that can be used to express requirements of timed automata.

Timed Properties. The timed temporal logic TCTL adapts the classical branching-time temporal logic CTL [15] with the means to express quantitative timing constraints on the system's timed behaviour. For example, the logic TCTL can express requirements such as "a response always follows a request within 5 microseconds". TCTL has been introduced in two main forms: in [4], the syntax of the "until" operator of CTL is augmented by a time constraint to obtain a "time-bounded until" path operator $\phi_1 \mathcal{U}_{\sim c} \phi_2$, where ϕ_1, ϕ_2 are TCTL formulae, $\sim \in \{<, \leq, =, \geq, >\}$ is a comparison operator and $c \in \mathbb{N}$. The path operator $\phi_1 \mathcal{U}_{\sim c} \phi_2$ is evaluated, not on a state, but on a path, and, if satisfied by the path, means intuitively "ϕ_2 is true at some point along the path, where the total time elapsed at this point satisfies $\sim c$, and ϕ_1 is true until this point". As with CTL, such a path operator must be nested immediately within an existential or universal path quantifier to obtain a CTL-like formula which is true or false in a state. The syntax of TCTL is stated formally as follows:

$$\phi ::= a \mid \phi \vee \phi \mid \neg \phi \mid \phi \exists \mathcal{U}_{\sim c} \phi \mid \phi \forall \mathcal{U}_{\sim c} \phi$$

where $a \in AP$ is an atomic proposition, and $\sim$ and $c \in \mathbb{N}$ are defined as above. In the sequel, we use occasionally $\Diamond_{\sim c} \phi$ to abbreviate the path formula $\mathtt{true} \mathcal{U}_{\sim c} \phi$ (where $\mathtt{true}$ denotes a formula which is trivially true in all states, such as $a \vee \neg a$).

A second syntax of TCTL is given in [26], and differs with respect to the manner in which time-constrained properties are specified. Here, a formula can refer explicitly to clocks of the system using clock constraints, and can also refer to a new set of *formula clocks* which can be set to 0 by special *reset quantifiers* of the formula. Although this second syntax can express all of the operators of the first, we do not consider it here for simplicity.

We return to the previous syntax of TCTL, and introduce the notion of position along a path to identify the exact state at any point along a path, including at a time point which falls during the time-elapse part of a transition. A *position* of a path $\omega = s_0 \xrightarrow{\delta_0} s_1 \xrightarrow{\delta_1}$ is a pair $(i, \delta) \in \mathbb{N} \times \mathbb{R}_{\geq 0}$ such that δ is not greater than the duration δ_i of the $(i + 1)$th transition $s_i \xrightarrow{\delta_i} s_{i+1}$ of ω. We define a total order $\prec$ on positions of a path by defining $(i, \delta) \prec (j, \delta')$ if and only if either $i < j$, or $i = j$ and $\delta < \delta'$. The *state at position* (i, δ) of ω, denoted by $\omega(i, \delta)$, is defined as $(l_i, val_i + \delta)$, where $(l_i, val_i) = s_i$ is the $(i + 1)$th state along ω. We define $\mathrm{duration}(\omega, i, \delta)$ along the path ω at position (i, δ) to be the sum $\delta + \sum_{0 \leq k < i} \delta_k$. Given a state (l, val) of TA, and a TCTL formula ϕ, the satisfaction relation $\models^{\mathrm{T}}$ of TCTL is defined inductively as follows. Note that we express the satisfaction relation for the "until" formulae $\phi_1 \exists \mathcal{U}_{\sim c} \phi_2$ and $\phi_1 \forall \mathcal{U}_{\sim c} \phi_2$ using the formula $\phi_1 \mathcal{U}_{\sim c} \phi_2$, the satisfaction of which relates to a path, rather than to a state.

$(l, val) \models^{\mathrm{T}} a$ iff $a \in lab(l, val)$

$(l, val) \models^{\mathrm{T}} \phi_1 \vee \phi_2$ iff $(l, val) \models^{\mathrm{T}} \phi_1$ or $(l, val) \models^{\mathrm{T}} \phi_2$

$(l, val) \models^{\mathrm{T}} \neg\phi$ iff   not $(l, val) \models^{\mathrm{T}} \phi$

$(l, val) \models^{\mathrm{T}} \phi_1 \exists \mathcal{U}_{\sim c} \phi_2$ iff   there exists a divergent path $\omega \in Path_{div}(l, val)$ such that $\omega \models^{\mathrm{T}} \phi_1 \mathcal{U}_{\sim c} \phi_2$

$(l, val) \models^{\mathrm{T}} \phi_1 \forall \mathcal{U}_{\sim c} \phi_2$ iff   for all divergent paths $\omega \in Path_{div}(l, val)$, we have $\omega \models^{\mathrm{T}} \phi_1 \mathcal{U}_{\sim c} \phi_2$

$\omega \models^{\mathrm{T}} \phi_1 \mathcal{U}_{\sim c} \phi_2$ iff   there exists a position (i, δ) of ω such that $\omega(i, \delta) \models^{\mathrm{T}} \phi_2$, $\mathrm{duration}(\omega, i, \delta) \sim c$, and for all positions (j, δ') of ω such that $(j, \delta') \prec (i, \delta)$, either $\omega(j, \delta') \models^{\mathrm{T}} \phi_1$ or both $\omega(j, \delta') \models^{\mathrm{T}} \phi_2$ and $\mathrm{duration}(\omega, j, \delta') \sim c$.

We say that the timed automaton TA with the initial location $\bar{l}$ satisfies the TCTL formula ϕ if and only if $(\bar{l}, \mathbf{0}) \models^{\mathrm{T}} \phi$.

In addition to TCTL, timed automata can themselves be used as a property specification language, given that they are suitably augmented with acceptance conditions, to provide an timed extension of linear-time specification mechanisms such as finite-state automata [5].

The Region Graph. The primary tool for obtaining decidability and complexity results for timed automata verification is clock equivalence, as presented in Section 2, which is used to define the key notion of *regions*. In the following definition of regions, we assume the timed automaton TA and the TCTL formula ϕ to be fixed, and use $\mathsf{TS} = (S, \bar{s}, Act, T, lab)$ to denote the underlying transition system of the timed automaton TA. Let max be the maximal natural constant against which any clock is compared in the invariant conditions and guards of TA, or which is included as a time bound of a subformula of ϕ. We define *region equivalence of* TA *and* ϕ to be the relation $\cong_{\max} \subseteq S \times S$, where $(l, val) \cong_{\max} (l', val')$ if and only if:

- $l = l'$ (the location components are equal);
- $val \equiv_{\max} val'$ (the valuations are clock equivalent with respect to max).

We define a *region* to be an equivalence class of $\cong_{\max}$. Region equivalent states satisfy the same TCTL formulae [4]. Note that region equivalence is a bisimulation of the underlying transition system of a timed automaton (not necessarily the coarsest), provided that the duration of time-elapse transitions is abstracted.

A fundamental property of region equivalence is that the number of regions of any timed automaton is *finite*. This allows us to construct a finite-state transition system which represents the behaviour of the timed automaton, where the states of the new system are regions, and transitions are obtained in a natural manner: if some state from region r_1 can reach a state in a region r_2 either by letting time elapse or by traversing an edge, then there is a sequence of transitions in the region graph from r_1 to r_2. Formally, given a timed automaton TA, its underlying transition system TS and its maximal constant max, we define the *region graph* to be the transition system $RG = (R, \bar{r}, RT, Rlab)$, where R is the finite set of regions, and the initial region $\{(\bar{l}, 0)\}$ is denoted by $\bar{r}$. The transition relation of the region graph $RT \subseteq R \times R$ is obtained in the following manner. We define RT to be the smallest relation such that $((l, \alpha), (l', \beta)) \in RT$ if either of the following conditions hold:

Time Transition: $l = l'$ and β is the $inv(l)$-satisfying time successor of the clock equivalence class α;

Discrete Transition: there exists an edge $(l, g, X, l') \in E$ such that $\alpha \models g$ and $\beta = \alpha[X := 0]$.

Finally, the labelling function $Rlab$ is derived from the labelling function of the timed automaton by letting $Rlab(l, \alpha) = \mathcal{L}(l)$ for all regions $(l, \alpha) \in R$. The significance of the region graph lies in the close correspondence between its paths and those of the underlying transition system of the timed automaton. More precisely, for any region r and any path ω of the region graph which commences in r, we can find a path of the transition system which commences in a state contained within the region r, and which passes through states contained in the regions of the path ω in the correct sequence. Conversely, for any state s of the transition system, and any divergent path ω' which commences in s, we can find a path of the region graph which commences in the region which contains s, and which passes through regions which contain the states of the path ω' in the correct sequence.

Model Checking Using Regions. The region graph RG obtained from a non-Zeno timed automaton TA and a TCTL formula ϕ can be used to establish whether $\mathsf{TA} \models^\mathsf{T} \phi$. The model-checking method is based on the classical labelling algorithms of [15, 44]: it proceeds up through levels of the parse tree of ϕ, from the smallest subformulae of ϕ to the largest (ϕ itself), labelling regions with the subformulae that they satisfy. The manner in which the labelling is done for the atomic propositions and the boolean combinations $\neg\phi$ and $\phi_1 \vee \phi_2$ follows straightforwardly from classical model checking. The case for formulae of the form $\phi_1 \exists \mathcal{U}_{\geq 0} \phi_2$ is also standard: we label a region with $\phi_1 \exists \mathcal{U}_{\geq 0} \phi_2$ if there exists a path from the region in which ϕ_2 is true at some point, and $\phi_1 \vee \phi_2$ is true at all preceding points. Such a labelling can be computed by iterating an "existential" predecessor operator on regions until a fixpoint is obtained. Formulae of the form $\phi_1 \forall \mathcal{U}_{\geq 0} \phi_2$ label a region if, for all paths from the region, we have ϕ_2 true at some point, and $\phi_1 \vee \phi_2$ is true at all preceding points. We can compute a labelling of such formulae again by iterating a "universal" predecessor operation on regions until a fixpoint is obtained.

Formulae of the form $\phi_1 \exists \mathcal{U}_{\sim c} \phi_2$ and $\phi_1 \forall \mathcal{U}_{\sim c} \phi_2$ are handled by introducing an extra clock $z \notin \mathcal{X}$ to the clock set of the timed automaton, and two extra atomic propositions a_b and $a_{\sim c}$ used to label the resulting region graph. More precisely, the atomic proposition a_b labels those "boundary" regions in which no time is allowed to pass (that is, when the value of at least one clock equals an integer), while the atomic proposition $a_{\sim c}$ labels those regions for which $val_z \sim c$, for all valuations val of that region. Then a region (l, α) is labelled with $\phi_1 \exists \mathcal{U}_{\sim c} \phi_2$ if there exists a path starting in the region $(l, \alpha[z := 0])$ in which $\phi_2 \wedge a_{\sim c} \wedge (a_b \vee \phi_1)$ is true at some point and ϕ_1 is true at all preceding points. The reason for the conjunct $(a_b \vee \phi_1)$ is that, if the region is not a boundary region, then some time may pass in the region, and, as we want ϕ_1 to hold at all points before ϕ_2 holds, then ϕ_1 must also hold in the region. A similar rule holds for $\phi_1 \forall \mathcal{U}_{\sim c} \phi_2$. These algorithms can also be implemented using fixpoint computations.

The model-checking algorithm will terminate, because the predecessor operations used to compute fixpoints return sets of regions, of which there are a finite number. The correctness of the algorithm follows from the fact that TA is non-Zeno, and because region-equivalent states satisfy the same TCTL formulae. Hence, we have a method for establishing whether TA satisfies ϕ. Combining this with a complexity-theoretic argument considering the size of the region graph yields the result that model checking of timed automata against TCTL specifications is PSPACE-complete [4].

Note that we can check whether a timed automaton is non-Zeno by checking that all states satisfy the formula $\exists \Diamond_{=1} \mathtt{true}$ (that is, all states allow time to progress by one time unit). A method for strengthening the invariant conditions of a timed automaton so that non-Zenoness is guaranteed is given in [26]. Furthermore, [47] gives syntactic (and thus easily checked) conditions for a timed automaton to be *strongly non-Zeno*, in which all loops of the model are required to let at least one time unit elapse. Strong non-Zenoness implies non-Zenoness.

Beyond Regions. Although regions are useful to supply the basic decidability and complexity results of timed automata model checking, in practice more sophisticated methods are used in the verification of timed automata, which we now summarize briefly. The main approach is to use clock constraints to define *sets* of regions, which are then manipulated during model checking algorithms. This approach was pioneered in the papers [26] for TCTL properties, in [50, 39] for reachability properties, and in [12] for mixed logic-automata properties. Both the TCTL and reachability algorithms have been implemented in the tool KRO-NOS [52], while algorithms for reachability and a subset of liveness properties have been implemented in the tool UPPAAL [6].

The problem of model checking a timed automaton using semantics based on discrete-time, rather than continuous-time, has been addressed by [27, 7, 13].

4   Real-Time Probabilistic Processes

4.1   Modelling with Real-Time Probabilistic Processes

In order to introduce the first model for *probabilistic* timed systems featured in this paper, we revisit the example of a light switch. In this section, we assume that the light switches off deterministically 3 minutes after being switched on, and that the light is switched on at a point in time given by a uniform distribution between 1 minute and 30 minutes. The real-time probabilistic process modelling this switch is shown in Figure 3. The model consists of locations and

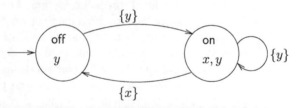

Fig. 3. A real-time probabilistic process modelling a switch

edges, and has two clocks: the clock y models the amount of time remaining before a person wishing to switch on the light arrives, and the clock x models the amount of time remaining before the light switches off automatically. If a clock annotates a location, then it is *scheduled* in that location. Upon entry to a location l, those clocks scheduled in l that were not scheduled in the previous location l' are reset to new, negative values. A scheduled clock may be used to trigger edge traversals from the location when its value reaches 0. An edge which can be triggered by a clock is labelled by the clock within { }. More than one clock may reach 0 at the same time, and therefore edges can also be labelled with sets of clocks, with the interpretation that the edge is triggered if all clocks within the set labelling the edge reach 0. In Figure 3, the clock y is scheduled in the initial location off, and is set initially to a value given by the amount of time before the light switch is used next. More precisely, if this amount of time is δ, then y is set to $-\delta$. Time then elapses and the values of all scheduled clocks increase, until the value of at least one clock becomes 0. In our example, the clock

y is the only clock scheduled in off, and will become 0 after δ time units. The clock(s) which are now 0 then trigger outgoing edges labelled with the clock(s). Thus, the edge from off to on is traversed in our example, indicating that a user has switched the light on. This alternation between letting time elapse until at least one clock reaches 0 continues in the subsequent location. In our example, when we newly enter the location on, it is possible that another user may arrive while the light is still on, so y takes a new value corresponding to the arrival time of the next user. Also, the clock x is deterministically set to the value -3 to denote that the light will turn off automatically in 3 minutes. In contrast to the example in Section 3, we assume that if another person arrives, he will not switch the light on again; this is denoted by the self-loop of the location on labelled by $\{y\}$. Naturally, if x reaches the value 0, the real-time probabilistic process will return to the location off.

The reader familiar with stochastic processes will note that real-time probabilistic processes have many similarities with generalized semi-Markov processes [49, 24].

4.2   Syntax of Real-Time Probabilistic Processes

Probability Distributions. A *continuous density function* on $\mathbb{R}$ is a function f such that $f(q) \geq 0$ for all $q \in \mathbb{R}$ and $\int_{-\infty}^{\infty} f(q)dq = 1$. Let $\mathsf{CDF}(\mathbb{R})$ be the set of continuous density functions on $\mathbb{R}$. The *support* of a continuous density function is the smallest closed subset of $\mathbb{R}$ which has probability 1. A continuous density function is *positive-bounded* if its support lies within an interval of $\mathbb{R}_{\geq 0}$ with natural numbered end-points. Let $PB \subseteq \mathsf{CDF}(\mathbb{R})$ be the set of positive-bounded continuous density functions.

A discrete probability *distribution* over a countable set Q is a function $\mu :$ $Q \to [0,1]$ such that $\sum_{q \in Q} \mu(q) = 1$. For a possibly uncountable set Q', let $\mathsf{Dist}(Q')$ be the set of distributions over countable subsets of Q'. For some element $q \in Q$, let $\{q \mapsto 1\} \in \mathsf{Dist}(Q)$ be the distribution which assigns probability 1 to q.

Syntax. A *real-time probabilistic process* $\mathsf{RTPP} = (L, \bar{p}, \mathcal{X}, sched, prob, delay, \mathcal{L})$ [2, 3] comprises:

- a finite set L of *locations*,
- an *initial distribution* $\bar{p} \in \mathsf{Dist}(L)$,
- a finite set $\mathcal{X}$ of *clocks*,
- a *scheduling function* $sched : L \to 2^{\mathcal{X}}$,
- a *probabilistic transition function* $prob : L \times 2^{\mathcal{X}} \to \mathsf{Dist}(L)$,
- a *delay distributions function* $delay : \mathcal{X} \to \mathsf{CDF}(\mathbb{R})$, and
- a *labelling function* $\mathcal{L} : L \to 2^{AP}$.

The initial distribution $\bar{p}$ is a discrete probability distribution which gives the probability of the real-time probabilistic process starting its behaviour in any location. A set $sched(l)$ of clocks from the set $\mathcal{X}$ are *scheduled* when the process is in a location l, with the intuition that only scheduled clocks can trigger a discrete transition from a location. The probabilistic transition function *prob* is a partial function which associates a discrete probability distribution with pairs comprising a location and a clock set. For any $l \in L$ and $X \subseteq \mathcal{X}$, we require that $prob(l, X)$ is defined only if X is a non-empty subset of $sched(l)$; then $prob(l, X)(l')$ is the probability of making a transition to location l' if the clock set X "expires" (reaches 0) in l. Finally, the delay distribution function *delay* associates a continuous density function with each clock, where we restrict the distributions to be of the following forms:

- distributions of the form $f_x(q) = 1$ for some $q \in \mathbb{N}$; we say that any clock $x \in \mathcal{X}$ is a *fixed-delay* clock if $delay(x)$ is of this form;
- positive bounded continuous density functions f_x with support $[a, b]$ where $a < b$ (therefore $\int_a^b f_x(q)\, dq = 1$); we say that any clock $x \in \mathcal{X}$ is a *variable-delay* clock if $delay(x)$ is of this form;
- exponential distributions f_x with the associated cumulative distribution function $F_x(t) = 1 - e^{-\lambda t}$ for some positive $\lambda \in \mathbb{R}_{\geq 0}$. Therefore, the interpretation of the distribution is that $F_x(t)$ gives the probability that the value selected by the distribution is less than or equal to t. We say that any clock $x \in \mathcal{X}$ is a *exponential-delay* clock if $delay(x)$ is of this form.

Note that [17] defines ways to obtain distributions with more permissive conditions from syntactic combinations of the above distributions. Finally, we note that the previous definitions of real-time probabilistic processes [2, 3], in addition to the literature on generalized semi-Markov processes [49, 24], refers to *events* where we have clocks in our definition above. This changes our *interpretation* of the apparatus used to trigger changes in location, but, as a unique clock is associated with each event in these previous studies, the change does not have a effect on the technicalities with regard to model checking. We therefore choose to include clocks explicitly in our presentation of real-time probabilistic processes in order to be consistent with the definitions of the other models in this paper. The reader more familiar with generalized semi-Markov processes may prefer to regard a clock as indicating an event.

The behaviour of a real-time probabilistic process RTPP takes the following form. The process commences in some location $l \in L$ such that $\bar{p}(l) > 0$, and schedules the set of clocks $sched(l)$ by choosing independently a real value d_x for each clock $x \in sched(l)$ according to the clock's delay distribution $delay(x)$. The value of the clock x is set to the chosen value $-d_x$. Time then passes, and the values of the clocks increase at the same rate, until one or more of the clocks reach the value 0. At this point, a discrete transition is made according to the probability transition function *prob* in the following manner: recalling that the current location is l, and using early $\subseteq sched(l)$ to denote the set of clocks which have reached 0, we use the discrete distribution $prob(l, \text{early})$ to choose probabilistically the location to which we make the transition. The set of scheduled clocks in the new location l' is determined by partitioning the set $sched(l')$ into two sets:

1. let $old(l, \text{early}, l') = sched(l') \cap (sched(l) \setminus \text{early})$ be the set of clocks that were scheduled in the previous location and did not elapse, and
2. let $new(l, \text{early}, l') = sched(l') \setminus old(l, \text{early}, l')$ be the set of clocks that are newly scheduled.

These sets are used to define the manner in which the clocks take values in the new location l'. Each clock in the set $old(l, \text{early}, l')$ retains its value from before the discrete transition. In contrast, for each clock x in the set $new(l, \text{early}, l')$, a value d_x is randomly and independently chosen from the clock's continuous density function $delay(x)$, and the value of x is set to $-d_x$. The behaviour of the process then proceeds anew with time elapsing until at least one clock reaches the value 0, at which point another discrete transition occurs.

A significant difference between real-time probabilistic processes and timed automata is that the former exhibit *only* probabilistic choice (which manifests itself within the probabilistic setting of clock values and the probabilistic branching between locations on discrete transitions), whereas the latter exhibit *only* nondeterministic choice (which manifests itself within the nondeterministic choice as to the exact time to select edges for choice, and the nondeterministic choice between simultaneously enabled edges).

4.3   Semantics of Real-Time Probabilistic Processes

Dense Markov Chains. A *dense Markov chain* DMC $= (S, \Sigma, \bar{\mathbf{p}}, \mathbf{P}, lab)$ consists of a set S of states, a sigma-field Σ over S, an initial distribution taking the form of a probability measure $\bar{\mathbf{p}} : \Sigma \to [0, 1]$, a probabilistic transition function $\mathbf{P} : S \times \Sigma \to [0, 1]$, where $\mathbf{P}(s, \cdot)$ is a probability measure for all states $s \in S$, and a labelling function $lab : S \to 2^{AP}$ such that $\{s \in S \mid lab(s) = a\} \in \Sigma$ for each atomic proposition $a \in AP$.

Semantics. Real-time probabilistic processes are expressed typically in terms of continuous-time Markov processes with continuous state spaces [2, 3, 17]. For consistency with Section 6, we instead present the formal semantics of real-time probabilistic processes using dense Markov chains. More precisely, the semantics of a real-time probabilistic process RTPP is expressed formally in terms of a dense Markov chain DMC $= (S, \Sigma, \bar{\mathbf{p}}, \mathbf{P}, lab)$. The state set S consists of location-valuation pairs (l, val), although with the following caveat: the valuations are defined only over the *scheduled* clocks of the location l. Hence, for a state $(l, val) \in S$ of the underlying dense Markov chain of a real-time probabilistic process, we have $val \in \mathbb{R}^{|sched(l)|}$. We note that this choice is made for notational convenience rather than from necessity.

The initial distribution of the dense Markov chain $\bar{\mathbf{p}} : \Sigma \to [0, 1]$ is derived from the initial distribution $\bar{p}$ of RTPP and the delay distribution function $delay$ in the following manner. First, for every location $l \in L$, we define the set $fix(l) \subseteq sched(l)$ to be the set of fixed-delay clocks scheduled in l, and $nonfix(l) = sched(l) \setminus fix(l)$ to be the set of variable- and exponential-delay clocks scheduled in l. Next, we choose an interval or point for each scheduled clock $x \in sched(l)$ in the following way:

1. for each non-fixed, scheduled clock $x \in nonfix(l)$, let $I_x \subseteq (-\infty, 0)$ be an interval of the negative real-line;
2. for each fixed, scheduled clock $x \in fix(l)$, let $I_x = [-q, -q]$, where $-q$ is the unique value for which $f_x(q) = 1$.

Let $(l, I) \subseteq S$ be the set of states with the location l and valuation val for which $val_x \in I_x$ for all scheduled clocks $x \in sched(l)$. Noting that $(l, I) \in \Sigma$, we let:

$$\bar{\mathbf{p}}(l, I) = \bar{p}(l) \cdot \prod_{x \in nonfix(l)} \int_{I_x} f_x \ .$$

An *initial state* of RTPP is a state $(l, val) \in S$ such that $\bar{p}(l) > 0$ and, for all clocks $x \in sched(l)$ scheduled in l, the following conditions hold:

- if $x \in fix(l)$ is a fixed-delay clock, then $-val_x = q$ for the unique value q for which $f_x(q) = 1$;
- if $x \in nonfix(l)$ is a variable- or exponential-delay clock, then $-val_x \in Int_x$, where Int_x is the interior of the support set of f_x.

The transition probability function $\mathbf{P} : S \times \Sigma \to [0, 1]$ is defined in a similar manner to the initial distribution. Consider the source state $(l, val) \in S$ of a transition, let early $\subseteq sched(l)$ be the set of clocks which have the greatest value in val, and let δ_{early} be this value multiplied by -1 (formally, let $\delta_{\text{early}} = -\max_{x \in sched(l)} val_x$). Now, for each clock $x \in \mathcal{X}$, let I_x be an interval of the real-line obtained in the following manner:

1. for each non-fixed, *newly*-scheduled clock $x \in nonfix(l) \cap new(l, \text{early}, l')$, let $I_x \subseteq (-\infty, 0)$ be an interval of the negative real-line;
2. for each fixed, *newly*-scheduled clock $x \in fix(l) \cap new(l, \text{early}, l')$, let $I_x = [-q, -q]$, where $-q$ is the unique value for which $f_x(q) = 1$;
3. for each *old* clock $x \in old(l, \text{early}, l')$ that continues to be scheduled in the new location, let $I_x = [val_x + \delta_{\text{early}}, val_x + \delta_{\text{early}}]$.

Again, let $(l', I) \subseteq S$ be the set of states with the location l' and valuation val for which $val_x \in I_x$ for all scheduled clocks $x \in sched(l)$. Noting that $(l, I) \in \Sigma$, we then define:

$$\mathbf{P}((l, val), (l', I)) = prob(l, \text{early})(l') \cdot \prod_{x \in nonfix(l) \cap new(l, \text{early}, l')} \int_{I_x} f_x \ .$$

We use the notation $(l, val) \xrightarrow{\delta_{\text{early}}} (l', val')$ to signify a transition from (l, val) to (l', val'), where $prob(l, \text{early})(l') > 0$ and the valuation val' satisfies the following properties; for each clock $x \in sched(l')$ scheduled in the new location l', we have:

- if $x \in \mathit{fix}(l) \cap \mathit{new}(l, \mathsf{early}, l')$ is a fixed, newly-scheduled clock, then $-\mathit{val}'_x = q$ for the unique value q for which $f_x(q) = 1$;
- if $x \in \mathit{nonfix}(l) \cap \mathit{new}(l, \mathsf{early}, l')$ is a non-fixed, newly-scheduled clock, then $-\mathit{val}'_x \in \mathit{Int}_x$, where Int_x is the interior of the support set of f_x;
- if $x \in \mathit{old}(l, \mathsf{early}, l')$ is an old clock, then $\mathit{val}'_x = \mathit{val}_x$.

Details of how the probability measure $Prob$ over sets of infinite sequences of a dense Markov chain of a real-time probabilistic process is obtained can be found in [34].

A computation of a real-time probabilistic process corresponds to an infinite path of its dense Markov chain. As in the case of timed automata, some of these computations correspond to unrealisable behaviour in which an infinite number of transitions occur in a finite amount of time. First we identify the duration of a transition $(l, \mathit{val}) \xrightarrow{\delta_{\mathsf{early}}} (l', \mathit{val}')$ as δ_{early} for (l, val). Then a path $s_0 \xrightarrow{\delta_0} s_1 \xrightarrow{\delta_1} \cdots$ of the dense Markov chain of a real-time probabilistic process is divergent if the infinite sum $\sum_{i \geq 0} \delta_i$ diverges. We say that a real-time probabilistic process is non-Zeno if, for each state, the probability measure over its divergent paths is 1. Unfortunately, it is possible to construct pathological examples of real-time probabilistic processes which are not non-Zeno (for example, a model with a reachable bottom strongly connected component in which only fixed-delay transitions of duration 0 are taken). However, it possible to test whether a real-time probabilistic process is non-Zeno using a property expressed as a deterministic timed Muller automaton.

4.4   Model Checking Real-Time Probabilistic Processes

Timed Properties. The property description language considered in [2] takes the form of a variant of the temporal logic TCTL. Our presentation of the syntax of the logic differs from [2], to emphasize the following point: in contrast to the traditional presentation of TCTL in the context of timed automata, the path quantifiers $\exists$ and $\forall$ of [2] do not refer to the satisfaction of some or all paths, but instead to the satisfaction of paths *with probability greater than 0*, and *with probability 1*, respectively, rather than to some or all paths. We prefer to make this distinction between the interpretation of TCTL on timed automata and on real-time probabilistic processes explicit, and, borrowing notation from quantitative probabilistic logics such as PCTL [25], replace $\exists$ with $\mathbb{P}_{>0}$ and $\forall$ with $\mathbb{P}_{=1}$. Formally, the syntax of the logic, denoted by $\overline{\text{TCTL}}$, is given by:

$$\phi ::= a \mid \phi \vee \phi \mid \neg \phi \mid \mathbb{P}_{>0}(\phi \mathcal{U}_{\sim c} \phi) \mid \mathbb{P}_{=1}(\phi \mathcal{U}_{\sim c} \phi)$$

where $a \in AP$ is an atomic proposition, $\sim \in \{<, \leq, =, \geq, >\}$ is a comparison operator, and $c \in \mathbb{N}$ is a natural constant.

Given a real-time probabilistic process $\text{RTPP} = (L, \bar{p}, \mathcal{X}, \mathit{sched}, \mathit{prob}, \mathit{delay}, \mathcal{L})$ and a $\overline{\text{TCTL}}$ formula ϕ, we define the satisfaction relation $\models^{\mathsf{T}}$ of $\overline{\text{TCTL}}$ as follows. As with TCTL, we express the satisfaction of the "until" formulae $\mathbb{P}_{>0}(\phi_1 \mathcal{U}_{\sim c} \phi_2)$ and $\mathbb{P}_{=1}(\phi_1 \mathcal{U}_{\sim c} \phi_2)$ using the path formula $\phi_1 \mathcal{U}_{\sim c} \phi_2$. We omit the semantics of the formulae a, $\phi_1 \vee \phi_2$, $\neg \phi$, and $\phi_1 \mathcal{U}_{\sim c} \phi_2$, as it is the same as that presented in Section 3 with $\models^{\overline{T}}$ substituted for $\models^T$. For the remaining operators, we define:

$$(l, val) \models^{\overline{T}} \mathbb{P}_{>0}(\phi_1 \mathcal{U}_{\sim c}\phi_2) \text{ iff } Prob\{\omega \in Path(l, val) \mid \omega \models^{\overline{T}} \phi_1 \mathcal{U}_{\sim c}\phi_2\} > 0$$
$$(l, val) \models^{\overline{T}} \mathbb{P}_{=1}(\phi_1 \mathcal{U}_{\sim c}\phi_2) \text{ iff } Prob\{\omega \in Path(l, val) \mid \omega \models^{\overline{T}} \phi_1 \mathcal{U}_{\sim c}\phi_2\} = 1 .$$

Note that we can use duality to write $\mathbb{P}_{=1}(\phi_1 \mathcal{U}_{\sim c}\phi_2)$ as

$$\neg[\mathbb{P}_{>0}(\square_{\sim c}\neg\phi_2) \vee \mathbb{P}_{>0}((\neg\phi_2)\mathcal{U}_{\sim c}(\neg\phi_1 \wedge \neg\phi_2))] ,$$

where the semantics of $\mathbb{P}_{>0}(\square_{\sim c}\phi)$ is expressed as:

$$(l, val) \models^{\overline{T}} \mathbb{P}_{>0}(\square_{\sim c}\phi) \text{ iff } Prob\{\omega \in Path(l, val) \mid \omega \models^{\overline{T}} \square_{\sim c}\phi\} > 0$$
$$\omega \models^{\overline{T}} \square_{\sim c}\phi \quad \text{iff for all positions } (i, \delta) \text{ of } \omega \text{ such that}$$
$$\text{duration}(\omega, i, \delta) \sim c, \text{ we have } \omega(i, \delta) \models^{\overline{T}} \phi .$$

That is, $\square_{\sim c}$ is a time-bounded version of the standard "globally" operator of temporal logic. Thus we can write any $\overline{\text{TCTL}}$ formula in terms of probabilistic quantifiers of the form $\mathbb{P}_{>0}$ only. Then we say that the real-time probabilistic process RTPP satisfies the $\overline{\text{TCTL}}$ formula ϕ if and only if there exists an initial state s of the underlying Markov process such that $s \models^{\overline{T}} \phi$.

Finally, we note that deterministic timed Muller automata can be used as property specifications of real-time probabilistic processes, rather than temporal logic properties, as described in [3]. This result is extended in [42] to consider nondeterministic timed Büchi automata (which have the same expressive power as nondeterministic timed Muller automata).

The Region Graph. In order the define a decidable model checking algorithm for real-time probabilistic processes against $\overline{\text{TCTL}}$ formulae, we resort to a version of clock equivalence, similar to that used for the definition of the region graph of timed automata. In this subsection, we show how a new definition of clock equivalence can be used to obtain a finite-state transition system which corresponds to the real-time probabilistic process. We consider a basic construction of the region graph, which suffices for $\overline{\text{TCTL}}$ formulae of the form $\mathbb{P}_{>0}(\phi_1 \mathcal{U}_{\geq 0}\phi_2)$ and $\mathbb{P}_{=1}(\phi_1 \mathcal{U}_{\geq 0}\phi_2)$ with trivial time bounds, and extend it to non-trivial time bounds in the next subsection.

Assume the real-time probabilistic process RTPP to be fixed. We use $\text{DMC} = (S, \Sigma, \bar{p}, P, lab)$ to denote the underlying dense Markov chain of RTPP. Let max be the maximum constant used in the definition of any fixed- or variable-delay clock distribution, and let $\text{min} = -\text{max}$. Given the clock set $X \subseteq \mathcal{X}$, we define *RTPP clock equivalence (with respect to X and min)* to be the relation $\equiv_X^{\min} \subseteq [\text{min}, 0]^{|X|} \times [\text{min}, 0]^{|X|}$, where $val \equiv_X^{\min} val'$ if and only if:

- for each clock $x \in X$,
  - $\lfloor val_x \rfloor = \lfloor val'_x \rfloor$ (clock values have the same integer part) and
  - $val_x - \lfloor val_x \rfloor = 0$ if and only if $val'_x - \lfloor val'_x \rfloor = 0$ (the fractional parts of the clocks' values are either all zero or all positive); and

- for each pair of clocks $x, y \in X$, we have $\lfloor val_x - val_y \rfloor = \lfloor val'_x - val'_y \rfloor$ (the integer part of the difference between clocks values is the same).

Observe that RTPP clock equivalence has several minor differences with clock equivalence presented in Section 2. Most obviously, RTPP clock equivalence partitions a section of *negative* clock valuation space, which is bounded from below by min and from above by 0. Instead, the clock equivalence of Section 2 partitions *positive* clock valuation space, and is unbounded from above. Note that we do not consider non-scheduled clocks in clock valuations in this section, and therefore it is not necessary for us to take into account the value of non-scheduled clocks in the clock equivalence.

Now we define *bounded-delay* clocks to be scheduled, non-exponential-delay clocks (therefore, they are either fixed- or variable-delay), and denote the set of bounded-delay clocks scheduled in a location $l \in L$ by $bounded(l)$. By definition, $bounded(l) \subseteq sched(l)$. We define *region equivalence for* RTPP *and* ϕ to be the relation $\asymp \subseteq S \times S$, where $(l, val) \asymp (l', val')$ if and only if:

- $l = l'$ (location components are equal);
- $val \equiv^{min}_{bounded(l)} val'$ (the valuations over bounded clocks scheduled in l are RTPP clock equivalent with respect to a minimum of min and the clock set $bounded(l)$).

We define a region to be an equivalence class of $\asymp$, and let R be the set of all regions.[2] Note that the equivalence relation $\asymp$ has a finite number of equivalence classes. We identify the regions which are reached just before an edge transition as a *discrete trigger*. Formally, a region $(l, \alpha) \in R$ is a discrete trigger if, for some clock $x \in bounded(l)$, the time successor of α includes a valuation val such that $val_x = 0$.

We are now able to define a finite-state transition system called the region graph $\mathsf{RG} = (R, \bar{R}, RT, Rlab)$ in the following manner. Note that we consider a variant of transition systems which have an initial *set* of states; hence, the initial set of regions is denoted by $\bar{R}$, where $(l, \alpha) \in \bar{R}$ if and only if there exists an initial state (l, val) such that $val \in \alpha$. With respect to the transition relation, let $RT \subseteq R \times R$ be the smallest set such that $(r, r') \in RT$ if either of the following conditions:

Time Transitions: we have $r = (l, \alpha)$ where r is not a discrete trigger, and $r' = (l, \beta)$ where β is the time successor of α;

---

[2] Some comments are in order, as our presentation is somewhat different to that of [2, 3, 17]. The minimum constant min is not used in these previous papers, as the idea there is to define an equivalence relation with an infinite number of equivalence classes, but for which the region graph can be constructed so that it navigates only a finite number of these classes. This can be seen by the fact that, although the partitioning of negative real-valued space in [2, 3, 17] extends towards $-\infty$, clocks will not be set to values lower than min, and therefore regions corresponding to clock values below min will never be considered as targets of the transition relation in the region graph.

Bounded-Delay Discrete Transitions: we have $r = (l, \alpha)$, where r is a discrete trigger, and $r' = (l', \beta)$ is such that:

1. $prob(l, \text{early})(l') > 0$;
2. for each newly-scheduled, fixed-delay clock $x \in fix(l') \cap new(l, \text{early}, l')$, we have $\beta \lfloor_x = -r$ where $f_x(r) = 1$;
3. for each newly-scheduled, variable-delay clock $x \in nonfix(l') \cap new(l, \text{early}, l')$, we have $\beta \lfloor_x \subseteq I_x$ such that $\beta \lfloor_x$ is not an integer, and there does not exist another clock $y \in sched(l')$ such that $val_x - val_y$ is an integer, for some $val \in \beta$;
4. for each old clock $x \in old(l, \text{early}, l')$, we have $\beta_x = \gamma_x$, where γ is the time successor of α.

Exponential-Delay Discrete Transitions: we have $r = (l, \alpha)$, and $r' = (l', \alpha)$ is such that there exists some $x \in sched(l) \setminus bounded(l)$ for which $prob(l, \{x\})(l') > 0$.

Finally, the labelling function $Rlab$ is obtained in the usual manner by letting $Rlab(l, \alpha) = \mathcal{L}(l)$ for all regions $(l, \alpha) \in R$.

As with the timed automaton case, we can identify a correspondence between the paths of the underlying dense Markov chain of a real-time probabilistic process and the associated region graph. That is, for any path of the region graph, we can find a path of the dense Markov chain, the states of which are contained within the regions of the region graph in an appropriate sequence. Similarly, for any path of the dense Markov chain, we can find a path of the region graph with the same kind of correspondence between states and regions.

Model Checking Using Regions. Given the construction of the region graph RG of a real-time probabilistic process RTPP and a $\overline{\text{TCTL}}$ formula ϕ, we are now in a position to consider whether RTPP satisfies ϕ. Without loss of generality, we assume that the $\overline{\text{TCTL}}$ formula ϕ is written using probabilistic quantifiers of the form $\mathbb{P}_{>0}$ only (potentially introducing the path formula $\square_{\sim c}$). As in Section 3, our strategy is to use the classical labelling algorithm to successively label the regions of RG with the subformulae of ϕ.

Despite the correspondence between paths of a real-time probabilistic process and the region graph, it is not immediate that RG can be used for model checking of RTPP against the $\overline{\text{TCTL}}$ formula ϕ. While the idea of the model checking techniques presented below is to perform graph-theoretical analysis on the region graph to infer probabilistic properties that are sufficient for establishing $\overline{\text{TCTL}}$ formulae, the probabilistic quantifiers $\mathbb{P}_{>0}$ of such formulae do not consider paths with probability 0, and therefore we have to rule out the possibility that a region graph path may correspond to a set of real-time probabilistic process paths which has probability 0. The case in which we have subformulae of the form $\mathbb{P}_{>0}(\phi_1 \mathcal{U}_{\sim c}\phi_2)$ is not problematic: if there exists a finite path in the region graph which satisfies $\phi_1 \mathcal{U}_{\sim c}\phi_2$ (given suitable adjustments to the region graph to cater for the time bound $\sim c$), then there will exist at least one finite path of the real-time probabilistic process which satisfies $\phi_1 \mathcal{U}_{\sim c}\phi_2$, and, as this is a finite path, its probability must be greater than 0. However, in the case of subformulae of the form $\mathbb{P}_{>0}(\square_{\sim c}\phi')$ we encounter the following problems. Recall that, in the case of model checking finite-state Markov chains [18], a subformula $\mathbb{P}_{>0}(\square_{\sim c}\phi')$ is established by checking for a reachable *bottom strongly connected component* (a strongly connected component of the graph of the Markov chain which cannot be exited), all of the states of which are labelled with ϕ'. All of the transition probabilities of states within a bottom strongly connected component of a finite-state Markov chain are fixed, and therefore are bounded from below. Hence, once the bottom strongly connected component is entered, each state within the component is visited infinitely often with probability 1. Instead, it is not immediate that the probabilities of transition between regions of RG obtained from RTPP are bounded from below: for example, if the region r is such that the fractional parts of the clocks x, y, z have the order $0 < x < y < z < 1$, and there occurs a transition to a region in which the fractional parts have the order $0 < y < x < z < 1$, then the probability of the transition is proportional to the difference $z - y$. Given that this transition of the region graph is visited infinitely often, if the value of $z - y$ converges to 0, then the probability of the transition also converges to 0. Therefore, it is established in [2,3] that this case can never occur: there is a positive probability that the clock values are separated by at least some fixed amount, and hence the graph-theoretical approach, applied to the region graph, can also be used to verify $\mathbb{P}_{>0}(\square_{\sim c}\phi')$. These results are also necessary in the case of properties expressed as deterministic timed Muller automata [3].

Now we turn to the labelling algorithms used to label RG. The cases of subformulae of the form $\mathbb{P}_{>0}(\phi_1\mathcal{U}_{\geq 0}\phi_2)$ and $\mathbb{P}_{>0}(\square_{\geq 0}\phi')$ follow from the model checking algorithms of [18]. Such classical labelling algorithms require adjustment when we consider subformulae of the general form $\mathbb{P}_{>0}(\phi_1\mathcal{U}_{\sim c}\phi_2)$ and $\mathbb{P}_{>0}(\square_{\sim c}\phi')$. We proceed by adding a single clock z to the clock set of the real-time probabilistic process, which has the role to measure global time, and to compare its value to c. The clock z is a fixed-delay clock, and is deterministically set to $-c$. We also introduce a special atomic proposition $a_{\sim c}$, which labels regions for which the value of z is ~ 0, and the atomic propositions a_b, which labels boundary regions in which no time can pass. Formally, we define the new real-time probabilistic process $\mathsf{RTPP}^{\sim c} = (L_1 \cup L_2, \bar{p}', \mathcal{X} \cup \{z\}, sched', prob', delay', \mathcal{L}')$. For each location $l \in L$, there are two copies $l_1 \in L_1$ and $l_2 \in L_2$ in the location set of $\mathsf{RTPP}^{\sim c}$; we then let $\bar{p}'(l_1) = \bar{p}'(l_2) = \frac{\bar{p}(l)}{2}$, $sched'(l_1) = sched(l) \cup \{z\}$ and $sched'(l_2) = sched(l)$. Hence, the new clock z is scheduled in L_1 but not in L_2. For each source location $l \in L$, non-empty clock set $X \subseteq \mathcal{X}$ and target location $l' \in L$ such that there exists a probabilistic transition from l to l' via the expiration of clocks in X (formally, $prob(l, X)(l') > 0$), then there exists probabilistic transitions from l_1 to l_1' and probabilistic transitions from l_2 to l_2', both of which are triggered by the expiration of clocks in X, where $prob'(l_1, X)(l_1') = prob'(l_1, X)(l_2') = prob'(l_2, X)(l_1') = prob'(l_2, X)(l_2') = \frac{prob(l, X)(l')}{2}$. Furthermore, for all $l \in L$, we also let $prob'(l_1, \{z\})(l_2) = 1$ (that is, if z expires, then the new real-time probabilistic process makes a transition from the copy of l in which z is scheduled to the other copy).

Our next task is to construct the region graph of $\mathsf{RTPP}^{\sim c}$, which we denote by $\mathsf{RG}^{\sim c} = (R', \bar{R}', RT', Rlab')$. Although the definition of $\mathsf{RG}^{\sim c}$ follows similarly to that presented in the previous subsection, the definition of the labelling function is somewhat different. For each region $r \in R'$, we let $Rlab'(r)$ be the set of labels as described in the previous definition of $Rlab$, but also including the atomic proposition a_b for all boundary regions and the atomic proposition $a_{\sim c}$ if:

- the location component of r is in the set L_1 and $(c + 1 - val_z) \sim c$ for some valuation val contained in the clock equivalence class component of r;
- the location component of r is in the set L_2 and $c + 1 \sim c$.

Then we can apply traditional model checking algorithms to establish which regions satisfy the time-bounded subformulae. For example, for $\mathbb{P}_{>0}(\phi_1 \mathcal{U}_{\sim c} \phi_2)$, we check for a path along which $\phi_2 \wedge a_{\sim c} \wedge (a_b \vee \phi_1)$ holds at some point and ϕ_1 holds at all preceding points (the justification of the conjunct $a_b \vee \phi_1$ is the same as in Section 3).

To conclude, we can establish whether RTPP satisfies ϕ by verifying whether the labelling algorithm labels an initial region in $\bar{R}$ with ϕ.

4.5   Related Work

An approach to the qualitative verification of reachability properties of stochastic automata is presented in [19]. Stochastic automata are similar to real-time probabilistic processes but have some nondeterministic choice over edges when clocks expire, and can permit parallel composition of stochastic automaton sub-components. The approach of D'Argenio is to transform each sub-component into a timed automaton, following which reachability analysis may be performed using timed automata model checking tools such as UPPAAL or KRONOS. The verification is conservative in the sense that if a set of states is reached in the timed automaton model then it is also reached with non-zero probability in the stochastic automaton model, but not necessarily vice versa.

We also report on methods for the verification of models similar to real-time probabilistic processes against *quantitative* probabilistic properties, in which probabilistic quantifiers are more general than those used in this section (that is, the probabilistic quantifiers are of the form $\mathbb{P}_{\bowtie \lambda}$ for $\bowtie \in \{<, \leq, \geq, >\}$ and $\lambda \in [0, 1]$, as introduced in the probabilistic temporal logic PCTL [25]). Model checking quantitative probabilistic properties of semi-Markov chains is considered in [41]. Algorithms for the verification of a restricted class of stochastic automata (for which the model's clocks must be reset on entry to a location, similarly to semi-Markov processes) against quantitative probabilistic properties, are presented in [14]. These algorithms assume that nondeterminism is resolved before they are executed. In [51], a verification method based on *sampling* paths of a generalized-semi Markov process is presented. The method can both be used to obtain an estimate of the probability that a path formula is satisfied and to obtain error bounds for the estimate.

5   Probabilistic Timed Automata

5.1   Modelling with Probabilistic Timed Automata

A probabilistic timed automaton can be used to model the light switch of Section 3 in the presence of faults. Suppose that the mechanism of the switch is faulty, so that there is a 1% chance that, when the switch is pressed, the light will not switch on. The resulting probabilistic timed automaton is shown in Figure 4. Like timed automata, the model consists of locations annotated by clock constraints. However, the edges have changed form: from off and on there exist bifurcating edges that represent probabilistic choice. For example, in off, the light may be switched on at any time, corresponding to the bifurcating edge leaving this location. Once the decision to leave the location has been made, then a probabilistic choice of one branch (for example, the switch fails and we return to off with probability 0.01) or of the other branch (the switch works and we move to on while assigning 0 to the clock x with probability 0.99) is performed.

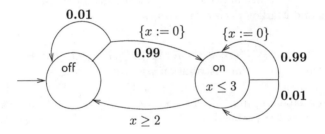

Fig. 4. A probabilistic timed automaton modelling a faulty switch

5.2   Syntax of Probabilistic Timed Automata

A *probabilistic timed automaton* PTA $= (L, \bar{l}, \mathcal{X}, inv, prob, \mathcal{L})$ [30, 35] is a tuple, the first four and the final elements of which are borrowed directly from the definition of timed automata: that is, L is a finite set of locations with the initial location $\bar{l}$, we have a finite set $\mathcal{X}$ of clocks, and inv and $\mathcal{L}$ are functions associating an invariant condition and a set of atomic propositions, respectively, with each location. In contrast to the definition of timed automata, the finite set $prob \subseteq L \times \Psi_{\mathcal{X}} \times \mathsf{Dist}(2^{\mathcal{X}} \times L)$ denotes a set of *probabilistic edges*. Each probabilistic edge $(l, g, p) \in prob$ is determined by (1) its source l location, (2) its guard g, and (3) its probability distribution p which assigns probability to pairs of the form (X, l') for some clock reset X and target location l'. Observe that p is a *discrete* probability distribution, because the sets $\mathcal{X}$ and L are finite.

The behaviour of a probabilistic timed automata takes a similar form to that of timed automata: in any location time can advance to an upper limit given by the restrictions of the invariant condition on clock values, and a probabilistic edge can be taken if its guard is satisfied. However, probabilistic timed automata generalize timed automata in the sense that, once a probabilistic edge is nondeterministically selected, then the choice of which clock reset to use and which target location to make the transition to is *probabilistic*. For example, in Figure 4 there is a probabilistic choice between the switch failing, resulting in a self-loop to the same location with no clock assignments, and the switch working, resulting in a transition to a location with an assignment of 0 to the clock x. Probabilistic timed automata extend timed automata with probabilistic branching, but are incomparable to the real-time probabilistic processes of Section 4, because they feature nondeterministic choice and are unable to set clocks according to continuous density functions.

Note that parallel composition for probabilistic timed automata can be defined by combining the parallel composition theories of timed automata [5] and probabilistic transition systems [46], given the extension of probabilistic timed automata with actions, as shown in [38].

5.3   Semantics of Probabilistic Timed Automata

The semantics of a probabilistic timed automaton is expressed in terms of Markov chains and Markov decision processes.

Markov Chains. A *Markov chain* MC $= (S, \bar{p}, P, lab)$ consists of a countable set S of states, an initial distribution $\bar{p} \in \mathsf{Dist}(S)$ over states, a probabilistic transition function $P : S \times S \to [0,1]$ such that $\sum_{s' \in S} P(s, s') = 1$ for all states $s \in S$, and a labelling function $lab : S \to 2^{AP}$. A probability measure *Prob* over infinite sequences of states of the Markov chain can be defined in the classical manner [31].

Markov Decision Processes. A *Markov decision process* MDP $= (S, \bar{s}, D, lab)$ comprises a set of states S with an initial state $\bar{s}$, a labelling function $lab : S \to 2^{AP}$, and a probabilistic-nondeterministic transition relation $D \subseteq S \times \mathsf{Dist}(S)$. The transitions from state to state of a Markov decision process are performed in two steps: the first step concerns a nondeterministic selection of a discrete probability distribution associated with the current state (that is, if the current state is s, then a distribution is chosen nondeterministically from the set $\{\mu \mid (s, \mu) \in D\}$); the second step comprises a probabilistic choice according to the chosen distribution (that is, given the choice of μ in the first step, we then make a transition to a state $s' \in S$ with probability $\mu(s')$). We often denote such a transition by $s \xrightarrow{\mu} s'$. Note that probabilistic timed automata extend Markov decision processes with the ability of resetting and testing the value of clocks, in the same way as timed automata extend finite automata with the ability of resetting and testing the value of clocks.

An infinite or finite *path* of MDP is defined as an infinite or finite sequence of transitions, respectively. We use $Path_{fin}$ to denote the set of finite paths of MDP, and $Path_{ful}$ the set of infinite paths of MDP. If ω is finite, we denote by $last(\omega)$ the last state of ω. We abuse notation by letting a state $s \in S$ to denote a path starting in s but consisting of no transitions, and by using $\omega \xrightarrow{\mu} s$ to refer to a path comprising the sequence of transitions of ω followed by the transition $last(\omega) \xrightarrow{\mu} s$. Finally, let $Path^A_{fin}(s)$ and $Path^A_{ful}(s)$ refer to the set of finite and infinite paths, respectively, commencing in state $s \in S$.

In contrast to a path, which corresponds to a resolution of nondeterministic and probabilistic choice, an *adversary* represents a resolution of nondeterminism *only*. Formally, an adversary of a Markov decision process MDP is a function A mapping every finite path $\omega \in Path_{fin}$ to a distribution $\mu \in \mathsf{Dist}(S)$ such that $(last(\omega), \mu) \in D$ [48]. Let Adv be the set of adversaries of MDP.

We now see how adversaries can be used to express probability measures of certain behaviours of a Markov decision process. For any adversary $A \in Adv$, let $Path^A_{fin}$ and $Path^A_{ful}$ denote the set of finite and infinite paths associated with A (more precisely, the paths resulting from the choices of distributions of A). Then, for a state $s \in S$, we define the probability measure $Prob^A_s$ over $Path^A_{ful}(s)$ in the following, standard way [48]. We define the Markov chain $\mathsf{MC}^A_s = (Path^A_{fin}(s), \{s \mapsto 1\}, \mathsf{A}, \_)$, where the set of states of the Markov chain is the set of finite paths generated by A starting in s, the initial distribution and the labelling function are arbitrary, and the probabilistic transition function function $\mathsf{A} : Path^A_{fin}(s) \times Path^A_{fin}(s) \to [0, 1]$ is defined such that:

$$\mathsf{A}(\omega, \omega') = \begin{cases} A(\omega)(s') & \text{if } \omega' = \omega \xrightarrow{A(\omega)} s' \\ 0 & \text{otherwise.} \end{cases}$$

Intuitively, $\mathsf{A}(\omega, \omega')$ refers to the probability of obtaining the finite path ω' when extending the finite path ω with one transition under the control of the adversary A. Now we are can define the probability measure $Prob^A_s$ over infinite paths which share the same finite prefixes, using the Markov chain MC^A_s obtained from the adversary A and classical methods [31].

Semantics. The semantics of the probabilistic timed automaton PTA $= (L, \bar{l}, \mathcal{X}, inv, prob, \mathcal{L})$ is the infinite-state Markov decision process MDP $= (S, \bar{s}, D, lab)$ derived in the following manner. The state set S comprises location-valuation pairs, where a valuation must satisfy the invariant of a location with which it is paired. As with timed automata, we let the initial state $\bar{s}$ be $(\bar{l}, \mathbf{0})$, and let the labelling function lab be such that $lab(l, val) = \mathcal{L}(l)$ for all states $(l, val) \in S$. The probabilistic transitions of MDP are defined in the following manner. We define D to be the smallest set such that $((l, val), \mu) \in D$ if both of the following conditions hold:

Time Elapse: there exists $\delta \in \mathbb{R}_{\geq 0}$ such that $val + \delta' \models inv(l)$ for all $0 \leq \delta' \leq \delta$, and

Edge Traversal: there exists a probabilistic edge $(l, g, p) \in E$ such that $val + \delta \models g$ and, for any $(l', val') \in S$, we have:

$$\mu(l', val') = \sum_{\substack{X \subseteq \mathcal{X} \,\& \\ val' = val[X:=0]}} p(X, l') \;.$$

We refer to the real δ used in the time-elapse part of a transition as the *duration* of the transition. For each $s' \in S$ such that $\mu(s') > 0$, we use $s \xrightarrow{\mu[\delta,p]} s'$ to denote the discrete transition from s to s' with the duration δ using the distribution p. The summation in the definition of discrete transitions is required for the cases in which multiple clock resets result in the same target state (l', val'). Note that we assume a well-formedness condition which stipulates that the invariant condition of a target location of a transition is always satisfied immediately after a transition is taken (for more details, see [38]).

As with our treatment of the semantics of timed automata and real-time probabilistic processes, a behaviour of a probabilistic timed automaton takes the form of an infinite sequence of transitions. We can also extend the idea of divergence and non-Zenoness to probabilistic timed automata: using the precedents of [45, 22], we associate a notion of divergence with *adversaries*. First we say that a path $s_0 \xrightarrow{\mu[\delta_0, p_0]} s_1 \xrightarrow{\mu[\delta_1, p_1]} \cdots$ of the underlying Markov decision process of a probabilistic timed automaton is *divergent* if the infinite sum $\sum_{\geq 0} \delta_i$ diverges. Then, for a state $s \in S$, the adversary $A \in Adv$ is a *divergent adversary from* s if $Prob_s^A \{\omega \in Path_{ful}^A(s) \mid \omega \text{ is divergent}\} = 1$. A probabilistic timed automaton is *non-Zeno* if there exists a divergent adversary from all states.

5.4   Model Checking Probabilistic Timed Automata

Probabilistic Timed Properties. We now proceed to define a *probabilistic, timed* temporal logic which can be used to specify properties of probabilistic timed automata. In particular, we note that the logic that we present in this section has the power to refer to *exact* probability values, in contrast to TCTL presented in Section 4, which only had the ability to refer to probability strictly greater than 0 or equal to 1. Our approach is to define a logic which draws ideas from TCTL, and the probabilistic temporal logic PCTL [25], in which a probabilistic quantifier of the form $\mathbb{P}_{\bowtie\lambda}$ is used in place of a path quantifier of CTL, where $\bowtie \in \{<, \leq, \geq, >\}$ is a comparison operator and $\lambda \in [0, 1]$ is a probability.

Formally, the syntax of PTCTL (Probabilistic Timed Computation Tree Logic) [35] is defined as follows:

$$\phi ::= a \mid \phi \vee \phi \mid \neg\phi \mid \mathbb{P}_{\bowtie\lambda}(\phi \mathcal{U}_{\sim c} \phi) \mid \mathbb{P}_{\bowtie\lambda}(\phi \mathcal{U}_{\sim c} \phi)$$

where $a \in AP$ is an atomic proposition, and $\bowtie$ and λ are the comparison operator and the probability defined above. Given a probabilistic timed automaton $\mathsf{PTA} = (L, \bar{l}, \mathcal{X}, inv, prob, \mathcal{L})$ and a PTCTL formula ϕ, we define the satisfaction relation $\models^{\mathrm{PT}}$ of PTCTL as follows. The semantics of the formulae a, $\phi_1 \vee \phi_2$, $\neg\phi$, and $\phi_1 \mathcal{U}_{\sim c} \phi_2$ is omitted, as it is the same as that presented for TCTL in Section 3 with $\models^{\mathrm{PT}}$ substituted for $\models^{\mathrm{T}}$. We then define:

$$(l, val) \models^{\mathrm{PT}} \mathbb{P}_{\bowtie \lambda}(\phi_1 \mathcal{U}_{\sim c} \phi_2) \text{ iff}$$
$$\text{for all adversaries } A \in Adv \text{ which are divergent from } (l, val),$$
$$\text{we have } Prob^A_{(l, val)}\{\omega \in Path^A_{ful}(l, val) \mid \omega \models^{\mathrm{PT}} \phi_1 \mathcal{U}_{\sim c} \phi_2\} \bowtie \lambda .$$

The probabilistic timed automaton PTA with the initial location $\bar{l}$ satisfies the PTCTL formula ϕ if and only if $(\bar{l}, \mathbf{0}) \models^{\mathrm{PT}} \phi$.

Probabilistic timed automata can also be verified against properties which refer to the maximal probability with which a Büchi condition is satisfied [10].

The Region Graph. The fundamental notion of clock equivalence, defined in Section 2, can be used to obtain a finite-state Markov decision process from a probabilistic timed automaton which contains sufficient information to establish or refute PTCTL properties. For the remainder of this section, assume that the probabilistic timed automaton PTA $= (L, \bar{l}, \mathcal{X}, inv, prob, \mathcal{L})$ and the PTCTL formula ϕ are fixed, and that we use MDP $= (S, \bar{s}, D, lab)$ to denote the underlying Markov decision process of PTA. Let max be the maximal constant against which any clock is compared in the invariant conditions and guards of PTA. Then we define *region equivalence for* PTA as the relation $\simeq \subseteq S \times S$, where $(l, val) \simeq (l', val')$ if and only if:

- $l = l'$ (the location components are equal);
- $val \equiv_{\max} val'$ (the valuations are clock equivalent with respect to max).

Observe that this is the same definition of region equivalence defined for timed automata in Section 3, now applied to the states of the underlying Markov decision process of a probabilistic timed automaton. We define a *region* to be an equivalence class of $\simeq$, and use R to denote the set of regions. Region equivalent states of a probabilistic timed automaton satisfy the same PTCTL formulae [35], and region equivalence is a probabilistic bisimulation (in the sense of [40, 46], but not necessarily the coarsest) of a probabilistic timed automaton, provided that the duration of transitions is abstracted.

The property that the set of clock equivalence classes is finite, and therefore the number of regions is finite, permits us to define a finite-state region graph. Many of the principles involved in the construction of the region graph are borrowed from the corresponding constructions that we have seen for timed automata and real-time probabilistic processes. However, in contrast with the definitions in Section 3 and Section 4, the region graph now incorporates probability values. More precisely, the *region graph of* PTA *and* ϕ is a Markov decision process $\mathcal{RG} = (R, \bar{r}, \mathcal{RT}, Rlab)$, where the initial region $\bar{r} = \{(\bar{l}, \mathbf{0})\}$ and region labelling function $Rlab(l, \alpha) = \mathcal{L}(l)$, for all regions $(l, \alpha) \in R$, are defined in a similar manner as in the case of the region graph of a timed automaton. The probabilistic transition relation $\mathcal{RT} \subseteq R \times \text{Dist}(R)$ is the smallest relation such that $((l, \alpha), \rho) \in \mathcal{RT}$ if either of the following conditions hold:

Time Elapse: $\rho = \{(l, \beta) \mapsto 1\}$, where β is an $inv(l)$-satisfying time successor of α;

Probabilistic Edge Traversal: there exists a probabilistic edge $(l, g, p) \in prob$ such that $\alpha \models g$ and, for any region $(l', \beta) \in R$, we have:

$$\rho(l', \beta) = \sum_{\substack{X \subseteq \mathcal{X} \ \& \\ \beta = \alpha[X:=0]}} p(X, l') \,.$$

The fact that the region graph defined above includes probabilistic information, derived from the probabilistic edges of the associated probabilistic timed automaton, allows us to strengthen the notion of correspondence between behaviours of our probabilistic timed system and its associated region graph. More precisely, in addition to showing that every path of the region graph corresponds to at least one path of the probabilistic timed automaton, and that every path of the probabilistic timed automaton corresponds to a path of the region graph, we can also make a stronger statement: every adversary of the probabilistic timed automaton has a corresponding adversary of the region graph, and that every adversary of the region graph corresponds to at least one adversary of the probabilistic timed automaton.

Model Checking Using Regions. Given the construction of the region graph $\mathcal{RG}$ of a probabilistic timed automaton PTA and a PTCTL formula ϕ, we can now proceed to model check $\mathcal{RG}$ to establish whether PTA satisfies ϕ. In analogy with the corresponding case of timed automata, we borrow ideas from finite-state model checking, in this case probabilistic model checking of finite-state Markov decision processes [11, 9]. The technique of extending PTA with an additional clock to count time relative to the time-bound subscripts of the PTCTL formula ϕ, and adding extra atomic propositions a_b and $a_{\sim c}$ to the region graph, is adopted directly from TCTL model checking. Then it remains to use probabilistic model checking techniques to calculate the probability of satisfying subformulae of ϕ in each region, progressing up the parse tree of ϕ until we reach ϕ itself. Then if the initial region $\bar{r}$ satisfies ϕ, we can conclude that PTA satisfies ϕ.

As in the case of model checking timed automata, the assumption of non-Zenoness of the probabilistic timed automaton guarantees that all regions have some successor. We can adapt strong non-Zenoness [47] to probabilistic timed automata to give syntactical conditions which guarantee the non-Zenoness property.

Finally, we note that a region-graph-based construction can also be used to obtain the maximum probability with which a probabilistic timed automaton satisfies a Büchi property [10].

Beyond Regions. As with timed automata, the region graph construction for probabilistic timed automata is expensive, and therefore efficient model checking techniques which do not rely explicitly on the region graph are required.

Adaptations of methods based on the manipulation of clock constraints are presented in [35, 36], which consider properties referring to the maximal probability of reaching a certain state set (that is, properties of the form $\mathbb{P}_{\leq c}(\Diamond a)$ or $\mathbb{P}_{< c}(\Diamond a)$). Furthermore, a discrete-time semantics for probabilistic timed automata is defined in [38], and permits direct use of the probabilistic model checker PRISM [32]. Probabilistic timed properties of the IEEE1394 FireWire root contention protocol have been verified using algorithms based on manipulation of clock constraints [38, 21] and the discrete-time semantics [38]. The discrete-time semantics has also been used to facilitate probabilistic model checking of properties of a IEEE802.11 wireless local area network [37] and the IPv4 link-local address configuration protocol [33]. The latter case study also illustrates how the discrete-time semantics may be used to refer to expected costs incurred before the probabilistic timed automaton reaches a certain set of states.

6   Continuous Probabilistic Timed Automata

6.1   Modelling with Continuous Probabilistic Timed Automata

We revisit the example of the light switch for a final time in order to introduce the model of continuous probabilistic timed automata. As in Section 4, we consider randomly distributed arrivals of people wanting to use the switch; more precisely, we consider arrivals uniformly distributed between 1 and 30 minutes. Furthermore, as in Section 5, we assume that the switch is faulty. The resulting continuous probabilistic timed automaton is shown in Figure 5. The model combines location annotations indicating clocks (such as y in off1) and clock constraints (such as $x \leq 3 \wedge y \leq 30$ in on). The first type of location annotation gives the clocks which must be reset on entry to the location, whereas the second indicates the location's invariant condition. The edges leaving a location can split into multiple branches, indicating probabilistic edges, such as those used for probabilistic timed automata in Section 5.

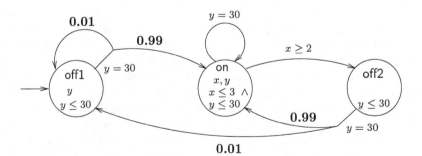

Fig. 5. A continuous probabilistic timed automaton modelling a faulty switch

In Figure 5, we always set the clock x to 0 on entry to a location with which it is annotated, whereas y is set uniformly to a real value in the interval $[0, 29]$. Then, if y reaches the value 30, we know that between 1 and 30 time units have elapsed since it was set. We use two locations off1 and off2 to model the light being off, as we need to distinguish between the case in which the clock y needs to be set to a new value (off1) and the case in which y should retain its old value (off2).

6.2   Syntax of Continuous Probabilistic Timed Automata

A *continuous probabilistic timed automaton* CPTA $= (L, \bar{l}, \mathcal{X}, inv, \underline{prob}, reset, \mathcal{L})$ [34] is a tuple, the first four and final elements of which are borrowed directly from timed automata. That is, L is a finite set of locations with initial location $\bar{l}$, the finite set $\mathcal{X}$ comprises clocks, and inv and $\mathcal{L}$ are functions associating an invariant condition and a set of atomic propositions, respectively, with each location. The finite set $\underline{prob} \subseteq L \times \Psi_{\mathcal{X}} \times \text{Dist}(L)$ is the set of probabilistic edges of the continuous probabilistic timed automaton. Observe that, for each probabilistic edge $(l, g, p) \in \underline{prob}$, the distribution component p is a distribution over the set of locations L *only*, rather than over the set $L \times 2^{\mathcal{X}}$ of pairs of locations and clock resets, as we saw in probabilistic timed automata.

The partial function $reset : L \times \mathcal{X} \to PB$, referred to as the *clock reset function*, associates a positive bounded density function with certain location-clock pairs. The interpretation of the clock reset function is as follows: if $reset(l, x)$ is defined for a location $l \in L$ and a clock $x \in \mathcal{X}$, then $reset(l, x)$ is the positive-bounded density function which captures the manner in which the clock x is reset upon entry into l. That is, in contrast to the case of probabilistic timed automata, we now do not only consider resetting the values of clocks to the value 0, but we also consider resetting to clocks to *exact, natural numbered values* (if $reset$ returns a distribution which assigns probability 1 to a natural number) or according to *random assignments* (if $reset$ returns a more general positive-bounded distribution function) in the manner of real-time probabilistic processes. Instead, if $reset(l, x)$ is undefined, then the clock x retains its old value upon entry to the location l. Thus, the behaviour of a continuous probabilistic timed automata derives aspects from all of the previous three models featured previously in this paper: it adopts the ability to select the exact time at which transitions are taken according to nondeterministic choice from timed automata, the discrete probability distributions which can be used to express relative likelihoods of transitions between locations from probabilistic timed automata, and the resetting of clocks to values taken from continuous density functions from real-time probabilistic processes. The invariant conditions and guards of probabilistic edges retain their usual interpretation from timed automata. That is, time can elapse in a location only for as long as the invariant condition is satisfied by the current values of the clocks, and a probabilistic edge can be selected (nondeterministically) for choice at all times when its guard is satisfied by the current values of the clocks. Furthermore, probabilistic edges retain their interpretation from probabilistic timed automata, in the sense that, once nondeterministically selected for choice, they involve a probabilistic choice made according to their distribution component. The manner in which clocks are reset bears more of a resemblance to the analogous mechanism of real-time probabilistic processes.

Note that we require a well-formedness condition which requires a clock to be set explicitly to a value before it is referred to in the invariant condition of a location or in the guards of one of the outgoing probabilistic edges. Once we have imposed this condition, we can adjust the density function $reset$ so that, in the initial location $\bar{l}$, the clocks that were not originally assigned a value are assigned an arbitrary value, for example 0. This final adjustment is made for purely technical reasons, and ensures that our clock valuations are always points in $\mathbb{R}_{\geq 0}^{|\mathcal{X}|}$ space, rather than in positive real-valued space of dimension smaller than $|\mathcal{X}|$, as we saw in the case of real-time probabilistic processes.

Observe that both real-time probabilistic processes and continuous probabilistic timed automata associate a set of clocks to each location (using the functions $sched$ and $reset$, respectively), and that these sets determine the manner in which clocks are reset. However, the interpretation of these clock sets is somewhat different for the two models. In real-time probabilistic processes, the set $sched(l)$ contains the clocks which are $scheduled$ in a location l; therefore, if such a clock has been scheduled in the previous location, it keeps its old value, otherwise it is reset to a new value. On the other hand, in continuous probabilistic timed automata, $sched$ defines the clocks that must be reset to a new value on entry to the location, regardless of the previous location, and all clocks not within this set retain their old values. (This type of mechanism is also witnessed in stochastic automata [20].)

It is important to note that continuous probabilistic timed automata are not a superclass of real-time probabilistic processes, as they do not support exponentially distributed functions. However, continuous probabilistic timed automata are a superclass of real-time probabilistic processes without such distributions (for example, the real-time probabilistic processes of [2]). A continuous probabilistic timed automaton which emulates a real-time probabilistic process can be constructed in the following way. First, we note that the clocks of continuous probabilistic timed automata take non-negative values only, whereas the clocks of real-time probabilistic processes take values in the interval [min, 0] only, where min is defined as in Section 4. Therefore, we "shift the axes" by adding the value $-$ min to each of the constant used in the description of the real-time probabilistic process. Note that this involves constructing invariants and guards using the constraints $x \leq -$ min and $x = -$ min, respectively, for all scheduled clocks and all locations. Our second task is to ensure that the continuous probabilistic timed automaton which represents the real-time probabilistic process resets the clocks in the correct manner, which can be achieved by adding extra copies of locations.

We can also observe the probabilistic timed automata are a subclass of continuous probabilistic timed automata. In probabilistic timed automata, all clock assignments can be regarded as fixed-delay distributions which set clocks to 0. Then, as in the previous example, we require some bookkeeping to require that the correct set of clocks are reset when the continuous probabilistic timed automaton makes a discrete transition, which is achieved by adding copies of locations.

6.3   Semantics of Continuous Probabilistic Timed Automata

Dense Markov Decision Processes. A *dense Markov decision process* DMDP $= (S, \Sigma, \bar{\mathbf{d}}, \mathbf{D}, lab)$ is a quintuple, the first three and the final elements of which are adopted directly from dense Markov chains. That is, DMDP comprises a set S of states, a sigma-field Σ over S, an initial distribution taking the form of a probability measure $\bar{\mathbf{d}} : \Sigma \to [0, 1]$, and a labelling function $lab : S \to 2^{AP}$ such that $\{s \in S \mid lab(s) = a\} \in \Sigma$ for each atomic proposition $a \in AP$. The difference between dense Markov chains and dense Markov processes resides in the probabilistic transition relation $\mathbf{D} \subseteq S \times (\Sigma \to [0, 1])$, which is defined such that $\mathbf{d}$ is a probability measure for each pair $(s, \mathbf{d}) \in \mathbf{D}$.

Semantics. We now describe formally the semantics of a continuous probabilistic timed automaton CPTA $= (L, \bar{l}, \mathcal{X}, inv, \underline{prob}, reset, \mathcal{L})$ in terms of a dense Markov decision process DMDP $= (S, \Sigma, \bar{\mathbf{d}}, \mathbf{D}, \overline{lab})$. As usual, the set S of states consists of pairs of the form (l, val) where $val \models inv(l)$, and the labelling function is defined as $lab(l, val) = \mathcal{L}(l)$ for each state $(l, val) \in S$.

Before proceeding to the definitions of the initial distribution $\bar{\mathbf{d}}$ and the probabilistic transition relation $\mathbf{D}$, we introduce some notation. For each location $l \in L$, let $entry(l) \subseteq \mathcal{X}$ be the set of clocks such that, for each clock $x \in entry(l)$, the function $reset(l, x)$ is defined (that is, $entry(l)$ comprises the clocks reset on entry to l). As in the case of real-time probabilistic processes, let $fix(l) \subseteq entry(l)$ be the set of fixed-delay clocks which are reset in l (formally, we have $x \in fix(l)$ if and only if $reset(l, x)$ is defined and equals $\{q \mapsto 1\}$ for some $q \in \mathbb{N}$). Then let $nonfix(l) = entry(l) \setminus fix(l)$ be the set of variable-delay clocks which are reset on entry to l.

The initial distribution $\bar{\mathbf{d}} : \Sigma \to [0, 1]$ is derived from the initial location $\bar{l}$ of CPTA and probability distributions $reset(\bar{l}, x)$, for each clock $x \in \mathcal{X}$ (recall that we assumed $entry(\bar{l}) = \mathcal{X}$). As we saw in the case of real-time probabilistic processes, we choose an interval or point for each clock $x \in \mathcal{X}$ in the following manner:

1. for each non-fixed clock $x \in nonfix(l)$, let $I_x \subseteq (0, \infty)$ be an interval of the positive real-line;
2. for each fixed clock $x \in fix(l)$, let $I_x = [q, q]$, where q is the unique value for which $f_x(q) = 1$.

Let $(\bar{l}, I) \subseteq S$ be the set of states with the initial location $\bar{l}$ and valuation val for which $val_x \in I_x$ for all scheduled clocks $x \in \mathcal{X}$. Noting that $(l, I) \in \Sigma$, we then let:

$$\bar{\mathbf{d}}(\bar{l}, I) = \prod_{x \in nonfix(\bar{l})} \int_{I_x} reset(\bar{l}, x) \,.$$

The probabilistic transition relation $\mathbf{D} \subseteq S \times (\Sigma \to [0, 1])$ is defined in a similar manner. As with all of the other models studied in this paper, transitions consist of a time-elapse phase and an edge-traversal phase. We let $\mathbf{D}$ be the smallest relation such that $((l, val), \mathbf{d}) \in \mathbf{D}$ if both of the following conditions hold:

Time Elapse: there exists $\delta \in \mathbb{R}_{\geq 0}$ such that $val + \delta' \models inv(l)$ for all $0 \leq \delta' \leq \delta$, and

Probabilistic Edge Traversal: there exists a probabilistic edge $(l, g, p) \in$ _prob_, such that $val + \delta \models g$ and, for each set (l', I) of states comprising some location $l' \in L$ and a set I of valuations satisfying the following conditions:

1. for each $x \in nonfix(l')$, the interval $I_x \subseteq (0, \infty)$ is an interval of the real-line;
2. for each $x \in fix(l')$, the interval I_x is such that $I_x = [q, q]$, where $q \in \mathbb{N}$ is the unique value for which $reset(l', x) = \{q \mapsto 1\}$;
3. for each $x \in \mathcal{X} \setminus entry(l')$, the interval I_x is such that $I_x = [val_x + \delta, val_x + \delta]$;

then we have:

$$\mathbf{d}(l', I) = p(l') \cdot \prod_{x \in nonfix(l')} \int_{I_x} reset(l', x) \,.$$

We denote a transition of the dense Markov process as $(l, val) \xrightarrow{\mathbf{d}[\delta, p]} (l', val')$ if $\delta \in \mathbb{R}_{\geq 0}$ is the duration and p the distribution used to define the transition above, and the target state (l', val') is such that the following conditions hold:

- for each defined clock $x \in nonfix(l')$ for which $reset(l', x)$ has support $[q_1, q_2]$, we have $val'_x \in (q_1, q_2)$;
- for each defined clock $x \in fix(l')$ for which $reset(l', x)$ has support $[q, q]$, we have $val'_x = q$;
- for each non-defined clock $x \in \mathcal{X} \setminus entry(l')$, we have $val'_x = val_x$.

As with probabilistic timed automata, we assume a well-formedness condition which guarantees that a clock never has a value which does not satisfy an invariant condition on entry to a location.

We can define notions of path and adversaries of dense Markov decision processes. An infinite or finite _path_ of DMDP is defined as an infinite or finite sequence of transitions, respectively. We use $Path_{fin}$ to denote the set of finite paths of DMDP, and $Path_{ful}$ the set of infinite paths of DMDP. If ω is finite, we denote by $last(\omega)$ the last state of ω, and let a state $s \in S$ denote a path consisting of no transitions. We use $\omega \xrightarrow{\mathbf{d}} s$ to refer to a path comprising the sequence of transitions of ω followed by the transition $last(\omega) \xrightarrow{\mathbf{d}} s$.

An adversary of a dense Markov decision process DMDP is a function A mapping every finite path $\omega \in Path_{fin}$ to a probability measure $\mathbf{d} : \Sigma \to [0, 1]$ such that $(last(\omega), \mathbf{d}) \in \mathbf{D}$. Let Adv be the set of adversaries of DMDP. For any adversary $A \in Adv$, let $Path_{fin}^A$ and $Path_{ful}^A$ denote the set of finite and infinite paths associated with A (more precisely, the paths resulting from the choices of dense distributions of A).

Given an adversary $A \in Adv$ of DMDP and a finite path $\bar{\omega} \in Path_{fin}^A$ of A, we define the probability measure $Prob_{\bar{\omega}}^A$ over $Path_{ful}^A$ in the following way. First we define the dense Markov chain $\mathsf{DMC}_{\bar{\omega}}^A = (Path_{fin}^A, \Sigma^A, \{\bar{\omega} \mapsto 1\}, \mathbf{A}, \_)$

obtained from A and $\bar{\omega}$. The set of states of $\mathsf{DMC}_{\bar{\omega}}^A$ are the set of finite paths of A. The sigma-field Σ^A is generated by all of the sets of the following form, where $\omega \in Path_{fin}^A$ is a finite path, $l \in L$ is a location and I is a set of clock valuations obtained from intervals $I_x \subseteq [0, \infty)$ for each defined clock $x \in entry(l)$:

$$\Xi[A, \omega, l, I] = \{\omega \xrightarrow{A(\omega)} (l, val) \in Path_{fin}^A \mid val_x \in I_x \text{ for all } x \in entry(l, x)\} \ .$$

The initial probability distribution $\{\bar{\omega} \mapsto 1\}$ assigns probability 1 to $\bar{\omega}$. The transition relation $\mathbf{A} : Path_{fin}^A \times \Sigma^A \to [0, 1]$ is defined such that:

$$\mathbf{A}(\omega, \Xi[A, \omega', l', I]) = \begin{cases} A(\omega)(l', I) & \text{if } \omega' = \omega \\ 0 & \text{otherwise.} \end{cases}$$

Intuitively, $\mathbf{A}(\omega, \Xi[A, \omega', l', I])$ refers to the probability of obtaining the set of finite paths $\Xi[A, \omega', l', I]$ when extending the finite path ω with one transition under the control of the adversary A. Now that we have defined the dense Markov chain $\mathsf{DMC}_{\bar{\omega}}^A$ obtained from A and $\bar{\omega}$, we are in a position to define the probability measure $Prob_{\bar{\omega}}^A$ over sets of its infinite paths: refer to [34] for details.

6.4   Model Checking Continuous Probabilistic Timed Automata

In this section, we consider how the notion of clock equivalence can be used to facilitate a model checking algorithm for continuous probabilistic timed automata. The property description language that we use is PTCTL, the syntax and semantics of which have been presented in Section 5. For simplicity, assume that $reset(\bar{l}, x)$ is defined as $\{[0, 0] \mapsto 1\}$ for all clocks $x \in \mathcal{X}$; this is not a serious restriction, as we can ensure that no time passes in $\bar{l}$ before a transition to the "proper" initial location which includes more liberal clock resetting functions is made. A continuous probabilistic timed automaton then satisfies a PTCTL formula ϕ if and only if $(\bar{l}, 0) \models^{\text{PT}} \phi$. We now proceed directly to the definition of the region graph of a continuous probabilistic timed automaton.

The Region Graph. Assume that the continuous probabilistic timed automaton $\mathsf{CPTA} = (L, \bar{l}, \mathcal{X}, inv, \underline{prob}, reset, \mathcal{L})$ and the PTCTL formula ϕ are fixed. We use $\mathsf{DMDP} = (S, \Sigma, \bar{\mathsf{d}}, \mathbf{D}, lab)$ to denote the underlying dense Markov decision process of CPTA. Let max be the maximal constant against which any clock is compared in the invariant conditions and guards of CPTA. We define *region equivalence for* CPTA as the relation $\approx \subseteq S \times S$, where $(l, val) \approx (l', val')$ if and only if:

  – $l = l'$ (the location components are equal);
  – $val \equiv_{\max} val'$ (the valuations are clock equivalent with respect to max).

This is the same definition of region equivalence defined for timed automata in Section 3, and for probabilistic timed automata in Section 5, applied in the context of continuous probabilistic timed automata. As usual, we define a *region* to be an equivalence class of $\approx$, and use R to denote the set of regions. As in the previous sections, the fact that region equivalence is finite permits us to define a finite-state region graph which represents the behaviour of the continuous probabilistic timed automaton.

The definition of the region graph is somewhat different to the definition presented in Section 5. The difference resides in the fact that regions do not encode enough information to calculate transition probabilities when clocks are reset according to continuous density functions. (Indeed, this problem also arises in the context of real-time probabilistic processes, meaning that it is in general not possible to obtain a Markovian continuous-time stochastic process from region equivalence.) Consider the following example, originally due to Alur: assume that a clock x is set to some value within the interval $(0, 1)$ on entry to the location l. Time then elapses in location l, and, before the value of x reaches the value 1, a transition is made to location l'. In this new location l', the clock x retains its existing value, whereas another clock y is set to some value in the interval $(0, 1)$. We have three possible relationships between the clocks x and y in location l': either (1) the value of x is less than that of y, (2) the value of x is greater than that of y, or (3) the values of x and y are equal. While we can encode the probability of the value of the clocks being equal is 0, it is impossible to use region equivalence to obtain the exact probabilities of outcomes (1) and (2). The problem is that region equivalence abstracts from the exact amount of time that elapses in location l, upon which the probabilities of outcomes (1) and (2) depends.

However, observe that we can identify the relationship between *only* the clocks that are reset on entry to a location (namely, those in the set $entry(l)$), simply by combining continuous density functions and performing integrations. The values of all other clocks remain unchanged by the change of location. Our problem then reduces to that of describing how the newly-reset clocks and the unchanged clocks relate to one another. Unfortunately, it is impossible to obtain information giving the relative orders the fractional parts of reset and unchanged clocks. Therefore, we let the choice as to which relative order is chosen in the region graph be *nondeterministic*. The presence of such a nondeterministic choice in the place of a probabilistic choice introduces the possibility of an error in the probabilistic model checking results.

To define formally these concepts, we introduce the following notation. For a subset $X \subseteq \mathcal{X}$ of clocks, let $\equiv_{\max}^{X} \subseteq \mathbb{R}_{\geq 0}^{|X|} \times \mathbb{R}_{\geq 0}^{|X|}$ be a variant of clock equivalence of Section 2 restricted to clocks in X. A *union region* is a triple of the form (l, α_u, α_r), where $l \in L$ is a location, α_u is a clock equivalence class of unchanged clocks in $\mathcal{X} \setminus entry(l)$, and α_r is a clock equivalence class of reset clocks in $entry(l)$. We let U denote the set of union regions, which is finite. The set of *combinations* of two clock equivalence classes, one α_X defined over a set $X \subseteq \mathcal{X}$ of clocks, the other $\alpha_{(\mathcal{X} \setminus X)}$ defined over the remainder $\mathcal{X} \setminus X$, is defined as the set of clock equivalence classes over $\mathcal{X}$ obtained by combining the values of the clocks from the two distinct sets X and $\mathcal{X} \setminus X$ in any way (that is, with any relative order on the fractional parts between the clocks in X and the clocks in $\mathcal{X} \setminus X$). Formally, let:

$$\mathsf{Comb}(\alpha_X, \alpha_{(\mathcal{X} \setminus X)}) = \{\alpha \mid \alpha \lfloor_X = \alpha_X \text{ and } \alpha \lfloor_{\mathcal{X} \setminus X} = \alpha_{(\mathcal{X} \setminus X)}\} .$$

Finally, for a location $l \in L$, we define μ_l to be a discrete probability distribution over the clock equivalence classes of $entry(l)$, such that $\mu_l(\alpha)$ is the probability that, when the location l is entered (and thus when the clocks in $entry(l)$ are reset), the clock equivalence class over $entry(l)$ is α. The probabilities can be obtained using standard integration.

We now proceed to define the *region graph of* CPTA *and* ϕ as the Markov decision process $\mathcal{RG} = (R \cup U, \bar{r}, \mathcal{RT}, lab)$. The initial region is defined as $\bar{r} = \{(\bar{l}, \mathbf{0})\}$, while the labelling function is defined by $lab(l, \alpha) = \mathcal{L}(l)$, for all regions $(l, \alpha) \in R$, and by $lab(l, \alpha_u, \alpha_r) = \mathcal{L}(l)$, for all union regions $(l, \alpha_u, \alpha_r) \in U$. Then the probabilistic transition relation $\mathcal{RT} \subseteq R \times \mathsf{Dist}(R)$ is defined in the following way. Let $\mathcal{RT}$ be the smallest set such that $(r, \rho) \in \mathcal{RT}$ if any of the following conditions hold:

Time Elapse: $r = (l, \alpha) \in R$ is a region, and $\rho = \{(l, \beta) \mapsto 1\}$, where β is an $inv(l)$-satisfying time successor of α;

Probabilistic Edge Traversal: $r = (l, \alpha) \in R$ is a region, there exists a probabilistic edge $(l, g, p) \in prob$ such that $\alpha \models g$ and, for each union region $(l', \beta_u, \beta_r) \in U$, we have $\rho(l', \beta_u, \beta_r) = p(l') \cdot \mu_{l'}(\beta_r)$, where $\beta_u = \alpha \lfloor_{\mathcal{X} \setminus entry(l)}$;

Combination: $r = (l, \alpha_u, \alpha_r) \in U$ is a union region, and $\rho = \{(l, \beta) \mapsto 1\}$, where $\beta \in \mathsf{Comb}(\alpha_u, \alpha_r)$ is a combination of α_u, α_r.

Model Checking Using Regions. Given that we have constructed a region graph of a continuous probabilistic timed automaton CPTA, with respect to a particular PTCTL property ϕ, we are now in a position to define an approximate model-checking algorithm to determine whether CPTA satisfies ϕ. The algorithm proceeds according to the same principles established for verifying PTCTL properties of probabilistic timed automata: we convert the PTCTL property into an untimed probabilistic property, and extend the region graph with an extra clock and atomic propositions. However, in the context of continuous probabilistic timed automata, we must also consider the *error* that may arise in the computation of the probabilities.

Note that the minimum probability of satisfying a property established using the region graph will be equal to or less than the minimum probability of satisfying a property in the continuous probabilistic timed automaton; similarly, the maximal probability of satisfying a property in the region graph will be equal to or greater than the corresponding maximum probability in the continuous probabilistic timed automaton. This follows from the fact that, in the definition of the region graph, we replace some probabilistic choice (the manner in which the values of certain clocks are ordered) with nondeterministic choice. Observe that we can envisage an adversary which, instead of selecting among nondeterministic alternatives *deterministically* (as in our previous definitions of adversaries) selects alternatives *probabilistically* according to some distribution. Then, considering that we have replaced a probabilistic choice in the continuous probabilistic timed automaton with a nondeterministic choice in the region graph, we must have one such randomized adversary that resolves the nondeterministic choice in the same manner as the original probabilistic choice (that is, using the same distribution). Now, recalling that for properties such as those of PTCTL the minimal and maximal probabilities of satisfying a property are given by non-randomized adversaries [11], it must be the case that the minimal and maximal probabilities of satisfying a property on the region graph using such non-randomized adversaries bound the minimal and maximal probabilities of satisfying the property using randomized adversaries, and then, as one of these randomized adversaries corresponds to the actual probabilistic choice made, they therefore bound the actual minimal and maximal probabilities of satisfying the property in the original continuous probabilistic timed automaton. Hence, the minimal and maximal probability bounds obtained using the region graph are *conservative*.

Our second point concerns the computation of the error. Note that we can identify regions which can potentially cause an error to arise as those in which there is doubt as to the relative ordering of at least a pair of clocks; that is, if we have at least one clock which keeps its old value after a transition, and at least one clock which is newly reset. Therefore, for any adversary, we can identify a bound on the error of satisfying an "until" property by that adversary by computing the probability of the adversary reaching such "doubtful" regions.

The third point concerns the manner in which it may be possible to obtain a more accurate estimation of the required probabilities. This can be achieved by *re-scaling* the time unit so that clock equivalence partitions the valuation space more finely. Consider the example in which the clock x keeps its old value in the interval $(d, d + 1)$, and the other clock y is newly set in the interval $(d, d + 1)$, and that two outgoing probabilistic edges of the current location have guards $x = d + 1$ and $y = d + 1$, respectively. Say that we re-scaled the time unit by a factor of two: then we would now consider the intervals $(d, d + \frac{1}{2})$ and $(d + \frac{1}{2}, d + 1)$. Hence, say the old clock x was in the interval $(d, d + \frac{1}{2})$ and the new clock y was probabilistically reset in the interval $(d + \frac{1}{2}, d + 1)$; then we would know that the value of y is greater than that of x, and that the probabilistic edge with the guard $y = d + 1$ will be enabled before the probabilistic edge with the guard $x = d + 1$. Note that there is still the possibility of error, as, say, clock y could nevertheless be probabilistically reset in the interval $(d + \frac{1}{2}, d + 1)$, in which case we would still have no knowledge about the relative order of x and y, and therefore no knowledge about which probabilistic edge would be enabled first. However, refining clock equivalence in such a manner will not increase the error, and may potentially lead to improvements in the computation of the probabilities (see [34] for an example).

Note that this third point relates to the complexity of the procedure, as the size of the region graph is exponential in the size of the maximal constant used multiplied by the scaling factor. Therefore, the aims of future research in the area of model checking continuous probabilistic timed automata must either concern the adaptation of algorithms based on clock constraints or the identification of useful subclasses of continuous probabilistic timed automata for which model checking is less expensive.

7   Conclusions

This overview of probabilistic extensions of timed automata has been concerned primarily with theoretical issues centred around the use of clock equivalence to define model-checking algorithms. Future work in this area should bridge the gap between such theory and practical algorithms and tools. Note that timed automata tools such as UPPAAL [6] and KRONOS [52] have been available for several years, and are developed to an advanced stage of maturity. Probabilistic model checkers such as PRISM [32], E ⊢ MC^2 [28] and RAPTURE [29] have also been developed, although more recently. Note that PRISM was used to probabilistically model check discrete-time probabilistic timed automata in [38, 37, 33]; this approach shows promise, although the size of the state space of the obtained discrete-time representation of a probabilistic timed automaton is sensitive to the maximal constant used in the description model, which can prove to be a problem particularly if the description of the system uses different time scales. Hence, a complementary approach based on manipulation of clock constraints is required [35, 36].

References

1. R. Alur. Timed automata. In N. Halbwachs and D. Peled, editors, *Proceedings of the 11th International Conference on Computer-Aided Verification (CAV'99)*, volume 1633 of *LNCS*, pages 8–22. Springer, 1999.
2. R. Alur, C. Courcoubetis, and D. L. Dill. Model-checking for probabilistic real-time systems. In J. Leach Albert, B. Monien, and M. Artalejo Rodríguez, editors, *Proceedings of the 18th International Conference on Automata, Languages and Programming (ICALP'91)*, volume 510 of *LNCS*, pages 115–136. Springer, 1991.
3. R. Alur, C. Courcoubetis, and D. L. Dill. Verifying automata specifications of probabilistic real-time systems. In J. W. de Bakker, C. Huizing, W. P. de Roever, and G. Rozenberg, editors, *Proceedings of Real-Time: Theory in Practice, REX Workshop*, volume 600 of *LNCS*, pages 28–44. Springer, 1991.
4. R. Alur, C. Courcoubetis, and D. L. Dill. Model-checking in dense real-time. *Information and Computation*, 104(1):2–34, 1993.
5. R. Alur and D. L. Dill. A theory of timed automata. *Theoretical Computer Science*, 126(2):183–235, 1994.
6. T. Amnell, G. Behrmann, J. Bengtsson, P. R. D'Argenio, A. David, A. Fehnker, T. Hune, B. Jeannet, K. G. Larsen, M. O. Möller, P. Pettersson, C. Weise, , and W. Yi. UPPAAL - Now, next, and future. In F. Cassez, C. Jard, B. Rozoy, and M. Ryan, editors, *Proceedings of the Summer School on Modelling and Verification of Parallel Processes (MOVEP'2k)*, volume 2067 of *LNCS*, pages 100–125. Springer, 2001.

7. E. Asarin, O. Maler, and A. Pnueli. On discretization of delays in timed automata and digital circuits. In R. de Simone and D. Sangiorgi, editors, *Proceedings of the 9th International Conference on Concurrency Theory (CONCUR'98)*, volume 1466 of *LNCS*, pages 470–484. Springer, 1998.
8. C. Baier, B. Haverkort, H. Hermanns, and J.-P. Katoen. Model-checking algorithms for continuous-time Markov chains. *IEEE Transactions on Software Engineering*, 29(6):524–541, 2003.
9. C. Baier and M. Kwiatkowska. Model checking for a probabilistic branching time logic with fairness. *Distributed Computing*, 11(3):125–155, 1998.
10. D. Beauquier. On probabilistic timed automata. *Theoretical Computer Science*, 292(1):65–84, 2003.
11. A. Bianco and L. de Alfaro. Model checking of probabilistic and nondeterministic systems. In P. S. Thiagarajan, editor, *Proceedings of the 15th Conference on Foundations of Software Technology and Theoretical Computer Science (FSTTCS'95)*, volume 1026 of *LNCS*, pages 499–513. Springer, 1995.
12. A. Bouajjani, S. Tripakis, and S. Yovine. On-the-fly symbolic model checking for real-time systems. In K.-J. Lin and S. H. Son, editors, *Proceedings of the 18th IEEE Real-Time Systems Symposium (RTSS'97)*. IEEE Computer Society Press, 1997.
13. M. Bozga, O. Maler, and S. Tripakis. Efficient verification of timed automata using dense and discrete time semantics. In L. Pierre and T. Kropf, editors, *Proceedings of the 10th IFIP WG 10.5 Advanced Research Working Conference on Correct Hardware Design and Verification Methods (CHARME'99)*, volume 1703 of *LNCS*, pages 125–141. Springer, 1999.
14. J. Bryans, H. Bowman, and J. Derrick. Model checking stochastic automata. To appear in *ACM Transactions on Computational Logic*, 2003.
15. E. M. Clarke and E. A. Emerson. The design and synthesis of synchronization skeletons using temporal logic. In D. Kozen, editor, *Proceedings of the Workshop on Logics of Programs*, volume 131 of *LNCS*, pages 52–71. Springer, 1981.
16. E. M. Clarke, O. Grümberg, and D. Peled. *Model checking*. MIT Press, 1999.
17. C. Courcoubetis and S. Tripakis. Probabilistic model checking: formalisms and algorithms for discrete and real-time systems. In M. K. Inan and R. P. Kurshan, editors, *Verification of Digital and Hybrid Systems*, pages 183–219. Springer, 2000.
18. C. Courcoubetis and M. Yannakakis. The complexity of probabilistic verification. *Journal of the ACM*, 42(4):857–907, 1995.
19. P. R. D'Argenio. A compositional translation of stochastic automata into timed automata. Technical Report CTIT 00-08, University of Twente, 2000.
20. P. R. D'Argenio, J.-P. Katoen, and E. Brinksma. An algebraic approach to the specification of stochastic systems (extended abstract). In D. Gries and W.-P. de Roever, editors, *Proceedings of the IFIP Working conference on Programming Concepts and Methods (PROCOMET'98)*, IFIP Series, pages 126–147. Chapman & Hall, 1998.
21. C. Daws, M. Kwiatkowska, and G. Norman. Automatic verification of the IEEE 1394 root contention protocol with KRONOS and PRISM. In R. Cleaveland and H. Garavel, editors, *Proceedings of the 7th International Workshop on Formal Methods for Industrial Critical Systems (FMICS 2002)*, volume 66(2) of *Electronic Notes in Theoretical Computer Science*, 2002.
22. L. de Alfaro. *Formal verification of probabilistic systems*. PhD thesis, Stanford University, Department of Computer Science, 1997.
23. C. Derman. *Finite-State Markovian Decision Processes*. Academic Press, 1970.

24. P. W. Glynn. A GSMP formalism for discrete-event systems. *Proceedings of the IEEE*, 77:14–23, 1989.

25. H. Hansson and B. Jonsson. A logic for reasoning about time and reliability. *Formal Aspects of Computing*, 6(5):512–535, 1994.

26. T. Henzinger, X. Nicollin, J. Sifakis, and S. Yovine. Symbolic model checking for real-time systems. *Information and Computation*, 111(2):193–244, 1994.

27. T.A. Henzinger, Z. Manna, and A. Pnueli. What good are digital clocks? In W. Kuich, editor, *Proceedings of the 19th International Colloquium on Automata, Languages and Programming (ICALP'92)*, volume 623 of *LNCS*, pages 545–558. Springer, 1992.

28. H. Hermanns, J.-P. Katoen, J. Meyer-Kayser, and M. Siegle. A Markov chain model checker. *Software Tools for Technology Transfer*, 4(2):153–172, 2003.

29. B. Jeannet, P. R. D'Argenio, and K. G. Larsen. RAPTURE: A tool for verifying Markov Decision Processes. In I. Cerna, editor, *Tools Day'02*, Brno, Czech Republic, Technical Report. Faculty of Informatics, Masaryk University Brno, 2002.

30. H. E. Jensen. Model checking probabilistic real time systems. In B. Bjerner, M. Larsson, and B. Nordström, editors, *Proceedings of the 7th Nordic Workshop on Programming Theory*, volume 86, pages 247–261. Chalmers Institute of Technology, 1996.

31. J. G. Kemeny, J. L. Snell, and A. W Knapp. *Denumerable Markov Chains*. Graduate Texts in Mathematics. Springer, 2nd edition, 1976.

32. M. Kwiatkowska, G. Norman, and D. Parker. PRISM: Probabilistic symbolic model checker. In T. Field, P. Harrison, J. Bradley, and U. Harder, editors, *Proceedings of the 12th International Conference on Modelling Techniques and Tools for Computer Performance Evaluation (TOOLS 2002)*, volume 2324 of *LNCS*, pages 200–204. Springer, 2002.

33. M. Kwiatkowska, G. Norman, D. Parker, and J. Sproston. Performance analysis of probabilistic timed automata using digital clocks. In K. G. Larsen and P. Niebert, editors, *Proceedings of the 1st International Workshop on Formal Modeling and Analysis of Timed Systems (FORMATS 2003)*, LNCS. Springer, 2003.

34. M. Kwiatkowska, G. Norman, R. Segala, and J. Sproston. Verifying quantitative properties of continuous probabilistic timed automata. In C. Palamidessi, editor, *Proceedings of the 11th International Conference On Concurrency Theory (CONCUR 2000)*, volume 1877 of *LNCS*, pages 123–137. Springer, 2000.

35. M. Kwiatkowska, G. Norman, R. Segala, and J. Sproston. Automatic verification of real-time systems with discrete probability distributions. *Theoretical Computer Science*, 286:101–150, 2002.

36. M. Kwiatkowska, G. Norman, and J. Sproston. Symbolic computation of maximal probabilistic reachability. In K. Larsen and M. Nielsen, editors, *Proceedings of the 13th International Conference on Concurrency Theory (CONCUR 2001)*, volume 2154 of *LNCS*, pages 169–183. Springer, 2001.

37. M. Kwiatkowska, G. Norman, and J. Sproston. Probabilistic model checking of the IEEE 802.11 wireless local area network protocol. In H. Hermanns and R. Segala, editors, *Proceedings of the 2nd Joint International Workshop on Process Algebra and Performance Modelling and Probabilistic Methods in Verification (PAPM-PROBMIV 2002)*, volume 2399 of *LNCS*, pages 169–187. Springer, 2002.

38. M. Kwiatkowska, G. Norman, and J. Sproston. Probabilistic model checking of deadline properties in the IEEE 1394 FireWire root contention protocol. *Formal Aspects of Computing*, 14(3):295–318, 2003.

39. K. G. Larsen, P. Pettersson, and W. Yi. Compositional and symbolic model-checking of real-time systems. In A. Burns, Y.-H. Lee, and K. Ramamritham, editors, *Proceedings of the 16th IEEE Real-Time Systems Symposium (RTSS'95)*, pages 76–87. IEEE Computer Society Press, 1995.
40. K. G. Larsen and A. Skou. Bisimulation through probabilistic testing. *Information and Computation*, 94(1):1–28, 1991.
41. G. Infante Lòpez, H. Hermanns, and J.-P. Katoen. Beyond memoryless distributions: Model checking semi-Markov chains. In L. de Alfaro and S. Gilmore, editors, *Proceedings of Process Algebra and Probabilistic Methods. Performance Modeling and Verification. Joint International Workshop (PAPM-PROBMIV 2001)*, volume 2165 of *LNCS*, pages 57–70. Springer, 2001.
42. A. V. Moura and G. A. Pinto. A note on the verification of automata specifications of probabilistic real-time systems. *Information Processing Letters*, 82(5):223–228, 2002.
43. M. L. Puterman. *Markov Decision Processes*. J. Wiley & Sons, 1994.
44. J.-P. Queille and J. Sifakis. Specification and verification of concurrent systems in CESAR. In M. Dezani-Ciancaglini and U. Montanari, editors, *Proceedings of the International Symposium on Programming*, volume 137 of *LNCS*, pages 337–351. Springer, 1982.
45. R. Segala. *Modeling and Verification of Randomized Distributed Real-Time Systems*. PhD thesis, Massachusetts Institute of Technology, 1995.
46. R. Segala and N. A. Lynch. Probabilistic simulations for probabilistic processes. *Nordic Journal of Computing*, 2(2):250–273, 1995.
47. S. Tripakis. Verifying progress in timed systems. In J.-P. Katoen, editor, *Proceedings of the 5th International AMAST Workshop on Real-Time and Probabilistic Systems (ARTS'99)*, volume 1601 of *LNCS*, pages 299–314. Springer, 1999.
48. M. Y. Vardi. Automatic verification of probabilistic concurrent finite-state programs. In *Proceedings of the 16th Annual Symposium on Foundations of Computer Science (FOCS'85)*, pages 327–338. IEEE Computer Society Press, 1985.
49. W. Whitt. Continuity of generalized semi-Markov processes. *Mathematics of Operations Research*, 5:494–501, 1980.
50. W. Yi, P. Pettersson, and M. Daniels. Automatic verification of real-time communicating systems by constraint-solving. In D. Hogrefe and S. Leue, editors, *Proceedings of the 7th International Conference on Formal Description Techniques*, pages 223–238. North–Holland, 1994.
51. H. L. S. Younes and R. G. Simmons. Probabilistic verification of discrete event systems using acceptance sampling. In E. Brinskma and K. G. Larsen, editors, *Proceedings of the 14th International Conference on Computer Aided Verification (CAV2002)*, volume 2404 of *LNCS*, pages 223–235. Springer, 2002.
52. S. Yovine. Kronos: A verification tool for real-time systems. *International Journal of Software Tools for Technology Transfer*, 1(1/2):123–133, 1997.
53. S. Yovine. Model checking timed automata. In G. Rozenberg and F. Vaandrager, editors, *Embedded Systems*, volume 1494 of *LNCS*, pages 114–152. Springer, 1998.

Serial Disk-Based Analysis of Large Stochastic Models

Rashid Mehmood

School of Computer Science, University of Birmingham,
Birmingham B15 2TT, United Kingdom
R.Mehmood@cs.bham.ac.uk

Abstract. The paper presents a survey of out-of-core methods available for the analysis of large Markov chains on single workstations. First, we discuss the main sparse matrix storage schemes and review iterative methods for the solution of systems of linear equations typically used in disk-based methods. Next, various out-of-core approaches for the steady state solution of CTMCs are described. In this context, serial out-of-core algorithms are outlined and analysed with the help of their implementations. A comparison of time and memory requirements for typical benchmark models is given.

1 Introduction

Computer and communication systems are being used in a variety of applications ranging from stock markets, information services to aeroplane and car control systems. Discrete-state models have proved to be a valuable tool in the analysis of these computer systems and communication networks. Modelling of such systems involves the description of the system's behaviour by the set of different states the system may occupy, and identifying the transition relation among the various states of the system. Uncertainty is an inherent feature of real-life systems and, to take account of such behaviour, probability distributions are associated with the possible events (transitions) in each state, so that the model implicitly defines a stochastic process. If the probability distributions are restricted to be either geometric or exponential, the stochastic process can be modelled as a discrete time (DTMC) or a continuous time (CTMC) Markov chain respectively. A Markov decision process (MDP) admits a number of discrete probability distributions enabled in a state which are chosen nondeterministically by the environment. We concentrate in this paper on continuous time Markov chains.

A CTMC may be represented by a set of states and a transition rate matrix Q containing state transition rates as coefficients. The matrix coefficients determine transition probabilities and state sojourn times. A CTMC is usually stored as a sparse matrix, where only the nonzero entries are stored. In general, when analysing CTMCs, the performance measure of interest corresponds to either the probability of being in a certain state at a certain time (*transient*) or the

C. Baier et al. (Eds.): Validation of Stochastic Systems, LNCS 2925, pp. 230–255, 2004.

long-run (*steady state*) probability of being in a state. Transient state probabilities can be determined by solving a system of ordinary differential equations. The computation of steady state probabilities involves the solution of a system of linear equations. We focus in this paper on the steady state solution of a CTMC.

The overall process of the state based analytical modelling for CTMCs involves the *specification* of the system, the *generation* of the state space, and the *numerical computation* of all performance measures of interest. Specification of a system at the level of a Markov chain, however, is difficult and error-prone. Consequently, a wide range of high-level formalisms have been developed to specify abstractions of such systems. These formalisms, among others, include queueing networks (QN) [13], stochastic Petri nets (SPN) [32], generalised SPNs (GSPN) [28, 29], stochastic process algebras (SPA), such as PEPA [20], EMPA [7] and TIPP [18], mixed forms such as queueing Petri nets (QPN) [4], and stochastic automata networks (SAN) [34, 35]. See for example [38], for a survey of model representations.

Once a system is specified using some suitable high-level formalism, the entire[1] state space needs to be generated from this specification. In this paper, we do not discuss state space generation algorithms and techniques.

The size of a system (number of states) is typically exponential in the number of parallel components in the system. This problem is known as the *state-space explosion* or the *largeness* problem, and has motivated researchers to tread a number of directions for the numerical solution of a Markov chain. These can be broadly classified into *explicit* and *implicit* techniques. The so-called explicit methods store a CTMC explicitly, using a data structure of size proportional to the number of states and transitions. The implicit methods, on the other hand, use some kind of symbolic data structure for the storage of a CTMC.

The focus of this paper is the serial disk-based performance analysis of CTMCs, and hence we will discuss here those *out-of-core*[2] techniques that store all or a part of the data structure on disk for the numerical solution of a CTMC. The term *in-core* is used when the data is in the main memory of a computer. The term "serial" indicates that the numerical computations are performed on a single processor, as opposed to in "parallel", where the computational task is distributed between a number of processors.

The paper is organised as follows. The numerical solution methods used in the serial disk-based approaches are discussed in Section 2. A compact sparse matrix representation is at the heart of the analysis techniques for large Markov chains, especially considering that the time required for disk read/write determines the overall solution time for out-of-core methods. The main storage schemes for sparse matrices are reviewed in Section 3. The out-of-core algorithms are presented in Section 4. These algorithms are further analysed in Section 5

---

[1] With the exception of product-form queueing networks [3]. *On-the-fly* techniques [16], to some extent, are also an exception.

[2] Algorithms that are designed to achieve high performance when their data structures are stored on disk are known as *out-of-core algorithms*; see [40], for example.

with the help of results of their implementations applied to typical benchmark models. We conclude with Section 6.

2   Numerical Methods

Performance measures for stochastic models are traditionally derived by generating and solving a Markov chain obtained from some high-level formalism. Let $Q \in \mathbb{R}^{n \times n}$ be the infinitesimal generator matrix; the order of Q equals the number of states in the CTMC. The off-diagonal elements of the matrix Q satisfy $q_{ij} \in \mathbb{R}_{\geq 0}$, and the diagonal elements are given by $q_{ii} = -\sum_{j \neq i} q_{ij}$. The matrix element q_{ij} gives the rate of moving from state i to state j. The matrix Q is usually sparse; further details about the properties of these matrices can be found in [39]. The steady state behaviour of a CTMC is given by:

$$\pi Q = 0, \quad \sum_{i=0}^{n-1} \pi_i = 1, \tag{1}$$

where π is the steady state probability vector. A sufficient condition for the unique solution of the equation (1) is that the CTMC is finite and irreducible. A CTMC is *irreducible* if every state can be reached from every other state [39]. For the remainder of the paper, we restrict attention to solving only irreducible CTMCs; for details of the solution in the general case see [39]. The equation (1) can be reformulated as $Q^T \pi^T = 0$, and hence well-known methods for the solution of systems of linear equations of the form $Ax = b$ can be used.

The numerical solution methods for linear systems of the form $Ax = b$ are broadly classified into direct methods and iterative methods. For large systems, direct methods become impractical due to the phenomenon of fill-in, caused by the generation of new matrix entries during the factorisation phase. Iterative methods do not modify matrix A; rather, they involve the matrix only in the context of the matrix-vector product (MVP). The term "iterative methods" refers to a wide range of techniques that use successive approximations to obtain more accurate solutions to a linear system at each step [2].

Before we move on to the next section, where we discuss the basic iterative methods for the solution of the system of equations $Ax = b$, we mention the *Power method*. Given the generator matrix Q, setting $P = I + Q/\alpha$ in Equation (1), where $\alpha \geq max_i \mid q_{ii} \mid$, leads to:

$$\pi P = \pi. \tag{2}$$

Using $\pi^{(0)}$ as the initial estimate, an approximation of the steady state probability vector after k transitions is given by $\pi^{(k)} = \pi^{(k-1)} P$. This method successively multiplies the steady state probability vector with the matrix P until convergence is reached. It is guaranteed to converge (for irreducible CTMCs), though convergence can be very slow. Below we consider alternative iterative methods, which may fail to converge for some models but, in practice, converge much faster than the Power method.

2.1   Basic Iterative Methods

We discuss, in this section, the so-called *stationary iterative methods* for the solution of the system of equations $Ax = b$, that is, the methods that can be expressed in the simple form $x^{(k)} = Bx^{(k-1)} + c$, where neither B nor c depend upon the iteration count k [2]. Beginning with a given approximate solution, these methods modify the components of the approximation in each iteration, until a required accuracy is achieved. In the k-th iteration of the *Jacobi method*, for example, we calculate element-wise:

$$x_i^{(k)} = a_{ii}^{-1} \left(b_i - \sum_{j \neq i} x_j^{(k-1)} a_{ij} \right), \quad 0 \leq i < n, \tag{3}$$

where a_{ij} denotes the element in row i and column j of matrix A. It can be seen in equation (3) that the new approximation of the iteration vector $(x^{(k)})$ is calculated using only the old approximation of the vector $(x^{(k-1)})$. This makes the Jacobi method suitable for parallelisation. However, Jacobi method exhibits slow convergence.

The *Gauss-Seidel* (GS) iterative method, which in practice converges faster than the Jacobi method, uses the most recently computed approximation of the solution:

$$x_i^{(k)} = a_{ii}^{-1} \left(b_i - \sum_{j<i} x_j^{(k)} a_{ij} - \sum_{j>i} x_j^{(k-1)} a_{ij} \right), \quad 0 \leq i < n. \tag{4}$$

Another advantage of the Gauss-Seidel method is that it can be implemented with a single iteration vector, whereas the Jacobi method requires two.

Finally, we mention the *successive over-relaxation* (SOR) method. An SOR iteration is of type

$$x_i^{(k)} = \omega \hat{x}_i^{(k)} + (1 - \omega) x_i^{(k-1)}, \quad 0 \leq i < n, \tag{5}$$

where $\hat{x}$ denotes a Gauss-Seidel iterate, and $\omega \in (0, 2)$ is a relaxation factor. The method is under-relaxed for $0 < \omega < 1$, and is over-relaxed for $\omega > 1$; the choice $\omega = 1$ reduces SOR to Gauss-Seidel. It is shown in [22] that SOR fails to converge if $\omega \notin (0, 2)$. For a good choice of ω, SOR can have considerably better convergence behaviour than Gauss-Seidel, but unfortunately a priori computation of an optimal value for ω is not feasible.

2.2   Block Iterative Methods

Consider a partitioning of the state space S of a CTMC into B contiguous partitions $S_0, \ldots, S_{B-1}$ of sizes $n_0, \ldots, n_{B-1}$, such that $n = \sum_{i=0}^{B-1} n_i$. Using this, the matrix A can be divided into B^2 blocks, $\{A_{ij} \mid 0 \leq i, j < B\}$, where the rows and columns of block A_{ij} correspond to the states in S_i and S_j, respectively, i.e., block A_{ij} is of size $n_i \times n_j$. We also define $n_{\max} = \max\{n_i \mid 0 \leq i < B\}$.

Using such a partitioning of the state space for $B = 4$, the system of equations $Ax = b$ can be partitioned as:

$$
\begin{pmatrix}
A_{00} & A_{01} & A_{02} & A_{03} \\
A_{10} & A_{11} & A_{12} & A_{13} \\
A_{20} & A_{21} & A_{22} & A_{23} \\
A_{30} & A_{31} & A_{32} & A_{33}
\end{pmatrix}
\begin{pmatrix}
X_0 \\
X_1 \\
X_2 \\
X_3
\end{pmatrix}
=
\begin{pmatrix}
B_0 \\
B_1 \\
B_2 \\
B_3
\end{pmatrix}
\tag{6}
$$

Given the partitioning introduced above, block iterative methods essentially involve the solution of B sub-systems of linear equations of sizes $n_0, \ldots, n_{B-1}$ within a *global* iterative structure, say Gauss-Seidel. Hence, from (4) and (6), the *block Gauss-Seidel* method for the solution of the system $Ax = b$ is given by:

$$
X_i^{(k)} = A_{ii}^{-1} \left(B_i - \sum_{j<i} A_{ij} X_j^{(k)} - \sum_{j>i} A_{ij} X_j^{(k-1)} \right), \quad 0 \le i < B, \tag{7}
$$

where $X_i^{(k)}$, $X_i^{(k-1)}$ and B_i are the i-th blocks of vectors $x^{(k)}$, $x^{(k-1)}$ and b respectively, and, as above, A_{ij} denotes the (i,j)-th block of matrix A. Hence, in each of the B phases of the k-th iteration of the block Gauss-Seidel iterative method, we solve the equation (7) for $X_i^{(k)}$. These sub-systems can be solved by either direct or iterative methods. It is not necessary even to use the same method for each sub-system. If iterative methods are used to solve these sub-systems then we have several *inner* iterative methods within a global or *outer* iterative method. Each sub-system of equations can receive either a fixed or varying number of *inner* iterations. Such block methods (with an inner iterative method) typically require fewer iterations, but each iteration requires more work (multiple inner iterations). Block iterative methods are well known and are an active area of research; see [39], for example.

2.3   Test for Convergence

Usually, a test for convergence is carried out in each iteration of an iterative method and the method is stopped when the convergence criterion is met; [2] discusses this subject in some details. In the context of the steady state solution of a CTMC, a widely used test is the relative error criterion:

$$
\max_i \left(\frac{|x_i^{(k)} - x_i^{(k-1)}|}{|x_i^{(k)}|} \right) < \varepsilon \ll 1. \tag{8}
$$

3   Matrix Storage Considerations

An $n \times n$ dense matrix is usually stored in a two-dimensional $n \times n$ array. For sparse matrices, in which most of the entries are zero, storage schemes are sought which can minimise the storage while keeping the computational costs to a minimum. Consequently, a number of sparse storage schemes exist which exploit various matrix properties, e.g., the *sparsity pattern* of a matrix. The sparse schemes we will discuss in this section are not exhaustive; for more schemes see, for instance, [2].

The simplest of sparse schemes which makes no assumption about the matrix is the so-called *coordinate format* [37, 19]. Figure 1 gives an 4×4 sparse matrix with a off-diagonal nonzero entries and its storage in coordinate format. The scheme uses three arrays. The first array Val (of size $a + n$ doubles) stores the matrix nonzero entries in any order, while the arrays Col and Row, both of size $a + n$ ints, store the column and row indices for these entries, respectively. Given an 8-byte floating point number representation (double) and a 4-byte integer representation (int), the coordinate format requires $16(a + n)$ bytes to store the whole sparse matrix, including diagonal and off-diagonal entries.

$$
n = 4, \quad a = 6 \quad
\begin{pmatrix}
-0.2 & 0 & 0.2 & 0 \\
0 & -0.9 & 0.4 & 0.5 \\
0.6 & 0.7 & -0.13 & 0 \\
0.9 & 0 & 0 & -0.9
\end{pmatrix}
$$

Val | -0.2 | 0.2 | -0.9 | 0.4 | 0.5 | 0.9 | 0.7 | -0.13 | 0.6 | -0.9 |

Col | 0 | 2 | 1 | 2 | 3 | 0 | 1 | 2 | 0 | 3 |

Row | 0 | 0 | 1 | 1 | 1 | 3 | 2 | 2 | 2 | 3 |

(a) A CTMC matrix   (b) The coordinate format

Fig. 1. A 4×4 sparse matrix and its storage in the coordinate format

Figure 2(a) illustrates the storage of the matrix in Figure 1(a) in the *compressed sparse row* (CSR) [37] format. All the $a + n$ nonzero entries are stored row by row in the array Val, while Col contains column indices of these nonzero entries; the elements within a row can be stored in any order. The i-th element of the array Starts (of size n ints) contains the index in Val (and Col) of the beginning of the i-th row. The CSR format requires $12a + 16n$ bytes to store the whole sparse matrix including the diagonal.

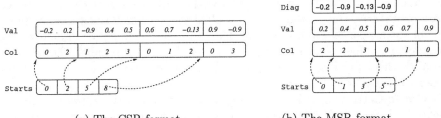

(a) The CSR format   (b) The MSR format

Fig. 2. Storage in the CSR and MSR formats of the matrix in Figure 1(a)

3.1   Separate Diagonal Storage

A second dimension for sparse schemes is added if we consider that many iterative algorithms treat diagonal entries of a matrix differently. This gives us the following two additional choices for sparse storage schemes; an alternative choice will be mentioned in Section 3.2.

Case A: The diagonal entries may be stored separately in an array of n doubles. Storage of column indices of diagonal entries in this case is not required, which gives us a saving of $4n$ bytes over the CSR format. This scheme is known as the *modified sparse row* (MSR) format [37] (a modification of CSR), see Figure 2(b). We note that the MSR scheme essentially is the same as the CSR format except that the diagonal elements are stored separately. The scheme requires $12(a + n)$ bytes to store the whole sparse matrix. Some computational advantage may be obtained by storing the diagonal entries as $1/a_{ii}$ instead of a_{ii}, which replaces n division operations with n multiplications.

Case B: In certain contexts, such as for the steady state solution of a Markov chain, it is possible to avoid the *in-core* storage of the diagonal entries. Given $R = Q^T D^{-1}$, the system $Q^T \pi^T = 0$ can be equivalently written as $Q^T D^{-1} D \pi^T = Ry = 0$, with $y = D\pi^T$. The matrix D is defined as the diagonal matrix with $d_{ii} = q_{ii}$, for $0 \leq i < n$. Consequently, the equivalent system $Ry = 0$ can be solved with all the diagonal entries of the matrix R being 1. The original diagonal entries can be stored on disk for computing π from y. This saves $8n$ bytes of the in-core storage, along with computational savings of n divisions per each step in an iterative method such as Gauss-Seidel.

3.2   Exploiting Matrix Properties

The number of distinct values in a generator matrix depends on the model. This characteristic can lead to significant memory savings if one considers indexing the nonzero entries in the above mentioned formats. Consider the MSR format. Let *MaxD* be the number of distinct values the off-diagonal entries of a matrix can take, where $MaxD \leq 2^{16}$; then *MaxD* distinct values can be stored as double Val[*MaxD*]. The indices to this array of distinct values cannot exceed 2^{16}, and, in this case, the array double Val[a] in MSR format can be replaced with short Val_i[a]. In the context of CTMCs, in general, the maximum number of entries per row of a generator matrix is also small, and is limited by the maximum number of transitions leaving a state. If this number does not exceed 2^8, the array int Starts[n] in MSR format can be replaced by the array char row_entries[n].

Consequently, in addition to the array of distinct values, Val[*MaxD*], the indexed variation of MSR mentioned above uses three arrays: the array Val_i[a] of length $2a$ bytes for the storage of a short (2-byte integer representation) indices to the *MaxD* entries, an array of length $4a$ bytes to store a column indices as int (as in MSR), and the n-byte long array row_entries[n] to store the number of entries in each row. The total in-core memory requirement for this scheme is $6a + n$ bytes plus the storage for the actual distinct values in the matrix. Since the storage for the actual distinct values is relatively small for large models, we do not consider it in future discussions. Such variations of the MSR format, based on indexing the matrix elements, have been used in the literature [15, 5, 23, 6, 24] under different names. We call it the *indexed MSR* format.

We note that, in general, for any of the above-mentioned formats, it is possible to replace the array double Val[a] with short Val_i[a], or with char Val_i[a], if $MaxD$ equals 2^{16}, or 2^8, respectively. In fact, $\lceil log_2(MaxD) \rceil$ bits suffice for each index. Similarly, it is also possible to index diagonal entries of a matrix provided the diagonal vector has relatively few distinct entries. This justifies an alternative choice for separate diagonal storage (see Section 3.1).

We describe here another variation of the MSR scheme. It has been mentioned that the indexed MSR format exploits the fact that the number of entries in a generator matrix is relatively small. We have also seen that the indexed MSR format (this is largely true for all the formats considered here) stores the column index of a nonzero entry in a matrix as an int. An int usually uses 32 bits, which can store a column index as large as 2^{32}. The size of the models which can be stored within the RAM of a modern workstation are much smaller than 2^{32}. For example, it is evident from Table 1 that using an in-core method such as Gauss-Seidel, 54 million states is a limit for solving a CTMC on a typical single workstation. The column index for a model with 54 million states requires at most 26 bits, leaving 6 bits unused. Even more bits can be made available if we consider that, for out-of-core and for parallel solutions, it is common practice (or, at least it is possible) to use local numbering for a column/row index inside each matrix block.

The *compact MSR* format [24] exploits the above mentioned facts and stores the column index of a matrix entry along with the index to the actual value of this entry in a single int. This is depicted in Figure 3. The storage and retrieval of these indices into, and from, an int can be carried out efficiently using bit operations. Since the operation $Q^T D^{-1}$ increases the number of distinct values in the resulting matrix R, matrix Q and the diagonal vector d can be stored separately. As mentioned earlier, the diagonal entries can be indexed, and the distinct entries can be stored as $1/a_{ii}$ to save n divisions per iteration; indices to these distinct values may be stored as short. The compact MSR scheme uses three arrays: the array Col_i[a] of length $4a$ bytes which stores the column positions of matrix entries as well as the indices to these entries, the n-byte long array row_entries[n] to store the number of entries in each row, and the $2n$-byte long array Diag_i[n] of short indices to the original values in the diagonal. We do not consider the storage for the original matrix entries. The total memory requirements in the compact MSR format is thus $4a + 3n$ bytes, around 30% more compact than the indexed MSR format.

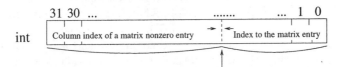

Fig. 3. Storage of an index to a matrix distinct value using available space in int

Table 1. Storage requirements for the FMS models in various formats

k	states (n)	off-diagonal nonzero (a)	a/n	memory required for Q (MB)			MB per π
				MSR format	Indexed MSR	Compact MSR	
8	4,459,455	38,533,968	8.64	489	225	160	34
9	11,058,190	99,075,405	8.96	1,260	577	409	84
10	25,397,658	234,523,289	9.23	2,962	1,366	967	194
11	54,682,992	518,030,370	9.47	6,554	3,016	2,133	417
12	111,414,940	1,078,917,632	9.68	13,563	6,279	4,433	850
13	216,427,680	2,136,215,172	9.87	32,361	15,149	10,580	1,651

Table 1 summarises a fact sheet for the model of a flexible manufacturing system (FMS) [11] comparing storage requirements in MSR, indexed MSR, and compact MSR formats. The first column in Table 1 gives the parameter k (number of tokens in the system); the second and third columns list the resulting number of reachable states and the number of transitions respectively. The number of states and the number of transitions increase with an increase in the parameter k. The largest model generated is FMS ($k = 13$) with over 216 million reachable states and 2.1 billion transitions. The fourth column (a/n) gives the average number of the off-diagonal nonzero entries per row, which serves as an indication of the matrix sparsity. Columns 5–7 give storage requirements for the matrices in MB for the MSR, the indexed MSR and the compact MSR schemes respectively. Finally, the last column lists the memory required to store a single iteration vector of doubles (8 bytes) for the solution phase.

3.3   Alternatives for Matrix Storage

We have discussed various explicit sparse matrix schemes for CTMCs. Another approach, which has been very successful in model checking, is the implicit storage of the matrix. The term implicit methods is used because these data structures do not require size proportional to the number of states. Implicit methods include multi-terminal binary decision diagrams (MTBDDs) [12, 1], the Kronecker approach [34], matrix diagrams (MDs) [9] and on-the-fly methods [16]. A brief explanation of MTBDDs, for which a vector out-of-core implementation will be discussed in 4.5, is given in the following paragraph; further discussion of the implicit methods is beyond the scope of this paper, and the interested reader is invited to consult the individual references, or see [31, 8] for recent surveys of such data structures.

Multi-Terminal Binary Decision Diagrams (MTBDDs)

MTBDDs [12, 1] are an extension of binary decision diagrams (BDDs). An MTBDD is a rooted, directed acyclic graph, which represents a function mapping Boolean variables to real numbers. MTBDDs can be used to encode real-valued matrices (and vectors) by encoding their indices as Boolean variables. The prime reason for using MTBDDs is that they can provide extremely compact storage for the generator matrices of very large CTMCs, provided that structure and regularity derived from their high-level description can be exploited. Here, we describe a slight variant of MTBDDs called *offset-labelled MTBDDs*

[27,33]. This data structure additionally allows information about which states of a CTMC are reachable. Techniques which use data structures based on BDDs are often called *symbolic* approaches.

$$\begin{pmatrix} 0 & 0.9 & 0 & 1.2 \\ 1.2 & 0 & 0 & 1.2 \\ 0 & 0 & 0 & 0 \\ 3.7 & 0.9 & 0 & 0 \end{pmatrix}$$

Entry of	Encoding				MTBDD Path			
the matrix	x_1	x_2	y_1	y_2	x_1	y_1	x_2	y_2
$(0,1) = 0.9$	0	0	0	1	0	0	0	1 → 0.9
$(0,3) = 1.2$	0	0	1	1	0	1	0	1 → 1.2
$(1,0) = 1.2$	0	1	0	0	0	0	1	0 → 1.2
$(1,3) = 1.2$	0	1	1	1	0	1	1	1 → 1.2
$(3,0) = 3.7$	1	1	0	0	1	0	1	0 → 3.7
$(3,1) = 0.9$	1	1	0	1	1	0	1	1 → 0.9

Fig. 4. An offset-labelled MTBDD representation of a matrix

Figure 4 shows a matrix, its representation as an offset-labelled MTBDD, and a table explaining how the information is encoded. To preserve structure in the symbolic representation of a CTMC's generator matrix, its diagonal elements are stored separately as an array. Hence, the diagonal of the matrix in Figure 4 is shown to be zero. The offset-labelled MTBDD in the figure represents a function over four Boolean variables x_1, y_1, x_2, x_2. For a given valuation of these variables, the value of this function can be computed by tracing a path from the top of the offset-labelled MTBDD to the bottom, taking the dotted edge if the Boolean variable on that level is 0 and the solid edge if the variable is 1. The value can be read from the label of the bottom (terminal) node reached. For example, if $x_1 = 1, y_1 = 0, x_2 = 1, y_2 = 1$, the function returns 0.9.

To represent the matrix, the offset-labelled MTBDD in Figure 4 uses the variables x_1, x_2 to encode row indices and y_1, y_2 to encode column indices. Notice that these are ordered in an interleaved fashion in the figure. This is a common heuristic in BDD-based representations to reduce their size. In the example, row and column indices are encoded using the standard binary representation of integers. For example, the row index 3 is encoded as 11 ($x_1 = 1, x_2 = 1$) and the column index 1 is encoded as 01 ($y_1 = 0, y_2 = 1$). To determine the value of the matrix entry, we read the value of the function represented by the MTBDD for $x_1 = 1, y_1 = 0, x_2 = 1, y_2 = 1$. Hence, the matrix entry (3,1) is 0.9.

The integer values on the nodes of the data structure are *offsets*, used to compute the *actual* row and column indices of the matrix entries (in terms of reachable states only). This is typically essential since the *potential state space* can be much larger than the *actual state space*. The actual row index is determined by summing the offsets on x_i nodes from which the solid edge is taken (i.e. if $x_i = 1$). The actual column index is computed similarly for y_i nodes. In the example in Figure 4, state 2 is not reachable. For the previous example of matrix entry (3,1), the actual row index is 2+0=2 and the column index is 1 (using only the offset on the y_2 level).

Note that each node of an MTBDD can be seen to represent a submatrix of the matrix represented by the whole MTBDD. Since an MTBDD is based on binary decisions, descending each level (each pair of x_i and y_i variables) of the data structure splits the matrix into 4 submatrices. Hence, descending l levels, gives a decomposition into $(2^l)^2$ blocks. For example, descending one level of the MTBDD in Figure 4, gives a decomposition of the matrix into 4 blocks and we see that each x_2 node represents a 2×2 submatrix of the 4×4 matrix. This allows convenient and fast access to individual submatrices. However, since the MTBDD actually encodes a matrix over its potential state space and the distribution of the reachable states across the state space is unpredictable, descending l levels of the MTBDD actually results in blocks of varying and uneven sizes.

Numerical solution of CTMCs can be performed purely using MTBDDs (see e.g. [17]). This is done by representing both the matrix and the vector as MTB-DDs and using an MTBDD-based matrix-vector multiplication algorithm ([12, 1]). However, this approach is often very inefficient because the MTBDD representation of the vector is irregular and grows quickly during solution. A better approach in general is to use offset-labelled MTBDDs to store the matrix and an array to store the vector ([27, 33]). Computing the *actual* row and column indices of matrix entries is important in this case because they are needed to access the elements of the array storing the vector. All matrix entries can be extracted from an offset-labelled MTBDD in a single pass using a recursive traversal of the data structure.

4   Out-of-Core Iterative Solution Methods

We survey here the serial out-of-core methods for the steady state solution of CTMCs, i.e., the methods which store whole or part of the data structure on disk. We begin with an in-core block Gauss-Seidel algorithm in the next section. We then present and explain a matrix and a complete out-of-core Gauss-Seidel algorithm in Sections 4.2 and 4.4, respectively. The matrix out-of-core approach of [15] is described in Section 4.3, and the symbolic out-of-core solution of [25] is discussed in Section 4.5.

4.1   The In-Core Approach

We give an in-core block Gauss-Seidel algorithm for the solution of the system $Ax = 0$, where $A = Q^T$ and $x = \pi^T$. Block iterative methods are described in Section 2.2 and the block Gauss-Seidel method was formulated in equation (7).

A typical iteration of the block Gauss-Seidel method is shown in Figure 5. The algorithm requires an array x of size n (number of states) to store the iteration vector, the i-th block of which is denoted X_i, and another array $\tilde{X}_i$ of size $n_{\max}$ to accumulate the sub-MVPs, $A_{ij}X_j$, i.e., the multiplication of a single matrix block by a single vector block; see Section 2.2. The subscript i of $\tilde{X}_i$ in the algorithm is used to make the description intuitive and to keep the vector block notation consistent; it does not imply that we have used B such arrays.

Each iteration of the algorithm is divided into B phases. In the i-th phase, the method updates elements in the i-th block of the iteration vector. The up-

1.   **for** $i = 0$ **to** $B - 1$
2. $\tilde{X}_i \leftarrow 0$
3.   **for** $j = 0$ **to** $B - 1$
4.   **if** $j \neq i$
5. $\tilde{X}_i = \tilde{X}_i - A_{ij} X_j$
6.   Solve $A_{ii} X_i = \tilde{X}_i$
7.   Test for convergence
8.   Stop if converged

Fig. 5. An iteration of the block Gauss-Seidel algorithm

date of the i-th block, X_i, only requires access to entries from the i-th row of blocks in A, i.e., A_{ij} for $0 \leq j < B$. This is illustrated in Figure 6 for $B = 4$ and $i = 1$; the matrix and vector blocks used in the calculation are shaded grey. In the figure, all blocks are of equal size but this is generally not the case. Line 5 of the algorithm in Figure 5 performs a unit of computation, a sub-MVP, and accumulates these products. Finally, line 6 corresponds to solving a system of equations, either by direct or iterative methods (see Section 2.2). We use the Gauss-Seidel iterative method to solve $A_{ii} X_i = \tilde{X}_i$. More precisely, we apply the following to update each of the n_i elements of the i-th vector block, X_i:

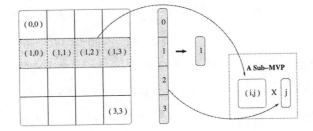

Fig. 6. Matrix vector multiplication at block level

for $p = 0$ **to** $n_i - 1$
$$A_{ii}[p, p]\, X_i[p] = \tilde{X}_i[p]\ -\ \sum_{q \neq p} A_{ii}[p, q]\, X_i[q],$$

where $X_i[p]$ is the p-th entry of the vector block X_i, and $A_{ii}[p, q]$ denotes the (p, q)-th element of the diagonal block (A_{ii}). We note that applying one *inner* Gauss-Seidel iteration for each X_i in the global Gauss-Seidel iterative structure reduces the block Gauss-Seidel method to the standard Gauss-Seidel method, although the method is based on block sub-MVPs.

4.2   A Matrix Out-of-Core Approach

In an implementation of the block Gauss-Seidel algorithm mentioned in the previous subsection, the matrix can be stored using some sparse storage scheme (see Section 3) and the vector can be stored as an array of doubles. However, this makes the total memory requirements for large CTMCs well above the size of the RAM available in standard workstations. One possible solution is to store the matrix on disk and read blocks of matrix into RAM when required. In each iteration of an iterative method, we can do the following:

while there is a block to read
    read a matrix block
    do computations using the matrix and vector blocks

We note that, in this disk-based approach, the processor will remain idle until a block has been read into RAM. We also note that the next disk read operation will not be initiated until the computations have been performed. It is not an efficient approach, particularly for large models and iterative methods, because they require a relatively large amount of data per floating point operation. We would like to use a two-process approach, where the disk I/O and the computation can proceed concurrently. We begin with a basic out-of-core algorithm and explain its working. In the coming sections we discuss the out-of-core approaches pursued in the literature.

Figure 7 presents a high-level description of a typical, matrix out-of-core algorithm which uses the block Gauss-Seidel method for the solution of the system $Ax = 0$. The term "matrix out-of-core" implies that only the matrix is stored out-of-core and the vector is kept in-core. The algorithm is implemented using two separate concurrent processes: the *DiskIO Process* and the *Compute Process*. The two processes communicate via shared memory and synchronise with semaphores.

Integer constant: B (*number of vector blocks*)
Semaphores: S_1, S_2: occupied
Shared variables: R_0, R_1 (*To read matrix A blocks into RAM*)

DiskIO Process	*Compute Process*
1. Local variables: $i, j, t = 0$	1. Local variables: $p, q, t = 0$
2. while not converged	2. while not converged
3. for $i \leftarrow 0$ to $B - 1$	3. for $i \leftarrow 0$ to $B - 1$
4. $R_t = $ read($A_{ij}, j = 0 : B - 1$)	4. Wait(S_1)
5. Signal(S_1)	5. Signal(S_2)
6. Wait(S_2)	6. $\bar{X}_i = $ subMVPs($-\sum_{j \neq i}^{0:B-1} A_{ij} X_j$, R_t)
7. $t = (t + 1) \bmod 2$	7. Solve($A_{ii} X_i = \bar{X}_i$, R_t)
	8. Test for convergence
	9. $t = (t + 1) \bmod 2$

Fig. 7. A matrix out-of-core block Gauss-Seidel algorithm

The algorithm of Figure 7 assumes that the (n-state) CTMC matrix to be solved is divided into B^2 blocks of size $n/B \times n/B$, and is stored on disk. For intuitive reasons, in Figure 7, the vector x is also shown to be divided into B blocks, although a single array of doubles is used to keep the whole vector. Another vector $\bar{X}_i$ of size n/B is required to accumulate the sub-MVPs (line 6). The algorithm assumes that, before it commences, the array x holds an initial approximation to the solution.

Each iteration of the algorithm given in Figure 7 is divided into B phases, where the i-th phase computes the next approximation for the i-th block of the iteration vector. To update the i-th block, X_i, the *Compute Process* requires access to entries from the i-th row of blocks in A, i.e., A_{ij} for $0 \leq j < B$; see Figure 6. The *DiskIO Process* helps the *Compute Process* with this task and fetches from disk (line 4) the required row of the blocks in matrix A. The algorithm uses two shared memory buffers, R_0 and R_1, to achieve the communication between the two processes. At a certain point in time during the execution, the *DiskIO Process* is reading a block of matrix into one shared buffer, say R_0, while the *Compute Process* is consuming a matrix block from the other buffer, R_1. Both processes alternate the value of a local variable t between 0 and 1, in order to switch between the two buffers R_0 and R_1. The two semaphores S_1 and S_2 are used to synchronise the two processes, and to prevent inconsistencies.

The high-level structure of the algorithm, given in Figure 7, is that of a producer-consumer problem. In each execution of its for loop, the i-th phase (lines $3 - 7$), the *DiskIO Process* reads the i-th row of blocks in A, into one of the shared memory buffers R_t, and issues a Signal($\cdot$) operation on S_1 (line 5). Since the two semaphores are occupied initially, the *Compute Process* has to wait on S_1 (line 4). On receiving this signal, the *Compute Process* issues a return signal on S_2 (line 5) and then advances to update the i-th vector block (lines $6 - 7$). The *DiskIO process*, on receiving this signal from the *Compute Process*, advances to read the next i-th row of blocks in A. This activity of the two processes is repeated until all of the vector blocks have been updated; the processes then advance to the next iteration.

Implementation

We have used the compact MSR storage scheme (see Section 3) to store the matrix in our implementation of the matrix out-of-core algorithm. The blocks have been stored on disk in the order they are required during the numerical computation. Hence, the *DiskIO Process* is able to read the file containing the matrix sequentially throughout an iteration of the algorithm.

4.3   The Deavours and Sanders' Approach

Deavours and Sanders [14] were the first to consider a matrix out-of-core technique for the steady state solution of Markov models. They used the block Gauss-Seidel method in their tool to reduce the amount of disk I/O. They partitioned the matrix into a number of sub-systems (or blocks) and applied multiple Gauss-Seidel *inner* iterations on each block (see Section 2.2 and 4.1). The number of inner iterations was a tunable parameter of their tool. They analysed the tool by presenting results for a fixed and varying number of inner iterations.

Deavours and Sanders reported the solution of models with up to 15 million states on a workstation with 128MB RAM. Later, in [15], they improved their earlier work by applying a compression technique on the matrix before storing it to disk, thus reducing the file I/O time. For our implementations of the explicit in-core and out-of-core methods, we have used the compact MSR scheme for matrix storage, which can also be considered as a compression technique. The compact MSR scheme provides 30% or more saving over conventional schemes. Furthermore, Deavours and Sanders report that the decompression time accounts for 50% of the computation time; therefore, in our case, we do not expect any overall gain from matrix compression.

The other notable papers on the matrix out-of-core solution are [23, 6]. However, the main emphasis of these papers is on the parallelisation of the out-of-core approach of Deavours and Sanders which is not the subject of this paper.

4.4   A Complete Out-of-Core Approach

The limitation of the Deavours and Sanders' approach is the in-core storage of the iteration vector. In [24], Kwiatkowska and Mehmood extend the earlier out-of-core methods and present the *complete out-of-core* method which stores the matrix as well the iteration vector on disk. They solved the system $\pi Q = 0$ using the block Gauss-Seidel method.

We reformulate the system $\pi Q = 0$ as $Ax = 0$, as in Section 4.2, and give the complete out-of-core block Gauss-Seidel algorithm for the solution in Figure 8. The algorithm assumes that the vector and the matrix are divided into B and B^2 blocks of equal size, respectively. It also assumes that, before it commences, an initial approximation for the probability vector x has been stored on disk and that the approximation for the last block, X_{B-1}, is already in-core.

The complete out-of-core algorithm of Figure 8 uses two shared memory buffers, R_0 and R_1, to read blocks of matrix into RAM from disk (line 9). Similarly, vector blocks are read (line 11) from disk into shared memory buffers, $Xbox_0$ and $Xbox_1$. Another array $\tilde{X}_i$, which is local to the *Compute Process*, is used to accumulate the sub-MVPs (line 12, *Compute Process*). As in Section 4.2, a local variable t can be used to switch between the pair of shared buffers, R_t and $Xbox_t$. Although the shared memory buffers are mentioned in the shared storage requirements of the algorithm, these buffers and the related local variables have been abstracted from the algorithm for the sake of simplicity.

The vector blocks corresponding to the empty matrix blocks are not read from disk (line 8, *DiskIO Process*). Moreover, to reduce disk I/O, the algorithm reads only the range of those elements in a vector block which are required for a sub-MVP. Once a vector block has been updated, this new approximation of the block must be updated on disk (line 15, *DiskIO Process*). The variable k (line 2, 18, *DiskIO Process*) is used to keep track of the index of the vector block to be written to disk during an inner iteration (line $7 - 17$, *DiskIO Process*). The variable j (line $5 - 6$ and $16 - 17$, *DiskIO Process*) is used to keep track of the indices of the matrix and vector blocks to be read from disk. The matrix blocks are stored on disk in such a way that a diagonal block (A_{ii}) follows all the off-diagonal blocks (all A_{ij}, $0 \le j < B$, $j \ne i$) in a row of matrix blocks, implying that the diagonal block will always be read and used for computation when $h = B - 1$ (or when $j = i$). Similarly, the *Compute Process* uses the variable j (line $4 - 5$ and $16 - 17$) to ensure that all the off-diagonal sub-MVPs are accumulated (line 12) before the sub-system $A_{ii}X_i = \tilde{X}_i$ (line 14) is solved.

Integer constant: B (*number of vector blocks*)
Semaphores: S_1, S_2: occupied
Shared variables: R_0, R_1 *(To read matrix A blocks into RAM)*
Shared variables: $Xbox_0$, $Xbox_1$ *(To read iteration vector x blocks into RAM)*

DiskIO Process	*Compute Process*
1. Local variable: h, i, j, k	1. Local variable: i, j, h
2. $k \leftarrow B - 1$	2. while not converged
3. while not converged	3. for $i \leftarrow 0$ to $B - 1$
4. for $i \leftarrow 0$ to $B - 1$	4. if $i = 0$ then $j \leftarrow B - 1$
5. if $i = 0$ then $j \leftarrow B - 1$	5. else $j \leftarrow i - 1$
6. else $j \leftarrow i - 1$	6. $\tilde{X}_i \leftarrow 0$
7. for $h \leftarrow 0$ to $B - 1$	7. for $h \leftarrow 0$ to $B - 1$
8. if not an *empty* block	8. Wait(S_1)
9. read A_{ij} from disk	9. Signal(S_2)
10. if $h \neq 0$	10. if $j \neq i$
11. read X_j from disk	11. if not an *empty* block
12. Signal(S_1)	12. $\tilde{X}_i \leftarrow \tilde{X}_i - A_{ij}X_j$
13. Wait(S_2)	13. else
14. if $h = 0$	14. Solve $A_{ii}X_i = \tilde{X}_i$
15. write X_k to disk	15. Test for convergence
16. if $j = 0$ then $j \leftarrow B - 1$	16. if $j = 0$ then $j \leftarrow B - 1$
17. else $j \leftarrow j - 1$	17. else $j \leftarrow j - 1$
18. $k \leftarrow k + 1 \mod B$	

Fig. 8. A complete out-of-core block Gauss-Seidel iterative algorithm

In the light of this discussion, and the explanation of the matrix out-of-core algorithm given in Section 4.2, the *Compute Process* is self-explanatory.

Implementation

We have used two separate files to store the matrix (in compact MSR) and the iteration vector on disk. As for the matrix out-of-core solution, the matrix blocks for the complete out-of-core method have been stored on disk in an order which enables the *DiskIO Process* to read the file sequentially throughout an iteration. However, the case for reading through the file which keeps the vector is more involved because, in this case, the *DiskIO Process* has to skip those vector blocks which correspond to empty blocks of the matrix.

Finally, the complete out-of-core implementation uses an array of size B^2 to keep track of the zero and nonzero matrix blocks. A sparse scheme may also be used to store this information. The number of blocks B for the complete out-of-core solution is small, usually less than 100, and therefore the memory required for the array is negligible.

4.5   A Symbolic Out-of-Core Solution Method

In the complete out-of-core method, the out-of-core scheduling of both the matrix and the iteration vector incurs a huge penalty in terms of disk I/O. Keeping the matrix in-core, in a compact representation, can significantly reduce this penalty while at the same time allowing for larger models to be analysed. This motivates the *symbolic out-of-core* method [25].

The idea of the symbolic out-of-core approach for the steady state solution of CTMCs is to keep the matrix in-core, in an appropriate symbolic data structure, and to store the probability vector on disk. The iteration vector is divided into a number of blocks. During the iterative computation phase, these blocks can be fetched from disk, one after another, into main memory to perform the numerical computation. Once a vector block has been updated with the next approximation, it is written back to disk. The symbolic out-of-core solution uses offset-labelled MTBDDs [27] to store the matrix, while the iteration vector for numerical computation is kept on disk as an array. An improved implementation of the symbolic out-of-core method is reported in [30].

Implementation
We have used a block iterative method to implement the symbolic out-of-core method. In this block method, we partition the system into a number of sub-systems and apply one (*inner*) iteration of the Jacobi iterative method[3] on each sub-system, in a *global* Gauss-Seidel iterative structure (see Section 2.2). This method is referred to in [33] as the *pseudo Gauss-Seidel method*.

The basic operation of the symbolic out-of-core block Gauss-Seidel method is the computation of the sub-MVPs, as was the case for the explicit out-of-core methods explained in earlier secions. The matrix is stored in an offset-labelled MTBDD. To implement each sub-MVP, we need to extract the matrix entries for a given matrix block from the MTBDD. We have seen in Section 3.3 that a matrix represented by an MTBDD can be decomposed into $(2^l)^2$ blocks by descending l levels of the MTBDD. Hence, in our implementation of the symbolic method, we select a value of l, take the number of blocks $B = 2^l$ and use the natural decomposition of the matrix given by the MTBDD. To access each block we simply need to store a pointer to the relevant node of the offset-labelled MTBDD. For large l, many of the matrix blocks may be empty. Therefore, instead of using a $B \times B$ array of pointers, a sparse scheme may be used to store the pointers to the nonzero matrix blocks; we select the compact MSR sparse scheme for this purpose. The extraction of matrix blocks, as required for each sub-MVP in the symbolic method, is therefore simple and fast; see [30] for further details on this implementation.

5   Results

In this section, we compare performance of the out-of-core solution methods discussed in Section 4. We have implemented the algorithms on an UltraSPARC-II

---

[3] The offset-labelled MTBDDs do not admit an efficient use of the Gauss-Seidel method, and therefore we use the Jacobi method to solve each sub-system.

440MHz CPU machine running SunOS 5.8 with 512MB RAM, and a 6GB local disk. We tested the implementations on three widely used benchmark models: a flexible manufacturing system (FMS) [11], a Kanban system [10] and a cyclic server Polling system [21]. These models were generated using PRISM [26], a probabilistic model checker developed at the University of Birmingham. More information about these models, a wide range of other PRISM case studies and the tool itself can be found at the PRISM web site [36].

The times to generate the files for the out-of-core solution phase are proportional to the times required to convert a model from BDD representation (see Section 3.3 on Page 238) to a sparse format. This file generation process can be optimised either for time or for memory. Optimising for memory can be achieved by allocating in RAM a data structure of the size of a submatrix of Q which is written to disk repeatedly as the conversion progresses. The generation process can be optimised for time by carrying out the entire process stated above in one step, i.e, converting the whole model into sparse format and then writing to file. We do not discuss the generation process any further, and hereon concentrate on the numerical solution phase.

The organisation of this section is as follows. In Section 5.1, we present and compare times per iteration for in-core and out-of-core versions for both explicit and symbolic approaches. In Section 5.2, we further analyse the out-of-core solutions with the help of performance graphs.

5.1   A Comparison of In-Core and Out-of-Core Solutions

Table 2 summarises results for the Kanban, FMS and Polling system case studies. The parameter k in column 2 denotes the number of tokens in the Kanban and FMS models, and the number of stations in the Polling system models. The resulting number of reachable states are given in column 3. Column 4 lists the average number of off-diagonal entries per row, giving an indication of the sparsity of the matrices.

Columns 5–7 in Table 2 list the time per iteration results for "Explicit" implementations: these include the standard in-core, where the matrix and the vector are kept in RAM; the matrix out-of-core (see Section 4.2), where only the matrix is stored on disk and the vector is kept in RAM; and the complete out-of-core (Section 4.4), where both the matrix and the vector are stored on disk. We have used the Gauss-Seidel method for the three reported explicit implementations and the resulting number of iterations are reported in column 8. The matrices for the explicit methods are stored in the compact MSR scheme, which requires $4a + 3n$ bytes to store the matrix. The entries "a/n" in column 4 can be used to calculate the memory required to store the matrices.

Table 2 reports the time per iteration results for "Symbolic" implementations in columns 9–10: both in-core, where both the matrix and the iteration vector are stored in RAM, and out-of-core (Section 4.5), where the matrix is kept in RAM and the vector is stored on disk. The matrix for these symbolic implementations has been stored using the offset-labelled MTBDD data structure, and the vector is kept as an array. We have used a block iterative method for the

Table 2. Comparing times per iteration for in-core and out-of-core methods

Model	k	States (n)	a/n	Time (seconds per iteration)						
				Explicit				Symbolic		
				In-core	Out-of-core		Iter.	In-core	Out-of-core	Iter.
					Matrix	Complete				
FMS	6	537,768	7.8	0.3	0.5	1.1	812	1.0	1.3	916
	7	1,639,440	8.3	1.1	1.7	3.8	966	3.0	3.2	1,079
	8	4,459,455	8.6	3.2	5.1	10.7	1,125	8.9	10.4	1,245
	9	11,058,190	8.9	–	24	51.8	1,287	39.6	35.9	1,416
	10	25,397,658	9.2	–	69	146	1,454	149	142	1,591
	11	54,682,992	9.5	–	–	374	1,624	–	708	1,770
	12	111,414,940	9.7	–	–	–	–	–	1,554	>50
	13	216,427,680	9.9	–	–	–	–	–	3,428	>50
Kanban system	4	454,475	8.8	0.3	0.5	1.0	323	0.5	0.8	373
	5	2,546,432	9.6	1.8	3.0	6.0	461	3.1	4.5	532
	6	11,261,376	10.3	–	30	68.6	622	15.6	22.6	717
	7	41,644,800	10.8	–	180	283	802	–	143	924
	8	133,865,325	11.3	–	–	–	–	–	601	1,151
Polling system	15	737,280	8.3	0.5	0.7	1.2	32	0.8	0.8	263
	16	1,572,864	8.8	1.1	1.9	2.9	33	1.7	2.0	276
	17	3,342,336	9.3	2.4	3.9	6.4	34	3.9	4.6	289
	18	7,077,888	9.8	5.5	15.8	20.4	34	8.3	10.5	302
	19	14,942,208	10.3	–	41	71	35	20.3	23.8	315
	20	31,457,280	10.8	–	101	162	36	–	52	328
	21	66,060,288	11.3	–	–	359	36	–	177	340
	22	138,412,032	11.8	–	–	–	–	–	374	353

symbolic implementations (see Section 4.5), and the respective number of iterations are reported in column 11. The run times for the FMS system ($k = 12, 13$) are taken for 50 iterations; we were unable to wait for their convergence, and hence the total numbers of iterations are not reported in the table.

The convergence criterion we have used in all our implementations is given by the equation (8) for $\varepsilon = 10^{-6}$. All reported run times are *wall clock* times.

We note, in Table 2, that the in-core explicit method provides the fastest run-times. However, pursuing this approach, the largest model solvable on a 512MB workstation is the Polling system with 7 million states. The in-core symbolic solution can solve larger models because, in this case, the matrix is stored symbolically. The largest model solvable with this symbolic in-core approach is the FMS system with 25.3 million states. We now consider the out-of-core approaches. The matrix out-of-core solution requires in-core storage for one iteration vector and two blocks of the matrix. The memory required for these matrix blocks can, in general, be reduced by increasing the number of blocks. However, in this case, the largest solvable model is limited by the size of the iteration vector. Pursuing the matrix out-of-core approach, the largest model solvable on the workstation is the Kanban system with 41.6 million states.

The out-of-core storage of both matrix and vector can solve even larger models. This is reported in column 7, and the largest model reported in this case is the Polling system with 66 million states. The limit in this case is the size of the available disk (here 6GB). Finally, the largest solvable model on the available machine is attributed to the symbolic out-of-core approach, i.e., the FMS system with 216 million states. This is possible because, in this case, only the vector is stored on disk and the symbolic data structure provides a compact in-core representation for the matrix.

The matrix out-of-core and the symbolic in-core methods both provide a solution to the matrix storage problem, and hence it is interesting to compare the run times for the two approaches. We note in Table 2, for the Polling and Kanban systems, that the run times per iteration for the symbolic in-core are faster than the marix out-of-core method. However, for the FMS system, matrix out-of-core method provides better run times. Similarly, we note that the symbolic out-of-core method provides faster run times for Polling and Kanban systems, but is slower than the complete out-of-core approach for the FMS system.

We observe that the results for explicit solutions are quite consistent for all three example models. However, the performance of symbolic solutions, both in-core and out-of-core, depends on the particular system under study. This is because the symbolic methods exploit model structure through sharing (subgraphs of the MTBDD). The FMS system is the least structured of the three models which equates to a large MTBDD to store it; the larger the MTBDD, the more time is required to perform its traversal. The Polling system, on the other hand, is very structured and therefore results in a smaller MTBDD. We conclude with our observation of Table 2 that, for large models, the symbolic out-of-core solution provides the best overall results for the examples considered.

5.2   Further Analysis of the Solution Methods

In this section, we investigate performance for the out-of-core solution methods by analysing the memory and time properties plotted against the number of blocks. All experiments reported in this section have been performed on the same machine as in Section 5.1. We begin by analysing the matrix out-of-core solution and then move on to the complete and the symbolic out-of-core solutions.

The Matrix Out-of-Core Method

A matrix out-of-core algorithm was given in Section 4.2 and the time per iteration results presented in Table 2. In Figure 9(a), we have plotted the memory requirements of the matrix out-of-core solution for three CTMCs, one from each case study. The plots display the total amount of memory used against the number of vector blocks B of equal size ($B \times B$ matrix blocks). Consider the plot for the Polling system. The memory required to store the vector and matrix in this case, if kept completely in RAM, is approximately 26MB and 135MB respectively. Decomposing the matrix into blocks, keeping it on disk, and reading one block at a time reduces the total in-core memory requirement of the solution. The memory required for the case $B = 4$ is 85MB. Increasing the number of blocks up to $B = 64$ reduces the memory requirements to nearly 30MB. This minimum is bounded by the storage required for the iteration vector. Similar properties are evident in the plots for the FMS and Kanban system CTMCs.

In Figure 9(b), we analyse the time per iteration characteristics for the same three CTMCs, plotted against the number of vector blocks. We note a slight decrease in the time per iteration for all three CTMCs. The reason for this slight

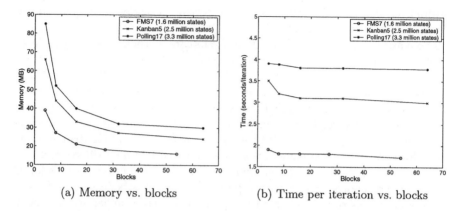

(a) Memory vs. blocks   (b) Time per iteration vs. blocks

Fig. 9. The matrix out-of-core method: effects of varying the number of blocks

decrease in time is the decrease in the memory requirement of the solution process, as demonstrated in the corresponding Figure 9(a). This effect would be more obvious for larger models. Another reason for this effect is that increasing the number of blocks typically results in smaller blocks, which possibly have less variations[4] in their sizes, consequently resulting in a better balance between the disk I/O and the computation. However, increasing the number of blocks to a very high number will possibly result in an increase in solution time due to the higher number of synchronisation points (see Section 4.2).

The Complete Out-of-Core Method

Figure 10(a) illustrates the total memory requirement of the complete out-of-core solution against the number of blocks for the same three CTMCs. The vector and the CTMC are partitioned into B and $B \times B$ blocks respectively, and, in this case, both are stored on disk. We note that the memory plots in the figure show a similar trend as in Figure 9(a).

The time per iteration properties of the complete out-of-core solution are plotted in Figure 10(b). In contrast to the matrix out-of-core method, we note a significant increase in solution time for the complete out-of-core method with an increase in the number of blocks. This is due to the fact that, for the complete out-of-core method, the iteration vector[5] must be read B times in each iteration. Consequently, an increase in the number of blocks generally results in an increase in the amount of disk I/O, and hence an increase in the solution time.

The Symbolic Out-of-Core Method

In Figure 11, we analyse the performance of the symbolic out-of-core method by plotting the memory and time properties against the number of blocks for

---

[4] Although the blocks have an equal number of rows, they can still have a varying and unequal number of nonzero entries.

[5] Of course, the vector blocks corresponding to the zero matrix blocks are not read.

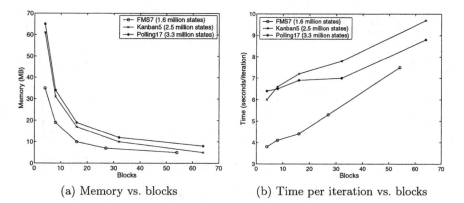

(a) Memory vs. blocks   (b) Time per iteration vs. blocks

Fig. 10. The complete out-of-core method: effects of varying the number of blocks

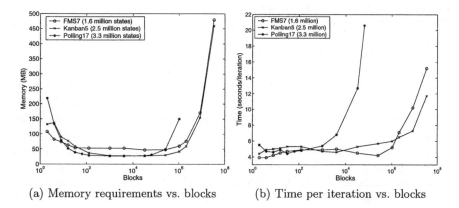

(a) Memory requirements vs. blocks   (b) Time per iteration vs. blocks

Fig. 11. The symbolic out-of-core method: effects of varying the number of blocks

the same three CTMCs as in Figure 10. To further explore the behaviour of the symbolic out-of-core method, in Figure 12, we plot similar properties for larger CTMCs. We note in the two figures that the range for the numbers of blocks is much higher compared to the earlier graphs for explicit solutions. The reason is that the MTBDD actually encodes a matrix over its *potential state space*, which typically includes many unreachable states (see Section 3.3). In the following, we explain the plots for Figure 12; in the light of this discussion, Figure 11 should be self-explanatory.

The total memory requirement of the symbolic out-of-core solution against the number of blocks for three CTMCs is plotted in Figure 12(a) (compare with Figure 11(a)). We explain the plot for the Kanban system ($k = 6$). The memory required for the case $B = 2$ is above 650MB. The increase in the number of blocks reduces the memory requirements for the Kanban system to nearly 140MB. A similar properties are evident for the other plots in Figure 12(a). For large numbers of blocks (i.e. the rightmost portions of the plots), we note an increase in

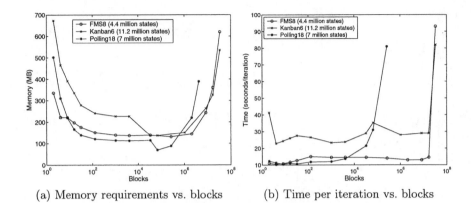

(a) Memory requirements vs. blocks   (b) Time per iteration vs. blocks

Fig. 12. The symbolic out-of-core method: effects of varying the number of blocks

the amount of memory. This is because the memory overhead required to store information about the blocks of the MTBDD dominates the overall memory in these cases.

The time per iteration properties of the symbolic out-of-core solution are analysed in Figure 12(b), plotted against the number of vector blocks. Consider the plot for the Kanban system. Initially, for $B = 2$, the memory required (see Figure 12(a)) for the iteration vector is more than the available RAM. This causes thrashing and results in a high solution time. An increase in the number of blocks removes this problem and explains the initial downward jump in the plot. From this point on, however, the times vary. The rationale for this is as follows. As we explained in Section 3.3, our decomposition of the MTBDD matrix into blocks can result in a partitioning such that the resulting matrix (and hence vector) blocks are of unequal sizes. This can affect the overlap of computation and disk I/O, effectively increasing the solution time. The sizes of the partitions are generally unpredictable, being determined both by the sparsity pattern of the matrix and by the encoding of the matrix into the MTBDD. Finally, we note that the end of the plot shows an increase in the solution time. This is due to the overhead of manipulating a large number of blocks of the matrix and the increased memory requirements that this imposes, as is partially evident from Figure 12(a). Note that, for the symbolic implementations, we use a block iterative method where we apply one Jacobi (inner) iteration on each sub-system, in a global Gauss-Seidel iterative structure (see Section 4.5). Increasing the number of blocks, therefore, typically causes a reduction in the required number of iterations for a CTMC to converge.

We conclude this section here with the observation that among the many factors which affect performance of the out-of-core solutions are the data structures that hold the matrix, and the number of blocks, B, that the matrix (and vector) are partitioned into. We have also found that, under different values of k (see Table 2) for each case study, similar patterns for the memory and time plots against the number of blocks are observed. This can, in fact, be useful for predicting good choices of B for larger values of k. A useful direction for future work would be to investigate more fully what constitutes a good choice for the number of blocks.

6   Conclusions

In this paper, serial out-of-core algorithms for the analysis of large Markov chains have been surveyed. Earlier work, in this context, has focussed on implicit and parallel explicit solutions. During the last five or so years, some progress has been made, beginning with the solution of a 15 million states system on a workstation with 128MB RAM, with no assumption on its structure [14, 15]. This approach was parallelised in 1999 [23], leading to the solution of 724 million states on a 26-node dual-processor cluster [6]. A number of solution techniques had been devised by the late 1990s to cope with the matrix storage problems. However, explicit storage of the solution vector(s) hindered further progress for both implicit and explicit methods. These limitations were later relaxed by the out-of-core storage of the vector. The complete out-of-core solution of a 41 million states system on a 128MB RAM machine was demonstrated in [24]. Furthermore, a combination of MTBDD-based in-core symbolic storage of matrix and out-of-core storage of the vector allowed the solution of a 216 million states system on a single workstation with 512MB RAM [25, 30].

We surveyed the out-of-core approaches in this paper. The algorithms were analyzed using tabular and graphical results of their implementations with the help of benchmark case studies. Our focus in this paper has been the steady state solution of an (irreducible) CTMC. We intend to generalise these techniques to other numerical computation problems, such as transient analysis of CTMCs and analysis of DTMCs and MDPs. Another direction for future research is to extend the complete out-of-core and the symbolic out-of-core approaches by employing parallelisation.

The Kronecker approach provides a space-efficient representation of a Markov chain. Representations based on such an approach have increasingly gained popularity. These methods, however, still require explicit storage of the solution vector(s). The out-of-core storage of the solution vector can also provide a solution in this case.

Further improvements in out-of-core techniques can be achieved with the help of redundant arrays of independent disks (RAID). In future, we anticipate that a combination of parallel and out-of-core techniques will play an important role in the analysis of large stochastic models.

Acknowledgment

I am grateful to Markus Siegle and Alexander Bell for their extremely valuable suggestions and many fruitful discussions during the write up of this manuscript. David Parker, Gethin Norman and Marta Kwiatkowska are all very kindly acknowledged for their support and helpful comments.

References

1. I. Bahar, E. Frohm, C. Gaona, G. Hachtel, E. Macii, A. Pardo, and F. Somenzi. Algebraic Decision Diagrams and their Applications. In *Proc. ICCAD'93*, pages 188–191, Santa Clara, 1993.
2. R. Barrett, M. Berry, T. F. Chan, J. Demmel, J. M. Donato, J. Dongarra, V. Eijkhout, R. Pozo, C. Romine, and H. van der Vorst. *Templates for the Solution of Linear Systems: Building Blocks for Iterative Methods.* Philadalphia: Society for Industrial and Applied Mathematics, 1994.
3. F. Baskett, K. M. Chandy, R. R. Muntz, and F. G. Palacios. Open, Closed, and Mixed Networks of Queues with Different Classes of Customers. *Journal of ACM*, 22(2), 1975.
4. F. Bause. Queueing Petri Nets: A Formalism for the Combined Qualitative and Quantitative Analysis of Systems. In *Proc. PNPM'93*. IEEE Computer Society Press, October 1993.
5. A. Bell. Verteilte Bewertung Stochastischer Petrinetze, Diploma thesis, RWTH, Aachen, Department of Computer Science, March 1999.
6. A. Bell and B. R. Haverkort. Serial and Parallel Out-of-Core Solution of Linear Systems arising from Generalised Stochastic Petri Nets. In *Proc. High Performance Computing 2001*, Seattle, USA, April 2001.
7. M. Bernardo and R. Gorrieri. Extended Markovian Process Algebra. In *Proc. CONCUR 1996, LNCS Volume 1119*, Italy, 1996. Springer-Verlag.
8. P. Buchholz and P. Kemper. Kronecker based Matrix Representations for Large Markov Models. In *this Proceedings*, 2003.
9. G. Ciardo and A. Miner. A Data Structure for the Efficient Kronecker Solution of GSPNs. In *Proc. PNPM'99*, Zaragoza, 1999.
10. G. Ciardo and M. Tilgner. On the use of Kronecker Operators for the Solution of Generalized Stochastic Petri Nets. ICASE Report 96-35, Institute for Computer Applications in Science and Engineering, 1996.
11. G. Ciardo and K. S. Trivedi. A Decomposition Approach for Stochastic Reward Net Models. *Performance Evaluation*, 18(1):37–59, 1993.
12. E. Clarke, M. Fujita, P. McGeer, J. Yang, and X. Zhao. Multi-Terminal Binary Decision Diagrams: An Effificient Data Structure for Matrix Representation. In *Proc. International Workshop on Logic Synthesis (IWLS'93)*, May 1993.
13. A. E. Conway and N. D. Georganas. *Queueing Networks - Exact Computational Algorithms: A Unified Theory based on Decomposition and Aggregation.* MIT Press, 1989.
14. D. D. Deavours and W. H. Sanders. An Efficient Disk-based Tool for Solving Very Large Markov Models. In *Proc. TOOLS'97*, volume 1245 of *LNCS*, pages 58–71. Springer-Verlag, 1997.
15. D. D. Deavours and W. H. Sanders. An Efficient Disk-based Tool for Solving Large Markov Models. *Performance Evaluation*, 33(1):67–84, 1998.
16. D. D. Deavours and W. H. Sanders. "On-the-fly" Solution Techniques for Stochastic Petri Nets and Extensions. *IEEE Transactions on Software Engineering*, 24(10):889–902, 1998.
17. H. Hermanns, J. Meyer-Kayser, and M. Siegle. Multi Terminal Binary Decision Diagrams to Represent and Analyse Continuous Time Markov Chains. In *Proc. NSMC'99*, Zaragoza, 1999.
18. H. Hermanns and M. Rettelbach. Syntax, Semantics, Equivalences, and Axioms for MTIPP. In *Proc. PAPM'94*, Germany, 1994.

19. M. Heroux.   A proposal for a sparse BLAS Toolkit, Technical Report TR/PA/92/90, Cray Research, Inc., USA, December 1992.
20. J. Hillston. *A Compositional Approach to Performance Modelling.* PhD thesis, University of Edinburgh, 1994.
21. O. Ibe and K. Trivedi. Stochastic Petri Net Models of Polling Systems. *IEEE Journal on Selected Areas in Communications*, 8(9):1649–1657, 1990.
22. W. Kahan. *Gauss-Seidel methods of solving large systems of linear equations.* PhD thesis, University of Toronto, 1958.
23. W. J. Knottenbelt and P. G. Harrison. Distributed Disk-based Solution Techniques for Large Markov Models. In *Proc. NSMC'99*, 1999.
24. M. Kwiatkowska and R. Mehmood. Out-of-Core Solution of Large Linear Systems of Equations arising from Stochastic Modelling. In *Proc. PAPM-PROBMIV'02*, July 2002. Available as Volume 2399 of *LNCS*.
25. M. Kwiatkowska, R. Mehmood, G. Norman, and D. Parker. A Symbolic Out-of-Core Solution Method for Markov Models. In *Proc. Parallel and Distributed Model Checking (PDMC'02)*, August 2002. Appeared in Volume 68, issue 4 of ENTCS (http://www.elsevier.nl/locate/entcs).
26. M. Kwiatkowska, G. Norman, and D. Parker. PRISM: Probabilistic Symbolic Model Checker. In *Proc. TOOLS'02*, volume 2324 of *LNCS*, 2002.
27. M. Kwiatkowska, G. Norman, and D. Parker. Probabilistic Symbolic Model Checking with PRISM: A Hybrid Approach. In *Proc. TACAS 2002, LNCS Volume 2280*, April 2002.
28. M. A. Marsan, G. Balbo, and G. Conte. A Class of Generalized Stochastic Petri Nets for the Performance Analysis of Multiprocessor Systems. *ACM Transactions on Computer Systems*, 2(2), 1984.
29. M. A. Marsan, G. Balbo, G. Conte, S. Donatelli, G. Franceschinis, and D. Kartson. *Modelling With Generalized Stochastic Petri Nets.* John Wiley & Son Ltd, 1995.
30. R. Mehmood. *On the Development of Techniques for the Analysis of Large Markov Models.* PhD thesis, University of Birmingham, 2003. To appear.
31. A. Miner and D. Parker. Symbolic Representations and Analysis of Large State Spaces. In *this Proceedings*, 2003.
32. M. K. Molloy. Performance Analysis using Stochastic Petri Nets. *IEEE Trans. Comput.*, 31:913–917, September 1982.
33. D. Parker. *Implementation of Symbolic Model Checking for Probabilistic Systems.* PhD thesis, University of Birmingham, August 2002.
34. B. Plateau. On the Stochastic Structure of Parallelism and Synchronisation Models for Distributed Algorithms. In *Proc. 1985 ACM SIGMETRICS Conference on Measurement and Modeling of Computer Systems*, 1985.
35. B. Plateau and K. Atif. Stochastic Automata Network for Modeling Parallel Systems. *IEEE Transactions on Software Engineering*, 17(10), 1991.
36. PRISM Web Page. http://www.cs.bham.ac.uk/~dxp/prism/.
37. Y. Saad. SPARSKIT: A basic tool kit for sparse matrix computations. Technical Report RIACS-90-20, NASA Ames Research Center, CA, 1990.
38. M. Siegle. Advances in Model Representations. In *Proc. PAPM/PROBMIV 2001, LNCS Volume 2165*, Aachen, Germany, 2001. Springer Verlag.
39. W. J. Stewart. *Introduction to the Numerical Solution of Markov Chains.* Princeton University Press, 1994.
40. S. Toledo. A Survey of Out-of-Core Algorithms in Numerical Linear Algebra. In *External Memory Algorithms and Visualization*, DIMACS Series in Discrete Mathematics and Theoretical Computer Science. American Mathematical Society Press, Providence, RI, 1999.

Kronecker Based Matrix Representations for Large Markov Models

Peter Buchholz[1]* and Peter Kemper[2]**

[1] Technische Universität Dresden, D-01062 Dresden, Germany
p.buchholz@inf.tu-dresden.de
[2] Universität Dortmund, D-44221 Dortmund, Germany
peter.kemper@udo.edu

Abstract. State-based analysis of discrete event systems notoriously suffers from the largeness of state spaces, which often grow exponentially with the size of the model. Since non-trivial models tend to be built by submodels in some hierarchical or compositional manner, one way to achieve a compact representation of the associated state-transition system is to use Kronecker representations that accommodate the structure of a model at the level of a state transition system. In this paper, we present the fundamental idea of Kronecker representation and discuss two different kinds of representations, namely modular representations and hierarchical representations. Additionally, we briefly outline how the different representations can be exploited in efficient analysis algorithms.

1 Introduction

The class of discrete event systems covers a wide variety of systems from different application areas such as communication systems, computer systems or manufacturing systems. All these different systems have in common that they can be mapped onto models that are characterized by a countable or finite set of states and a set of atomic transitions that move the system from one state to another and mimic the dynamic behavior of the system. Transitions are often assigned a label and/or a numerical value. The label gives additional information concerning each transition typically related to the event that performs it in the discrete event system. Numerical values that are associated with transitions quantify costs, times, rates, or probabilities. Based on this description, it is possible to analyze systems according to a wide variety of qualitative and quantitative aspects. In this overview paper, we consider quantitative analysis of systems that are described by continuous time Markov chains with a finite state space. However, the presented concepts can be used in a much wider context for the analysis of discrete event systems.

---

* This research is supported by DFG, collaborative research center 358 'Automatic Systems Design'
** This research is supported by DFG, collaborative research center 559 'Modeling of Large Logistic Networks'

C. Baier et al. (Eds.): Validation of Stochastic Systems, LNCS 2925, pp. 256–295, 2004.

The presentation is structured as follows. We start in Section 2 with some basic definitions to describe a Markov process by its generator matrix and introduce numerical analysis for its transient (time-bounded) behavior and steady state (long-range) behavior. Furthermore, we introduce stochastic automata networks (SANs) to describe large systems by a set of relatively small automata and composition operations that turn out to match operators of a specific matrix algebra. In particular, a Kronecker algebra with operators Kronecker product and sum are defined that is suitable to express the generator matrix of a SAN in a compositional manner. In the last subsection, the approach is illustrated by an example and certain difficulties are identified and discussed in subsequent sections.

The descriptor matrix as given by Kronecker algebra is the basis for the techniques introduced in Section 3. We consider algorithms to perform a multiplication of a vector with a Kronecker representation, which is a fundamental operation for many analysis techniques. Two weaknesses of a Kronecker representation of a generator matrix require further attention: One is that transition rates in a model may depend on the overall state of a model, for instance in case of service times at shared resources; the other is that a Kronecker representation provides a generator matrix for a state space that is only potentially reachable and that contains the state space of the considered Markov chain as a part. If the set of potentially reachable states is significantly larger than the reachable set, the whole approach loses its efficiency. For the problem of defining dependencies due to shared resources, we introduce state-dependent transition rates and generalized Kronecker sums/products that are able to handle those state dependencies. The problem of unreachable states is handled by an additional data structure that helps algorithms to restrict "themselves" to reachable states.

Section 4 introduces a hierarchical approach which represents the generator and not the descriptor. The difference between descriptor and generator is that the former describes transitions in the potentially reachable state space, whereas the latter considers only transitions among reachable states which is sufficient for the analysis. The idea of the representation of the generator is to define a hierarchical view of the system such that macro states are described and each macro state represents a set of micro states characterized by the cross-product of subsets of automata states. We show that this presentation is very natural for queuing network like models, but can also be derived automatically from a SAN description. The resulting matrix has a block structure where each block is represented as a sum of Kronecker products of automata matrices.

In Section 5, we very briefly describe analysis methods usable in conjunction with the Kronecker representation. However, we mainly give references to relevant literature and do not introduce the techniques in detail in order to keep the paper concise. The paper ends with the conclusions where we briefly outline current research efforts in the area and mention some open research problems. Table 1 gives a brief overview of the notation used throughout the paper.

Table 1. Notation

$\mathcal{RS}, \|\mathcal{RS}\| = n$	reachable set of states and its cardinality
$\mathcal{RS}^{(i)}, \|\mathcal{RS}^{(i)}\| = n^{(i)}$	reachable set of states of component i and its cardinality
$\mathcal{PS}, \|\mathcal{PS}\| = \bar{n} = \prod_{i=1}^{J} n^{(i)}$	potential set of states and its cardinality
$\mathcal{T} \supset \mathcal{TS}$	set of transitions and subset of synchronized transitions
$\mathcal{T}^{(i)} \supset \mathcal{TL}^{(i)}$	transitions of component i and subset of local transitions
i, j	numbers of automata in a network
x, y, z	state indices
$\mathbf{x}, \mathbf{y}$	state vectors

2   Basic Definitions

The first section collects the basic definitions which are required to present the Kronecker representation of generator matrices. We begin with the introduction of stochastic automata and Markov processes. Afterwards networks of stochastic automata are defined and the basic operations of Kronecker algebra are presented. The last two subsections contain a definition of the Kronecker representation of the descriptor for a network of stochastic automata and an example which serves as a running example for the rest of the paper.

2.1   Stochastic Automata and Markov Processes

In our framework, a Markov process is described as a stochastic automaton with a finite state space $\mathcal{RS}$ including n states. Since $\mathcal{RS}$ is isomorphic to $\{0, ..., n-1\}$, states can be characterized by their number. We assume that transitions among states are labeled with labels from a finite set $\mathcal{T}$ and additionally we associate a positive real value with each transition. Labels are used to distinguish among transitions, numerical values to describe the rate λ of a transition, i.e., the transition takes place after a delay whose length has a negative exponential distribution with mean value $1/\lambda$. A given automaton of this kind can be formalized in two ways: as a node and arc labeled graph (a labeled transition system) or as a matrix which is the adjacency matrix of the graph. If we want to distinguish transitions according to their labels we may employ a formal sum of adjacency matrices with one matrix per label. The graph representation naturally leads to graph exploration algorithms like depth-first search or breadth-first search, while the matrix representation yields to a mathematical treatment in terms of matrix operations. Since the matrix representation is more appropriate for the realization of analysis algorithms and it furthermore allows us to introduce an algebra to describe networks of components, we consider here a matrix description of automata.

An automaton is characterized by $|\mathcal{T}|$ matrices of dimension $n \times n$. Let $\mathbf{E}_t$ be the matrix including numerical values for transitions labeled with $t \in \mathcal{T}$. The case of several transitions between two states s, s' with the same label t can be handled in different ways: either transitions shall be considered as individual transitions so that they need to obtain different labels t and t', or transitions shall not be treated individually and in that case they are represented by a single entry $\mathbf{E}_t(s, s')$ which is defined as the sum of the two values. For Markov processes, the sum characterizes a race situation since the minimum of two random variables with exponential distribution with rates λ, resp. λ' is exponentially distributed with rate $\lambda + \lambda'$. Furthermore let λ_t be the basic rate of t-labeled transitions. This rate is introduced to normalize the elements in matrix $\mathbf{E}_t$. The value of λ_t is chosen in such a way that all elements of $\mathbf{E}_t$ are in the interval $[0, 1]$. Thus, $\lambda_t \mathbf{E}_t(x, y)$ for $0 \le x, y < n$ describes the rate of t-labeled transitions between states x and y. Define $\mathbf{D}_t = diag(\mathbf{E}_t \mathbf{e}^T)$ where $\mathbf{e}$ is the unit row vector of length n and $diag(\mathbf{a})$ for column vector $\mathbf{a}$ is a diagonal matrix with $\mathbf{a}(i)$ in position (i, i). Usually the specification of an automaton additionally contains the initial state. In stochastic automata, it is sometimes convenient to allow a set of initial states instead of a single state. The set of initial states is characterized by a probability distribution $\mathbf{p}_0 \in \mathbb{R}^n$; a state i is an initial state with probability $\mathbf{p}_0(i)$. We use the convention that $i = 0$ is the only initial state if no initial distribution is explicitly defined.

Matrices $\mathbf{E}_t$ contain the full description of the functional behavior of the automaton, i.e., a behavior that is described in terms of sequences of transitions where transitions are only distinguished according to their label. In that case, we need to distinguish only zero and non-zero entries in $\mathbf{E}_t$ since the latter describe transitions. Additionally, the matrices describe the Continuous Time Markov Chain (CTMC) specified by the automaton. Let

$$\mathbf{E} = \sum_{t \in \mathcal{T}} \lambda_t \mathbf{E}_t \text{ and } \mathbf{D} = \sum_{t \in \mathcal{T}} \lambda_t \mathbf{D}_t \ .$$

The generator matrix of the underlying CTMC is given by $\mathbf{Q} = \mathbf{E} - \mathbf{D}$. From $\mathbf{Q}$ and the initial distribution $\mathbf{p}_0$ the distribution at time τ is computed as

$$\mathbf{p}_\tau = \mathbf{p}_0 exp(\mathbf{Q}\tau) \ . \tag{1}$$

If $\mathbf{Q}$ is irreducible, then the stationary distribution $\boldsymbol{\pi}$ is uniquely defined as the solution of

$$\boldsymbol{\pi}\mathbf{Q} = \mathbf{0} \text{ and } \boldsymbol{\pi}\mathbf{e}^T = 1 \ . \tag{2}$$

Many different methods exist for the solution of (1) and (2). Usually iterative methods based on sparse matrix data structures are used for the solution of the equations. After the stationary or transient distribution has been computed, different quantitative measures can be derived, e.g., the throughput of type t transitions equals $\lambda_t \boldsymbol{\pi} \mathbf{E}_t \mathbf{e}^T$. Similarly other measures can be computed. For a detailed introduction of Markov processes and their analysis, we refer to [62].

So far, we have presented automata as the basic model class. However, for specification purposes, this basic notation is usually too low level and other more abstract specification paradigms are used. Examples are queueing networks, stochastic Petri nets, or stochastic process algebras. In this paper, we restrict the models to those with a finite state space. The modeling formalisms allow us to describe more compact models from which the underlying automaton can be generated with the help of some well-known algorithms. Thus, in principle, the steps from model specification to analysis are well defined and analysis algorithms are known for each step. However, the major problem of the approach is the *state space explosion* which means that the size of the state space that results from a queueing network or stochastic Petri net tends to grow exponentially with the size of the model. This implies that even models that appear simple may have state spaces with several millions of states. Thus, handling of state space explosion is the major challenge of state-based analysis techniques.

2.2   Networks of Stochastic Automata

Here, we present an approach which relies on the idea that exponentially growing problems often can be handled efficiently by a divide and conquer approach. Thus, instead of a single large automaton, a number of small automata is considered. The resulting model is a network of communicating automata, which is a well known paradigm to specify communicating processes [50, 58]. In this context, we focus on stochastic automata networks (SANs) as the basic paradigm. SANs originate from the work of Plateau [60], a recent textbook presentation is given in [62], Chapter 9. A SAN is defined on a set of automata of the kind introduced above. Those automata are numbered consecutively from 1 to J. Automaton i is characterized by state space $\mathcal{RS}^{(i)}$ containing $n^{(i)}$ states, the set of transitions $\mathcal{T}^{(i)}$ (with corresponding matrices $\mathbf{E}_t^{(i)}$ and $\mathbf{D}_t^{(i)}$), and by the initial distribution $\mathbf{p}_0^{(i)}$.

In a SAN, different automata are connected via synchronizing their transitions. Thus, we have to distinguish between synchronized and local transitions. Local transitions occur independently in the automata, whereas on a synchronized transition at least two automata participate and the transition can only occur if all participating automata agree that it occurs. Let $\mathcal{J} = \{1, \ldots, J\}$ be the set of automata in the SAN. Let $\mathcal{TS} \subseteq \cup_{i \in \mathcal{J}} \mathcal{T}^{(i)}$ be the set of synchronized transitions and assume that for each $t \in \mathcal{TS}$ at least two automata i and j exist such that $t \in \mathcal{T}^{(i)} \cap \mathcal{T}^{(j)}$. The set of local transitions for each automaton i is defined as $\mathcal{TL}^{(i)} = \mathcal{T}^{(i)} \setminus \mathcal{TS}$.

For a SAN, we can define a potential state space $\mathcal{PS} = \times_{j=1}^{J} \mathcal{RS}^{(i)}$ that includes all potentially reachable states. The number of states in $\mathcal{PS}$ equals $\bar{n} = \prod_{i=1}^{J} n^{(i)}$. However, due to synchronization constraints, not all of these states must be reachable and we define the set of reachable states as $\mathcal{RS} \subseteq \mathcal{PS}$. A state of the SAN can be described by a vector $\mathbf{x} = (\mathbf{x}(1), \ldots, \mathbf{x}(J))$ with $\mathbf{x}(i) \in \mathcal{RS}^{(i)}$. This description can be "linearized" using a so called mixed radix number representation [36] such that state $\mathbf{x} \in \mathcal{PS}$ is mapped onto integer x from the set $\{0, \ldots, \bar{n} - 1\}$. The mapping is computed as

$$x = \sum_{i=1}^{J} \mathbf{x}(i) \prod_{j=i+1}^{J} n^{(j)} . \tag{3}$$

We will use the vector and linearized representation interchangeably. Observe that the state space of order $\bar{n}$ is expressed by J sets of size $n^{(i)}$ each. Memory requirements to store the automata state spaces seperately is in $O(\sum_{i=1}^{J} n^{(i)})$ whereas memory requirements to store all states are in $O(\prod_{i=1}^{J} n^{(i)})$. This relation holds as long as the size of $\mathcal{RS}$ is in the same order of magnitude as the size of $\mathcal{PS}$. The isolated storage of automata state spaces instead of the overall state space eliminates the state space explosion to a large extent.

2.3   Definitions Related to Kronecker Representations

The idea of representing a large state space as a cross product of small state spaces is now extended to matrices. This step requires matrix operations that are able to express the behavior that results from synchronized transitions and the behavior that results from local transitions. The former will be handled by a Kronecker product, the latter by a Kronecker sum, which are defined as follows.

Definition 1. *The Kronecker (tensor) product of two matrices $\mathbf{A} \in \mathbb{R}^{r_1 \times c_1}$ and $\mathbf{B} \in \mathbb{R}^{r_2 \times c_2}$ is defined as $\mathbf{C} = \mathbf{A} \otimes \mathbf{B}$, $\mathbf{C} \in \mathbb{R}^{r_1 r_2 \times c_1 c_2}$, where*

$$\mathbf{C}(x_1 \cdot r_2 + x_2, y_1 \cdot c_2 + y_2) = \mathbf{A}(x_1, y_1)\mathbf{B}(x_2, y_2)$$

($1 \leq x_l \leq r_l$, $1 \leq y_l \leq c_l$, $l \in \{1, 2\}$).
    The Kronecker (tensor) sum of two square matrices $\mathbf{A} \in \mathbb{R}^{n_1 \times n_1}$ and $\mathbf{A} \in \mathbb{R}^{n_2 \times n_2}$ is defined as

$$\mathbf{A} \oplus \mathbf{B} = \mathbf{A} \otimes \mathbf{I}_{n_2} + \mathbf{I}_{n_1} \otimes \mathbf{B}$$

where $\mathbf{I}_n$ is the identity matrix of order n.

We consider as a simple example the Kronecker product and sum of two small matrices.

$$\mathbf{A} = \begin{pmatrix} \mathbf{A}(0,0) & \mathbf{A}(0,1) \\ \mathbf{A}(1,0) & \mathbf{A}(1,1) \end{pmatrix}, \quad \mathbf{B} = \begin{pmatrix} \mathbf{B}(0,0) & \mathbf{B}(0,1) \\ \mathbf{B}(1,0) & \mathbf{B}(1,1) \end{pmatrix}$$

The Kronecker product can be used to combine matrices of arbitrary dimensions.

$$\mathbf{C} = \mathbf{A} \otimes \mathbf{B} = \begin{pmatrix} \mathbf{A}(0,0)\mathbf{B} & \mathbf{A}(0,1)\mathbf{B} \\ \mathbf{A}(1,0)\mathbf{B} & \mathbf{A}(1,1)\mathbf{B} \end{pmatrix}$$

Note that the resulting matrix $\mathbf{C}$ basically contains all possible products of single entries in $\mathbf{A}$ with single entries in $\mathbf{B}$. A product matches the behavior of a synchronized transition, because a synchronized transition takes place if and only if all involved components perform the transition together and a value like $\mathbf{C}(0,0)$ is nonzero if and only if all entries $\mathbf{A}(0,0), \mathbf{B}(0,0)$ are nonzero. For a specific application, we need to define numerical entries of $\mathbf{A}$ and $\mathbf{B}$ in a way such that the resulting product appropriately describes the behavior of the synchronized transition.

The Kronecker sum is based on the Kronecker product. The role of identity matrices in $\mathbf{A} \otimes \mathbf{I}_{n_2} + \mathbf{I}_{n_1} \otimes \mathbf{B}$ (+ is the "normal" elementwise addition of elements) is that of a neutral component in a synchronized transition, i.e., the neutral component accepts only those transitions that do not alter its own

state by $\mathbf{I}(x, x) = 1$ and the numerical entry 1 is the neutral element of multiplication such that it does not modify the outcome of the Kronecker product.
$\mathbf{D} = \mathbf{A} \oplus \mathbf{B} =$

$$\begin{pmatrix} \mathbf{A}(0,0) + \mathbf{B}(0,0) & \mathbf{B}(0,1) & \mathbf{A}(0,1) & 0 \\ \mathbf{B}(1,0) & \mathbf{A}(0,0) + \mathbf{B}(1,1) & 0 & \mathbf{A}(0,1) \\ \mathbf{A}(1,0) & 0 & \mathbf{A}(1,1) + \mathbf{B}(0,0) & \mathbf{B}(0,1) \\ 0 & \mathbf{A}(1,0) & \mathbf{B}(1,0) & \mathbf{A}(1,1) + \mathbf{B}(1,1) \end{pmatrix}$$

So the Kronecker sum can be used to describe the behavior of local transitions that take place independently from other components since each term in $\mathbf{A} \otimes \mathbf{I}_{n_2} + \mathbf{I}_{n_1} \otimes \mathbf{B}$ describes the behavior of one component while the other remains inactive. Clearly, the summation over these terms must be compatible with the definition of dynamic behavior in the considered modeling formalism. Summation behaves like a logical OR from a functional point of view, an entry in the resulting matrix is non-zero if at least one term in the sum contains a non-zero entry at that position. As for the Kronecker product, the numerical values must be chosen such that their sum represents the correct value according to the stochastic automaton of the overall system. For instance, in a Markovian process, summation is appropriate since we assume a race situation among transitions where the fastest transition becomes effective with a rate that is the sum over all rates. In the following, we focus on Markov processes and will define numerical values accordingly.

Kronecker product and sum have some nice algebraic properties [44] which hold if the involved matrix products and inverse matrices are defined:

1. $(\mathbf{A} \otimes \mathbf{B}) \otimes \mathbf{C} = \mathbf{A} \otimes (\mathbf{B} \otimes \mathbf{C})$ and $(\mathbf{A} \oplus \mathbf{B}) \oplus \mathbf{C} = \mathbf{A} \oplus (\mathbf{B} \oplus \mathbf{C})$ (associativity)
2. $(\mathbf{AB}) \otimes (\mathbf{CD}) = (\mathbf{A} \otimes \mathbf{C})(\mathbf{B} \otimes \mathbf{D})$ (compatibility with multiplication)
3. $(\mathbf{A} + \mathbf{B}) \otimes (\mathbf{C} + \mathbf{D}) = \mathbf{A} \otimes \mathbf{C} + \mathbf{A} \otimes \mathbf{D} + \mathbf{B} \otimes \mathbf{C} + \mathbf{B} \otimes \mathbf{D}$ (distributivity over addition)
4. $(\mathbf{A} \otimes \mathbf{B})^T = \mathbf{A}^T \otimes \mathbf{B}^T$ (compatibility with transposition)
5. $(\mathbf{A} \otimes \mathbf{B})^{-1} = \mathbf{A}^{-1} \otimes \mathbf{B}^{-1}$ (compatibility with inversion)

Since Kronecker products and sums are associative, the following generalizations are well-defined.

$$\bigotimes_{j=1}^{J} \mathbf{A}_j = \mathbf{A}_1 \otimes \left(\bigotimes_{j=2}^{J} \mathbf{A}_j \right) \text{ and } \bigoplus_{j=1}^{J} \mathbf{A}_j = \mathbf{A}_1 \oplus \left(\bigoplus_{j=2}^{J} \mathbf{A}_j \right)$$

Kronecker products and sums can be transformed into some standardized form by applying the above properties of the operations several times.

$$\bigotimes_{j=1}^{J} \mathbf{A}_j = \prod_{j=1}^{J} \mathbf{I}_{l_j} \otimes \mathbf{A}_j \otimes \mathbf{I}_{u_j} \text{ and } \bigoplus_{j=1}^{J} \mathbf{A}_j = \sum_{j=1}^{J} \mathbf{I}_{l_j} \otimes \mathbf{A}_j \otimes \mathbf{I}_{u_j} ,$$

where $\mathbf{A}_j \in \mathbb{R}^{r_j, c_j}$, $l_j = \prod_{i=1}^{j-1} c_i$ and $u_j = \prod_{i=j+1}^{J} r_i$. In that form, the naming Kronecker product and Kronecker sum becomes more clear.

For square matrix $\mathbf{A}_j \in \mathbb{R}^{n_j,n_j}$ $(j = 1, ..., J)$ and a permutation σ on $\{1, ..., J\}$, a pseudo commutativity of the Kronecker product holds, since

$$\bigotimes_{j=1}^{J} \mathbf{A}_j = \mathbf{P}_\sigma \left(\bigotimes_{\sigma(j)=1}^{J} \mathbf{A}_j \right) \mathbf{P}_\sigma^T$$

for a $\prod_{j=1}^{J} n_j \times \prod_{j=1}^{J} n_j$ permutation matrix $\mathbf{P}_\sigma$ with

$$\mathbf{P}_\sigma((x_1, ..., x_J), (y_1, ..., y_J)) = \begin{cases} 1 & \text{if } y_{\sigma(k)} = x_k \text{ for all } k = 1, 2, ..., J \\ 0 & \text{otherwise.} \end{cases}$$

We can equivalently define $\mathbf{P}_\sigma(x, y) = 1$ if $x = \sum_{j=1}^{J} x_j \prod_{k=j+1}^{J} n_k$ and $y = \sum_{j=1}^{J} x_j \prod_{\sigma(k)=j+1}^{J} n_k$ and 0 otherwise by help of equation (3), in case one prefers a linearized indexation of $\mathbf{P}_\sigma$.

The concept of permuting rows and columns of a matrix can be extended to define order-preserving mappings into a subset of the set of rows and columns. With a slight misuse of notation, let $\bar{n} = \prod_{j=1}^{J} n_j$. Let $\mathbf{A}_j \in \mathbb{R}^{n_j,n_j}$ $(j = 1, ..., J)$, then $\otimes_{j=1}^{J} \mathbf{A}_j \in \mathbb{R}^{\bar{n},\bar{n}}$. Furthermore let $\mathcal{M} \subset \{0, ..., \bar{n}-1\}$ be a subset of the set of states described by the Kronecker product. We assume $|\mathcal{M}| = m < \bar{n}$ and define a mapping $\psi_\mathcal{M} : \{0, ..., \bar{n}-1\} \rightarrow \{0, ..., m\}$ such that $\psi_\mathcal{M}(x) = m$ for $x \notin \mathcal{M}$ and $\psi_\mathcal{M}(x) \in \{0, ..., m-1\}$ for $x \in \mathcal{M}$. Mapping $\psi_\mathcal{M}$ is order preserving if $\psi_\mathcal{M}(x) < \psi_\mathcal{M}(y)$ for $x, y \in \mathcal{M}$ and $x < y$. Order-preserving mappings are very important in Kronecker based analysis approaches since they allow the restriction of transitions to specific parts of the product space without losing the compact representation of the matrix as a sum of Kronecker products. For this purpose, projection matrices are defined that are similar to the permutation matrices introduced above. Let $\mathbf{P}_{\psi_\mathcal{M}}$ be a $m \times \bar{n}$ matrix with $\mathbf{P}_{\psi_\mathcal{M}}(x, y) = 1$ if $x = \psi_\mathcal{M}(y)$ and 0 otherwise. Then the matrix

$$\mathbf{P}_{\psi_\mathcal{M}} \left(\bigotimes_{j=1}^{J} \mathbf{A}_j \right) \mathbf{P}_{\psi_\mathcal{M}}^T$$

describes transitions between states in $\mathcal{M}$ [52, 20].

2.4   Kronecker Representations of SAN Models

If $\mathbf{A}$ and $\mathbf{B}$ are two generator matrices of CTMCs, then $\mathbf{A} \oplus \mathbf{B}$ equals the generator of the independently combined CTMC[1].

We first consider local transitions in the automata. Assume that automaton i changes its state from x_i to y_i via some transition labeled with $t \in \mathcal{TL}^{(i)}$ with rate μ. Now assume that the state of the SAN is $\mathbf{x}$ (with $\mathbf{x}(i) = x_i$) before

---

[1] In a discrete time setting, if $\mathbf{A}$ and $\mathbf{B}$ are two transition matrices of discrete time Markoc chains (DTMCs) the transition matrix of the combined DTMC equals $\mathbf{A} \otimes \mathbf{B}$.

the transition takes place, then $\mathbf{y}$ with $\mathbf{y}(j) = \mathbf{x}(j)$ for $j \neq i$ and $\mathbf{y}(i) = y_i$ is the state after the transition occured. Since the generator matrix $\mathbf{Q}$ of the complete SAN is in $\mathbb{R}^{\bar{n} \times \bar{n}}$, we have to use the integer representation computed via (3) instead of the vector representation for the representation of states. For a state transition $x \overset{t}{\to} y$, the value of y is given as $y = x + (y_i - x_i) \cdot \prod_{j=i+1}^{J} n^{(j)}$. This simple addressing scheme can be formulated by the following matrix term

$$\lambda_t \left(\mathbf{I}_{l^{(i)}} \otimes \mathbf{E}_t^{(i)} \otimes \mathbf{I}_{u^{(i)}} \right) ,$$

where $l^{(i)} = \prod_{j=1}^{i-1} n^{(j)}$ and $u^{(i)} = \prod_{j=i+1}^{J} n^{(j)}$. Since local transitions are independent and in a CTMC with probability 1 at most one transition occurs, the matrices of different local transitions can be added to build the part of the generator matrix that considers local transitions only.

$$\sum_{i=1}^{J} \sum_{t \in T\mathcal{L}^{(i)}} \lambda_t \left(\mathbf{I}_{l^{(i)}} \otimes \mathbf{E}_t^{(i)} \otimes \mathbf{I}_{u^{(i)}} \right)$$

By defining a separate matrix of local transitions

$$\mathbf{E}_l^{(i)} = \sum_{t \in T\mathcal{L}^{(i)}} \lambda_t \mathbf{E}_t^{(i)}$$

we obtain the following representation for local transitions:

$$\mathbf{E}_l = \sum_{i=1}^{J} \left(\mathbf{I}_{l^{(i)}} \otimes \mathbf{E}_l^{(i)} \otimes \mathbf{I}_{u^{(i)}} \right) = \bigoplus_{i=1}^{J} \mathbf{E}_l^{(i)} \tag{4}$$

In a similar way, the diagonal elements of this matrix can be represented in a diagonal matrix $\mathbf{D}_l$.

$$\mathbf{D}_l = \sum_{i=1}^{J} \left(\mathbf{I}_{l^{(i)}} \otimes \mathbf{D}_l^{(i)} \otimes \mathbf{I}_{u^{(i)}} \right) = \bigoplus_{i=1}^{J} \mathbf{D}_l^{(i)} \tag{5}$$

where

$$\mathbf{D}_l^{(i)} = \sum_{t \in T\mathcal{L}^{(i)}} \lambda_t \mathbf{D}_t^{(i)} .$$

Synchronized transitions can be handled using Kronecker products instead of Kronecker sums. Assume that the state of the system is $\mathbf{x}$, then transition $t \in TS$ is enabled, if it is enabled in all automata which synchronize at t. If $t \notin T^{(i)}$, then t cannot be disabled by automaton i and the occurrence of t cannot change the state of i. Consequently, we define $\mathbf{E}_t^{(i)} = \mathbf{I}_{n^{(i)}}$ for $t \notin T^{(i)}$. In state $\mathbf{x}$, t is enabled if all rows $\mathbf{E}_t^{(i)}(\mathbf{x}(i), *)$ are non-zero and state $\mathbf{y}$ is reachable from $\mathbf{x}$ by occurrence of t, if $\mathbf{E}_t^{(i)}(\mathbf{x}(i), \mathbf{y}(i)) > 0$ for all $i = 1, ..., J$. In that case, the transition rate is $\lambda_t \prod_{i=1}^{J} \mathbf{E}_t^{(i)}(\mathbf{x}(i), \mathbf{y}(i))$. The whole matrix for transition $t \in \mathcal{T}^{(i)}$ is then generated via the Kronecker product of the automata matrices.

$$\mathbf{E}_t = \lambda_t \bigotimes_{i=1}^{J} \mathbf{E}_t^{(i)} \tag{6}$$

Similarly, the diagonal elements due to transition t are collected in the following matrix.

$$\mathbf{D}_t = \lambda_t \bigotimes_{i=1}^{J} \mathbf{D}_t^{(i)} \tag{7}$$

Thus we have generated matrices with rates of global transitions in the state space of the SAN from the small automata matrices of dimension $n^{(i)}$. The descriptor matrix $\mathbf{Q}$ of the complete CTMC is generated from the matrices for the transitions as follows.

$$
\begin{aligned}
\mathbf{Q} = \mathbf{E} - \mathbf{D} &= \left(\mathbf{E}_l + \sum_{t \in \mathcal{TS}} \mathbf{E}_t \right) - \left(\mathbf{D}_l + \sum_{t \in \mathcal{TS}} \mathbf{D}_t \right) \\
&= \left(\bigoplus_{i=1}^{J} \mathbf{E}_l^{(i)} + \sum_{t \in \mathcal{TS}} \lambda_t \bigotimes_{i=1}^{J} \mathbf{E}_t^{(i)} \right) - \left(\bigoplus_{i=1}^{J} \mathbf{D}_l^{(i)} + \sum_{t \in \mathcal{TS}} \bigotimes_{i=1}^{J} \mathbf{D}_t^{(i)} \right) \\
&= \sum_{i=1}^{J} \mathbf{I}_{l^{(i)}} \otimes \mathbf{E}_l^{(i)} \otimes \mathbf{I}_{u^{(i)}} + \sum_{t \in \mathcal{TS}} \lambda_t \prod_{i=1}^{J} \mathbf{I}_{l^{(i)}} \otimes \mathbf{E}_t^{(i)} \otimes \mathbf{I}_{u^{(i)}} \\
&\quad - \sum_{i=1}^{J} \mathbf{I}_{l^{(i)}} \otimes \mathbf{D}_l^{(i)} \otimes \mathbf{I}_{u^{(i)}} - \sum_{t \in \mathcal{TS}} \lambda_t \prod_{i=1}^{J} \mathbf{I}_{l^{(i)}} \otimes \mathbf{D}_t^{(i)} \otimes \mathbf{I}_{u^{(i)}} \, .
\end{aligned}
\tag{8}
$$

This matrix equals the generator matrix of the CTMC if $\mathcal{RS} = \mathcal{PS}$. The interesting aspect of the above representation is that the compact description of the matrix can be exploited in a wide variety of analysis algorithms such that the overall matrix never needs to be generated as a whole. The diagonal part of the matrix, which corresponds to the last two sums, and might be stored in a vector of length n, thus requiring more space but allows a more efficient access.

2.5   Kronecker Representations for High-Level Modeling Formalisms

The automata specification is relatively low level and most times a model results from the specification of some higher level formalism. Fortunately, many models in high level formalisms provide compositional or hierarchical structures that can be employed to derive a Kronecker representation.

For instance, stochastic process algebras [7, 9, 46, 49] are a very convenient family of modeling formalisms that can be used for the compositional specification of Markov models. The composition via synchronized transitions that we use for our automata models corresponds to the parallel composition operator, which is available in most stochastic process algebras. A direct mapping of the process algebra MPA to a SAN and a Kronecker representation of the underlying CTMC is given in [9]. For other process algebras, the correspondence might not be so straightforward due to different semantics for the parallel composition, but it usually can be achieved. The subclass of process algebra models which yield a SAN are those where, at the highest level, processes are composed in parallel. This structure is very common when modeling parallel processes.

Another class of models which can be easily mapped on SANs are Superposed Generalized Stochastic Petri Nets (SGSPNs) [41, 40, 52], a specific class of stochastic Petri Nets with a partition on the set of places defining a set of subnets which communicate via shared transitions. In this way, a SGSPN consists of a number of components interacting via synchronized transitions. A specific difference to SANs is that SGSPNs include additional immediate transitions which have a strict priority over exponentially timed transitions. However, if immediate transitions are local to the components, they can be eliminated locally using standard means [2] and the resulting matrices are exactly the same as the matrices presented above for SANs. If immediate transitions are non-local, then a Kronecker representation of the matrix can still be achieved even for different levels of priority [42]. In the sequel of this paper, we will use SGSPNs to specify models, because they have a nice graphical representation that helps to illustrate the dynamics of a model much better than a set of matrices.

Apart from process algebras, other specification techniques exist that are based on networks of communicating automata. The specification techniques UML and SDL may serve as examples that both might be and have been extended by some stochastic timing [5, 61]. The resulting models may as well be mapped on SANs. The difference is usually that communication is asynchronous rather than synchronous. Asynchronous communication can be modeled by transitions with preconditions in the initiating automaton only. This implies that for the automata that receives a message asynchronously by some transition t, the corresponding matrices $\mathbf{E}_t^{(i)}$ are stochastic matrices. For other specification paradigms that also describe a Markov model, a mapping onto SANs is not so straightforward. In particular, queuing networks or some types of stochastic Petri nets consider the flow of entities through a network of resources. This view has not a natural interpretation as a network of interacting automata and especially in that case some hierarchical description as presented in Section 4 is often more appropriate. However, simple queuing network models have been successfully modeled as SANs [62].

2.6   Example

The example is a simplified version of a unidirectional communication protocol with one sender and one receiver. A more detailed version of the protocol and the related Petri net model can be found in [65] and we refer the interested reader to that paper for a detailed explanation. Only a brief overview of the model is given in the following. Figure 1 presents the model as a stochastic Petri net. The whole system consists of four components or automata. Automaton 1 models the upper level protocol layers of the sender. Automata 1 and 2 communicate via synchronized transition ts_1. Automaton 2 models the network layer at the sender side, which splits messages into packets that are transferred to the receiver. The number of packets is not fixed. The postset of p_1^2 contains two transitions to

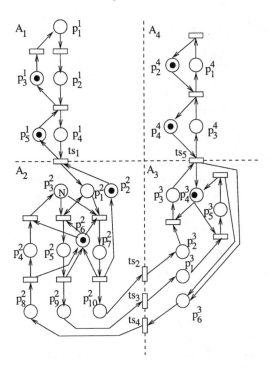

Fig. 1. SPN model of a simple communication protocol

model this effect, one handles all packets but the last one, the other transition handles the last packet and removes a token from p_1^2. The transfer protocol uses an acknowledgment scheme that acknowledges the receipt of each single packet by the receiver. For flow control, the sender uses a window of size N. Automata 3 and 4 describe the network layer and the upper level layers of the receiver. We will not describe the function of the model in detail, which can be seen by considering the dynamics of the Petri net and which is, of course, a very abstract description of a communication protocol. We rather focus on the state spaces and transition matrices in some more detail. The size of the state space is scalable by parameter N, the window size. Table 2 contains the sizes of the component state spaces $n^{(i)}$ ($i = 1, \ldots, 4$), the size of $\mathcal{RS}$, and the size of $\mathcal{PS}$ ($= \prod_{i=1}^{4} n^{(i)}$) for several values of N. Component state spaces are generated from the SPN specification of the components. For the first and the fourth component, state spaces and matrices can be generated directly from the SPN component. For components two and three, the isolated SPN for each component is unbounded after removing the synchronization restrictions due to other components. However, bounded state spaces that include all reachable states of the component as part of the SGSPN can be achieved by first computing place invariants for the whole net and then introduce place capacities resulting from the invariants (see [52] for details).

Table 2. State space sizes for different versions of the example model

N	$n^{(1)}$	$n^{(2)}$	$n^{(3)}$	$n^{(4)}$	$|\mathcal{RS}|$	$|\mathcal{PS}|$
1	6	13	6	4	456	1872
2	6	49	18	4	2760	21168
3	6	129	40	4	10584	123840

The sizes of the state spaces for the first and last component are independent of the window size. For the remaining two components, the sizes of their state spaces grow with a growing window size. This has the effect that the sizes of the reachable and potentially reachable state spaces grow very rapidly with parameter N. Due to the synchronization in the model, not all states from $\mathcal{PS}$ are reachable and the percentage of unreachable states in $\mathcal{PS}$ increases with increasing the size of the state space. This is a typical property of many SAN models. On the other hand, it can also be clearly seen that the automata state spaces are much smaller than $\mathcal{RS}$, let alone $\mathcal{PS}$. Thus, the representation by means of automata matrices is really compact, but contains unreachable states. This aspect will be addressed in the following sections. At this point, we present the matrices in some more detail and consider the smallest configuration with $N = 1$. States of each automaton are numbered consecutively from 0 through $n^{(i)} - 1$. We introduce a detailed description of the reachable states and the transition matrices for each automaton. Let the places of the Petri net for automaton i be numbered p_1^i to $p_{k_i}^i$ and let k_i be equal to the number of places in the corresponding Petri net, then we can characterize states by vectors of length k_i that contain the marking of the places in automaton i. Set $\mathcal{RS}^{(1)}$ contains 6 states as given below. Two matrices, namely one for local transitions and one for synchronized transition ts_1 describe the behavior of the first automaton. Matrices for the remaining synchronized transitions are identity matrices. Since we are not interested in the transition rates, we assume that all local transitions have rate μ and we do not distinguish between different local transitions.

$$\mathcal{RS}^{(1)} = \left\{ \begin{matrix} 0 = (0,0,1,0,1), \\ 1 = (1,0,0,0,1), \\ 2 = (0,1,0,0,1), \\ 3 = (0,0,1,1,0), \\ 4 = (1,0,0,1,0), \\ 5 = (0,1,0,1,0) \end{matrix} \right\}, \mathbf{E}_l^{(1)} = \begin{pmatrix} 0 & \mu & 0 & 0 & 0 & 0 \\ 0 & 0 & \mu & 0 & 0 & 0 \\ 0 & 0 & 0 & \mu & 0 & 0 \\ 0 & 0 & 0 & 0 & \mu & 0 \\ 0 & 0 & 0 & 0 & 0 & \mu \\ 0 & 0 & 0 & 0 & 0 & 0 \end{pmatrix}, \mathbf{E}_{ts_1}^{(1)} = \begin{pmatrix} 0 & 0 & 0 & 0 & 0 & 0 \\ 0 & 0 & 0 & 0 & 0 & 0 \\ 0 & 0 & 0 & 0 & 0 & 0 \\ 1 & 0 & 0 & 0 & 0 & 0 \\ 0 & 1 & 0 & 0 & 0 & 0 \\ 0 & 0 & 1 & 0 & 0 & 0 \end{pmatrix}.$$

For the second automaton, we show the state space, the matrix of local transition rates and the non-zero elements in the matrices $\mathbf{E}_{ts_k}^{(2)}$.

$$
\mathcal{RS}^{(2)} = \left\{ \begin{array}{ll}
0 = & (0,1,1,0,0,1,0,0,0,0), \\
1 = & (1,0,1,0,0,1,0,0,0,0), \\
2 = & (0,0,0,0,0,0,1,0,0,0), \\
3 = & (1,0,0,0,1,0,0,0,0,0), \\
4 = & (0,1,0,0,0,1,0,0,0,1), \\
5 = & (1,0,0,0,0,1,0,0,1,0), \\
6 = & (0,1,0,0,0,1,0,0,0,0), \\
7 = & (1,0,0,0,0,1,0,0,0,1), \\
8 = & (1,0,0,0,0,1,0,0,0,0), \\
9 = & (0,1,0,0,0,1,0,1,0,0), \\
10 = & (1,0,0,0,0,1,0,1,0,0), \\
11 = & (0,1,0,1,0,0,0,0,0,0), \\
12 = & (1,0,0,1,0,0,0,0,0,0)
\end{array} \right\}, \mathbf{E}_l^{(2)} =
\begin{pmatrix}
0 & 0 & 0 & 0 & 0 & 0 & 0 & 0 & 0 & 0 & 0 & 0 & 0 \\
0 & 0 & \mu & \mu & 0 & 0 & 0 & 0 & 0 & 0 & 0 & 0 & 0 \\
0 & 0 & 0 & 0 & \mu & 0 & 0 & 0 & 0 & 0 & 0 & 0 & 0 \\
0 & 0 & 0 & 0 & 0 & \mu & 0 & 0 & 0 & 0 & 0 & 0 & 0 \\
0 & 0 & 0 & 0 & 0 & 0 & 0 & 0 & 0 & 0 & 0 & 0 & 0 \\
0 & 0 & 0 & 0 & 0 & 0 & 0 & 0 & 0 & 0 & 0 & 0 & 0 \\
0 & 0 & 0 & 0 & 0 & 0 & 0 & 0 & 0 & 0 & 0 & 0 & 0 \\
0 & 0 & 0 & 0 & 0 & 0 & 0 & 0 & 0 & 0 & 0 & 0 & 0 \\
0 & 0 & 0 & 0 & 0 & 0 & 0 & 0 & 0 & 0 & 0 & 0 & 0 \\
0 & 0 & 0 & 0 & 0 & 0 & 0 & 0 & 0 & 0 & 0 & \mu & 0 \\
0 & 0 & 0 & 0 & 0 & 0 & 0 & 0 & 0 & 0 & 0 & 0 & \mu \\
\mu & 0 & 0 & 0 & 0 & 0 & 0 & 0 & 0 & 0 & 0 & 0 & 0 \\
0 & \mu & 0 & 0 & 0 & 0 & 0 & 0 & 0 & 0 & 0 & 0 & 0
\end{pmatrix},
$$

$$
\mathbf{E}_{ts_1}^{(2)}(0,1) = \mathbf{E}_{ts_1}^{(2)}(4,7) = \mathbf{E}_{ts_1}^{(2)}(6,5) = \mathbf{E}_{ts_1}^{(2)}(9,10) = \mathbf{E}_{ts_1}^{(2)}(11,12) =
$$
$$
\mathbf{E}_{ts_2}^{(2)}(4,6) = \mathbf{E}_{ts_2}^{(2)}(7,8) = \mathbf{E}_{ts_3}^{(2)}(5,8) = \mathbf{E}_{ts_4}^{(2)}(6,9) = \mathbf{E}_{ts_4}^{(2)}(8,10) = 1
$$

All remaining elements in these matrices are zero and matrix $\mathbf{E}_{ts_5}^{(2)}$ equals the identity matrix. The third automaton has the following state space.

$$
\mathcal{RS}^{(3)} = \left\{ \begin{array}{l}
0 = (0,0,0,1,0,0), \; 2 = (0,1,0,1,0,0), \; 4 = (0,0,1,0,0,0), \\
1 = (1,0,0,1,0,0), \; 3 = (0,0,0,0,1,0), \; 5 = (0,0,0,1,0,1)
\end{array} \right\}
$$

The following non-zero elements exist in the matrices.

$$
\mathbf{E}_l^{(3)}(1,3) = \mathbf{E}_l^{(3)}(2,4) = \mathbf{E}_l^{(3)}(3,5) = \mu \; ,
$$
$$
\mathbf{E}_{ts_2}^{(3)}(0,2) = \mathbf{E}_{ts_3}^{(3)}(0,1) = \mathbf{E}_{ts_4}^{(3)}(5,0) = \quad \mathbf{E}_{ts_5}^{(2)}(4,5) = 1
$$

The last automaton is the simplest one with only four states.

$$
\mathcal{RS}^{(4)} = \left\{ 0 = (0,1,0,1), \; 1 = (0,1,1,0), \; 2 = (1,0,0,1), \; 3 = (1,0,1,0) \right\}
$$

The automaton synchronizes only on ts_5 and the matrices for local transitions and synchronized transition ts_5 contain the following nonzero elements.

$$
\mathbf{E}_l^{(4)}(1,2) = \mathbf{E}_l^{(4)}(2,0) = \mathbf{E}_l^{(4)}(3,1) = \mu
$$
$$
\mathbf{E}_{ts_5}^{(4)}(0,1) = \mathbf{E}_{ts_5}^{(4)}(2,3) = 1
$$

As we have seen in the example, the introduced basic approach of representing models by interacting components is very convenient from a specification point of view, if we consider processes which run in parallel and interact via synchronization and communication. In the following section, we consider extensions to the Kronecker representation presented so far.

3   Modular Kronecker Representations

A modular Kronecker representation uses the representation of $\mathbf{Q}$ as given in (8), which we recall for convenience.

$$\mathbf{Q} = \sum_{t \in \mathcal{TS}} \lambda_t \bigotimes_{i=1}^{J} \mathbf{E}_t^{(i)} + \bigoplus_{i=1}^{J} \mathbf{E}_l^{(i)} - \mathbf{D}$$

Its key advantage is its space efficiency, since the right side of the equation requires to store only a set of matrices that are relatively small compared to the dimensions of $\mathbf{Q}$. In this section, we will discuss different ways to use this representation to perform a vector matrix multiplication $\mathbf{y} = \mathbf{x} \cdot \mathbf{Q}$. A vector matrix multiplication is a fundamental step in many analysis algorithms for CTMC analysis, as will be discussed in Section 5. Obviously, a multiplication of a vector with a modular Kronecker representation includes $\mathbf{x}^{(t)} = \mathbf{x} \cdot \bigotimes_{i=1}^{J} \mathbf{E}_t^{(i)}$ as a basic operation; a subsequent accumulation $\sum_{t \in \mathcal{TS}} \lambda_t \mathbf{x}^{(t)}$ and the analogous treatment of $\bigoplus_{i=1}^{J} \mathbf{E}_l^{(i)} - \mathbf{D}$ are then clear. Hence, we identify the crucial step for a vector matrix multiplication as

$$\mathbf{y} = \mathbf{x} \cdot \bigotimes_{i=1}^{J} \mathbf{E}_t^{(i)} \qquad (9)$$

The following subsection introduces a basic algorithm to perform the iteration (9), then we consider the situation where transitions in one automaton are functions of the state of other automata and the grouping of different automata into one automaton. The last two subsections describe an extended representation which avoids unreachable states by introduction of a mapping from the potential state space to the reachable state space and show how vector matrix products can be computed for this representation.

3.1   The Shuffle Algorithm

We begin with a sophisticated multiplication method that considers $\mathcal{PS}$ and constant transition rates. The shuffle multiplication algorithm [60, 62] requires $\bar{n} \cdot \sum_{i=1}^{J} n^{(i)}$ multiplications to perform (9) which is significantly less than the naive approach performed for $\mathbf{Q}$ with $\bar{n}^2$ multiplications. The shuffle algorithm exploits two observations. Namely, $\bigotimes_{i=1}^{J} \mathbf{E}_t^{(i)} = \prod_{i=1}^{J} \mathbf{I}_{l(i)} \otimes \mathbf{E}_t^{(i)} \otimes \mathbf{I}_{u(i)}$ which allows to perform (9) as a sequence of multiplications over i with intermediate vectors $\mathbf{x}^{(i)} = \mathbf{x}^{(i-1)} \cdot \mathbf{I}_{l(i)} \otimes \mathbf{E}_t^{(i)} \otimes \mathbf{I}_{u(i)}$ with $\mathbf{x}^{(0)} = \mathbf{x}$ and $\mathbf{x}^{(J)} = \mathbf{y}$. Multiplications of that type are specifically simple if $i = J$, because in that case $\mathbf{I}_{u(i)} = 1$ and $\mathbf{I}_{l(i)} \otimes \mathbf{E}_t^{(i)}$ has a peculiar structure: it is simply the matrix $\mathbf{E}_t^{(i)}$ repeated $\bar{n}/n^{(J)}$ times over the diagonal. The second observation is that this special case can be also achieved by a specific permutation, a perfect shuffle permutation. Let $\sigma(x, y)$ denote a permutation on $\{1, \ldots, J\}$ that switches positions of x and y. According to [36], we have

$$\bigotimes_{i=1}^{J} \mathbf{E}_t^{(i)} = \prod_{i=1}^{J} \mathbf{P}_{\sigma(i,J)} \cdot (\mathbf{I}_{\bar{n}/n^{(i)}} \otimes \mathbf{E}_t^{(i)}) \mathbf{P}_{\sigma(i,J)}^T ,$$

so the shuffle permutation rearranges vector entries appropriately to allow for a simple multiplication of a vector with a Kronecker matrix. Note that the above equation is a mathematical formulation; the shuffle algorithm, which implements the approach, expresses the shuffle permutation as simple index transformations that are performed whenever needed to read appropriate elements of $\mathbf{x}^{(i-1)}$ and write resulting elements of $\mathbf{x}^{(i)}$ for the $\bar{n}/n^{(i)}$ matrix vector multiplications with $\mathbf{E}_t^{(i)}$. The shuffle algorithm performs J multiplications of a vector with a Kronecker matrix with $O(\bar{n}/n^{(i)} \cdot (n^{(i)})^2)$ operations each (assuming a matrix $\mathbf{E}_t^{(i)}$ with $(n^{(i)})^2$ elements). This yields $O(\sum_{i=1}^{J} \bar{n} \cdot n^{(i)}) = O(\bar{n} \sum_{i=1}^{J} n^{(i)})$ as mentioned above.

3.2   State-Dependent Transition Rates

So far, we have considered only constant transition rates. However, modeling shared resources often implies that activities slow down in case of high utilization. Hence a transition rate becomes state-dependent, more formally matrix entries in $\mathbf{E}_t^{(i)}$ are functions $f : \times_{i=1}^{J} \mathcal{RS}^{(i)} \to \mathbb{R}$. In practice, functional transition rates often depend only on a subset of components $\{1, \ldots, J\}$. Clearly, two simple solutions are at hand: for the extreme case of local functional transitions where the entry in $\mathbf{E}_t^{(i)}$ only depends on s_i, the rate is constant in $\mathbf{E}_t^{(i)}$, and for the general case, we can use a sparse, explicit storage for matrix entries in $\mathbf{Q}$ that result from functional transition rates. In the following, we consider a third solution and discuss to what extend the shuffle algorithm is able to handle functional transition rates by generalized Kronecker products. A generalized Kronecker product contains matrices whose entries are functions $f : \times_{i=1}^{J} \mathcal{RS}^{(i)} \to \mathbb{R}$. The shuffle algorithm is able to accommodate functions that depend only on a subset of states at a component k, $f_k : \times_{i=k}^{J} \mathcal{RS}^{(i)} \to \mathbb{R}$ in $\mathbf{E}_t^{(k)}$. That condition is called order condition. If the order condition holds for all $k = 1, \ldots, J$, one can perform $\mathbf{x} \cdot \prod_{i=1}^{J} \mathbf{I}_{l^{(i)}} \otimes \mathbf{E}_t^{(i)} \otimes \mathbf{I}_{u^{(i)}}$ in a sequence that starts with $i = 1$. We only informally discuss why this is possible and refer to [43] for details. An informal explanation is that by computation of $\mathbf{y} = \mathbf{x} \cdot \bigotimes_{i=1}^{J} \mathbf{E}_t^{(i)}$, each entry $\mathbf{y}(s')$ obtains a weighted sum of values from $\mathbf{x}$ due to transition t, i.e., $\mathbf{y}(s') = \sum_{s \in \mathcal{PS}} \mathbf{x}(s) \cdot \prod_{i=1}^{J} f_i(s)$. If that product is computed as a sequence of Kronecker-vector multiplications with intermediate vectors $\mathbf{x}^{(i)}$, then the effect of transition t on the state of all components takes place as a sequence of local state changes in one component per step. So, if there is a state transition $s \xrightarrow{t} s'$ then informally $\mathbf{x}^{(i-1)}(s'_1, \ldots, s'_{i-1}, s_i, s_{i+1}, \ldots, s_J)$ contains the numerical value from $\mathbf{x}(s)$ after multiplication with the first $i - 1$ component matrices for t. Consequently, an evaluation of function $f(s)$ at $(s'_1, \ldots, s'_{i-1}, s_i, s_{i+1}, \ldots, s_J)$ is only possible for the remaining $(s_i, \ldots, s_J)$ fraction of the state description that is still consistent with s. Since permutations allow the reordering of components in each term of $\sum_{t \in \mathcal{TS}} \lambda_t \bigotimes_{i=1}^{J} \mathbf{E}_t^{(i)}$ in (8), the restriction applies for each term individually which makes the approach fairly powerful. In some cases, functional transition rates allow for simpler models with matrices $\mathbf{E}_t^{(i)}$ that tend to be more dense. This advantage comes at the price of more complex computations to evaluate functions at $\mathbf{E}_t^{(i)}$ instead of constant values. If for a single t, there are at least two components i and j with functional transition rates $f()$ and $g()$ that depend on both s_i and s_j, this is called cyclic dependency and in that case, the shuffle algorithm is not applicable. Therefore, in [62], explicit storage of matrix entries $\mathbf{E}_t$ is suggested, if evaluation of functions is too costly or if the condition on the possible ordering of components cannot be fulfilled. Alternatively, grouping may resolve cyclic dependencies in certain cases.

3.3   Grouping of Automata

Grouping means that we merge several components into a single new one. This reduces the number of components for the price of usually larger components. In the extreme case of merging all components, we obtain a conventional sparse matrix representation of $\mathbf{Q}$. Since we can always permute the order of components, it is sufficient to formally consider only the case that a subset of components $j, j + 1, \ldots, J$ shall be grouped into a new last component k. By grouping, we get a representation of $\mathbf{Q} = \sum_{t \in \mathcal{TS}} \lambda_t \bigotimes_{i=1}^{j-1} \mathbf{E}_t^{(i)} \otimes \mathbf{E}_t^{(k)} + \bigoplus_{i=1}^{j-1} \mathbf{E}_l^{(i)} \oplus \mathbf{R}_l^{(k)} - \mathbf{D}$ where the new $\mathbf{E}_t^{(k)} = \bigotimes_{i=j}^{J} \mathbf{E}_t^{(i)}$ and $\mathbf{R}_l^{(k)} = \bigoplus_{i=j}^{J} \mathbf{E}_l^{(i)}$. In many cases, we can select a subset of components, such that grouping results in a Kronecker representation where functions of state-dependent transitions fulfill the order condition required by the shuffle algorithm. Grouping also implies that the set of synchronized transitions is reduced, i.e., let $\mathcal{TS}'$ be the set of synchronized transitions after grouping, then $\mathcal{TS}' = \mathcal{TS} \cap (\cup_{i=1}^{j-1}) \mathcal{T}^{(i)}$. The latter case implies that for transitions that became local by grouping, we can add $\lambda_t \mathbf{E}_t^{(k)}$ to $\mathbf{R}_l^{(k)}$ and thus reduce the number of synchronized transitions. So grouping helps to reduce dependencies for functional transition rates, but also the amount of synchronization between components, respectively the number of synchronized transitions. The price for it is the increase of component state space sizes. For the example system, grouping components A_2 and A_3 into a single new component A_{23} results in component state space $\mathcal{RS}^{(23)}$ with less states than $n^{(2)} \cdot n^{(3)}$ and in that special case $\mathcal{RS} = \mathcal{RS}^{(1)} \times \mathcal{RS}^{(23)} \times \mathcal{RS}^{(4)}$ such that we obtain an ideal decomposition. However, this is not possible in general.

3.4   Computing Reachable States Among Potential States

One drawback of the modular Kronecker representation so far is that only a subset of states may be reachable from the initial state, i.e., only a submatrix of $\mathbf{Q}$ is in fact required to represent the generator matrix of a CTMC. Unreachable states in $\mathbf{Q}$ can be present even if for each state s_i of a component

i holds that there is a state $\mathbf{s} \in \mathcal{RS}$ with $\mathbf{s}(i) = s_i$. The reason is that the Kronecker product is defined on the cross product of component state spaces while synchronization of transitions restricts reachability. Table 2 shows that $|\mathcal{RS}| << |\mathcal{PS}|$ for our example model and the given partition. The problem of unreachable states can be handled in different ways. One is to keep the Kronecker representation and matrix vector multiplication algorithms unchanged but ensure that for $\mathbf{y} = \mathbf{xQ}$ only reachable states in $\mathbf{x}$ contain non-zero values. In this case, $\mathbf{y}$ is correctly computed; unreachable states are only a source of inefficiency in space, since vectors $\mathbf{y}$ and $\mathbf{x}$ are of dimension $\mathcal{PS}$, and a source of inefficiency in time since multiplications involving unreachable states are useless.

However, a Kronecker representations contains sufficient information to compute $\mathcal{RS}$ [52, 53]. We define the set of successor states of a given state $\mathbf{s} = (s_1, \ldots, s_J)$ as $succ(\mathbf{s}) = \{(s'_1, \ldots, s'_J)| \; \exists t \in T \; \forall j \in \{1, \ldots, J\} \; \mathbf{E}_t^{(i)}(s_i, s'_i) \neq 0\}$ which immediately matches the semantics of the dynamics in our representation[2]. Set $\mathcal{RS}$ is then explored by an exploration algorithm as follows.

Reachability_analysis

$U = S = \{\mathbf{s}_{init}\}$;
while $U \neq \emptyset$ do
$\quad U' = \cup_{s \in U} succ(s)$;
$\quad U = U' \backslash S$;
$\quad S = S \cup U'$;
od

Set U contains the so-far-reached, but yet-unexplored states, S contains the reached and explored states. $\mathbf{s}_{init}$ denotes the initial state as given in the model specification. This rather general scheme can be realized with much variety. In [52], a bit vector of dimension $\mathcal{PS}$ is employed to represent $\mathcal{RS}$ by hashing with (3) as a perfect hash function. In [24, 26], $succ(s)$ is extended to consider sequences of transitions that either start with a synchronized transition and that are then followed by local transitions [26] or vice versa [24]. This speeds up the exploration and reduces the number of elements in set U which is represented by a stack.

An alternative approach is to use a symbolic state space exploration with multi-valued decision diagrams (MDDs) [31]. Decision diagrams for boolean functions are well known in the field of model checking (hence the term "symbolic model checking"). In our context, it is more useful to allow for more than binary decisions, hence we consider MDDs. MDD-like data structures appear in several publications under various names, however, the key idea to gain efficiency by representing isomorphic (equal) substructures only once is common to all those approaches [31–33, 55, 59]. MDDs in theory differ slightly from MDDs as a data structure in an implementation. From a theoretical point of view, an MDD encodes a function $f : \times_{i=1}^{J} \mathcal{RS}^{(i)} \to \{true, false\}$ and is a directed acyclic graph with terminal nodes $true$ and $false$ and non-terminal nodes.

---

[2] Note that T contains local and synchronized transitions.

Any non-terminal node v has a value $level(v) \in \{1, 2, \ldots, J\}$, a vector of edges of length $n^{(i)}$ with $i = level(v)$ that lead either to a terminal node or a non-terminal node. If edges of a node v do not lead to nodes v' with $level(v') \leq level(v)$, the MDD is called *ordered*. Only ordered MDDs are of interest and throughout the rest of the paper we consider only ordered MDDs. Figure 2 shows an example with $J = 3$ components that encodes a set of vectors $\mathcal{RS} = \{(0, 1, 0), (0, 1, 1), (0, 1, 3), (1, 0, 2), (1, 2, 0), (1, 2, 1), (1, 2, 3)\}$. In Fig. 2 (a), vectors are encoded as paths in an unreduced MDD that has the shape of a tree. Any path that starts at the root node and ends at the bottom node *true* encodes an element of $\mathcal{RS}$. Paths that finally end at *false* correspond to unreachable states. An MDD gains its space efficiency from the fact that it can be reduced in the sense that isomorphic subgraphs are shared. In Fig. 2 (a), two of the level-3 nodes are equal and can be shared by adjacent level-2 nodes as shown in (b). Note that both structures represent the same boolean function. Sharing is the key and principal idea in MDDs.

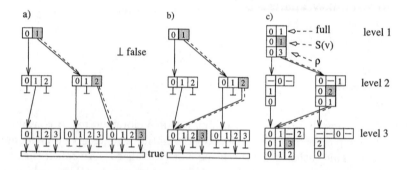

Fig. 2. $\mathcal{RS}$ as (a) an unreduced MDD, (b) a quasi-reduced MDD with edges to terminal nodes *true* and *false*, and (c) an MDD as a data structure

For applications, we need to enhance and modify the concept of MDD a bit further to make operations on MDDs more efficient. Clearly, any modification should not destroy or reduce the amount of sharing in an MDD. Required operations are

- a recursive enumeration of reachable states only,
- access to a single state that is an element of $\mathcal{RS}$, and
- an order-preserving mapping from $\mathcal{RS}$ to index set $\{0, 1, \ldots, |\mathcal{RS}| - 1\}$.

Finally, only required information should be stored in the data structure to save on space.

Since paths of unreachable states are not of interest for algorithms that analyze $\mathcal{RS}$ they can safely be removed from the structure. The remaining paths all end at terminal node *true*, so the terminal nodes are not necessary and can be removed. This leads to a "sparse" MDD structure like the one in Fig. 2 (c) which allows for an efficient enumeration of reachable states based on $S(v)$ using the definition below. Note that both variants contain the same amount of information with respect to paths, however Fig. 2 (c) contains additional information at each node that still needs to be defined. The dotted lines in the graphs illustrate paths that correspond to states $(1, 2, 0)$, $(1, 2, 1)$, and $(1, 2, 3)$. The shaded elements in all figures of Fig. 2 highlight state $(1, 2, 3)$.

With a slight misuse in notation, we define an MDD according to the "sparse" structure in Fig. 2 (c), namely as a special directed acyclic graph with a set of nodes V and edges E. Each node v of the graph contains a set of attributes:

- $level(v) \in \{1, 2, \ldots, J\}$ indicates to which component the node belongs,
- a finite index set $S(v) \subseteq \{0, 1, \ldots, n^{(i)} - 1\}$ with $i = level(v)$, where each element corresponds to an element of $\mathcal{RS}^{(i)}$ and $S(v) \neq \emptyset$. $S(v)$ indicates a set of possible local states and for each element $s \in S(v)$ there is an edge $v \xrightarrow{s} v'$ to a node v' with $level(v') = level(v) + 1$ if $level(v) < J$; at $level(v) = J$, there are no edges.
- a weight $\rho(v, s) \in \mathbf{N}$ for each $s \in S(v)$, $\rho(v, s)$ will be used to define the order-preserving mapping for the set of reachable states.

There is a dedicated root node with no incoming arcs that we denote by $mdd()$ and that is the only node of level 1. Any node v that has no outgoing arcs must have $level(v) = J$ by construction. Since the graph is directed, acyclic and has a unique root node, we obtain an identifier for each node by following a path from the root node by selecting appropriate edges, which we denote by $mdd(s_1, \ldots, s_j)$.

Sharing subgraphs requires a notion of equality. We define two nodes v and w being equal if all four of the following conditions hold: 1) $level(v) = level(w)$, 2) $S(v) = S(w)$, 3) for all $s \in S(v)$ with $v \xrightarrow{s} v'$ holds $w \xrightarrow{s} w'$ and $v' = w'$, and 4) $\rho(v, s) = \rho(v', s)$ for all $s \in S(v)$.

An MDD is quasi-reduced, if it contains no nodes that are equal. A node is redundant, if all its outgoing arcs point to the same element. For being reduced, theory of decision diagrams requires redundant nodes to be removed also, which implies that a path may visit less than J levels and represent a reachable state. However, this is not common in the MDD literature for this context [32, 59], so we consider only quasi-reduced MDDs in the following. An algorithm to reduce any given MDD proceeds bottom up, compares nodes of equal level, and keeps of any two equal nodes v and w only one, say v, and redirects incoming arcs of w to v. The set $S(v)$ of a node v with $level(v) = i$ and $mdd(s_1, \ldots, s_{i-1})$ gives the fraction of states in $\mathcal{RS}^{(i)}$ that are reachable under the condition that the states of components $1, \ldots, i - 1$ are $s_1, \ldots, s_{i-1}$, i.e. $S(v) = \{s_i \in \mathcal{RS}^{(i)} | v = mdd(s_1, \ldots, s_{i-1}), \exists(s_1, \ldots, s_{i-1}, s_i, x_{i+1} \ldots, x_J) \in \mathcal{RS}\}$. In Fig. 2 (c) the node at the lower left is shared. The graph in Fig. 2 (c) is the "sparse" MDD variant of the MDD in Fig. 2 (b). The reduction algorithm yields the MDD in Fig. 2 (b) for the unreduced MDD in (a) – provided we abstract from the differences between the two types of MDDs.

So far, we considered requirements with respect to space efficiency and efficient enumeration of $\mathcal{RS}$. We have not yet considered the order-preserving mapping which is established by adding appropriate weights to all nodes on a path. Weights are also denoted as offsets in the literature since they are used in numerical analysis to determine the position in a vector of length $\mathcal{RS}$. For instance, in Fig. 2 vector $(1, 2, 3)$ is the largest element of $\mathcal{RS}$ according to lexicographical order and thus should obtain a value 6, i.e., $\psi_\mathcal{M}(1, 2, 3) = 6$. In order to define weights $\rho(v, s)$, we need to count the number of paths that start in v for all $s' < s$, so we define for a node $v = mdd(s_1, \ldots, s_{i-1})$ a function[3] $count(v, s) = 1$ if $level(v) = J$ and

$$count(v, s) = \sum_{x \in S(mdd(s_1, \ldots, s_{i-1}, s))} count(mdd(s_1, \ldots, s_{i-1}, s), x)$$

otherwise. We define weights $\rho(v, s) = \sum_{x \in S(v), x < s} count(v, x)$. Weights are represented by the last row of values in each node of the MDD in Fig. 2 (c), e.g., following the shaded elements and the dotted line in the figure we have $\psi_\mathcal{M}(1, 2, 3) = 3 + 1 + 2 = 6$.

Finally, we consider access to a single state of $\mathcal{RS}$. Searching for a specific path and using only the information of $S(v)$ would require a binary search to determine which edge should be followed for $s \in S(v)$. To avoid that effort, we add a "full" vector to a node v with $level(v) = i$ that indicates whether $s \in \mathcal{RS}^{(i)}$ is an element of $S(v)$ or not. If $s \in S(v)$ the full vector provides the index position in vector $S(v)$. Fig. 2 (c) shows nodes where the "full" vector is the top row of values in each node.

MDDs can be employed for a symbolic reachability analysis that follows the same scheme as above, but represents sets S and U by MDDs. The main innovation is to compute the successor function not on a single state (a path in the MDD) but on an MDD, i.e., on a set of states at once. For a symbolic exploration we need to define $succ_t$ which takes a MDD U that encodes a set of states and generates a new MDD that encodes the set of successor states. For the algorithmic details of this construction we refer to [33]. The saturation technique in [31] speeds up state space exploration by considering sequences of transitions instead of a single transition. In that respect, it is related to the revised definition of $succ(s)$ in [24, 26] as stated above.

3.5   Multiplication Schemes

Once $\mathcal{RS}$ is known, several approaches enhance Kronecker representations with respect to reachability. They improve on the following, basic multiplication scheme that performs (9) with vectors $\mathbf{x}, \mathbf{y}$ of length $\mathcal{PS}$.

---

[3] Function *count* is like the SAT-count function in the theory of boolean functions that determines the cardinality of the set of parameter values that make a boolean function evaluate to true.

for all $s_1 \in \mathcal{RS}^{(1)}$

$\dots$

for all $s_J \in \mathcal{RS}^{(J)}$

for all $s'_1 \in \mathcal{RS}^{(1)}$

$\dots$

for all $s'_J \in \mathcal{RS}^{(J)}$

$$\mathbf{y}(s') \quad += \quad \mathbf{x}(s) \cdot \prod_{i=1}^{J} \mathbf{E}_t^{(i)}(s_i, s'_i)$$

where s is determined by linearizing $(s_1, \dots, s_J)$ using (3) and s' analogously. Informally, it enumerates all rows and for each row it multiplies a vector entry with all matrix elements to accumulate values on all possible positions in $\mathbf{y}$.

In order to take reachability into account, we use an order-preserving mapping $\psi_{\mathcal{RS}}$ to translate between indices of reachable states in vectors $\mathbf{x}, \mathbf{y} \in \mathbb{R}^{|\mathcal{RS}|}$ and their corresponding indices in $\times_{i=1}^{J} \mathcal{RS}^i$, so that we can perform

$$\mathbf{y} = \mathbf{x} \cdot \mathbf{P}_{\psi_{\mathcal{RS}}} \left(\otimes_{i=1}^{J} \mathbf{E}_t^{(i)} \right) \mathbf{P}_{\psi_{\mathcal{RS}}}^T \tag{10}$$

The advantage in terms of space is, that vectors $\mathbf{x}$ and $\mathbf{y}$ are of length $\mathcal{RS}$ (and not $\mathcal{PS}$) although the Kronecker representation implicitly describes an overall matrix of dimension $\mathcal{PS} \times \mathcal{PS}$. This is possible by help of permutation matrices $\mathbf{P}_{\psi_{\mathcal{RS}}}$, $\mathbf{P}_{\psi_{\mathcal{RS}}}^T$ in (10). It remains to discuss which data structures are suitable to represent $\mathbf{P}_{\psi_{\mathcal{RS}}}$ in a space efficient manner and how a multiplication is performed. Note that the following approaches realize $\psi_{\mathcal{RS}}$ only for reachable states, for unreachable states the mapping is undefined. A first approach in [52] suggests to use a single integer vector $\mathbf{v} \in \{0, 1, \dots, n-1\}^{|\mathcal{RS}|}$ of length $\mathcal{RS}$ that contains indices of reachable states in $\times_{i=1}^{J} \mathcal{RS}^{(i)}$, i.e., $\mathbf{v}_{\psi(x)} = x$. Entries are ordered with respect to the lexicographical order in $\times_{i=1}^{J} \mathcal{RS}^{(i)}$, their position in the vector gives the result of the mapping in $0, 1, \dots, |\mathcal{RS}| - 1$. Equation (3) is used to decode an entry s of $\mathbf{v}$ into $(s_1, \dots, s_J)$ and back. The multiplication scheme is:

for all $s = \mathbf{v}_k, k \in \{0, 1, \dots, |\mathcal{RS}| - 1\}$, with $(s_1, \dots, s_J) = s$

for all $s'_1 \in \mathcal{RS}^{(1)}$ with $\mathbf{E}_t^{(1)}(s_1, s'_1) \neq 0$

$\dots$

for all $s'_J \in \mathcal{RS}^{(J)}$ with $\mathbf{E}_t^{(J)}(s_J, s'_J) \neq 0$

$$\mathbf{y}(\psi_M(s')) += \mathbf{x}(s) \cdot \prod_{i=1}^{J} \mathbf{E}_t^{(i)}(s_i, s'_i)$$

where $\psi_M(s')$ is determined by linearizing $(s'_1, \ldots, s'_J)$ into s' and binary search on $\mathbf{v}$. This representation serves its purpose but shows several disadvantages. For instance, it takes a significant amount of space ($\mathcal{RS}$ integer values) and entries of $\mathbf{v}$ are subject to overflow domains of integer values (4 Bytes, 8 Bytes). Finding a corresponding value for ψ_x is possible in $O(1)$ since the vector position is known, but for $x \in \mathcal{PS}$ it requires a binary search with effort $O(\log(\mathcal{RS}))$. However, a permutation vector has been integrated into a recent variant of the shuffle algorithm [1] to restrict data structures to the size of $\mathcal{RS}$. That variant of the shuffle algorithm avoids binary search on $\mathbf{v}$ but requires sorting resulting values of the Kronecker-vector multiplication before its accumulation on $\mathbf{y}$.

MDDs improve on this situation. Mapping ψ_M is encoded in an MDD by help of $\rho()$. Let $\psi_M(s_1, \ldots, s_J) = \sum_{i=1}^J \rho(mdd(s_1, \ldots, s_{i-1}), s_i)$ for an MDD that represents $\mathcal{RS}$.

It is clear that an MDD as a directed acyclic graph allows for simple recursive functions to visit all nodes through all paths in the graph. Hence it is straightforward to formulate algorithms that perform a Kronecker matrix vector multiplication if the levels of an MDD are ordered in the same way as the terms in a Kronecker product. Figure 3 illustrates that a path in the MDD leads to a position in vector $\mathbf{x}$ via summation of $\rho()$ and at each level i refers to a row of a matrix $\mathbf{E}_t^{(i)}$, which may also be an identity matrix or which may contain an empty row for a state s_i. Note that nodes at the same level can be of different dimension since fixing states at higher levels restricts or conditions the reachable fraction at lower levels towards leaf nodes. The MDD allows us to restrict the ranges of s_i, s'_i in the basic scheme above and achieve in this way an algorithm for (10):

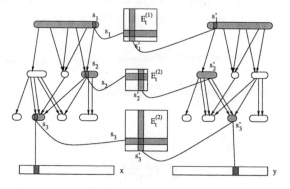

Fig. 3. Kronecker matrix vector multiplication directed by an MDD

for all $s_1 \in mdd()$, $s'_1 \in mdd()$ with $\mathbf{E}_t^{(i)}(s_1, s'_1) \neq 0$

$\ldots$

for all $s_J \in mdd(s_1, \ldots, s_{J-1})$, $s'_J \in mdd(s'_1, \ldots, s'_{J-1})$ with
$\mathbf{E}_t^{(i)}(s_J, s'_J) \neq 0$

$\mathbf{y}(\psi_M(s')) {+}= \mathbf{x}(\psi_M(s)) \cdot \prod_{i=1}^J \mathbf{E}_t^{(i)}(s_i, s'_i)$

where $\psi_M(s) = \sum_{i=1}^{J} \rho(mdd(s_1, \ldots, s_{i-1}), s_i)$ and $\psi_M(s')$ analogously. Since the order of s_i, s'_i matches the order of levels that a recursive enumeration of all paths in the MDD actually sees, one can simply use MDDs to represent $\mathbf{P}_{\psi(\mathcal{RS})}$ and direct the control flow in a multiplication algorithm. Note that for instance $\rho(mdd(s'_1, \ldots, s'_{i-1}), s'_i)$ either requires a binary search on the set $S(mdd(s'_1, \ldots, s'_{i-1}))$ or an enhanced MDD structure that contains at each node at level i an additional vector of length $\mathcal{RS}^{(i)}$ whose entries indicate the indices of elements in $S(mdd(s'_1, \ldots, s'_{i-1})) \subset \mathcal{RS}^{(i)}$. Note that partial sums like $\sum_{i=1}^{j} \rho(mdd(s_1, \ldots, s_{i-1}), s_i)$ and products like $\prod_{i=1}^{j} \mathbf{E}_t^{(i)}(s_i, s'_i)$ for some $j < J$ can be memorized and reused due to the nested for all operators. The given scheme considers all elements of a matrix in an order that is imposed by the order of the enumeration, which does not fit the typical mathematical notations that proceed by rows or columns. However, the order is compatible with e.g., the power method, the method of Jacobi, or randomization. For a Gauss-Seidel iteration, one needs to proceed by columns, i.e., one value of $\mathbf{y}$ must be completely computed before the computation of another value could be started. This is accomplished by the following scheme that goes by columns.

For every path $(s'_1, \ldots, s'_J)$ in the MDD we consider

for all $s_1 \in mdd()$ with $\mathbf{E}_t^{(i)}(s_1, s'_1) \neq 0$

$\ldots$

for all $s_J \in mdd(s_1, \ldots, s_{J-1})$ with $\mathbf{E}_t^{(i)}(s_J, s'_J) \neq 0$

$$\mathbf{y}(\psi_M(s')) + = \mathbf{x}(\psi_M(s)) \cdot \prod_{i=1}^{J} \mathbf{E}_t^{(i)}(s_i, s'_i)$$

and consider all synchronized transitions t and local transitions before we can consider the next path. As above, $\psi_M(s) = \sum_{i=1}^{J} \rho(mdd(s_1, \ldots, s_{i-1}), s_i)$ and $\psi_M(s')$ analogously.

Multiplication schemes with different levels of priorities. The order preserving mapping and MDDs are useful for much more general purposes than just reachability issues. For instance in [42], this concept is used to reflect the impact of different levels of priority among transitions. Let $\alpha : \mathcal{T} \to \{0, 1, \ldots, K\}$ be the priority function. α imposes a partition on $\mathcal{T} = \cup_{k=0}^{K} \mathcal{T}_k$ with $\mathcal{T}_k$ being the set of transitions that have priority k. Priorities sharpen the enabling conditions for transition t, i.e., if a transition t is enabled at state s then this implies that all transitions t' with $\alpha(t') < \alpha(t)$ are disabled at state s. So, in any state only transitions of one priority can take place, which implies a partition on $\mathcal{RS}$ according to the priority of the transitions that are enabled in a state $s \in \mathcal{RS}$. Let $\mathcal{RS}_k = \{s \in \mathcal{RS}$ such that $\exists t \in \mathcal{T}_k$ and t is enabled at $s\}$. A partition of $\mathcal{RS}$ can be performed on a corresponding MDD as well. It requires a subtraction operation "\" that implements subtraction of sets in addition to intersection and union. The construction uses $mdd()$ of $\mathcal{RS}$ and for each transition t an $mdd_t()$ that represents all states in $\mathcal{PS}$ that enable t assuming that there are no priorities. We can sequentially construct $mdd_K() = mdd() \cap (\cup_{t \in \mathcal{T}_K} mdd_t())$, $mdd_{k-1} = mdd() \backslash mdd_k()$ for $k = K - 1, \ldots, 1$. Note that it is important, that all $mdd_k()$ share the same function $\rho()$ in order to implement the same order preserving mapping from $\mathcal{PS}$ to $\mathcal{RS}$.

Parallel multiplication schemes. Finally, MDDs support parallelization of algorithms [54]. We define a partition on $\mathcal{RS}$ into P non-empty subsets $\{1,\ldots,P\}$ such that for each $p \in \{1,\ldots,P-1\}$ and any $s_p \in \mathcal{RS}_p, s'_{p+1} \in \mathcal{RS}_{p+1}$ holds that $s_p < s'_{p+1}$. The $mdd()$ of $\mathcal{RS}$ needs to be partitioned into P MDDs accordingly, which is simple if $P \leq |\mathcal{RS}^{(1)}|$ because in that case it requires only to split the root node of the $mdd()$ for $\mathcal{RS}$. The following scheme for $p \in \{1,\ldots,P\}$ writes only values on positions in $\mathbf{y}$ that belong to the corresponding set $\mathcal{RS}_p$ such that all P parts can be computed in parallel on a shared memory architecture. As for the priorities above, it is important that all MDDs share the same function $\rho()$ to compute the same order preserving mapping.

for all $s_1 \in mdd()$, $s'_1 \in mdd_p()$ with $\mathbf{E}_t^{(i)}(s_1, s'_1) \neq 0$

$\ldots$

for all $s_J \in mdd(s_1, \ldots, s_{J-1})$, $s'_J \in mdd_p(s'_1, \ldots, s'_{J-1})$ with
$\mathbf{E}_t^{(i)}(s_J, s'_J) \neq 0$

$$\mathbf{y}(\psi_M(s'))+ = \mathbf{x}(\psi_M(s)) \cdot \prod_{i=1}^{J} \mathbf{E}_t^{(i)}(s_i, s'_i)$$

where $\psi_M(s)$ and $\psi_M(s')$ are determined as above. In an iterative algorithms that performs several Kronecker-matrix vector multiplication of the kind of (10), a barrier synchronization between complete multiplications is necessary if vectors $\mathbf{x}, \mathbf{y}$ exchange values before subsequent multiplications.

4   Hierarchical Kronecker Representations

The motivation for a hierarchical Kronecker representation originally came from the modeling of closed queueing networks as communicating submodels [17] and has been subsequently extended to other classes of models.

Here we present first a hierarchical representation of the state space, followed by the introduction of an algorithm to generate the hierarchy. The last two subsections describe the resulting hierarchical representation of the generator matrix $\mathbf{Q}$ and present the hierarchical representation of the running example.

4.1   Hierarchical State Space Description

If one considers a single class queueing network with k stations with exponential service time distributions and population m, then the state space of the underlying Markov chains is described by all integer vectors $\mathbf{x} \in \mathbf{N}^k$ with $\sum_{l=1}^{k} \mathbf{x}(l) = m$ where $\mathbf{x}(l)$ describes the number of customers at station l. To apply the Kronecker based approach, we can define submodels that act as automata. Assume that the first l ($< k$) stations form submodel 1 and the remaining stations $l+1,\ldots,k$ build a second submodel. The state spaces $\mathcal{RS}^{(i)}$ ($i = 1, 2$) are defined by integer vectors of dimension l and $k - l$, respectively. Since the submodels interchange customers, a vector describes a valid state if its elements sum up to some value $c \leq m$. For the submodels, the matrices can be easily generated. The movement of a customer between stations of one submodel describes a local transition and the movement of a customer from one submodel to another describes a synchronized transition. The model can be analyzed like a SAN by interpreting submodels as automata. However, in that case $|\mathcal{RS}| << |\mathcal{PS}|$, because $\mathcal{PS}$ contains all state vectors where the elements sum up to some value $c \leq 2m$ but only vectors with $c = m$ describe reachable states.

A natural way to avoid unreachable states is to define macro states according to the population in the subnets. Thus, $\mathcal{RS}^{(i)}$ is decomposed into $m + 1$ subsets $\mathcal{RS}^{(i)}[0], \ldots, \mathcal{RS}^{(i)}[m]$ such that $\mathcal{RS}^{(i)}[r]$ contains all states with r customers at the stations of the subnet. Let $\mathcal{RS}^{(0)} = \{(0, m), \ldots, (m, 0)\}$ be the macro state space. Each macro state of the global system defines a macro state for each component. According to the macro states, the global state space is decomposed into subsets $\mathcal{RS}[(r, m - r)] = \mathcal{RS}^{(1)}[r] \times \mathcal{RS}^{(2)}[m - r]$. As long as the number of macro states is small compared to the size of the state space, the representation remains compact. According to the structure of the state space, the generator matrix can be described in a compact form by considering matrices for submodels. Those matrices describe transitions inside and between macro states such that the generator matrix of the complete model is block structured and each block is represented as a sum of Kronecker products of small matrices. Details of the matrix representation will be given below. The hierarchical representation obviously can be generalized for all kinds of closed multi-class queueing networks that are decomposed into subnets. Additionally, it is possible to define hierarchies of more than two levels. However, experience shows that usually two levels are sufficient for a very compact representation of large matrices.

Hierarchical representations for other modeling formalisms. Apart from queueing network models, there is no straightforward way to define a hierarchical representation for modeling formalisms in general, for instance Stochastic Petri Nets or Stochastic Automata. It is not obvious for these formalisms how to define macro states and even if macro states are defined it is not necessarily the case that $\mathcal{RS}$ can be represented as $\cup \times_{i=1}^{J} \mathcal{RS}^{(i)}[x]$. The hierarchical representation for queueing networks relies on the notion of customers which forms an invariant in closed queueing networks. Such an invariant is not present in Petri nets per se since tokens can be arbitrarily generated or deleted. A first approach that introduces a hierarchical Kronecker representation for Stochastic Petri Nets requires the definition of macro states by the modeler at the net level [8]. Although such a net level description can be supported by an appropriate modeling formalism [6], the use of the hierarchical specification does not assure that the resulting state space contains no unreachable states and, additionally, it requires some knowledge of the user to define an appropriate structure. Consequently, it is important to find algorithmic methods to generate automatically a hierarchical structure for a Stochastic Petri Net. Based on earlier results for marked graphs [29], a restricted subclass of Petri nets, it could be shown that for this class of nets a hierarchical representation can be generated at the net level by introducing implicit places and the resulting representation contains no unreachable states [21]. For nets that are more general than marked graphs, the approach usually yields a compact representation of the state space which, unfortunately, may contain unreachable states [30]. Nevertheless, it is possible to extend the approach by defining an algorithm that partitions a Stochastic Petri Nets into regions forming subnets and an upper level model which defines the macro states [22, 27].

4.2   Algorithmic Generation of Hierarchies

We now present an approach for SANs that generates a hierarchical description without unreachable states and with a minimal number of macro states for a given partitioning of automata. The approach has been presented in [12] and has been integrated in the APNN-toolbox [23]. We start with a SAN with potential state space $\mathcal{PS}$. The first step is the computation of the set of reachable states $\mathcal{RS}$ which has already been considered in this paper. If $\mathcal{RS} = \mathcal{PS}$, then the flat SAN representation is appropriate since (8) completely characterizes the generator matrix and can be used in different analysis approaches. Thus we assume that $\mathcal{RS} \subset \mathcal{PS}$ and generate a hierarchical representation such that

$$\mathcal{RS} = \cup_{\tilde{\mathbf{x}} \in \mathcal{RS}^{(0)}} \mathcal{RS}[\tilde{\mathbf{x}}] = \cup_{\tilde{\mathbf{x}} \in \mathcal{RS}^{(0)}} \times_{i=1}^{J} \mathcal{RS}^{(i)}[\tilde{\mathbf{x}}(i)] \ , \qquad (11)$$

where $\mathcal{RS}^{(0)}$ is the set of macro states and $\mathcal{RS}^{(i)}[\tilde{\mathbf{x}}(i)]$ is the set of micro states that belongs to macro state $\tilde{\mathbf{x}}(i)$ of component i. We use a tilde to distinguish macro- and micro states and assume that $\mathcal{RS}^{(i)}$ is decomposed into disjoint subsets $\mathcal{RS}^{(i)}[\tilde{x}]$ (i.e., $\mathcal{RS}^{(i)} = \cup \mathcal{RS}^{(i)}[\tilde{x}]$ and $\mathcal{RS}^{(i)}[\tilde{x}] \cap \mathcal{RS}^{(i)}[\tilde{y}] = \emptyset$ for $\tilde{x} \neq \tilde{y}$). Let $n^{(i)}(\tilde{x}) = |\mathcal{RS}^{(i)}[\tilde{x}]|$ be the number of micro states in macro state $\tilde{x}$. Each state $\tilde{\mathbf{x}} \in \mathcal{RS}^{(0)}$ is characterized by a J-dimensional vector where component $\tilde{\mathbf{x}}(i)$ contains the macro state for automaton i. The above representation exists for an arbitrary SAN since $\mathcal{RS}^{(0)} = \mathcal{RS}$ gives a trivial but valid hierarchy. However, in that case the representation has no advantages since it does not reduce the complexity. The goal is to find a representation with a smaller or, ideally, minimal number of macro states for a given SAN.

A necessary condition for two states to belong to the same macro state is that they are both reachable in combination with the same states of the other automata in the SAN. We denote this as identical reachability of states. Formally $x_i, y_i \in \mathcal{RS}^{(i)}$ are identically reachable if and only if

$$(x_1, \ldots, x_{i-1}, x_i, x_{i+1}, \ldots, x_J) \in \mathcal{RS} \Leftrightarrow (x_1, \ldots, x_{i-1}, y_i, x_{i+1}, \ldots, x_J) \in \mathcal{RS} \ .$$

If this condition is not observed, then a state $(x_1, \ldots, x_{i-1}, x_i, x_{i+1}, \ldots, x_J) \in \mathcal{RS}$ exists such that $(x_1, \ldots, x_{i-1}, y_i, x_{i+1}, \ldots, x_J) \notin \mathcal{RS}$ or vice versa. In both cases, (11) cannot hold since x_i and y_i are combined with different states. On the other hand, identical reachability is also sufficient for a combination of states into macro states such that (11) holds. This follows from the fact that identical reachable states x_i, y_i can be substituted by $\tilde{x}_i$ and whenever a state is reachable where the i-th component equals $\tilde{x}_i$, then $\tilde{x}_i$ can be substituted by x_i or y_i and both resulting states belong to $\mathcal{RS}$. Thus, the substitution of identically reachable states by macro states defines a macro state space with a minimal number of macro states for the given partitioning of automata. Obviously, a change of the partitioning by grouping automata implies changes on the macro state space, since different macro states are generated.

An algorithmic approach to generate the macro state space for a SAN is given in [12] using a bit vector representation of $\mathcal{RS}$. We will briefly introduce this approach here and show afterwards how the approach can be realized if $\mathcal{RS}$ is stored as an MDD.

First assume that $\mathcal{RS}$ is described as a bit-vector of length $\bar{n}$ that contains a 1 if a state in $\mathcal{PS}$ is reachable and 0 if it is not reachable. Generation of macro states is done consecutively for the automata 1 through J. For the first automaton, we can interpret the bit-vector to represent $\mathcal{RS}$ as a matrix with $n^{(1)}$ rows and $\bar{n}^{(1)} = \prod_{i=2}^{J} n^{(i)}$ columns. Let $\mathbf{b}^x$ be row x of this matrix for some $0 \leq x < n^{(1)}$. Each vector $\mathbf{b}^x$ can be mapped to an integer

$$b^x = \sum_{i=0}^{\bar{n}^{(1)}-1} \mathbf{b}^x(i) \cdot 2^i \ .$$

The relation $b^x = b^y \Leftrightarrow \mathbf{b}^x = \mathbf{b}^y$ holds. States x and y are identically reachable if $b^x = b^y$. To build macro states, values b^x are computed and sorted and afterwards macro states are generated for states with identical values b^x. Let $\tilde{n}^{(1)}$ be the number of macro states and $\widetilde{\mathcal{RS}}^{(1)}$ be the set of macro states for the first automaton. Since micro states that belong to the same macro state have identical rows in $\mathbf{b}^x$, it is sufficient to store one row per macro state such that the resulting matrix has a dimension of $\tilde{n}^{(1)} \times \prod_{i=2}^{J} n^{(i)}$. The effort for this step is linear in the number of states in $\mathcal{PS}$. In the next step, states of the second automaton are combined into macro states. To perform this step, a matrix of dimension $n^{(2)} \times (\tilde{n}^{(1)} \cdot \prod_{i=3}^{J} n^{(i)})$ is generated by a shuffle operation [36]. The corresponding algorithm is described in [28] and requires an effort linear in the number elements in the matrix which is less or equal to the size of $\mathcal{PS}$. After the shuffle operation has been performed, macro states of automaton 2 are computed as described above and with subsequent shuffle and reordering operations, macro states for all automata are computed. As a result of the computation for each automaton a set of macro states is available. Furthermore, the matrix resulting from building macro states for the last automaton contains $\prod_{i=1}^{J} \tilde{n}^{(i)}$ elements and each element belongs to a macro state $\tilde{\mathbf{x}}$ $(\tilde{\mathbf{x}}(i) \in \widetilde{\mathcal{RS}}^{(i)})$ of the SAN.

We denote a macro state $\tilde{\mathbf{x}}$ is globally reachable if $\mathbf{x} \in \mathcal{RS}$ for $\mathbf{x}(i) \in \mathcal{RS}^{(i)}[\tilde{\mathbf{x}}(i)]$. Observe that the conditions for building macro states assure that for a globally reachable macro state $\tilde{\mathbf{x}}$ all $\mathbf{x}$ with $\mathbf{x}(i) \in \mathcal{RS}^{(i)}[\tilde{\mathbf{x}}(i)]$ belong to $\mathcal{RS}$. Let $\widetilde{\mathcal{RS}}$ be the set of globally reachable macro states and $\tilde{n}$ be its cardinality, then $\tilde{\mathbf{x}} \in \widetilde{\mathcal{RS}}$ if the corresponding value in the matrix that results from the computation of macro states is 1. Thus, the algorithm implicitly computes the set of reachable macro states.

The major drawback of the approach is the size of the bit vector which determines space and time requirements of the algorithms. For models where $\mathcal{PS}$ is much larger than $\mathcal{RS}$, the approach becomes impractical. In [12], a method is described how to reduce the size of $\mathcal{PS}$ a priori by first reducing the sizes of the subsets $\mathcal{RS}^{(i)}$ by combining states which are identically reachable in all possible environments of an automaton. Identically reachable states in an automaton can be found by computation of an equivalence relation of the bisimulation type and representing equivalence classes of the relation by single states. Afterwards the reduced automata can be used to build an aggregated state space of the SAN which still contains enough information to recreate the complete reachable state space $\mathcal{RS}$. We refer to [12, 26] for further details.

If $\mathcal{RS}$ is stored as an MDD, macro states can be built with an effort linear in the number of reachable states which is much more efficient. Two states $x_i, y_i \in \mathcal{RS}^{(i)}$ are identically reachable, if they appear in the same nodes and have the same successors in each node they appear in. Due to the structure of MDDs, states in one node of a MDD have the same predecessors, such that it is sufficient to consider only successors to prove identical reachability. An algorithm to compute macro states starts in the first node and combines states with identical successor nodes. Then the algorithm continues level by level and combines states which are in the same nodes at the current level and have the same successors. The comparison of states to find those with identical successors can be done using hashing. Assuming that no collisions in the hash function occur, which is a reasonable assumption here, the effort to compute the hash function and sort the states according to the value of the hash function is $O(n^{(i)})$ for each node of the MDD. Assuming that all $n^{(i)}$ are of a similar size and $\tilde{n} \gg n^{(i)}$, the overall effort is in $O(\tilde{n} \log n^{(i)}) \approx O(\tilde{n})$ since the MDD contains only reachable states. Of course, the approach can be combined with a reduction of the automata state spaces according to reachability preserving equivalence [26].

4.3   Hierarchical Matrix Representation

After the generation of macro states for each automaton, states in the subsets $\mathcal{RS}^{(i)}$ have to be reordered such that states that belong to the first macro state are followed by states that belong to the second macro state and so on. The same permutation is applied to the matrices $\mathbf{E}_t^{(i)}$ and $\mathbf{D}_t^{(i)}$ which can be done by using an appropriate permutation matrix. Afterwards each matrix is block structured into $\tilde{n}^{(i)} \times \tilde{n}^{(i)}$ blocks.

$$
\mathbf{E}_t^{(i)} = \begin{pmatrix} \mathbf{E}_t^{(i)}[0,0] & \cdots\cdots & \mathbf{E}_t^{(i)}[0, \tilde{n}^{(i)} - 1] \\ \vdots & \ddots & \vdots \\ \vdots & \ddots & \vdots \\ \mathbf{E}_t^{(i)}[\tilde{n}^{(i)} - 1, 0] & \cdots\cdots & \mathbf{E}_t^{(i)}[\tilde{n}^{(i)} - 1, \tilde{n}^{(i)} - 1] \end{pmatrix},
$$

$$
\mathbf{D}_t^{(i)} = \begin{pmatrix} \mathbf{D}_t^{(i)}[0,0] & \mathbf{0} & \cdots & & \mathbf{0} \\ \mathbf{0} & \ddots & \ddots & & \vdots \\ \vdots & \ddots & \ddots & & \mathbf{0} \\ \mathbf{0} & & \cdots & \mathbf{0} & \mathbf{D}_t^{(i)}[\tilde{n}^{(i)} - 1, \tilde{n}^{(i)} - 1] \end{pmatrix}.
$$

Matrices $\mathbf{D}_t^{(i)}[\tilde{x}, \tilde{x}]$ are diagonal matrices.

The generator matrix $\mathbf{Q}$ is structured into $\tilde{n}^2$ submatrices describing transitions between macro states. Each diagonal block $\mathbf{Q}[\tilde{x}, \tilde{x}]$ can be represented as

$$
\mathbf{Q}[\tilde{x}, \tilde{x}] = \left(\bigoplus_{i=1}^{J} \mathbf{E}_l^{(i)}[\tilde{x}(i), \tilde{x}(i)] + \sum_{t \in TS} \lambda_t \bigotimes_{i=1}^{J} \mathbf{E}_t^{(i)}[\tilde{x}(i), \tilde{x}(i)] \right) \\
- \left(\bigoplus_{i=1}^{J} \mathbf{D}_l^{(i)}[\tilde{x}(i), \tilde{x}(i)] + \sum_{t \in TS} \lambda_t \bigotimes_{i=1}^{J} \mathbf{D}_t^{(i)}[\tilde{x}(i), \tilde{x}(i)] \right) . \tag{12}
$$

This representation is very similar to the basic representation for SANs as shown in (8). In fact, if $\widetilde{RS}$ includes only a single macro state, then (8) and (12) are identical and the generator matrix can be represented as a sum of Kronecker products since $PS = RS$ in this case. For the non-diagonal blocks $\mathbf{Q}[\tilde{x}, \tilde{y}]$ ($\tilde{x} \neq \tilde{y}$), we distinguish two cases. If $\tilde{x}$ and $\tilde{y}$ differ only in one component i (i.e., $\tilde{x}(j) = \tilde{y}(j)$ for $j \neq i$ and $\tilde{x}(i) \neq \tilde{y}(i)$), then the corresponding transitions result from a synchronized transition or a local transition in automaton i and the submatrices are described as

$$
\mathbf{Q}[\tilde{x}, \tilde{y}] = \left(\bigotimes_{j=1}^{i-1} \mathbf{I}_{n^{(i)}(\tilde{x}(j))} \right) \otimes \mathbf{E}_l^{(i)}[\tilde{x}(i), \tilde{y}(i)] \otimes \left(\bigotimes_{j=i+1}^{J} \mathbf{I}_{n^{(i)}(\tilde{x}(j))} \right) \\
+ \sum_{t \in TS} \lambda_t \bigotimes_{j=1}^{J} \mathbf{E}_t^{(j)}[\tilde{x}(j), \tilde{y}(j)] . \tag{13}
$$

For the remaining non-diagonal blocks, where $\tilde{x}$ and $\tilde{y}$ differ in more than one position, only synchronized transitions have to be considered, because local transitions modify the state of one automaton only. Therefore the matrices have the following simple representation.

$$
\mathbf{Q}[\tilde{x}, \tilde{y}] = \left(\sum_{t \in TS} \lambda_t \bigotimes_{j=1}^{J} \mathbf{E}_t^{(j)}[\tilde{x}(j), \tilde{y}(j)] \right) . \tag{14}
$$

The hierarchical representation is obviously a generalization of the basic representation for SANs. This representation remains compact as long as $|RS| \ll$

$|\mathcal{RS}|$ which is usually the case. If too many macro states are generated, grouping of automata often helps to reduce this number. Like in the basic SAN approach, the number of automata should not exceed 10 for an efficient analysis.

The hierarchical approach can be easily extended for models with functional transitions rates. In this case, the Kronecker products are substituted by generalized Kronecker products as in the SAN approach [60, 62, 43]. If functional transition rates depend only on the macro state, the use of generalized Kronecker products is not necessary since the macro state determines the rate of the functional transitions and the values are fixed in a submatrix of the generator. Thus, for SANs with a small number of functional rates, macro states can be artificially introduced to generate submatrices that can be represented using ordinary instead of generalized Kronecker products.

The realization of vector matrix product computations with a block structured matrix $\mathbf{Q}$ is straightforward. The vector $\mathbf{p}$ of length n is decomposed in subvectors $\mathbf{p}[\tilde{\mathbf{x}}]$ according to the structure of the state space into macro states. The length of $\mathbf{p}[\tilde{\mathbf{x}}]$ is $|\mathcal{RS}[\tilde{\mathbf{x}}]|$ and computation $\mathbf{pQ}$ is done at a block level. Thus, for each macro state $\tilde{\mathbf{x}} \in \widetilde{\mathcal{RS}}$ vector $\mathbf{p}[\tilde{\mathbf{x}}]$ is computed as

$$\mathbf{p}[\tilde{\mathbf{x}}] = \sum_{\tilde{\mathbf{y}} \in \widetilde{\mathcal{RS}}} \mathbf{p}[\tilde{\mathbf{y}}]\mathbf{Q}[\tilde{\mathbf{y}}, \tilde{\mathbf{x}}]$$

where each vector matrix product can be computed using the shuffle algorithm or any other method to compute the product of a vector with a Kronecker product of matrices [18, 43]. Furthermore, the block structure of the matrix defines a natural partition of the matrix which is useful for distributing the matrix for parallel processing.

4.4   Hierarchical Representation of the Example

The MDD for the reachable state space of the example with window size 1 ($N = 1$) is shown in figure 4 on the left side. From the structure of the reachable state space it becomes clear that for automata 1 and 4 only a single macro state needs to be generated. In our example, the interesting automata that introduce unreachable states in the potential state space are 2 and 3. For each automaton,

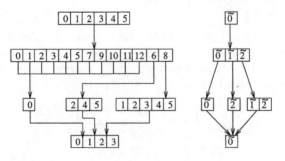

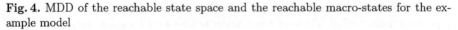

Fig. 4. MDD of the reachable state space and the reachable macro-states for the example model

3 macro states are generated. For A_2, the states 6 and 8 define their own macro states ($\tilde{1}$ and $\tilde{2}$) and the remaining states are collected in one macro state ($\tilde{0}$). In state 6 (macro state $\tilde{1}$) the places p_2^2 and p_6^2 contain a token which means that no packets have to be sent, but the sender waits for an acknowledgment for the last packet that was sent. For that context, automaton 3 can only be in state 2, 4, or 5 for the processing of the last packet. State 8 (macro state $\tilde{2}$) describes the situation that the places p_1^2 and p_6^2 contain a token which means that there are packets to send, while the sender has to wait for the acknowledgment. For that context, automaton 3 can be in any of the states $\{1, 2, 3, 4, 5\}$, because it can process one of the intermediate packets $\{1, 3, 5\}$ or the last packet $\{2, 4, 5\}$; the latter happens if A_1 triggers A_2 to send packets (firing of ts_1) and A_2 was waiting for an acknowledgement of the last packet sent (A_2 was in state 6). In the remaining states (macro state $\tilde{0}$), the sender has no outstanding packets or acknowledgment. For A_3, we have also 3 macro states. Macro state $\tilde{0}$ contains state 0 which is the initial state of this automaton. In this state, A_2 has to process no packets or acknowledgment. In macro state $\tilde{1}$, consisting of the states 1 and 3, the automaton processes one packet which is not the last packet of a message. The last macro state $\tilde{2}$ contains all states where the last packet or the acknowledgment of a message is processed. The MDD representing the set of reachable macro states is shown on the right side of figure 4.

The set of reachable macro states equals

$$\widetilde{\mathcal{RS}} = \{(\tilde{0}, \tilde{0}, \tilde{0}, \tilde{0}), (\tilde{0}, \tilde{1}, \tilde{2}, \tilde{0}), (\tilde{0}, \tilde{2}, \tilde{1}, \tilde{0}), (\tilde{0}, \tilde{2}, \tilde{2}, \tilde{0})\} .$$

From this representation, the state spaces for A_2 and A_3 can be reordered and the automata (sub-)matrices can be generated and combined for the generator matrix. Table 3 contains the number of macro states ($\tilde{n}$), the number of non-zeros in matrix $\mathbf{Q}$ ($nz(\mathbf{Q})$) excluding diagonal elements, the number of non-zeros to be stored for the hierarchical Kronecker representation ($nz(\sum \mathbf{Q}_i)$), and the number of states in $\mathcal{PS}$ and $\mathcal{RS}$ for different values of N. It can be seen that the hierarchical representation remains compact without introducing unreachable states.

5   Analysis Algorithms

A wealth of analysis algorithms has been developed for both Kronecker representations we presented. In this brief overview of analysis techniques, we start with quantitative analysis techniques and outline subsequently qualitative analysis approaches based on the Kronecker representation.

5.1   Quantitative Analysis

The basic approach for quantitative analysis is the computation of the stationary or transient solution vector of the CTMC as the solution of (1) and (2), respectively. For transient analysis, usually the randomization approach is used [62] which represents the vector $\mathbf{p}_\tau$ as

Table 3. State space sizes and numbers of non-zero elements for the hierarchical representation

N	$|\mathcal{RS}|$	$|\mathcal{PS}|$	$\bar{n}$	$nz(\mathbf{Q})$	$nz(\sum \mathbf{Q}_i)$
1	456	1872	4	1238	38
2	2760	21168	7	9578	147
3	10584	123840	10	42354	425
4	31368	502200	13	138070	986
5	78360	1605744	16	369434	1972

$$\mathbf{p}_\tau = \mathbf{p}_0 \sum_{k=l}^{r} (\mathbf{Q}/\alpha + \mathbf{I})^k \frac{(\alpha\tau)^k}{k!} e^{\alpha\tau} ,$$

where $\alpha \geq \max |\mathbf{Q}(s,s)|$ and l, r are appropriate constants. Computation of $\mathbf{p}_\tau$ requires the computation of r products of a vector with the matrix $\mathbf{Q}$. This operation can be performed on the compact matrix representation using an algorithm for vector matrix product computation which is available for all matrix representations we have introduced.

For stationary analysis a large number of different analysis approaches exist [62]. Among these approaches all iterative algorithms that do not modify the non-zero structure of the generator matrix can be realized in conjunction with a Kronecker representation of the generator matrix. The Power method, the method of Jacobi, projection methods [63, 13] and SOR [18, 64] are examples for methods which have been used in this context. Since Kronecker based approaches are mainly used for very large state spaces with up to several millions of states, solutions methods need to be efficient to compute the solution with an acceptable accuracy in an acceptable time. This requires that basic steps like the multiplication of a vector with a Kronecker representation needs to be as fast as possible and that a numerical method needs to provide a good rate of convergence. For the latter requirement, preconditioned projection techniques, block-SOR and aggregation/disaggregation techniques have been considered for the numerical analysis of Markov chains [37, 62]. However, not all of the methods can be adopted for SANs without significant modifications.

Research Directions. Currently, several research directions can be observed. In [39], the *separation of numerical method and matrix representation* is proposed. For different matrix representations including the modular Kronecker representation, an interface is described that allows to perform many numerical methods independently of the encapsulated, internal matrix representation. This separation of concerns has the potential to reduce the complexity of Kronecker-based methods and to simplify a more general usage of Kronecker representations. A different research direction aims at new numerical methods that exploit the Kronecker structure for a solution. Those approaches can be roughly classified in methods that use the structure to define new preconditioners and methods that use the structure for the definition of aggregation/disaggregation steps.

Preconditioning. First approaches for preconditioning were based on a series expansion of the generator matrix and its approximation by a Kronecker representation [63, 13]. However, although these preconditioners reduce the number of iterations when combined with some iterative method, they often do not reduce the solution time because the smaller number of iterations is outweighed by the effort for preconditioning. A preconditioner that results from the "nearest" Kronecker product of a matrix is a promising alternative, which opens a direction for future research [57].

Aggregation. The idea of aggregation/disaggregation is to compute solutions for smaller systems of equations that result from the original system by keeping the distribution of some dimensions (i.e., automata) fixed. The solution for the smaller dimension is afterwards projected on the original state space. This basic idea is very similar to the ideas underlying multigrid methods which are extremely efficient solvers for partial differential equations. For a hierarchical system with J components, $2^J - 1$ reduced systems can be defined by considering or not considering different components in detail. This shows that the resulting methods allow a great flexibility in how to define aggregation/disaggregation steps, but finding a good strategy is a non-trivial task. Aggregation/disaggregation methods have been proposed in different forms for SANs [14, 15, 45]. First experiences with these methods are very promising since the methods were able to compute solution vectors in an extremely efficient manner, but additional research work seems to be necessary to find really robust and efficient algorithms for a wide class of models.

Since the Kronecker structure of the generator matrix preserves the model structure, it is easy to substitute components by other components as long as the interface behavior, realized by synchronized transitions, remains the same. In this case only the matrices for one submodel need to be modified in the Kronecker representation and the generator matrix of the modified system is generated. This feature can be exploited in modeling by refining components or testing different realizations of components and it can be exploited for analysis purposes by first reducing components due to some equivalence relation that preserves the relevant behavior. Typical equivalence relations that preserve the stochastic behavior are exact and ordinary performance equivalence or bisimulation [10, 48, 49] which can be efficiently computed for a given component. Based on these equivalences, smaller matrices are built and substitute the original matrices in the Kronecker expressions yielding an exact aggregation of the Markov chain which means that stationary and transient results are not modified.

Vector sizes. The major problem of computing solution vectors for really large Markov chains is today the size of the iteration vectors. Some work has been done to reduce the vector size by storing also the vector in a symbolic form [25,

47], but the break-through is still missing since all approaches we are aware of result in an enormous overhead for the computation of vector matrix products and are sometimes even not really compact. Furthermore numerical stability might become a problem when using symbolic storage schemes for vectors.

Currently two possibilities seem to exist, when iteration vectors do not fit into main memory, namely disk-based methods and approximation methods. In disk-based methods, fast disks are exploited to enlarge the available memory. Available disk-based approaches [38, 56] use sparse storage schemes for the matrix such that matrix rows and vectors have to be reloaded from disk. Considering the efficiency of Kronecker based vector matrix computations (e.g., [14, 19]) it is very unlikely that disk based methods using sparse matrices can be more efficient, if the Kronecker approach can be applied which means if the model can be appropriately structured into components. However, as already pointed out in [56], disk-based methods can be combined with symbolic representations by keeping the compact matrix representation in main memory and storing only solution vectors on disk which might result in an increase in the size of solvable models. However, the disadvantage of disk-based methods is that implementations often become hardware dependent and therefore less flexible.

Apart from an exact analysis, the Kronecker structure of a model can also be used to realize approximate solution techniques which often drastically reduce the effort and introduce only small approximation errors. A general class of approximate solution techniques use a fixed point approach [35]. In these techniques, parts of a system are considered in detail and parts are aggregated. In the context of Kronecker based analysis this means that some components are aggregated and some are left in detailed representation. Since the aggregation of component i in a model reduces the state space size roughly by a factor of $n^{(i)}$, even the aggregation of a few components often yields much smaller state spaces to be analyzed. Different realization of the fixed point approach for the Kronecker representation have been proposed recently [11, 34, 16]. Current research activities have to consider algorithmic methods supporting the use of the these methods in modeling tools.

5.2   Qualitative Analysis

Apart from quantitative analysis also qualitative analysis profits from a Kronecker representation of the transitions matrix which can in this case be used for Boolean matrices indicating a transition or no transition between two states. Based on this compact representation of the transitions matrix, standard methods for qualitative analysis like model checking of CTL-formulas can be efficiently realized [26]. The Kronecker based approach can be seen as a complementary representation to graph based structures like BDDs. Experience shows that for problems which do not result from Boolean functions, the Kronecker representation often is more compact and convenient for the realization of efficient analysis algorithms. However, much more experience is required to decide which representation to use for which problem.

The combination of quantitative and qualitative analysis is in some sense realized by model checking approaches for Markov chains. Very popular in this context is the logic CSL [3] which forms the basis for different model checking algorithms [4, 51]. As shown in [20], these model checking algorithms profit from the exploitation of the Kronecker structure of the generator matrix.

6   Conclusion

This paper surveys the representation of transition matrices resulting from different modeling formalisms using Kronecker algebra. The key idea is to represent a huge matrix as a sum of Kronecker products of small matrices which describe local and synchronized transitions. This approach alleviates the state space explosion problem of state-based analysis. It defines a mathematically sound algebra to represent discrete-event systems, which reflects the compositional or hierarchical structure of the system at the state transition level. Furthermore, the matrix representation is well suited for the efficient realization of analysis algorithms for quantitative and qualitative analysis.

The current state of Kronecker based analysis for discrete-event systems is that the representation is well understood and methods exist to generate a Kronecker representation without unreachable states for automata networks or Petri nets with partitions on the set of places. Additionally, several analysis algorithms have been developed and first tool implementations are available. However, especially for finding efficient and robust analysis algorithms which exploit the Kronecker representation, considerable research work still needs to be done.

References

1. A. Benoit, B. Plateau, W.J. Stewart. Memory-efficient iterative methods for stochastic automata networks. Technical Report, Rapport de recherche INRIA n. 4259, France, Sept. 2001.
2. M. Ajmone-Marsan, G. Balbo, G. Conte, S. Donatelli, and G. Franceschinis. *Modelling with generalized stochastic Petri nets*. Wiley, 1995.
3. A. Aziz, K. Sanwal, V. Singhal, and R. Brayton. Model checking continuous time Markov chains. *ACM Trans. on Computational Logic*, 1(1):162–170, 2000.
4. C. Baier, B. Haverkort, H. Hermanns, and J. P. Katoen. Model checking continuous-time Markov chains by transient analysis. In E. A. Emerson and A. P. Sistla, editors, *Computer Aided Verification (CAV'00)*, pages 358–372. Springer LNCS 1855, 2000.
5. F. Bause and P. Buchholz. Protocol analysis using a timed version of SDL. In J. Quemada, J. Manas, and E. Vazquez, editors, *Formal Description Techniques 90*, pages 239–254. North Holland, 1990.
6. F. Bause, P. Buchholz, and P. Kemper. QPN-tool for the specification and analysis of hierarchically combined queueing Petri nets. In H. Beilner and F. Bause, editors, *Quantitative Evaluation of Computing and Communication Systems*, pages 224–238. Springer LNCS 977, 1995.

7. M. Bernado and R. Gorrieri. A tuturial of EMPA: A theory of concurrent process-
    es with nondeterminism, priorities, probabilities and time. *Theoretical Computer
    Science*, 202:1–54, 1998.
8. P. Buchholz. A hierarchical view of GCSPNs and its impact on qualitative and
    quantitative analysis. *Journal of Parallel and Distributed Computing*, 15(3):207–
    224, 1992.
9. P. Buchholz. Markovian process algebra: composition and equivalence. In U. Her-
    zog and M. Rettelbach, editors, *Proc. of the 2nd Work. on Process Algebras and
    Performance Modelling*, pages 11–30. Arbeitsberichte des IMMD, University of Er-
    langen, no. 27, 1994.
10. P. Buchholz. Equivalence relations for stochastic automata networks. In W. J.
    Stewart, editor, *Computations with Markov Chains*, pages 197–216. Kluwer Aca-
    demic Publishers, 1995.
11. P. Buchholz. Iterative decomposition and aggregation of labelled GSPNs. In J. De-
    sel and M. Silva, editors, *Application and Theory of Petri Nets 1998*, pages 226–245.
    Springer LNCS 1420, 1998.
12. P. Buchholz. Hierarchical structuring of superposed GSPNs. *IEEE Transactions
    on Software Engineering*, 25(2):166–181, 1999.
13. P. Buchholz. Projection methods for the analysis of stochastic automata networks.
    In B. Plateau, W. J. Stewart, and M. Silva, editors, *Numerical Solution of Markov
    Chains (NSMC'99)*, pages 149–168. Prensas Universitarias de Zaragoza, 1999.
14. P. Buchholz. Structured analysis approaches for large Markov chains. *Applied
    Numerical Mathematics*, 31(4):375–404, 1999.
15. P. Buchholz. Multi-level solutions for structured Markov chains. *SIAM Journal
    on Matrix Methods and Applications*, 22(2):342–357, 2000.
16. P. Buchholz. An adaptive decomposition approach for the analysis of stochas-
    tic Petri nets. In *Proc. Int. Conf. on Dependable Systems and Networks*, pages
    647–656. IEEE CS-Press, 2002.
17. P. Buchholz. A class of hierarchical queueing networks and their analysis. *Queueing
    Systems*, 15(1):1994, 59-80.
18. P. Buchholz, G. Ciardo, S. Donatelli, and P. Kemper. Complexity of Kronecker
    operations and sparse matrices with applications to the solution of Markov models.
    INFORMS Journal on Computing, 12(3):203–222, 2000.
19. P. Buchholz and T. Dayar. Block SOR for Kronecker structured representations.
    Proc. of the Int. Workshop on the Numerical Solution of Markov Chains, 2003, to
    appear as a revised version in *Journal Linear Algebra and Applications*.
20. P. Buchholz, J. P. Katoen, P. Kemper, and C. Tepper. Model-checking large struc-
    tured Markov chains. *Journal of Logic and Algebraic Programming*, 56(1/2):69–97,
    2003.
21. P. Buchholz and P. Kemper. Numerical analysis of stochastic marked graph nets.
    In *Proc. of the 6th Int. Workshop on Petri Nets and Performance Models*, pages
    32–41. IEEE CS-Press, 1995.
22. P. Buchholz and P. Kemper. On generating a hierarchy for GSPN analysis. *ACM
    Performance Evaluation Review*, 26(2):5–14, 1998.
23. P. Buchholz and P. Kemper. A toolbox for the analysis of discrete event dynam-
    ic systems. In N. Halbwachs and D. Peled, editors, *Computer Aided Verification
    (CAV'99)*, pages 483–486. Springer LNCS 1633, 1999.
24. P. Buchholz and P. Kemper. Efficient computation and representation of large
    reachability sets for composed automata. In R. Boel and G. Stremersch, edi-
    tors, *Discrete Event Systems Analysis and Control*, pages 49–56. Kluwer Academic,
    2000.

25. P. Buchholz and P. Kemper. Compact representations of probability distributions in the analysis of superposed GSPNs. In R. German and B. Haverkort, editors, *Proc. Petri Nets and Performance Models'01*, pages 81–90. IEEE CS-Press, 2001.

26. P. Buchholz and P. Kemper. Efficient computation and representation of large reachability sets for composed automata. *Discrete Event Dynamic Systems Theory and Applications*, 12(3):265–286, 2002.

27. P. Buchholz and P. Kemper. Hierarchical rechability graph generation for Petri nets. *Formal Methods in System Design*, 21(3):281–315, 2002.

28. P. E. Buis and W. R. Dyksen. Efficient vector and parallel manipulation of tensor products. *ACM Trans. Math. Software*, 22(1):18–23, 1996.

29. J. Campos, J.M. Colom, Jungnitz H, and M. Silva. Approximate throughput computation of stochastic marked graphs. *IEEE Transactions on Software Engineering*, 20(7):525–535, 1994.

30. J. Campos, M. Silva, and S. Donatelli. Structured solution of asynchronously communicating stochastic modules. *IEEE Transactions on Software Engineering*, 25(2):147–165, 1999.

31. G. Ciardo, G. Luettgen, and R. Siminiceanu. Saturation: an efficient iteration strategy for symbolic state-space generation. In T. Margaria and W. Yi, editors, *Tools and Algorithms for the Construction and Analysis of Systems (TACAS 2001)*, pages 328–342. Springer LNCS 2031, 2001.

32. G. Ciardo and A. Miner. A data structure for the efficient Kronecker solution of GSPNs. In P. Buchholz and M. Silva, editors, *Proc. 8th int. Workshop Petri Nets and Performance Models*, pages 22–31. IEEE CS Press, 1999.

33. G. Ciardo and A. Miner. Efficient reachability set generation and storage using decision diagrams. In S. Donatelli and J. Kleijn, editors, *Proc. 20th int. Conf. Application and Theory of Petri Nets*. Springer, LNCS 1639, 1999.

34. G. Ciardo, A. S. Miner, and S. Donatelli. Using the exact state space of a model to compute approximate stationary measures. In J. Kurose and P. Nain, editors, *Proc. ACM Sigmetrics*, pages 207–216. ACM Press, 2000.

35. G. Ciardo and K. Trivedi. A decomposition approach for stochastic reward net models. *Performance Evaluation*, 18:37–59, 1994.

36. M. Davio. Kronecker products and shuffle algebra. *IEEE Transactions on Computers*, 30:116–125, 1981.

37. T. Dayar and W. J. Stewart. Comparison of partitioning techniques for two-level iterative solvers on large, sparse Markov chains. *SIAM Journal on Scientific Computing*, 21:1691–1705, 2000.

38. D. D. Deavours and W. H. Sanders. An efficient disk-based tool for solving very large Markov models. In R. Marie, B. Plateau, M. Calzarossa, and G. Rubino, editors, *Computer Performance Evaluation - Modelling Techniques and Tools 1997*, pages 58–71. Springer LNCS 1245, 1997.

39. S. Derivasi, P. Kemper, and W.H. Sanders. The Möbius state-level abstract functional interface. In *Proc. of 12th Int. Conf. on Modelling Tools and Techniques for Computer and Communication System Performance Evaluation (Performance TOOLS 2002)*, LNCS 2324. Springer, 2001.

40. S. Donatelli. Superposed generalized stochastic Petri nets: definition and efficient solution. In R. Valette, editor, *Application and Theory of Petri Nets 1994*, pages 258–277. Springer LNCS 815, 1994.

41. S. Donatelli. Superposed stochastic automata: a class of stochastic Petri nets amenable to parallel solution. *Performance Evaluation*, 18:21–36, 1994.

42. S. Donatelli and P. Kemper. Integrating synchronization with priority into a Kronecker representation. *Performance Evaluation*, 44(1-4):73–96, 2001.

43. P. Fernandes, B. Plateau, and W. J. Stewart. Efficient descriptor-vector multiplication in stochastic automata networks. *Journal of the ACM*, 45(3):381–414, 1998.
44. A. Graham. *Kronecker products and matrix calculus with applications*. Ellis Howard, 1981.
45. O. Gusak and T. Dayar. Iterative aggregation-disaggregation versus block Gauss-Seidel on stochastic automata networks with unfavorable partitionings. In M. S. Obaidat and F. Davoli, editors, *Proc. Int. Symp. on Perf. Eval. of Comp. and Telecomm. Sys.*, pages 617–623. SCS-Press, 2001.
46. H. Hermanns, U. Herzog, and V. Mertsiotakis. Stochastic process algebras – between LOTOS and Markov chains. *Computer Networks and ISDN Systems*, 30(9/10):901–924, 1998.
47. H. Hermanns, M. Kwiatkowska, G. Norman, D. Parker, and M. Siegle. On the use of MTBDDs for performability analysis and verification of stochastic systems. *Journal of Logic and Algebraic Programming*, 56:23–67, 2003.
48. H. Hermanns and M. Rettelbach. Syntax, semantics, equivalences, and axioms for MTIPP. In U. Herzog and M. Rettelbach, editors, *Proc. of the 2nd Work. on Process Algebras and Performance Modelling*. Arbeitsberichte des IMMD, University of Erlangen, no. 27, 1994.
49. J. Hillston. A compositional approach for performance modelling. Phd thesis, University of Edinburgh, Dep. of Comp. Sc., 1994.
50. C. Hoare. *Communicating sequential processes*. Prentice Hall, 1985.
51. J. P. Katoen, M. Kwiatkowska, G. Norman, and D. Parker. Faster and symbolic CTMC model checking. In L. de Alfaro and S. Gilmore, editors, *Process Algebras and Probabilistic Methods*, LNCS 2165, pages 123–38. Springer, 2001.
52. P. Kemper. Numerical analysis of superposed GSPNs. *IEEE Transactions on Software Engineering*, 22(9):615–628, 1996.
53. P. Kemper. Reachability analysis based on structured representations. In J. Billington and W. Reisig, editors, *Application and Theory of Petri Nets 1996*, pages 269–288. Springer LNCS 1091, 1996.
54. P. Kemper. Parallel randomization for large structured Markov chains. In *Proc. of the 2002 int. Conf. om Dependable Systems and Networks*, pages 657–666. IEEE CS Press, 2002.
55. P. Kemper and R. Lübeck. Model checking based on Kronecker algebra. Forschungsbericht 669, Fachbereich Informatik, Universität Dortmund (Germany), 1998.
56. M. Kwiatkowska and R. Mehmood. Out-of-core solution of large linear systems of equations arising from stochastic modelling. In *Proc. PAPM/PROBMIV'02*, pages 135–151. Springer LNCS 2399, 2002.
57. A. N. Langville and W. J. Stewart. A Kronecker product approximate preconditioner for SANs. *Numerical Linear Algebra with Applications* (to appear).
58. R. Milner. *Communication and concurrency*. Prentice Hall, 1989.
59. A. Miner. Efficient solution of GSPNs using canonical matrix diagrams. In R. German and B. Haverkourt, editors, *Proc. 9th int. Workshop Petri Nets and Performance Models*, pages 101–110. IEEE CS Press, 2001.
60. B. Plateau. On the stochastic structure of parallelism and synchronisation models for distributed algorithms. *Performance Evaluation Review*, 13:142–154, 1985.
61. R. Pooley and P. King. The unified modeling language and performance enginering. *IEE Proceedings - Software*, 146(1):2–10, 1999.
62. W. J. Stewart. *Introduction to the numerical solution of Markov chains*. Princeton University Press, 1994.

63. W.J. Stewart, K. Atif, and B. Plateau. The numerical solution of stochastic automata networks. *European Journal of Operational Research*, 86:503–525, 1995.
64. E. Uysal and T. Dayar. Iterative methods based on splittings for stochastic automata networks. *European Journal of Operational Research*, 110(1):166–186, 1998.
65. C. M. Woodside and Y. Li. Performance Petri net analysis of communications protocol software by delay equivalent aggregation. In *Proc. 4th Int. Workshop on Petri Nets and Performance Models*, pages 64–73. IEEE CS-Press, 1991.

Symbolic Representations and Analysis of Large Probabilistic Systems

Andrew Miner[1]* and David Parker[2]**

[1] Dept. of Computer Science, Iowa State University, Ames, Iowa, 50011
[2] School of Computer Science, University of Birmingham, UK

Abstract. This paper describes symbolic techniques for the construction, representation and analysis of large, probabilistic systems. Symbolic approaches derive their efficiency by exploiting high-level structure and regularity in the models to which they are applied, increasing the size of the state spaces which can be tackled. In general, this is done by using data structures which provide compact storage but which are still efficient to manipulate, usually based on binary decision diagrams (BDDs) or their extensions. In this paper we focus on BDDs, multi-valued decision diagrams (MDDs), multi-terminal binary decision diagrams (MTBDDs) and matrix diagrams.

1 Introduction

This paper provides an overview of *symbolic* approaches to the validation of stochastic systems. We focus on those techniques which are based on the construction and analysis of finite-state probabilistic models. These models comprise a set of states, corresponding to the possible configurations of the system being considered, and a set of transitions which can occur between these states, labeled in some way to indicate the likelihood that they will occur. In particular, this class of models includes discrete-time Markov chains (DTMCs), continuous-time Markov chains (CTMCs) and Markov decision processes (MDPs).

Analysis of these models typically centers around either computation of transient or steady-state probabilities, which describe the system at a particular time instant or in the long-run, respectively, or computation of the probability that the system will reach a particular state or class of states. These are the key constituents of several approaches to analyzing these models, including both traditional performance or dependability evaluation and more recent, model checking based approaches using temporal logics such as PCTL or CSL.

As is well known, one of the chief practical problems plaguing the implementations of such techniques is the tendency of models to become unmanageably large, particularly when they comprise several parallel components, operating concurrently. This phenomenon is often referred to as 'the state space explosion problem', 'largeness' or 'the curse of dimensionality'. A great deal of work

* Supported in part by fellowships from the NASA Graduate Student Researchers Program (NGT-1-52195) and the Virginia Space Grant Consortium.
** Supported in part by QinetiQ and EPSRC grant GR/S11107.

C. Baier et al. (Eds.): Validation of Stochastic Systems, LNCS 2925, pp. 296–338, 2004.
© Springer-Verlag Berlin Heidelberg 2004

has been put into developing space and time efficient techniques for the storage and analysis of probabilistic models. Many of the recent approaches that have achieved notable success are *symbolic* techniques, by which we mean those using data structures based on binary decision diagrams (BDDs). Also known as *implicit* or *structured* methods, these approaches focus on generating compact model representations by exploiting structure and regularity, usually derived from the high-level description of the system. This is possible in practice because systems are typically modeled using structured, high-level specification formalisms such as Petri nets and process algebras. In contrast, *explicit* or *enumerative* techniques are those where the entire model is stored and manipulated explicitly. In the context of probabilistic models, sparse matrices are the most obvious and popular explicit storage method.

The tasks required to perform analysis and verification of probabilistic models can be broken down into a number of areas, and we cover each one separately in this paper. In Section 2, we consider the storage of sets of states, such as the reachable state space of a model. In Section 3, we look at the storage of the probabilistic model itself, usually represented by a real-valued transition matrix. We also discuss the generation of each of these two entities. In Section 4, we describe how the processes of analysis, which usually reduce to numerical computation, can be performed in this context. Finally, in Section 5, we discuss the relative strengths and weaknesses of the various symbolic approaches, and suggest some areas for future work.

2   Representing Sets of States

We begin by focusing on the problem of representing a set of states. The most obvious task here is to represent the entire set of states which the system can possibly be in. Usually, some proportion of the potential configurations of the system are impossible. These are removed by computing the set of *reachable* states, from some initial state(s), and removing all others. Note that in some verification technologies, non-probabilistic variants in particular, this *reachability* computation may actually constitute the verification itself. Here, though, it is seen as part of the initial construction of the model since probabilistic analysis cannot be carried out until afterwards.

In addition to storing the set of reachable states of the model, it may often be necessary to represent a particular class of states, e.g. those which satisfy a given specification. For model checking, in particular, this is a fundamental part of any algorithm. Clearly, in both cases, there is a need to store sets of states compactly and in such a way that they can be manipulated efficiently. In this section, we consider two symbolic data structures for this purpose: *binary decision diagrams* (BDDs) and *multi-valued decision diagrams* (MDDs).

2.1   Binary Decision Diagrams

Binary decision diagrams (BDDs) are rooted, directed, acyclic graphs. They were originally proposed by Lee [55] and Akers [3], but were later popularized by Bryant [13], who refined the data structure and presented a number of algorithms for their efficient manipulation. A BDD is associated with a finite set of Boolean variables and represents a Boolean function over these variables. We denote the function represented by the BDD B over K variables $x_K, \ldots, x_1$ as $f_B : \mathbb{B}^K \to \mathbb{B}$.

The vertices of a BDD are usually referred to as *nodes*. A node m is either *non-terminal*, in which case it is labeled with a Boolean variable $var(\mathsf{m}) \in \{x_K, \ldots, x_1\}$, or *terminal*, in which case it is labeled with either 0 or 1. Each non-terminal node m has exactly two children, $then(\mathsf{m})$ and $else(\mathsf{m})$. A terminal node has no children. The value of the Boolean function f_B, represented by BDD B, for a given valuation of its Boolean variables can be determined by tracing a path from its root node to one of the two terminal nodes. At each node m, the choice between $then(\mathsf{m})$ and $else(\mathsf{m})$ is determined by the value of $var(\mathsf{m})$: if $var(\mathsf{m}) = 1$, $then(\mathsf{m})$ is taken, if $var(\mathsf{m}) = 0$, $else(\mathsf{m})$ is taken. Every BDD node m corresponds to some Boolean function f_m. The terminal nodes correspond to the trivial functions $f_0 = 0$, $f_1 = 1$.

For a function f, variable x_k, and Boolean value b, the *cofactor* $f|_{x_k=b}$ is found by substituting the value b for variable x_k:

$$f|_{x_k=b} = f(x_K, \ldots, x_{k+1}, b, x_{k-1}, \ldots, x_1).$$

An important property of BDDs is that the children of a non-terminal node m correspond to cofactors of function f_m. That is, for every non-terminal node m, $f_{then(\mathsf{m})} = f_\mathsf{m}|_{var(\mathsf{m})=1}$, and $f_{else(\mathsf{m})} = f_\mathsf{m}|_{var(\mathsf{m})=0}$. We will also refer to the cofactor of a BDD node m, with the understanding that we mean the BDD node representing the cofactor of the function represented by node m.

A BDD is said to be *ordered* if there is a total ordering of the variables such that every path through the BDD visits nodes according to the ordering. In an ordered BDD (OBDD), each child m' of a non-terminal node m must therefore either be terminal, or non-terminal with $var(\mathsf{m}) > var(\mathsf{m}')$.

A *reduced* OBDD (ROBDD) is one which contains no *duplicate* nodes, i.e. non-terminal nodes labeled with the same variable and with identical children, or terminal nodes labeled with the same value; and which contains no *redundant* nodes, i.e., non-terminal nodes having two identical children. Since the function represented by a redundant node m does not depend on the value of variable $var(\mathsf{m})$, redundant nodes are sometimes referred to as *don't care* nodes.

Any OBDD can be reduced to an ROBDD by repeatedly eliminating, in a bottom-up fashion, any instances of duplicate and redundant nodes. If two nodes are duplicates, one of them is removed and all of its incoming pointers are redirected to its duplicate. If a node is redundant, it is removed and all incoming pointers are redirected to its unique child.

A common variant of the above reduction scheme is to allow redundant nodes but not duplicate nodes. In particular, an OBDD is said to be *quasi-reduced* if it contains no duplicate nodes and if all paths from the root node (which must have label x_K) to a terminal node visit exactly one node for each variable. Note that a quasi-reduced OBDD (QROBDD) can be obtained from any ROBDD by inserting redundant nodes as necessary.

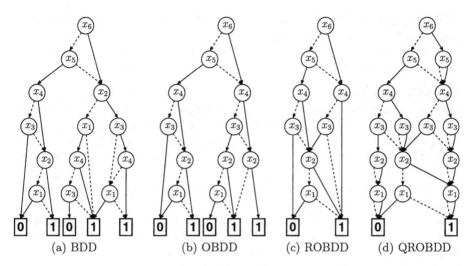

Fig. 1. Example BDDs for the same Boolean function

Figure 1 shows four equivalent data structures, a BDD, an OBDD, an ROB-DD and a QROBDD, each representing the same Boolean function, f. Tracing paths from the root node to the terminal nodes of the data structures, we can see, for example, that $f(0,0,1,0,0,1) = 1$ and $f(0,1,0,1,1,1) = 0$. The most commonly used of these four variants is the ROBDD and this will also be the case in this paper. For simplicity, and by convention, from this point on we shall refer to ROBDDs simply as BDDs. On the occasions where we require QROBDDs, this will be stated explicitly.

It is important to note that the reduction rules for BDDs described in the previous paragraphs have no effect on the function being represented. They do, however, typically result in a significant decrease in the number of BDD nodes. More importantly still, as shown by Bryant [13], for a fixed ordering of the Boolean variables, BDDs are a *canonical* representation. This means that there is a one-to-one correspondence between BDDs and the Boolean functions they represent. Similarly, it can also be shown that QROBDDs are a canonical representation for a fixed variable ordering.

The canonical nature of BDDs has important implications for efficiency. For example, it makes checking whether or not two BDDs represent the same function very easy. This is an important operation in many situations, such as the implementation of iterative fixed-point computations. In practice, these reductions are taken one step further. Many BDD packages will actually store all BDDs in a single, multi-rooted graph structure, known as the *unique-table*, where no two nodes are duplicated. This means that comparing two BDDs for equality is as simple as checking whether they are stored in the same place in memory.

It is also important to note that the choice of an ordering for the Boolean variables of a BDD can have a tremendous effect on the size of the data structure, i.e. its number of nodes. Finding the optimal variable ordering, however, is known to be computationally expensive [10]. For this reason, the efficiency of

BDDs in practice is largely reliant on the development of application-dependent *heuristics* to select an appropriate ordering, e.g. [38]. There also exist techniques such as *dynamic variable reordering*, which can be used to change the ordering for an existing BDD in an attempt to reduce its size, see e.g. [66].

BDD Operations. One of the main appeals of BDDs is the efficient algorithms for their manipulation which have been developed, e.g. [13, 14, 12]. A common BDD operation is the ITE (IfThenElse) operator, which takes three BDDs, B_1, B_2 and B_3, and returns the BDD representing the function "if f_{B_1} then f_{B_2} else f_{B_3}". The ITE operator can be implemented recursively, based on the property $ITE(B_1, B_2, B_3)|_{x_k=b} = ITE(B_1|_{x_k=b}, B_2|_{x_k=b}, B_3|_{x_k=b})$.

The algorithm to perform this is shown in Figure 2. Lines 1–8 cover the trivial base cases. In lines 14 and 15, the algorithm splits recursively, based on the cofactors of the three BDD operands. The variable, x_k, used to generate these cofactors is the top-most variable between the three BDDs. The resulting BDD, H, is generated by attaching the BDDs from the two recursive calls to a node labeled with x_k. In line 16, reduction of H is performed. Assuming that the operands to the algorithm were already reduced, the only two checks required here are that: (a) *then*(H) and *else*(H) are distinct; and (b) the root node of H does not already exist.

ITE(in: B_1, B_2, B_3)

- B_1, B_2, B_3 are BDD nodes
1: **if** $B_1 = 0$ **then**
2:     Return B_3;
3: **else if** $B_1 = 1$ **then**
4:     Return B_2;
5: **else if** $(B_2 = 1) \wedge (B_3 = 0)$ **then**
6:     Return B_1;
7: **else if** $B_2 = B_3$ **then**
8:     Return B_2;
9: **else if** $\exists$ computed-table entry (B_1, B_2, B_3, H) **then**
10:     Return H;
11: **end if**
12: $x_k \leftarrow$ top variable of B_1, B_2, B_3;
13: H $\leftarrow$ new non-terminal node with label x_k;
14: *then*(H) $\leftarrow$ ITE($B_1|_{x_k=1}, B_2|_{x_k=1}, B_3|_{x_k=1}$);
15: *else*(H) $\leftarrow$ ITE($B_1|_{x_k=0}, B_2|_{x_k=0}, B_3|_{x_k=0}$);
16: Reduce(H);
17: Add entry (B_1, B_2, B_3, H) to computed-table;
18: Return H;

Fig. 2. Algorithm for the BDD operator ITE

A crucial factor in the efficiency of the ITE algorithm is the *computed-table*, which is used to cache the result of each intermediate call to the algorithm. Notice how, in line 9, entries in the cache are checked and reused if possible. In practice, many recursive calls would typically be repeated without this step. Furthermore, this means that the computational complexity of the algorithm is bounded by the maximum number of distinct recursive calls to ITE, $\mathcal{O}(|\mathsf{B}_1| \cdot |\mathsf{B}_2| \cdot |\mathsf{B}_3|)$, where $|\mathsf{B}|$ denotes the number of nodes in BDD B.

Another common operation is Apply, which takes two BDDs, B_1 and B_2, plus a binary Boolean operator op, such as $\wedge$ or $\vee$, and produces the BDD which represents the function $f_{\mathsf{B}_1} \, op \, f_{\mathsf{B}_2}$. For convenience we often express such operations in infix notation, for example, $\mathsf{B}_1 \vee \mathsf{B}_2 \equiv \mathsf{Apply}(\vee, \mathsf{B}_1, \mathsf{B}_2)$. The Apply operator can be implemented using a recursive algorithm similar to the one for ITE, described above, again making use of the computed-table. Alternatively, any Apply operator can be expressed using ITE; e.g. $\mathsf{Apply}(\wedge, \mathsf{B}_1, \mathsf{B}_2) \equiv \mathsf{ITE}(\mathsf{B}_1, \mathsf{B}_2, 0)$. The latter has the advantage that it is likely to increase the hit-rate in the computed-table cache since only one operation is required, as opposed to several operations with their own computed-table caches (one for each binary Boolean operator op).

2.2 Multi-Valued Decision Diagrams

Multi-valued decision diagrams (MDDs) are also rooted, directed, acyclic graphs [47]. An MDD is associated with a set of K variables, $x_K, \ldots, x_1$, and an MDD M represents a function $f_{\mathsf{M}} : \mathbb{N}_K \times \cdots \times \mathbb{N}_1 \to \mathbb{M}$, where $\mathbb{N}_k$ is the finite set of values that variable x_k can assume, and $\mathbb{M}$ is the finite set of possible function values. It is usually assumed that $\mathbb{N}_k = \{0, \ldots, N_k - 1\}$ and $\mathbb{M} = \{0, \ldots, M - 1\}$ for simplicity. Note that BDDs are the special case of MDDs where $\mathbb{M} = \mathbb{B}$ and $\mathbb{N}_k = \mathbb{B}$ for all k. MDDs are similar to the "shared tree" data structure described in [71].

Like BDDs, MDDs consist of *terminal* nodes and *non-terminal* nodes. The terminal nodes are labeled with an integer from the set $\mathbb{M}$. A non-terminal node m is labeled with a variable $var(\mathsf{m}) \in \{x_K, \ldots, x_1\}$. Since variable x_k can assume values from the set $\mathbb{N}_k$, a non-terminal node m labeled with variable x_k has N_k children, each corresponding to a cofactor $\mathsf{m}|_{x_k=c}$. We refer to child c of node m as $child(\mathsf{m}, c)$, where $f_{child(\mathsf{m},c)} = f_{\mathsf{m}}|_{var(\mathsf{m})=c}$. Every MDD node corresponds to some integer function.

The BDD notion of ordering can also be applied to MDDs, to produce ordered MDDs (OMDDs). A non-terminal MDD node m is redundant if *all* of its children are identical, i.e., $child(\mathsf{m}, i) = child(\mathsf{m}, j)$ for all $i, j \in \mathbb{N}_{var(\mathsf{m})}$. Two non-terminal MDD nodes m_1 and m_2 are duplicates if $var(\mathsf{m}_1) = var(\mathsf{m}_2)$ and $child(\mathsf{m}_1, i) = child(\mathsf{m}_2, i)$ for all $i \in \mathbb{N}_{var(\mathsf{m})}$. Based on the above definitions, we can extend the notion of reduced and quasi-reduced BDDs to apply also to MDDs. It can be shown [47] that reduced OMDDs (ROMDDs) are a canonical representation for a fixed variable ordering. The same can also be shown for quasi-reduced OMDDs (QROMDDs). Finally, like BDDs, the number of ROMDD nodes required to represent a function may be sensitive to the chosen variable ordering.

Example MDDs are shown in Figure 3, all representing the same function over three variables, x_3, x_2, x_1 with $N_3 = N_2 = N_1 = 4$ and $M = 3$. The value

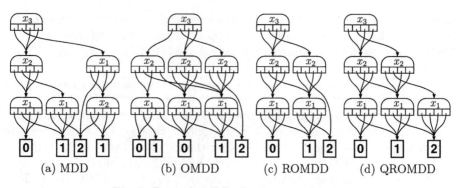

(a) MDD   (b) OMDD   (c) ROMDD   (d) QROMDD

Fig. 3. Example MDDs for the same function

of the function is zero if none of the variables has value 1, one if exactly one of
the variables has value 1, and two if two or more of the variables have value 1.
Figure 3(a) shows an MDD that is not ordered, Figure 3(b) shows an OMDD
that is not reduced, and Figure 3(c) and Figure 3(d) show the ROMDD and
QROMDD for the function, for the given variable ordering. Unless otherwise
stated, the remainder of the paper will assume that all MDDs are ROMDDs.

MDD Operations. Like BDDs, MDDs can be manipulated using various operators. One such operator is the Case operator, defined in [47] as:

$$\mathsf{Case}(\mathsf{F}, \mathsf{G}^0, \dots, \mathsf{G}^{M-1})(x_K, \dots, x_1) = \mathsf{G}^{f_\mathsf{F}(x_K, \dots, x_1)}(x_K, \dots, x_1).$$

Thus, Case selects the appropriate MDD G^i based on the value of F. Note that
Case is a generalization of ITE for BDDs: $\mathsf{ITE}(\mathsf{A}, \mathsf{B}, \mathsf{C}) \equiv \mathsf{Case}(\mathsf{A}, \mathsf{C}, \mathsf{B})$. A recursive algorithm to implement Case for MDDs is presented in [47], and is shown
in Figure 4(a). It is based on the recurrence:

$$\mathsf{Case}(\mathsf{F}, \mathsf{G}^0, \dots, \mathsf{G}^{M-1})|_{x_k=i} = \mathsf{Case}(\mathsf{F}|_{x_k=i}, \mathsf{G}^0|_{x_k=i}, \dots, \mathsf{G}^{M-1}|_{x_k=i})$$

with appropriate terminal conditions (e.g. if F is a constant). The Case algorithm is quite similar to the ITE algorithm: both algorithms first handle the
trivial cases, then handle the already-computed cases by checking entries in the
computed-table. The computation is performed for the other cases, and the resulting node is reduced. The main difference is that, since Case operates on
MDDs, a loop is required to generate the recursive calls to Case, one for each
possible value of top variable x_k.

As a simple example, the MDD produced by $\mathsf{Case}(\mathsf{F}, 1, 1, 0)$ is shown in
Figure 4(b), where F is the MDD shown in Figure 3(c).

As with ITE, a simple recursive implementation of Case without using a
computed-table can be computationally expensive: each call to Case with top
variable x_k will generate N_k recursive calls to Case. This leads to a computational complexity of $\mathcal{O}(\prod_{k=1}^{K} N_k)$ (assuming constant time for node reduction),
which is often intractable. Again, the number of *distinct* recursive calls to Case

Case(in: $F, G^0, \ldots, G^{M-1}$)
- $F, G^0, \ldots, G^{M-1}$ are MDD nodes
1: **if** $F = c, c \in \mathbb{M}$ **then**
2:      Return G^c;
3: **else if** $(G^0 = 0) \wedge \cdots \wedge (G^{M-1} = M - 1)$ **then**
4:      Return F;
5: **else if** $G^0 = G^1 = \cdots = G^{M-1}$ **then**
6:      Return G^0;
7: **else if** $\exists$ computed-table entry $(F, G^0, \ldots, G^{M-1}, H)$ **then**
8:      Return H;
9: **end if**
10: $x_k \leftarrow$ Top variable of $F, G^0, \ldots, G^{M-1}$;
11: $H \leftarrow$ new non-terminal node with label x_k;
12: **for each** $i \in \mathbb{N}_k$ **do**
13:      $child(H, i) \leftarrow$ Case$(F|_{x_k=i}, G^0|_{x_k=i}, \ldots, G^{M-1}|_{x_k=i})$;
14: **end for**
15: Reduce(H);
16: Add entry $(F, G^0, \ldots, G^{M-1}, H)$ to computed-table;
17: Return H;

(a) Algorithm   (b) Example

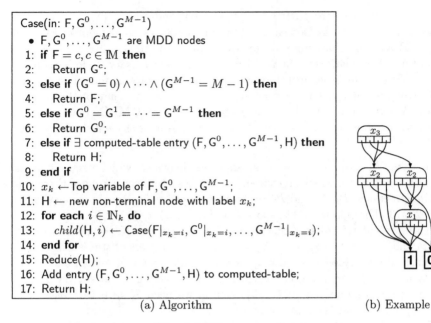

Fig. 4. MDD operator Case

is bounded by $|F| \cdot |G^0| \cdots |G^{M-1}|$; thus, the use of the computed-table bounds the worst-case computational complexity of Case by:

$$\mathcal{O}(|F| \cdot |G^0| \cdots |G^{M-1}| \cdot \max\{N_K, \ldots, N_1\}),$$

since each non-trivial call to Case with top variable x_k has computational cost of $\mathcal{O}(N_k)$. Note that the resulting MDD will have at most $|F| \cdot |G^0| \cdots |G^{M-1}|$ nodes, regardless of implementation.

2.3 State Set Representation and Generation

We now describe how the data structures introduced in the previous two sections, BDDs and MDDs, can be used to represent and manipulate sets of states of a probabilistic model. We also consider the problem of generating the set of reachable states for a model using these data structures.

Representing Sets of States. To represent a set of states using BDDs or MDDs, each state s must be expressible as a collection of K state variables, $\mathbf{s} = (x_K, \ldots, x_1)$, where each state variable can assume a finite number of values. If $\mathbb{N}_k$ is the set of possible values for state variable x_k, then the set of all possible states is $\mathcal{S} = \mathbb{N}_K \times \cdots \times \mathbb{N}_1$. Thus, any set of states will be a subset of $\mathcal{S}$.

The basic idea behind BDD and MDD state set representations is to encode the *characteristic function* of a subset $\mathcal{S}'$ of the set of states $\mathcal{S}$, i.e. the function

$\chi_{\mathcal{S}'} : \mathcal{S} \to \{0, 1\}$ where $\chi_{\mathcal{S}'}(x_K, \ldots, x_1) = 1$ if and only if $(x_K, \ldots, x_1) \in \mathcal{S}'$. If states are encoded as Boolean vectors, i.e. a state is an element of $\mathbb{B}^K$, then $\mathbb{N}_k = \mathbb{B}$ and the characteristic function can be encoded using a BDD in the usual way. To use BDDs when $\mathbb{N}_k \neq \mathbb{B}$, it becomes necessary to derive some encoding of the state space into Boolean variables. This process must be performed with care since it can have a dramatic effect on the efficiency of the representation. We will return to this issue later. Alternatively, MDDs can encode the characteristic function in a straightforward way as long as each set $\mathbb{N}_k$ is finite.

Example. We now introduce a small example which will be reused in subsequent sections. Consider an extremely simple system consisting of three cooperating processes, P_3, P_2 and P_1, each of which can be in one of four local states, 0, 1, 2 or 3. The (global) state space $\mathcal{S}$ is $\{0, 1, 2, 3\} \times \{0, 1, 2, 3\} \times \{0, 1, 2, 3\}$.

To represent sets of states using MDDs, we can use $K = 3$ state variables, x_3, x_2 and x_1, with $N_3 = N_2 = N_1 = 4$. To represent sets of states using BDDs, we must adopt a Boolean encoding. In this simple example, we can allocate two Boolean variables to represent each local state, and use the standard binary encoding for the integers $\{0, 1, 2, 3\}$. Hence, we need $K = 6$ variables, where x_{2n}, x_{2n-1} represent the state of process P_n.

Let us suppose that when a process in local state 1, it is modifying shared data and requires exclusive access to do so safely. Consider the set of states $\mathcal{S}'$, in which at most one process is in local state 1. The MDD and BDD representing $\mathcal{S}'$ can be seen in Figure 4(b) and Figure 1(c), respectively.

Manipulating and Generating Sets of States. Basic manipulation of state sets can easily be translated into BDD or MDD operations. For example, the union and intersection of two sets, $\mathcal{S}_1$ and $\mathcal{S}_2$, represented by BDDs, S_1 and S_2, can be computed using $\mathsf{Apply}(\vee, S_1, S_2)$ and $\mathsf{Apply}(\wedge, S_1, S_2)$, respectively. Similarly, sets $\mathcal{S}_1$ and $\mathcal{S}_2$, represented by MDDs S_1 and S_2, can be manipulated using Case; for example, $\mathcal{S}_1 \setminus \mathcal{S}_2$ and $\mathcal{S}_1 \cup \mathcal{S}_2$ can be computed using $\mathsf{Case}(S_2, S_1, 0)$ and $\mathsf{Case}(S_1, S_2, 1)$, respectively.

Of particular interest are the operations required to compute the BDD or MDD representing the set of reachable states of a model. To do so effectively requires a *next-state* function, which reports the states that are reachable from a given state in a single step. The next-state function must be represented in a format that is suitable for BDD and MDD manipulation algorithms.

One approach is to represent the next-state function using another BDD or MDD. Since the next-state function is a relation R over pairs of states from a set $\mathcal{S}$, a BDD or MDD can encode its characteristic function $\chi_R : \mathcal{S} \times \mathcal{S} \to \{0, 1\}$ where $\chi_R(\mathbf{s}, \mathbf{s}') = 1$ if and only if $(\mathbf{s}, \mathbf{s}') \in R$ for all $\mathbf{s}, \mathbf{s}' \in \mathcal{S}$. Note that this BDD or MDD R requires two sets of variables, $\{x_K, \ldots, x_1\}$ and $\{y_K, \ldots, y_1\}$, where $f_R(x_K, \ldots, x_1, y_K, \ldots, y_1) = \chi_R(\mathbf{s}, \mathbf{s}')$; variables $\{x_K, \ldots, x_1\}$ represent the "source" states, while variables $\{y_K, \ldots, y_1\}$ represent the "target" states.

Given a BDD or MDD S' in variables $\{x_K, \ldots, x_1\}$, representing a set of states $\mathcal{S}'$, the BDD or MDD S'' for the set of states $\mathcal{S}''$ reachable in exactly one step from any state in $\mathcal{S}'$ can be computed relatively easily, a process sometimes known as *image computation*. With BDDs, for example, we can use the following:

$$\mathsf{S}'' = \exists\{x_K \ldots, x_1\}(\mathsf{S}' \wedge \mathsf{R})$$

where $\exists x_k.\mathsf{B} = \mathsf{B}|_{x_k=0} \vee \mathsf{B}|_{x_k=1}$ and $\exists\{x_K, \ldots, x_1\}.\mathsf{B} = \exists x_K \ldots \exists x_1.\mathsf{B}$. Note that S'', the resulting BDD, will be in variables $\{y_K, \ldots, y_1\}$ but can easily be converted back to $\{x_K, \ldots, x_1\}$ if required.

Returning to our running example from the previous section, suppose that a process can asynchronously change from local states 1 to 2, from 2 to 3, or from 3 to 0. Processes can also change from local state 0 to 1, but only if no other process is already in its local state 1. The portion of the next-state function that characterizes the possible local state changes of process P_3, represented by a BDD over variables $\{x_6, \ldots, x_1\}, \{y_6, \ldots, y_1\}$, is shown in Figure 5(a), where terminal node 0 and all of its incoming arcs are omitted for clarity. The overall next-state function can be obtained by constructing similar BDDs for processes P_2 and P_1, and combining them using $\mathsf{Apply}(\vee, \cdot, \cdot)$. Note that the variable ordering of the BDD in Figure 5(a) interleaves the variables for "source" and "target" states. This is a well-known heuristic for reducing the size of BDDs which represent transition relations [36].

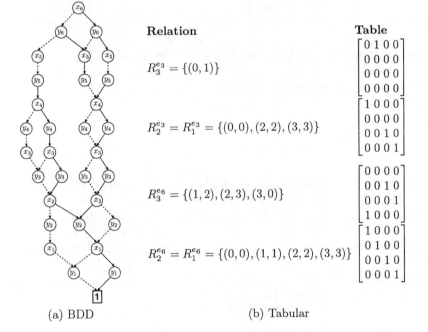

(a) BDD   (b) Tabular

Fig. 5. Next-state function representations

Another approach is to decompose the next-state function and represent its parts using tables. In practice, models are usually described using some high-level formalism; this approach requires to represent a separate next-state function for each model "event" (e.g. a Petri net transition) [60]. This implies that the next-state relation R can be expressed as $\chi_R(\mathbf{s}, \mathbf{s}') = \bigvee_{e \in \mathcal{E}} \chi_{R^e}(\mathbf{s}, \mathbf{s}')$, where R^e represents the next-state relation due to event e, and $\mathcal{E}$ is the (finite) set of all possible events. To represent the relation R^e efficiently, it is required that R^e can be expressed in a "product-form"

$$\chi_{R^e}((x_K, \ldots, x_1), (y_K, \ldots, y_1)) = \chi_{R_K^e}(x_K, y_K) \wedge \cdots \wedge \chi_{R_1^e}(x_1, y_1)$$

of "local" relations R_k^e. Each relation R_k^e can be stored using a Boolean table of size $N_k \times N_k$ (or in a sparse format).

Returning again to the running example, the next-state relation can be divided into events $e_1, \ldots, e_6$, where events e_k correspond to process P_k changing from local state 0 to 1, and events e_{3+k} correspond to process P_k changing from local state 1 to 2, 2 to 3, or 3 to 0. The local relations required to represent R^{e_3} and R^{e_6}, and their corresponding Boolean tables, are shown in Figure 5(b). Representations for relations R^{e_1} and R^{e_2} are similar to R^{e_3}, and the representations for R^{e_4} and R^{e_5} are similar to R^{e_6}.

Techniques based on tabular representation of the next-state function are typically applied to MDD-based storage of sets, rather than BDD-based storage (often the product-form requirement does not allow decomposition of the next-state function into Boolean components). Given an MDD S' representing a set of states $\mathcal{S}'$, and the "product-form" representation for relation R^e, the set of states reachable in exactly one step via event e from any state in $\mathcal{S}'$ can be determined using recursive algorithm Next, shown in Figure 6. Note that the algorithm should be invoked with $k = K$ at the top level. The overall set of states reachable in one step from $\mathcal{S}'$ can be determined by calling Next once for each event and taking the union of the obtained sets.

Using either of the two approaches outlined above for determining the states reachable in one step from a given set of states, it is relatively trivial to compute the set of all reachable states of a model. Starting with the BDD or MDD for some set of initial states, we iteratively compute all the states reachable in one step from the set of reachable states already discovered. These newly explored states are then added to the set of discovered states using set union. The algorithm can be terminated when the latter set ceases to increase.

This process equates to performing a breadth-first search of the model's state space. BDDs and MDDs are often well suited to this purpose for several reasons. Firstly, they can be extremely good at storing and manipulating *sets* of states, and exploiting regularity in these sets. Algorithms which require manipulation of *individual* states, such as depth-first search are less well suited. Secondly, checking for convergence of the algorithm, which requires verification that two sets are identical, reduces to testing two BDDs or MDDs for equality, a process which is efficient due to their canonicity.

It is important to emphasize that the state space generation techniques presented here are actually quite simplistic. Since the bulk of this paper focuses on issues specific to *probabilistic* models, we provide only an introductory presentation to this area. A wide range of much more sophisticated approaches,

---

Next(in: k, R^e, S')
- R^e is the product of local relations, $R^e_K \times \cdots \times R^e_1$.
- Returns states reachable from S' in one step via event e.
1: **if** $S' = 0$ **then**
2:    Return 0;
3: **else if** $k = 0$ **then**
4:    Return 1;
5: **else if** computed-table has entry (k, R^e, S', H) **then**
6:    Return H;
7: **end if**
8: H $\leftarrow$ new non-terminal node with label x_k;
9: **for each** $i \in \mathbb{N}_k$ **do**
10:    $child(H, i) \leftarrow 0$;
11: **end for**
12: **for each** $(i, j) \in R^e_k$ **do**
13:    D $\leftarrow$ Next$(k-1, R^e, S'|_{x_k=i})$;
14:    $child(H, j) \leftarrow$ Case$(child(H, j), D, 1)$;
15: **end for**
16: Reduce(H);
17: Add entry (k, R^e, S', H) to computed-table;
18: Return H;

Fig. 6. Image computation using a decomposed next-state relation

particularly those based on BDDs, have been presented in the literature. For instance, more sophisticated BDD-based state space generation techniques for Petri net models are described in [63, 64, 32]. Furthermore, BDDs have proved extremely successful in more complex applications than simple state space generation, such as in formal verification [19, 30]. The well-known (non-probabilistic) symbolic model checker SMV [56], for example, is based on BDD technology.

MDD-based approaches have also proved very successful in practice and, again, a number of enhanced techniques have been developed. Examples are those that exploit the concept of *event locality*: the fact that some events will only affect certain state variables [23, 57, 20]. In particular, [21] introduces the concept of *node saturation*: when a new MDD node is created corresponding to submodel k, it is immediately *saturated* by repeatedly applying all events that affect only submodels k through 1. Recently, more flexible next-state representations have been investigated that allow MDD-based reachability set generation algorithms to handle complex priority structure and immediate events [59], allowing for elimination of vanishing states both on-the-fly and after generation.

2.4   Indexing and Enumerating States

Once the set of reachable states of a model has been determined, it may be used for a variety of purposes. Of particular interest in the case of probabilistic models is the fact that it is usually required when performing numerical computation to analyze the model. Three operations that are commonly required are: (a) determining if a particular state is reachable or not; (b) enumerating all reachable states in order; and (c) determining the *index* of a given state in the set of all states. The latter, for example, is needed when a vector of numerical values, one for each state, is stored explicitly in an array and the index of a state is needed to access its corresponding value.

To perform these three operations efficiently, the BDD or MDD representation of the state space needs to be augmented with additional information. This process is described in [24, 57] for MDDs and in [62] for BDDs. In this section, we will describe how it can be done using the quasi-reduced variants of the data structures (i.e. QROBDDs and QROMDDs). Similar techniques can also be applied to fully reduced decision diagrams, but these require more complex algorithms to correctly handle cases where levels are "skipped" due to redundant nodes. In the following, we will assume that S denotes the set of all possible states of the model and that $S' \subseteq S$ is the set of all reachable states.

Given a K-variable BDD or MDD representation for S', determining if an arbitrary state $(s_K, \ldots, s_1)$ is contained in the set S' can be done by traversing the data structure, as described earlier: if the given variable assignments produce a path leading to terminal node 1, then the state is contained in the set. Note that, if each node has direct access to the downward pointers (i.e., an arbitrary child can be found in constant time), then the computational cost to determine if a state is reachable is exactly $\mathcal{O}(K)$ for reachable states, and at worst $\mathcal{O}(K)$ for unreachable states.

In order to enumerate the set of states S', one possibility is to go through all the states in S, and for each one, determine if it is in the set S' as described above. However, this has a computational cost of $\mathcal{O}(K \cdot |S|)$, which is often unacceptable since it is possible for the set S to be several orders of magnitude larger than S'. A more efficient approach is to follow all paths in the BDD or MDD for S' that can lead to terminal node 1. That is, for each node, we visit all children except those whose paths lead only to terminal node 0. Note that nodes whose paths lead only to terminal node 0 represent the constant function 0, and thus can be detected (and passed over) thanks to the canonicity property.

Figure 7(a) shows the recursive algorithm Enumerate, which enumerates all the states represented by QROMDD S', as just described. For the enumeration to be efficient, we must be able to visit only the non-zero children of each node; this can be done quickly if sparse storage is used for the children in each node, or if a list of the non-zero children is kept for each node. Assuming constant time access to the non-zero children (i.e. the ability to find the first or next non-zero child in constant time), it can be shown that algorithm Enumerate has a computational cost of $\mathcal{O}(|S'|)$ in the best case (when each non-terminal node has several non-zero children) and $\mathcal{O}(K \cdot |S'|)$ in the worst case (when most non-terminal nodes have a single non-zero child).

Note that, if the values s_k are selected in order in line 5 of Figure 7(a), then the states will be visited in lexicographical order. An example QROMDD, corresponding to the set of reachable states with process 2 in its local state 2 for our running example, is shown in Figure 7(b). Nodes corresponding to the

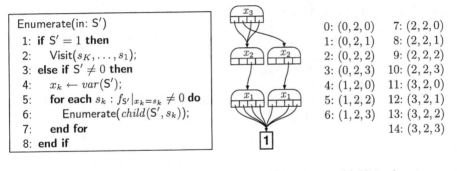

(a) Enumeration algorithm   (b) QROMDD   (c) Visited states

Fig. 7. Enumerating all states in a QROMDD-encoded set

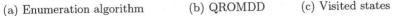

constant function 0 are omitted from the figure for clarity. The states visited by Enumerate for this QROMDD are shown in Figure 7(c), where the integer to the left of the state indicates the order in which states are visited.

For numerical solution, it is often necessary to assign a unique index to each state using some one-to-one indexing function $\psi : \mathcal{S}' \to \{0, \ldots, |\mathcal{S}'| - 1\}$. A commonly-used function ψ is to assign indices in lexicographical order; i.e., $\psi(\mathbf{s})$ is the number of states in $\mathcal{S}'$ that precede $\mathbf{s}$ in lexicographical order. Using an MDD to encode ψ will lead to a rather large MDD, one with $|\mathcal{S}'|$ terminal nodes (this is essentially the multi-level structure described in [23]). An alternative is to use an edge-valued MDD (EV$^+$MDD), a data structure described in [25], to represent the function ψ. An EV$^+$MDD is an MDD where each edge in the MDD from a non-terminal node m to child c has a corresponding value, $value(\mathrm{m}, c)$. The edge values are summed along a path to obtain the function value. With certain restrictions, EV$^+$MDDs are also a canonical representation for functions [25]. The edge values are sometimes referred to as *offsets* [24, 57, 62].

Given a QROMDD representing a set $\mathcal{S}'$, the EV$^+$MDD for ψ, corresponding to the lexicographical indexing function for states, can be constructed by assigning appropriate edge values in a bottom-up fashion, using algorithm BuildOffsets shown in Figure 8. The algorithm is based on the property that the index of a state is equal to the number of paths "before" the path of that state (in lexicographical order), which can be counted by enumerating the paths. Indeed, algorithm BuildOffsets is quite similar to algorithm Enumerate, except that once the paths from a given node to terminal node 1 have been counted, they do not need to be counted again. A computed-table is used to keep track of the number of paths from each non-terminal node to node 1. For the QROMDD shown in Figure 7(b), the resulting MDD with offset values (i.e., the EV$^+$MDD) produced by algorithm Enumerate is shown in Figure 9(a), where the state indices match those shown in Figure 7(c). For instance, the index for state $(3, 2, 1)$ is found by following the appropriate path through the EV$^+$MDD and summing the edge values $11 + 0 + 1 = 12$. Note that the edge values are unnecessary for children corresponding to the constant function 0, since it is impossible to reach terminal node 1 following these paths. As such, edge values for these children can be set

```
BuildOffsets(in: S')
 • Returns the number of paths from S' to terminal 1.
 1: if S' = 0 then
 2:    Return 0;
 3: else if S' = 1 then
 4:    Return 1;
 5: else if computed-table contains entry (S', n) then
 6:    Return n;
 7: end if
 8: x_k ← var(S');
 9: n ← 0;
10: for each s_k : f_{S'}|_{x_k=s_k} ≠ 0 do
11:    value(S', s_k) ← n;
12:    n ← n + BuildOffsets(child(S', s_k));
13: end for
14: Add (S', n) to computed-table;
15: Return n;
```

Fig. 8. Constructing "offset" edge-values

to any value, including "undefined", since the values will never be used; these are represented by dashes in Figure 9(a). Note that the pointers in the bottom-level nodes become unnecessary in this case: if the edge value is undefined, then the child must be terminal node 0, otherwise the child is terminal node 1. Other implementation tricks, including how to combine offsets with direct access and non-zero-only access, are described in [24, 57].

The QROBDD version of the set S' and indexing function ψ is shown in Figure 9(b). In this case, the MDD-encoded state $(3, 2, 1)$ is encoded as the binary state $(1, 1, 1, 0, 0, 1)$, and the state index is given by $7+4+0+0+0+1 = 12$. However, in the special case of BDDs, a simplification is possible. Since child 0 will always have either an edge value of zero or an undefined edge value (this is true for both MDDs and BDDs), the edge value for the *else* child in a BDD does not need to be explicitly stored in the node. Instead, it is sufficient to store only the edge value for the *then* child. This produces the QROBDD shown in Figure 9(c), where the label for each non-terminal node is the edge value for the *then* child (the variable for each node is shown to the left, for all nodes with the same variable). Note that the dashed values of Figure 9(b) have been changed to zeroes in Figure 9(c) by choice [62]. In this case, when tracing a path through the BDD corresponding to a particular state, the index can be computed by summing the offsets on nodes from which the *then* child was taken. For the example, the index for state $(1, 1, 1, 0, 0, 1)$ is given by $7 + 4 + 0 + 1 = 12$.

3   Model Representation

In the previous section, we have seen ways of storing and manipulating sets of states of a probabilistic model, in particular, the set of all of its reachable states.

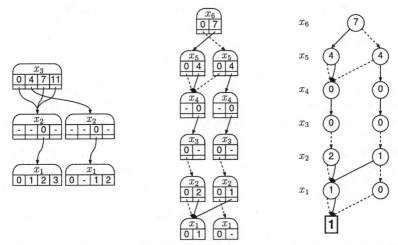

(a) MDD with offset values (b) BDD with offset values (c) Offset-labeled BDD

Fig. 9. Decision diagrams with offset information

The latter can provide some useful information about the model, for example, whether or not it is possible to reach a state which constitutes an error or failure. Typically, though, we are interested in more involved properties, for example, the *probability* of reaching such a state. Studying properties of this kind will generally require numerical calculation, which needs access to the model itself, i.e. not just the set of reachable states, but the transitions between these states and the probabilistic information assigned to them.

In this section, we concern ourselves with the storage of the model. For two of the most common types of probabilistic models, DTMCs and CTMCs, this representation takes the form of a real-valued square matrix. In the following sections we will see how symbolic data structures such as BDDs and MDDs can be extended to achieve this task. We will also mention extensions for more complex models such as MDPs. The two principal data structures covered are *multi-terminal binary decision diagrams* (MTBDDs) and *matrix diagrams*. The latter is based on the *Kronecker representation*. We also provide an introduction to this area. In addition, we will discuss two further data structures: decision node binary decision diagrams (DNBDDs) and probabilistic decision graphs (PDGs).

3.1   Multi-Terminal Binary Decision Diagrams (MTBDDs)

Multi-terminal binary decision diagrams (MTBDDs) are an extension of BDDs. The difference is that terminal nodes in an MTBDD can be labeled with arbitrary values, rather than just 0 and 1 as in a BDD. In typical usage, these values are real (or in practice, floating point). Hence, an MTBDD M over K Boolean variables $x_K, \ldots, x_1$ now represents a function of the form $f_M : \mathbb{B}^K \to \mathbb{R}$.

The basic idea behind MTBDDs was originally presented by Clarke et al. in [28]. They were developed further, independently, by Clarke et al. [27] and

Bahar et al. [5], although in the latter they were christened algebraic decision diagrams (ADDs).

These papers also proposed the idea of using MTBDDs to represent vectors and matrices. Consider a real-valued vector $\underline{v}$ of size 2^K. This can be represented by a mapping from integer indices to the reals, i.e. $f_{\underline{v}} : \{0, \ldots, 2^K - 1\} \to \mathbb{R}$. Given a suitable encoding of these integer indices into K Boolean variables, this can instead be expressed as a function of the form $f_V : \mathbb{B}^K \to \mathbb{R}$, which is exactly what is represented by an MTBDD V over K Boolean variables. Similarly, a real-valued matrix $\mathbf{M}$ of size $2^K \times 2^K$ can be interpreted as a mapping from pairs of integer indices to the reals, i.e. $f_M : \{0, \ldots, 2^K - 1\} \times \{0, \ldots, 2^K - 1\} \to \mathbb{R}$, and hence also as $f_M : \mathbb{B}^K \times \mathbb{B}^K \to \mathbb{R}$, which can represented by an MTBDD M over $2K$ Boolean variables, K of which encode row indices and K of which encode column indices.

Figure 10 shows an example of an MTBDD R which represents a 4×4 matrix $\mathbf{R}$. Since we are introducing MTBDDs in the context of a representation for probabilistic models, the matrix $\mathbf{R}$ is actually the transition rate matrix for a 4 state CTMC. This CTMC is also shown in Figure 10. Note that structure of an MTBDD is identical to a BDD, except for the presence of multiple terminal nodes labeled with real values. The function f_R represented by the MTBDD R can also be read off in identical fashion to a BDD. This process is illustrated by the rightmost five columns of the table in Figure 10.

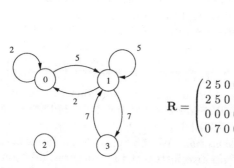
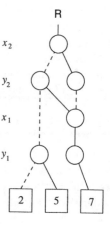

$$\mathbf{R} = \begin{pmatrix} 2 & 5 & 0 & 0 \\ 2 & 5 & 0 & 7 \\ 0 & 0 & 0 & 0 \\ 0 & 7 & 0 & 0 \end{pmatrix}$$

Entry in $\mathbf{R}$	x_2 x_1	y_2 y_1	x_2 y_2 x_1 y_1	f_R
$(0,0) = 2$	0 0	0 0	0 0 0 0	2
$(0,1) = 5$	0 0	0 1	0 0 0 1	5
$(1,0) = 2$	0 1	0 0	0 0 1 0	2
$(1,1) = 5$	0 1	0 1	0 0 1 1	5
$(1,3) = 7$	0 1	1 1	0 1 1 1	7
$(3,1) = 7$	1 1	0 1	1 0 1 1	7

Fig. 10. A CTMC, its transition rate matrix $\mathbf{R}$ and an MTBDD R representing it

The table also demonstrates how the matrix $\mathbf{R}$ is represented by the MTBDD. We use four Boolean variables: x_2 and x_1, which are used to encode row indices; and y_2 and y_1, which are used to encode column indices. In both cases, we have used the standard binary encoding of integers. For the matrix entry $(3, 1)$, for example, $x_2 = 1$ and $x_1 = 1$ encode the row index 3 and $y_2 = 0$ and $y_1 = 1$ encode the column index 1. Note that, in the MTBDD, the variables for rows and columns are interleaved, i.e. $x_2 > y_2 > x_1 > y_1$. This is a well-known heuristic for MTBDDs which represent transition matrices (or analogously, as mentioned previously in Section 2.3, for BDDs which represent transition relations) and typically gives a significant improvement in the size of the data structure. Therefore, to establish the value of entry $(3, 1)$ in $\mathbf{R}$, we trace the path $1, 0, 1, 1$ through R and determine that it is equal to 7.

MTBDD Operations. Operations on MTBDDs can be defined and implemented on MTBDDs in a similar way to BDDs. The Apply operator, for example, extends naturally to this data structure. In this case, the operation applied can be a real-valued function, such as addition or multiplication, for example Apply$(+, \mathsf{M}_1, \mathsf{M}_2)$ produces the MTBDD representing the function $f_{\mathsf{M}_2} + f_{\mathsf{M}_2}$. The MTBDD version of Apply can be implemented in essentially identical fashion to the BDD version, based on a recursive descent of the data structures and using a computed-table to cache intermediate results. The only difference is the handling of the terminal cases. Its time complexity is $\mathcal{O}(|\mathsf{M}_1| \cdot |\mathsf{M}_2|)$, where $|\mathsf{M}|$ denotes the number of nodes in MTBDD M.

   We are particularly interested in operations which can be applied to MTBDDs representing matrices and vectors. The Apply operation can be used to perform point-wise operations such as addition of two matrices or scalar multiplication of a matrix by a real constant. Of more interest are operations specifically tailored to matrices and vectors. The most obvious example is the matrix-vector or matrix-matrix multiplication operation. Crucially, because matrix-based multiplications can be expressed in a recursive fashion, they can be efficiently implemented using a similar approach as for the Apply operator. Three slight variants have been presented in [28, 27, 5]. In [5], empirical results are presented to compare the efficiency of the three algorithms.

Model Representation with MTBDDs. Some classes of probabilistic models, such as DTMCs and CTMCs, are described simply by a real valued matrix and can hence be represented by an MTBDD using the techniques described above. These observations have been made in numerous places, e.g. [39, 40, 8, 44]. We have already seen an example of the representation for a CTMC in Figure 10. In addition, we describe here how a third type of model, Markov decision processes (MDPs), can be represented as MTBDDs. This was initially proposed in [7, 6], the first concrete implementation of the idea was presented in [34] and the ideas have since been extended in [62].

   The chief difference between MDPs and DTMCs or CTMCs is the presence of nondeterminism. A DTMC or CTMC is specified by the likelihood of making a transition from each state to any other state. This is a real value, a discrete probability for a DTMC or a parameter of an exponential distribution for a CTMC. In either case, the values for all pairs of states can be represented by a square, real-valued matrix. In an MDP, each state contains several nondeterministic choices, each of which specifies a discrete probability for every state. In terms of a matrix, this equates to each state being represented by several different rows. This can be thought of either as a non-square matrix or as a three-dimensional matrix. In any case, we can treat non-determinism as a third index, meaning that an MDP is effectively a function of the form $f : S \times \{1, \ldots, n\} \times S \rightarrow [0, 1]$, where S is the set of states and n is the maximum number of nondeterministic choices in any state. By encoding the set $\{1, \ldots, n\}$ with Boolean variables, we can store the MDP as an MTBDD over three sets of variables, one for source states, one for destination states and one to distinguish between nondeterministic choices.

Model Generation with MTBDDs. The process of generating the MTBDD which represents a given probabilistic model is particularly important as it can have a huge impact on the efficiency of the MTBDD as a representation. This is an issue considered in [44], which presents a number of heuristics for this purpose. The most important of these is that one should try to exploit *structure* and *regularity*. In practice, a probabilistic model will be described in some high-level specification formalism, such as a stochastic process algebra, stochastic Petri nets or some other, custom language. Typically, this high-level description is inherently structured. Therefore, the most efficient way to construct the MTBDD representing the model is via a direct translation from the high-level description. This is demonstrated in [44] on some simple examples using formalisms such as process algebras and queuing networks. These findings have been confirmed by others, e.g. [34, 49, 62, 43] on a range of formalisms. Like BDDs, attention should also be paid to the ordering of the Boolean variables in the MTBDD. Discussions can be found in [44, 62, 43].

Direct translation of a model from its high-level description usually results in the introduction of unreachable states. This necessitates the process of reachability: computing all reachable states. This can be done with BDDs, as discussed in Section 2, and then the unreachable states can be removed from the MTBDD with a simple Apply operation. This process is facilitated by the close relation between BDDs and MTBDDs.

In practice, it has been shown that MTBDDs can be used to construct and store extremely large probabilistic models [39, 44, 34]. This is possible by exploitation of high-level structure in the description of the model. Often, the construction process itself is also found to be efficient. This is partly due to the fact that the symbolic representation is constructed directly from the high-level description and partly because the process can integrate existing efficient techniques for BDD-based reachability.

3.2   Kronecker Algebra

A well-accepted compact representation for certain types of models (usually CTMCs) is based on Kronecker algebra. Relevant well-known properties of Kronecker products are reviewed here; details can be found (for example) in [33]. The Kronecker product of two matrices multiplies every element of the first matrix by every element of the second matrix. Given square matrices $\mathbf{M}_2$ of dimension N_2 and $\mathbf{M}_1$ of dimension N_1, their Kronecker product

$$
\mathbf{M}_2 \otimes \mathbf{M}_1 =
\begin{bmatrix}
\mathbf{M}_2[0,0] \cdot \mathbf{M}_1 & \cdots & \mathbf{M}_2[0, N_2 - 1] \cdot \mathbf{M}_1 \\
\vdots & \ddots & \vdots \\
\mathbf{M}_2[N_2 - 1, 0] \cdot \mathbf{M}_1 & \cdots & \mathbf{M}_2[N_2 - 1, N_2 - 1] \cdot \mathbf{M}_1
\end{bmatrix}
$$

is a square matrix of dimension $N_2 N_1$. This can be generalized to the Kronecker product of K matrices $\mathbf{M} = \mathbf{M}_K \otimes \cdots \otimes \mathbf{M}_1$ since the Kronecker product is associative. In this case, the square matrix $\mathbf{M}$ has dimension $\prod_{k=1}^{K} N_k$, where N_k is the dimension of square matrix $\mathbf{M}_k$.

Kronecker algebra has been fairly successful in representing large CTMCs generated from various high-level formalisms [16, 35, 37, 46, 65]. The key idea behind Kronecker-based approaches is to represent the portion of the transition rate matrix for the CTMC due to a single event e, denoted as $\mathbf{R}^e$, as the Kronecker product of matrices $\mathbf{W}_k^e$, which describe the contribution to event e due to each model component k. The overall transition rate matrix is the sum of each $\mathbf{R}^e$ over all events e. Note that each matrix $\mathbf{W}_k^e$ is a square matrix of dimension N_k, and the dimension of the represented matrix is $N_K \cdots N_1$. Thus, just like MDDs, to use a Kronecker representation, a model state must be expressible as a collection of K state variables $(s_K, \ldots, s_1)$, where $s_k \in \mathbb{N}_k$. Unless matrix elements are allowed to be functions [37], the state variables and events must be chosen so that the rate of transition from state $(s_K, \ldots, s_1)$ to state $(s'_K, \ldots, s'_1)$ due to event e can be expressed as the product

$$
\lambda^e((s_K, \ldots, s_1), (s'_K, \ldots, s'_1)) = \lambda_K^e(s_K, s'_K) \cdots \lambda_1^e(s_1, s'_1)
$$

which means that the transition rates cannot arbitrarily depend on the global state. When the above product-form requirement is met, all matrix elements are constants: $\mathbf{W}_k^e[s_k, s'_k] = \lambda_k^e(s_k, s'_k)$.

We now return to the running example. Let us assume that the model is a CTMC, i.e. each transition of the model is associated with a rate. Let the rate of transition from local state 2 to local state 3 be 2.3, from local state 3 to local state 0 be 3.0, from local state 0 to local state 1 be $5.0 + 0.1 \cdot k$ for process k, and from local state 1 to local state 2 be $1.5 + 0.1 \cdot k$ for process k. The corresponding Kronecker matrices for $\mathbf{R}^{e_3}$ and $\mathbf{R}^{e_6}$ are shown in Figure 11. For this example, according to $\mathbf{R}^{e_6}$, the rate of transition from state $(1, 2, 1)$ to state $(2, 2, 1)$ is $1.8 \cdot 1.0 \cdot 1.0 = 1.8$. Note that the matrices $\mathbf{W}_2^{e_6}$ and $\mathbf{W}_1^{e_6}$ are identity matrices; this occurs whenever an event does not affect, and is not affected by, a particular component. Also note the similarities between the matrices of Figure 11 and the tables in Figure 5(b). The tabular next-state representation is in fact a Kronecker representation, where the table for relation R_k^e corresponds to a Boolean matrix $\mathbf{W}_k^e$.

$$\mathbf{W}_3^{e_3} = \begin{bmatrix} 0 & 5.3 & 0 & 0 \\ 0 & 0 & 0 & 0 \\ 0 & 0 & 0 & 0 \\ 0 & 0 & 0 & 0 \end{bmatrix} \quad \mathbf{W}_2^{e_3} = \begin{bmatrix} 1 & 0 & 0 & 0 \\ 0 & 0 & 0 & 0 \\ 0 & 0 & 1 & 0 \\ 0 & 0 & 0 & 1 \end{bmatrix} \quad \mathbf{W}_1^{e_3} = \begin{bmatrix} 1 & 0 & 0 & 0 \\ 0 & 0 & 0 & 0 \\ 0 & 0 & 1 & 0 \\ 0 & 0 & 0 & 1 \end{bmatrix}$$

$$\mathbf{W}_3^{e_6} = \begin{bmatrix} 0 & 0 & 0 & 0 \\ 0 & 0 & 1.8 & 0 \\ 0 & 0 & 0 & 2.3 \\ 3.0 & 0 & 0 & 0 \end{bmatrix} \quad \mathbf{W}_2^{e_6} = \begin{bmatrix} 1 & 0 & 0 & 0 \\ 0 & 1 & 0 & 0 \\ 0 & 0 & 1 & 0 \\ 0 & 0 & 0 & 1 \end{bmatrix} \quad \mathbf{W}_1^{e_6} = \begin{bmatrix} 1 & 0 & 0 & 0 \\ 0 & 1 & 0 & 0 \\ 0 & 0 & 1 & 0 \\ 0 & 0 & 0 & 1 \end{bmatrix}$$

Fig. 11. Kronecker matrices for the running example

For in-depth discussion of Kronecker representations and algorithms for efficient numerical solution of Markov chains expressed using Kronecker algebra, see for example [15, 18, 69].

3.3 Matrix Diagrams

Matrix diagrams (abbreviated as MDs or MxDs) are, like BDDs, MDDs and MTBDDs, rooted directed acyclic graphs. A matrix diagram is associated with K pairs of variables, $(x_K, y_K), \ldots, (x_1, y_1)$, and a matrix diagram M represents a matrix (or function) $f_M : (\mathbb{N}_K \times \cdots \times \mathbb{N}_1)^2 \to \mathbb{R}$. Note that both variables in a pair have the same possible set of values: $x_k, y_k \in \mathbb{N}_k$. Matrix diagrams consist of terminal nodes labeled with 1 and 0, and non-terminal nodes m labeled with variable pairs $var(\mathsf{m}) \in \{(x_K, y_K), \ldots, (x_1, y_1)\}$. A non-terminal node with label (x_k, y_k) is sometimes called a *level-k* node. According to the original definition [24], for a given node and pair of variable assignments, there is an associated *set* of children nodes, and each child node has a corresponding real edge value. We denote this as a set of pairs, $pairs(\mathsf{m}, i, j) \subset \mathbb{R} \times \mathcal{M}$, where (i, j) are the variable assignments for $var(\mathsf{m})$, and $\mathcal{M}$ is the set of matrix diagram nodes. If a set contains more than one pair, then a given group of variable assignments can lead to multiple paths through the matrix diagram. The value of a matrix element corresponding to the appropriate variable assignments is found by taking the product of the real values encountered along each path, and summing those products over all paths.

Like decision diagrams, the concept of ordering can be applied to obtain ordered MDs (OMDs). Two non-terminal matrix diagram nodes m_1 and m_2 are duplicates if $var(\mathsf{m}_1) = var(\mathsf{m}_2)$ and if $pairs(\mathsf{m}_1, i, j) = pairs(\mathsf{m}_2, i, j)$, for all possible $i, j \in \mathbb{N}_{var(\mathsf{m}_1)}$. The usual notion of quasi-reduction can be applied (using the above definition for duplicates) to obtain quasi-reduced OMDs (QROMDs). An element of the matrix encoded by a QROMD m can be computed using the recurrence:

$$f_\mathsf{m}(x_k, \ldots, x_1, y_k, \ldots, y_1) = \sum_{\forall (r, \mathsf{m}') \in pairs(\mathsf{m}, x_k, y_k)} r \cdot f_{\mathsf{m}'}(x_{k-1}, \ldots, x_1, y_{k-1}, \ldots, y_1)$$

with terminating conditions $f_1 = 1$ and $f_0 = 0$. However, QROMDs are *not* a canonical representation. Additional rules can be applied to matrix diagrams to obtain a canonical form, as described in [58]: sets of pairs are eliminated, so that each group of variable assignments corresponds to exactly one path through the matrix diagram, and the possible edge values are restricted. For the remainder of the paper, we assume that all matrix diagrams are QROMDs (with sets of pairs).

A simple example of a matrix diagram is shown in Figure 12(a). Each matrix element within a non-terminal node is either the empty set (represented by blank space) or a set of pairs (represented by stacked boxes). At the bottom level, the only possible child is terminal node 1 (pairs with terminal 0 children or zero edge values can be removed); thus, the pointers can be omitted. The matrix encoded by the matrix diagram is shown in Figure 12(b). The large blocks of zeroes are due to the blank spaces in the (x_3, y_3) matrix diagram node.

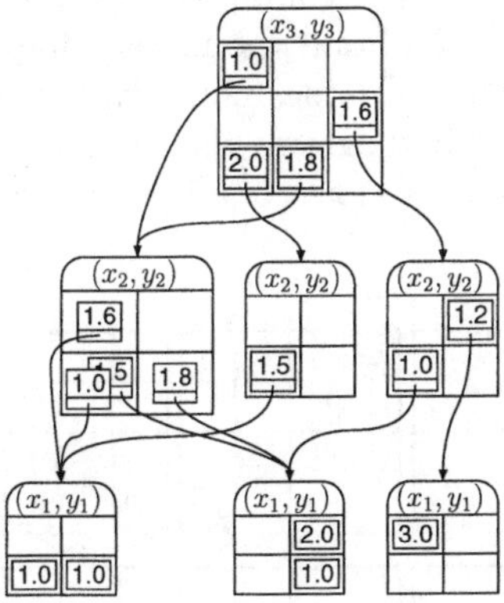

$$
\begin{bmatrix}
0 & 0 & 0 & 0 & 0 & 0 & 0 & 0 & 0 & 0 & 0 & 0 \\
1.6 & 1.6 & 0 & 0 & 0 & 0 & 0 & 0 & 0 & 0 & 0 & 0 \\
0 & 3.0 & 0 & 3.6 & 0 & 0 & 0 & 0 & 0 & 0 & 0 & 0 \\
1.0 & 2.5 & 0 & 1.8 & 0 & 0 & 0 & 0 & 0 & 0 & 0 & 0 \\
0 & 0 & 0 & 0 & 0 & 0 & 0 & 0 & 0 & 0 & 5.7 & 0 \\
0 & 0 & 0 & 0 & 0 & 0 & 0 & 0 & 0 & 0 & 0 & 0 \\
0 & 0 & 0 & 0 & 0 & 0 & 0 & 0 & 0 & 3.2 & 0 & 0 \\
0 & 0 & 0 & 0 & 0 & 0 & 0 & 0 & 0 & 1.6 & 0 & 0 \\
0 & 0 & 0 & 0 & 0 & 0 & 0 & 0 & 0 & 0 & 0 & 0 \\
0 & 0 & 0 & 0 & 2.8 & 2.8 & 0 & 0 & 0 & 0 & 0 & 0 \\
0 & 0 & 0 & 0 & 0 & 5.4 & 0 & 6.4 & 0 & 0 & 0 & 0 \\
3.0 & 3.0 & 0 & 0 & 1.8 & 4.5 & 0 & 3.2 & 0 & 0 & 0 & 0
\end{bmatrix}
$$

(a) Matrix diagram   (b) Encoded matrix

Fig. 12. Example matrix diagram

Matrix Diagram Operations. In this section, we describe three useful operations for matrix diagrams: constructing a matrix diagram from a Kronecker product, addition of matrix diagrams and selection of a submatrix from a matrix diagram.

Building a Kronecker product: Given matrices $\mathbf{M}_K, \ldots, \mathbf{M}_1$, we can easily construct a matrix diagram representation of their Kronecker product $\mathbf{M}_K \otimes \cdots \otimes \mathbf{M}_1$. The matrix diagram will have one non-terminal node for each of the K matrices, where the level-k node m_k corresponds to matrix $\mathbf{M}_k$, and the entries of node m_k are determined as follows. If $\mathbf{M}_k[i, j]$ is zero, then $pairs(\mathsf{m}_k, i, j)$ is the empty set; otherwise, $pairs(\mathsf{m}_k, i, j)$ is the set containing only the pair $(\mathbf{M}_k[i, j], \mathsf{m}_{k-1})$. The level-0 node m_0 is terminal node 1. The matrix diagrams obtained for the Kronecker products for the running example are shown in Figure 13. Since the matrix diagrams are QROMDs, the variable pairs for each node are omitted from the figure to save space.

Addition of two matrix diagrams: Given two level-k matrix diagram nodes m_1 and m_2, the level-k matrix diagram node m encoding their sum, i.e., $f_m = f_{m_1} + f_{m_2}$, can be easily constructed by set union

$$pairs(m, i, j) = pairs(m_1, i, j) \cup pairs(m_2, i, j)$$

for each possible $i, j \in \mathbb{N}_k$. For example, the sum of matrix diagrams for $\mathbf{R}^{e_1}$ and $\mathbf{R}^{e_2}$ from Figure 13 is shown in Figure 14(a). While this produces a correct matrix diagram, a more compact representation can sometimes be obtained. For instance, if a set contains pairs (r_1, m) and (r_2, m) with the same child m, the two pairs can be replaced with the single pair $(r_1 + r_2, m)$, since $r_1 \cdot \mathbf{M} + r_2 \cdot \mathbf{M} = (r_1 + r_2) \cdot \mathbf{M}$. Note that this implies that level-1 nodes never need to have sets with more than one pair. Reduction may be possible when a set contains pairs (r, m_1) and (r, m_2) with the same edge value r, by replacing the two pairs with the pair (r, m), where m encodes the sum of m_1 and m_2 (computed recursively). When this is done for the sum of $\mathbf{R}^{e_1}$ and $\mathbf{R}^{e_2}$, we obtain the matrix diagram shown in Figure 14(b). Note that this replacement rule does not always reduce the size of the matrix diagram. For canonical matrix diagrams that do not allow sets of pairs [58], addition is still possible but requires more complex algorithms.

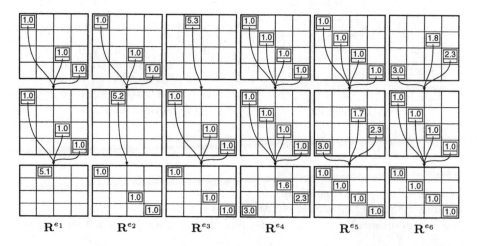

Fig. 13. Matrix diagrams built from Kronecker products for running example

Selecting a submatrix from a matrix diagram: Given a matrix diagram representation of a matrix, we can select a submatrix with specified rows and columns by using MDD representations for the sets of desired rows and columns. Using a QROMDD R over variables $x_K, \ldots, x_1$ for the set of rows, a QROMDD C over variables $y_K, \ldots, y_1$ for the set of columns, and a QROMD M over variable pairs $(x_K, y_K), \ldots, (x_1, y_1)$ allows for a straightforward recursive algorithm, shown in Figure 15(a), to construct $M' = \mathsf{Submatrix}(M, R, C)$. Note that, in addition to checking for duplicate nodes, Reduce also replaces nodes containing no

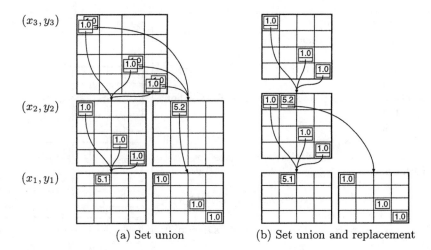

(x_3, y_3)

(x_2, y_2)

(x_1, y_1)

(a) Set union   (b) Set union and replacement

Fig. 14. Addition of $\mathbf{R}^{e_1}$ and $\mathbf{R}^{e_2}$ from Figure 13

pairs (i.e., $pairs(\mathsf{M}, i, j)$ is the empty set for all i, j) with terminal node 0. The Submatrix operator can also be implemented using ROMDD representations for R and C, using a slightly more complex algorithm. Note that it is also possible to physically remove the undesired rows and columns (rather than setting them to zero, as done by Submatrix); this may produce matrices with different dimensions within matrix diagram nodes with the same labels, and leads to more complex implementation [24, 57]. Figure 15(b) shows the matrix diagram obtained using $\mathbf{R}^{e_6}$ (from Figure 13) for the input matrix and the MDD encoding of the set of reachable states (from Figure 4(b)) for the desired rows and columns. Note that the rate of transition from state $(1, 2, 1)$ to state $(2, 2, 1)$ is $1.8 \cdot 1.0 \cdot 0 = 0$ in the obtained matrix diagram, since state $(1, 2, 1)$ is unreachable according to the MDD shown in Figure 4(b).

Model Representation and Construction with Matrix Diagrams. Given a high-level formalism that supports construction of the low-level model (e.g., CTMC) as a sum of Kronecker products, as described in Section 3.2, it is straightforward to construct the corresponding matrix diagram representation. Matrix diagrams can be constructed for each individual Kronecker product; these can then be summed to obtain the overall model. Note that both the overall matrix diagram and the Kronecker representation describe a model with states $\mathcal{S}$. The matrix diagram corresponding to the "reachable" portion of the model can be obtained using the Submatrix operator, using the set of reachable states $\mathcal{S}'$ as the set of desired rows and columns. This operation is important particularly for numerical solution, especially when $|\mathcal{S}| \gg |\mathcal{S}'|$. Note that this requires construction of the MDD representing the reachability set $\mathcal{S}'$, as described in Section 2.3. Summing the matrix diagrams in Figure 13 (using the replacement rules) and selecting the reachable portion of the matrix using Submatrix produces the matrix diagram shown in Figure 16.

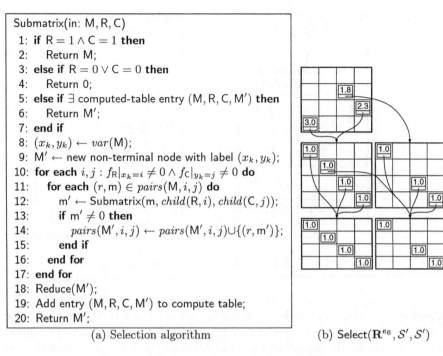

```
Submatrix(in: M, R, C)
 1: if R = 1 ∧ C = 1 then
 2:     Return M;
 3: else if R = 0 ∨ C = 0 then
 4:     Return 0;
 5: else if ∃ computed-table entry (M, R, C, M') then
 6:     Return M';
 7: end if
 8: (x_k, y_k) ← var(M);
 9: M' ← new non-terminal node with label (x_k, y_k);
10: for each i, j : f_R|_{x_k=i} ≠ 0 ∧ f_C|_{y_k=j} ≠ 0 do
11:     for each (r, m) ∈ pairs(M, i, j) do
12:         m' ← Submatrix(m, child(R, i), child(C, j));
13:         if m' ≠ 0 then
14:             pairs(M', i, j) ← pairs(M', i, j)∪{(r, m')};
15:         end if
16:     end for
17: end for
18: Reduce(M');
19: Add entry (M, R, C, M') to compute table;
20: Return M';
```

(a) Selection algorithm   (b) Select($\mathbf{R}^{e6}, S', S'$)

Fig. 15. Submatrix selection for matrix diagrams

Another method to construct a matrix diagram is to do so explicitly, by adding each non-zero entry of the matrix to be encoded into the matrix diagram, one at a time [58]. This approach essentially visits each non-zero entry, constructs a matrix diagram representing a matrix containing only the desired non-zero entry, and sums all the matrix diagrams to obtain the representation for the overall matrix. Overhead is kept manageable by mixing unreduced and reduced nodes in the same structure: in-place updates are possible for the unreduced nodes, but may contain duplicates. The unreduced portion is periodically merged with the reduced portion.

3.4   Other Data Structures

MTBDDs and matrix diagrams are not the only symbolic representations proposed for probabilistic models. There are a number of others, typically all extensions in some fashion of the basic BDD data structure. We describe two notable examples here: decision node binary decision diagrams (DNBDDs) and probabilistic decision graphs (PDGs).

Decision node binary decision diagrams (DNBDDs) were proposed by Siegle [67, 68] to represent CTMCs. While MTBDDs extend BDDs by allowing multiple terminal nodes, DNBDDs do so by adding information to certain *then* and *else* edges. Again, the purpose of this information is to encode the real values which the function being represented can take. This approach has the advantage that

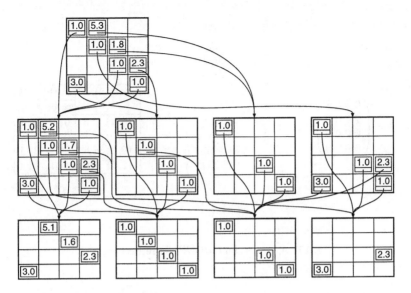

Fig. 16. Matrix diagram for the rate matrix of the running example

the DNBDD for a CTMC is identical in structure to the BDD which represents its transition relation, as might be used for example during reachability computation.

Figure 17 shows a simple example of a 4×4 matrix and the DNBDD which represents it. As for our MTBDD representation of a matrix of this size in Figure 10, the DNBDD uses two (interleaved) pairs of Boolean variables: x_2, x_1 to encode row indices and y_2, y_1 to encode column indices. The path through the DNBDD M which corresponds to each entry of matrix **M** is shown in the table in Figure 17. The nodes of the DNBDD which are shaded are *decision nodes*. Each node which has both *then* and *else* edges eventually leading to the 1 terminal node is a decision node. The value corresponding to a given path through the DNBDD is equal to the value which labels the edge from the last decision node along that path. For the matrix entry $(2, 3)$, for example, the row index is encoded as $1, 0$ and the column index as $1, 1$. Tracing the path $1, 1, 0, 1$ through the DNBDD, we see that the last labeled edge observed is 9, which is the value of the entry. In cases, where a single path corresponds to more than one matrix entry, the distinct values are differentiated by labeling the edge with an ordered list of values, rather than a single value.

In [45], DNBDDs were used to implement bisimulation algorithms for CT-MCs. They proved to be well suited to this application, since it can be seen as an extension of the non-probabilistic case, for which BDD-based algorithms had already been developed. It has not been shown, though, how DNBDDs could be used to perform analysis of CTMCs requiring numerical computation, as we consider in the next section.

Probabilistic decision graphs (PDGs) were proposed by Bozga and Maler in [11]. A PDG is a BDD-like data structure, designed specifically for storing vec-

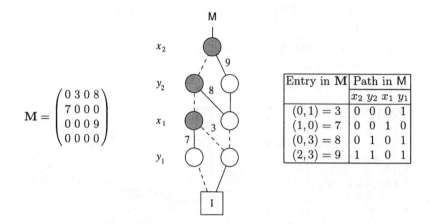

$$M = \begin{pmatrix} 0 & 3 & 0 & 8 \\ 7 & 0 & 0 & 0 \\ 0 & 0 & 0 & 9 \\ 0 & 0 & 0 & 0 \end{pmatrix}$$

Entry in **M**	Path in M			
	x_2	y_2	x_1	y_1
$(0,1) = 3$	0	0	0	1
$(1,0) = 7$	0	0	1	0
$(0,3) = 8$	0	1	0	1
$(2,3) = 9$	1	1	0	1

Fig. 17. A matrix and a DNBDD representing it

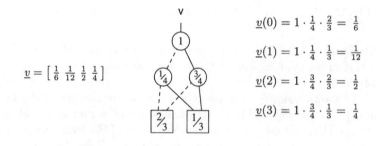

$$\underline{v} = \begin{bmatrix} \frac{1}{6} & \frac{1}{12} & \frac{1}{2} & \frac{1}{4} \end{bmatrix}$$

$$\underline{v}(0) = 1 \cdot \tfrac{1}{4} \cdot \tfrac{2}{3} = \tfrac{1}{6}$$

$$\underline{v}(1) = 1 \cdot \tfrac{1}{4} \cdot \tfrac{1}{3} = \tfrac{1}{12}$$

$$\underline{v}(2) = 1 \cdot \tfrac{3}{4} \cdot \tfrac{2}{3} = \tfrac{1}{2}$$

$$\underline{v}(3) = 1 \cdot \tfrac{3}{4} \cdot \tfrac{1}{3} = \tfrac{1}{4}$$

Fig. 18. A vector and the PDG representing it

tors and matrices of probabilities. Figure 18 shows a simple example of a PDG with two levels of nodes representing a vector of length 4. Nodes are labeled with probabilities and the values for all the nodes on each level must sum to one. Intuitively, these represent conditional probabilities: the actual value of each vector element can be determined by multiplying the conditional probabilities along the corresponding path, as illustrated in the figure. Like in a BDD, duplicate nodes are merged to save space. The hope is that some functions which have no compact representation as an MTBDD, perhaps because of an excessive number of distinct values, can be stored more efficiently as a PDG.

Buchholz and Kemper extended this work in [17]. They modified the PDG data structure to allow more than two children in each node, like in an MDD. They then used PDGs to store vectors and Kronecker representations to store matrices, combining the two to perform numerical solution of CTMCs. It was found, though, that the additional work required to manipulate the PDG data structure slowed the process of numerical computation.

4   Numerical Solution

In the preceding two sections, we have seen how symbolic data structures can be used to construct and store probabilistic models and their state spaces. In this section, we discuss how to actually perform analysis of the models when they are represented in this way. In the probabilistic setting, the most commonly required types of analysis will be those that require some form of numerical solution to be performed. Hence, this is the problem which we shall focus upon.

The exact nature of the work that needs to be performed will depend on several factors, for example, the type of probabilistic model being studied and the high-level formalisms which are used to specify both the model and the properties of the model which are to be analyzed. In *probabilistic model checking*, for example, properties are typically expressed in probabilistic extensions of temporal logics, such as PCTL [41] and CSL [4, 9]. This allows concise specifications such as "the probability that the system fails within 60s is at most 0.01" to be expressed. Model checking algorithms to verify whether or not such specifications are satisfied can be found in the literature (for surveys, see for example [26, 29]). In more traditional *performance analysis* based approaches, one might be interested in computing properties such as throughput, the average number of jobs in a queue, or average time until failure. In this presentation, we will avoid discussing the finer details of these various approaches. Instead, we will focus on some of the lower level numerical computations that are often required, regardless of the approach or formalism used. One of the most common tasks likely to be performed is the computation of a vector of probabilities, where one value corresponds to each state of the model. If the model is a continuous-time Markov chain (CTMC), this vector might, for example, contain *transient* or *steady-state* probabilities, which describe the model at a particular time instant or in the long-run, respectively. For the latter, the main operation required is the solution of a linear system of equations. For the former, a technique called uniformization provides an efficient and numerically stable approach. Another common case is where the values to be computed are the probabilities of, from each state in the model, reaching a particular class of states. If the model is a discrete-time Markov chain (DTMC), the principal operation is again solution of a linear system of equations; if the model is a Markov decision process (MDP), it is the solution of a linear optimization problem.

An important factor that all these types of computation have in common is that they can be (and often are) performed *iteratively*. Consider, for example, the problem of solving a linear equation system. This is, of course, a well-studied problem for which a range of techniques exist. For analysis of very large models, however, which is our aim, common *direct* methods such as Gaussian elimination are impractical because they do not scale up well. Fortunately, iterative techniques, such as the Power, Jacobi and Gauss-Seidel methods, provide efficient alternatives. There is a similar situation for the linear optimization problems required for analysis of MDPs. One option would be to use classic linear programming techniques such as the Simplex algorithm. Again, though, these are poorly suited to the large problems which frequently occur in these applications.

Fortunately, in this instance, there exist alternative, iterative algorithms to solve the problem. Lastly, we note that the aforementioned uniformization method for transient analysis of CTMCs is also an iterative method.

There is an important feature common to the set of iterative solution techniques listed in the previous paragraph. The computation performed depends heavily, of course, on the probabilistic model: a real matrix for DTMCs and CTMCs or a non-square matrix for MDPs. However, in all of these techniques, with the possible exception of some initialization steps, the model or matrix does not need to be modified during computation. Each iteration involves extracting the matrix entries, performing some computation and updating the solution vector. This is in contrast to some of the alternatives such as Gaussian elimination or the Simplex method which are based entirely on modifications to the matrix.

In the remainder of this section, we consider the issue of performing such numerical solution techniques, when the model is being represented either as an MTBDD or as a matrix diagram.

4.1   MTBDDs

Numerical solution using MTBDDs was considered in some of the earliest papers describing the data structure: [27, 5] gave algorithms for L/U decomposition and Gaussian elimination. The experimental results presented in the latter paper illustrated that, in comparison to more conventional, explicit approaches, MTBDDs were relatively poor for these tasks. They also identified the reasons for this. Methods such as Gaussian elimination are based on access to and modification of individual elements, rows or columns of matrices. MTBDDs, though are an inherently recursive data structure, poorly suited to such operations. Moreover, compact MTBDD representations rely on high-level structure in the matrix. This is typically destroyed by many such operations, increasing MTBDD sizes and time requirements.

Subsequently, Hachtel at al. [39, 40] and Xie and Beerel [70] presented MTBDD-based algorithms for iterative solution of linear equation systems, finding them better suited to symbolic implementation than direct methods. We have already observed that such techniques are preferable in our circumstances anyway because of the size of problems involved.

The basic idea of an iterative method is that a vector containing an estimate to the solution is repeatedly updated. This update is based on operations which use the matrix. Often, the main operation required is matrix-vector multiplication. For example, when solving the linear equation system $\mathbf{A} \cdot \underline{x} = \underline{b}$ using the Jacobi method, the kth iteration, which computes the solution vector $\underline{x}^{(k)}$ from the previous approximation $\underline{x}^{(k-1)}$ is as follows:

$$\underline{x}^{(k)} := \mathbf{D}^{-1} \cdot (\mathbf{L} + \mathbf{U}) \cdot \underline{x}^{(k-1)} + \mathbf{D}^{-1} \cdot \underline{b}$$

where $\mathbf{D}$, $\mathbf{L}$ and $\mathbf{U}$ are diagonal, lower- and upper-triangular matrices, respectively, such that $\mathbf{A} = \mathbf{D} - \mathbf{L} - \mathbf{U}$. As we saw earlier, matrix multiplication is well suited to MTBDDs because it can be implemented in a recursive fashion. The other operations required here, i.e. matrix addition, vector addition and inversion of a diagonal matrix can all be implemented using the Apply operator.

Hachtel et al. applied their implementation to the problem of computing steady-state probabilities for DTMCs derived from a range of large, benchmark circuits. The largest of these DTMCs handled had more than 10^{27} states. MT-BDDs proved invaluable in that an explicit representation of a matrix this size would be impossible on the same hardware. The MTBDDs were able to exploit a significant amount of structure and regularity in the DTMCs. However, it was observed that the MTBDD representation of the solution vector was more problematic. Hachtel et al. were forced to adopt techniques such as rounding all values below a certain threshold to zero.

Following the discovery that MTBDDs could be applied to iterative numerical solution, and inspired by the success of BDD-based model checking implementations, numerous researchers proposed symbolic approaches to the verification and analysis of probabilistic models. These included PCTL model checking of DTMCs [8, 42, 7, 6], PCTL model checking of MDPs [6, 53, 34, 31], computation of steady-state probabilities for CTMCs [44] and CSL model checking for CTMCs [9, 48]. While some of these papers simply presented algorithms, others implemented the techniques and presented experimental results. Their conclusions can be summarized as follows.

Firstly, MTBDDs can be used, as mentioned earlier, to construct and store extremely large, probabilistic models, where high-level structure can be exploited. Furthermore, in some instances, MTBDD-based numerical solution is also very successful. For example, [54] presented results for symbolic model checking of MDPs with more than 10^{10} states. This would be impossible with explicit techniques, such as sparse matrices, on the same hardware. However, in general, it was found that MTBDD-based numerical computation performed poorly in comparison to explicit alternatives, in terms of both time and space efficiency. The simple reason for this is that, despite compact MTBDD-based storage for probabilistic models, the same representation for solution vectors is usually inefficient. This is unsurprising since these vectors will usually be unstructured and contain many distinct values. Both of these factors generally result in greatly increased MTBDD sizes which are not only more expensive to store, but are slower to manipulate. By contrast, in an explicit (e.g. sparse matrix based) implementation, solution vectors are typically stored in arrays which are fast to access and manipulate, regardless of structure.

4.2   Offset-Labeled MTBDDs

To combat the problems described above, modifications to the MTBDD-based approach have been proposed [52, 62]. The basic idea is to combine the compact model representation afforded by symbolic techniques with the fast and efficient numerical solution provided by explicit approaches. This is done by performing numerical computation using an MTBDD for matrix storage but an array for the solution vector.

As above, the most important operation to be implemented is matrix-vector multiplication. A crucial observation is that this requires access to each of the non-zero entries in the matrix exactly once and that the order in which they are obtained is unimportant. The non-zero entries of a matrix stored by an MTBDD can be extracted via a recursive depth-first traversal of the data structure. Essentially, this traces every path from the root node of the MTBDD to a terminal node, each path corresponding to a single non-zero matrix entry.

During this process, we need to be able to compute the row and column index of each matrix entry. This is done by augmenting the MTBDD with additional information: we convert the MTBDD into an *offset-labeled MTBDD*, each node of which is assigned an integer offset. This is essentially the same as the indexing scheme we described for BDDs and MDDs in Section 2.4. Like in the simpler case, when tracing a path through an MTBDD, the required indices can be computed by summing the values of the offsets. More specifically, the row and column index are computed independently: the row index by summing offsets on nodes labeled with x_k variables from which the *then* edge was taken; and the column index by summing those on nodes labeled with y_k variables where the *then* edge was taken.

Figure 19 shows an example of an offset-labeled MTBDD R', the transition rate matrix **R** it represents, and a table explaining the way in which the information is represented. The matrix **R** is the same as used for the earlier example in Figure 10. State 2 of the CTMC which the matrix represents (see Figure 10) is unreachable. All entries of the corresponding row and column of **R** are zero. In Figure 19, this is emphasized by marking these entries as '–'.

Let us consider a path through the offset-labeled MTBDD: 1, 0, 1, 1. This is the same as used for the MTBDD example on page 313. The path leads to the 7 terminal, revealing that the corresponding matrix entry has value 7. To compute the row index of the entry, we sum the offsets on x_k nodes from which the *then* edge was taken. For this path, the sum is $2 + 0 = 2$. For the column index, we perform the same calculation but for y_k nodes. The *then* edge was only taken from the y_1 node, so the index is 1, the offset on this node. Hence, our matrix entry is $(2, 1) = 7$. Notice that the corresponding matrix entry last time was $(3, 1) = 7$, i.e. the offsets encode the fact that the unreachable state is not included in the indexing.

Compare the MTBDD R from Figure 10 with the offset-labeled variant R' in Figure 19, both of which represent the same matrix. Firstly, R' has been converted to its quasi-reduced form so that offsets can be added on each level (as was the case for BDDs in Section 2.4). Secondly, note that two additional nodes have been added (rightmost x_1 and y_1 nodes). These nodes would be duplicates (and hence removed) if not for the presence of offsets. This situation occurs when there are two paths through the MTBDD passing through a common node and the offset required to label that node is different for each path. Empirical results show that, in practice, this represents only a small increase in MTBDD size. For further details in this area, see [62].

Each row of the table in Figure 19 corresponds to a single path through R' which equates to a single non-zero entry of **R**. The entries are listed in the order

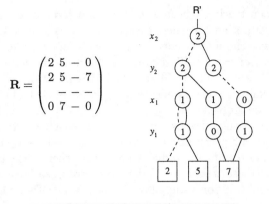

$$\mathbf{R} = \begin{pmatrix} 2 & 5 & - & 0 \\ 2 & 5 & - & 7 \\ - & - & - \\ 0 & 7 & - & 0 \end{pmatrix}$$

Path					Offsets				Entry of $\mathbf{R}$
x_2	y_2	x_1	y_1	$f_{R'}$	x_2	y_2	x_1	y_1	
0	0	0	0	2	-	-	-	-	$(0,0) = 2$
0	0	0	1	5	-	-	-	1	$(0,1) = 5$
0	0	1	0	2	-	-	1	-	$(1,0) = 2$
0	0	1	1	5	-	-	1	1	$(1,1) = 5$
0	1	1	1	7	-	2	1	0	$(1,2) = 7$
1	0	1	1	7	2	-	0	1	$(2,1) = 7$

Fig. 19. An offset-labeled MTBDD R′ representing a matrix **R**

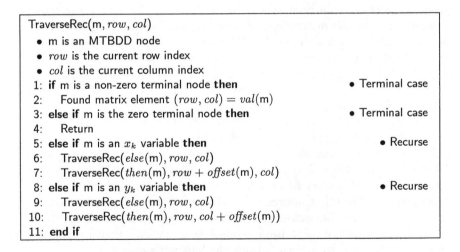

Fig. 20. Offset-labeled MTBDD traversal algorithm

in which they would be extracted by a recursive traversal of the offset-labeled MTBDD. Figure 20 shows the recursive algorithm, TraverseRec, used to perform such a traversal. Its first argument is the current MTBDD node m. Initially, this is the root node of the MTBDD being traversed. The other two arguments, *row* and *col*, keep track of the row and column indices respectively. Initially, these are both zero. Line 2 is where each matrix entry is found. In practice, this would actually use the matrix entry to perform a matrix-vector multiplication. We denote by *offset*(m) the offset labeling the node m. Note that, in essence, this algorithm is very similar to Enumerate in Figure 7(a).

Empirical results for the application of offset-labeled MTBDDs to several numerical solution problems can be found in [52, 62]. In practice, a number of optimizations are applied. For example, since TraverseRec is typically performed many times (once for every iteration of numerical solution), it is beneficial to cache some of the results it produces and reuse them each time. See [52, 62] for the details of these techniques. With these optimizations in place, that the speed of numerical solution using offset-labeled MTBDDs almost matches that of explicit methods based on sparse matrices. More importantly, because of the significant savings in memory made by storing the matrix symbolically, offset-labeled MTBDDs can usually handle much larger problems. An increase of an order of magnitude is typical.

4.3 Matrix Diagrams

In [24, 57], algorithms were given to compute the steady-state probabilities of a CTMC using the Gauss-Seidel iterative method, when the CTMC is represented with matrix diagrams and the steady-state probabilities are stored explicitly. To perform an iteration of Gauss-Seidel, we must be able to efficiently enumerate the columns of the transition rate matrix $\mathbf{R}$. This can be done using matrix diagrams, provided the matrix for each node is stored using a sparse format that allows for efficient column access [24, 57]. Column construction proceeds from the bottom level nodes of the matrix diagram to the top node. A recursive algorithm to perform the computation is shown in Figure 21. The algorithm parameters include the current "level", k, a level-k matrix diagram node M which encodes a matrix $\mathbf{M}$, a QROMDD node R with offsets which encodes the desired set of rows $\mathcal{R}$ (i.e., an EV$^+$MDD encoding the lexicographical indexing function ψ for $\mathcal{R}$), and the specification of the desired column $(y_k, \ldots, y_1)$. Using an ROMDD with offsets is also possible for $\mathcal{R}$, but leads to a slightly more complex algorithm to handle removed redundant nodes. Algorithm GetColumn returns the desired column of matrix $\mathbf{M}$ over the desired rows only, which is a vector of dimension $|\mathcal{R}|$. The offset information of R is used to ensure that an element at row $(x_K, \ldots, x_1)$ in the desired column is returned in position $\psi(x_K, \ldots, x_1)$ of the column vector. The algorithm assumes that the undesired rows have been removed from the matrix already using the Submatrix operator.

Since probabilistic systems tend to produce sparse matrices, the algorithm assumes that the desired column will also be sparse, i.e. that the number of non-zero elements it contains will be much less than $|\mathcal{R}|$. Hence, a sparse structure is used for the columns. Since the maximum number of non-zeros present in any column can be determined in a preprocessing step, we can bound the size of the sparse vector that is necessary for any call to GetColumn, and use an efficient array-based structure to represent the sparse columns. This allows for efficient

---

GetColumn(k, M, R, $y_k, \ldots, y_1$)
- R is an MDD with offsets, encoding the set of rows $\mathcal{R}$ to consider.
- M is a matrix diagram, where rows $\notin \mathcal{R}$ of the encoded matrix are zero.
- $(y_k, \ldots, y_1)$ is the desired column to obtain.

1: **if** $k = 0$ **then**   • Terminal case
2:    Return [1];
3: **else if** $\exists$ computed-table entry $(k, \text{M}, \text{R}, y_k, \ldots, y_1, \underline{col})$ **then**
4:    Return $\underline{col}$;
5: **end if**
6: $\underline{col} \leftarrow 0$;   • $\underline{col}$ is a sparse vector
7: **for each** $x_k : pairs(\text{M}, x_k, y_k) \neq \emptyset$ **do**
8:    **for each** $(r, \text{M}') \in pairs(\text{M}, x_k, y_k)$ **do**
9:       $\underline{d} \leftarrow$ GetColumn($k - 1$, M$'$, $child(\text{R}, x_k), y_{k-1}, \ldots, y_1$);
10:       **for each** $i : \underline{d}[i] \neq 0$ **do**
11:          $\underline{col}[i + value(\text{R}, x_k)] \leftarrow \underline{col}[i + value(\text{R}, x_k)] + r \cdot \underline{d}[i]$;
12:       **end for**
13:    **end for**
14: **end for**
15: Add entry $(k, \text{M}, \text{R}, y_k, \ldots, y_1, \underline{col})$ to computed-table;
16: Return $\underline{col}$;

Fig. 21. Recursive algorithm to select a matrix diagram column

implementation of the loop in lines 10 through 12. Note that the algorithm works correctly for dense matrices, but may not be the most efficient approach.

As with most recursive algorithms for decision diagrams, it is possible (and in fact likely) that a single top-level call to GetColumn will produce several *identical* recursive calls to GetColumn. Thus, GetColumn utilizes a *computed-table*, like many of the algorithms for decision diagram operators. The computed-table is also useful when performing a sequence of column operations, such as during numerical solution, which requires obtaining each reachable column in turn during a single iteration. However, results cannot be stored in the computed-table indefinitely; doing so would require storage space comparable to explicit sparse storage of the entire transition rate matrix once all columns have been computed. The solution to this problem adopted in [24, 57] is to save only the most recent column result for each node. Note that this allows simplification to lines 3 and 15 in Figure 21, since each matrix diagram node will have at most one computed-table entry. Other benefits of this approach are that memory usage is not only relatively small, but fixed: by traversing the matrix diagram once in a preprocessing step, we can determine the largest column that will be produced by each node, and allocate a fixed-size computed-table of the appropriate size. Also, the algorithm GetColumn can be implemented to use the computed-table space directly, thus avoiding the need to copy vectors into the table.

As an example, the computed-table entries after obtaining column $(2, 2, 1)$ for the running example are shown in Figure 22(b). Note that the matrix diagram

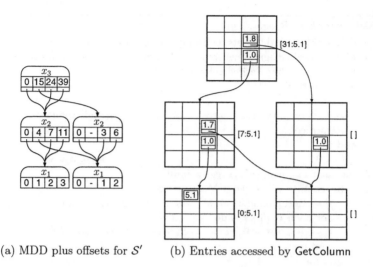

(a) MDD plus offsets for $\mathcal{S}'$ (b) Entries accessed by GetColumn

Fig. 22. Obtaining column $(2, 2, 1)$ for the running example

shown corresponds to the relevant portion of the matrix diagram of Figure 16. The reachable rows are encoded using the MDD with offset values shown in Figure 22(a). Note that column 1 of the rightmost level-1 matrix diagram node is empty; thus, the resulting column has no non-zero entries. The only transition to state $(2, 2, 1)$ is from state $(2, 2, 0)$, with rate $5.1 \cdot 1.0 \cdot 1.0 = 5.1$. Since state $(2, 2, 0)$ has index $24 + 7 + 0 = 31$ according to the MDD, the resulting column has a single non-zero entry at position 31, with value 5.1.

The computed-table must be invalidated at level k whenever the specified column changes for any component less or equal to k, since columns are constructed in a bottom-up fashion. For instance, if we just computed the column $(y_K, \ldots, y_1)$ and we want to compute the column $(y_K, \ldots, y_{k+1}, y'_k, y_{k-1}, \ldots, y_1)$ then the computed-table columns at levels K through k must be cleared. Thus, to improve our chances of hits in the computed-table (by reducing the number of times we must clear columns), after visiting the reachable column for state $(y_K, \ldots, y_1)$, we should next visit the reachable column for the state that follows $(y_K, \ldots, y_1)$ in lexicographic order *reading the strings in reverse*; that is, the order obtained when component K is the least significant (fastest changing) and component 1 is the most significant (slowest changing). Note that this order is *not* reverse lexicographic order. As described in [24], this order can be quickly traversed by storing the set of columns using an "upside-down" variable order. That is, if the reachable rows are stored using an MDD R with variable ordering $x_K, \ldots, x_1$, and the matrix diagram M is stored using ordering $(x_K, y_K), \ldots, (x_1, y_1)$, then the reachable columns should be stored in an MDD C with variable ordering $y_1, \ldots, y_K$. Enumerating the reachable columns of C as described in Section 2.4 will treat component K as the fastest changing component.

Numerical solution can then proceed, using explicit storage for solution vectors (with dimension $|\mathcal{R}|$), MDDs to represent the reachable states for the rows and columns, and a matrix diagram to represent the transition rate matrix $\mathbf{R}$. Often, the diagonal elements of $\mathbf{Q}$ must be easily accessible during numerical solution; these can be stored either explicitly in a vector (at a cost of increased storage requirements), or implicitly using a second matrix diagram (at a cost of increased computational requirements) [24].

Using matrix diagrams allows for the numerical solution of large models [24, 57]. Storage requirements for the required MDDs and for the matrix diagram itself are often negligible compared to the requirements for the solution vector; essentially, all available memory can be used for the solution vector. In practice, using matrix diagrams instead of explicit sparse matrix storage allows for solution of models approximately one order of magnitude larger.

5   Discussion and Conclusion

In this paper, we have surveyed a range of symbolic approaches for the generation, representation and analysis of probabilistic models. We conclude by summarizing some of their relative strengths and weaknesses and highlighting areas for future work.

In Section 2, we presented techniques for the representation and manipulation of state sets using BDDs and MDDs. These are at the heart of non-probabilistic symbolic model checking approaches and have proved to be extremely successful. They have also been examined at great length in the literature. Of more interest here are the extensions of these techniques to handle problems which are specific to the validation of probabilistic models.

In Sections 3 and 4, we focused on methods to construct, store and manipulate probabilistic models. We have seen that symbolic data structures such as MTBDDs and matrix diagrams can be successfully applied to this problem, allowing large, structured models to be constructed quickly and stored compactly. Furthermore, we have seen that, with suitable adaptations and algorithms, these data structures can be used to perform efficient numerical computation, as required for analysis of the probabilistic models. It is important to note that such results are reliant on high-level structure and regularity in these models. Furthermore, heuristics for state space encoding and variable ordering may need to be employed, especially where a binary encoding must be chosen. Implementations of both techniques are available in the tools PRISM [51, 1] and SMART [22, 2], respectively. These tools have been applied to a wide range of case studies. Overall, the conclusion is that we can increase by approximately an order of magnitude the size of problems which can be handled.

While the symbolic techniques discussed here have been successful in this respect, there remain important challenges for future work. One interesting area would be to investigate more closely the similarities and differences between the various data structures. This paper was broken into two distinct threads, describing the use of first BDDs and MTBDDs and then MDDs and matrix diagrams. It is evident that there are many similarities between the two approaches. There are also important differences though, leaving room for experimentation and exchanges of ideas.

The two principal issues are as follows. Firstly, the variables in BDDs and MTBDDs are Boolean, whereas in MDDs and matrix diagrams, they are multi-valued. For BDDs and MTBDDs, this simplifies both the storage of the data structure and the operations which are performed on it. However, it is not clear whether forcing everything to be encoded as Boolean variables is a good idea. It may be more intuitive or even more efficient to treat models as a small number of fixed-size submodels, as is the case for MDDs and matrix diagrams. Furthermore, larger variables will generally imply fewer levels to encode a given model, possibly implying less overhead during manipulation.

The second key difference is that matrix diagrams are implicitly tied to the Kronecker representation, where large matrices are stored as Kronecker-algebraic expressions of smaller matrices. This may well produce a more compact representation than MTBDDs, particularly where it results in many distinct values being stored. However, this may lead to more overhead in terms of actually computing each matrix entry than is necessary for MTBDDs, since floating-point multiplications are required.

Given these points, it is tempting to compare the two approaches experimentally. However, we feel that including experimental results here is inappropriate for the following reasons. First, since we are unaware of any single software tool that incorporates both approaches, it would be necessary to use different tools for each technique, introducing numerous uncertainties. For example, observed differences in performance could be due to the implementation quality of the respective tools, or to the fact that one tool handles a more general (but more computationally expensive) class of models. Second, without a thorough study of the differences between the two approaches, it would be unclear whether experimental differences apply in general, for a particular type of model, or simply for the models used in the experiments. Finally, results presented here would essentially duplicate those already published elsewhere. Clearly, further work is necessary to study the differences between the approaches in detail, in order to understand how the techniques can best be used in practice (to date, we are unaware of *any* work in this area).

There is one issue which unites the two schemes. We have seen that both usually allow an increase of approximately an order of magnitude in model size over explicit alternatives. While this is useful, improvement beyond this range is difficult. This is because, despite compact storage of the state space using BDDs or MDDs and of the model itself using MTBDDs or matrix diagrams, solution vectors proportional to the size of the model's state space must still be stored explicitly to allow acceptable speed of solution. This generally represents the limiting factor in terms of memory requirements. Hence, MTBDDs and matrix diagrams can be used to construct and store much larger models than can at present feasibly be solved. Attempts to store vectors symbolically using either

MTBDDs or PDGs have met with only limited success. It appears that the efficient representation for these cases is a much more difficult problem. In fact, it may be asked whether there is even structure there to exploit. Further research is needed to clarify this issue.

One possibility for alleviating the above difficulties, though, is by using parallel, distributed or out-of-core approaches, as in e.g. [50]. Also, it is also worth mentioning that these problems apply only to the computation of an *exact* solution. The ability to represent and manipulate a large model exactly using symbolic techniques may lead to new and intriguing *approximation* algorithms. Indeed, initial efforts in this area have been promising [61]. Much work is needed in this new area.

References

1. PRISM web site. www.cs.bham.ac.uk/~dxp/prism.
2. SMART web site. http://www.cs.wm.edu/~ciardo/SMART/.
3. S. Akers. Binary decision diagrams. *IEEE Transactions on Computers*, C-27(6):509–516, 1978.
4. A. Aziz, K. Sanwal, V. Singhal, and R. Brayton. Verifying continuous time Markov chains. In R. Alur and T. Henzinger, editors, *Proc. 8th International Conference on Computer Aided Verification (CAV'96)*, volume 1102 of *LNCS*, pages 269–276. Springer, 1996.
5. I. Bahar, E. Frohm, C. Gaona, G. Hachtel, E.Macii, A. Pardo, and F. Somenzi. Algebraic decision diagrams and their applications. In *Proc. International Conference on Computer-Aided Design (ICCAD'93)*, pages 188–191, 1993. Also available in *Formal Methods in System Design*, 10(2/3):171–206, 1997.
6. C. Baier. On algorithmic verification methods for probabilistic systems. Habilitation thesis, Fakultät für Mathematik & Informatik, Universität Mannheim, 1998.
7. C. Baier and E. Clarke. The algebraic mu-calculus and MTBDDs. In *Proc. 5th Workshop on Logic, Language, Information and Computation (WOLLIC'98)*, pages 27–38, 1998.
8. C. Baier, E. Clarke, V. Hartonas-Garmhausen, M. Kwiatkowska, and M. Ryan. Symbolic model checking for probabilistic processes. In P. Degano, R. Gorrieri, and A. Marchetti-Spaccamela, editors, *Proc. 24th International Colloquium on Automata, Languages and Programming (ICALP'97)*, volume 1256 of *LNCS*, pages 430–440. Springer, 1997.
9. C. Baier, J.-P. Katoen, and H. Hermanns. Approximate symbolic model checking of continuous-time Markov chains. In J. Baeten and S. Mauw, editors, *Proc. 10th International Conference on Concurrency Theory (CONCUR'99)*, volume 1664 of *LNCS*, pages 146–161. Springer, 1999.
10. B. Bollig and I. Wegner. Improving the variable ordering of OBDDs is NP-complete. *IEEE Transactions on Computers*, 45(9):993–1006, 1996.
11. M. Bozga and O. Maler. On the representation of probabilities over structured domains. In N. Halbwachs and D. Peled, editors, *Proc. 11th International Conference on Computer Aided Verification (CAV'99)*, volume 1633 of *LNCS*, pages 261–273. Springer, 1999.

12. K. Brace, R. Rudell, and R. Bryant. Efficient implementation of a BDD package. In *Proc. 27th Design Automation Conference (DAC'90)*, pages 40–45. IEEE Computer Society Press, 1990.

13. R. Bryant. Graph-based algorithms for Boolean function manipulation. *IEEE Transactions on Computers*, C-35(8):677–691, 1986.

14. R. Bryant. Symbolic Boolean manipulation with ordered binary-decision diagrams. *ACM Computing Surveys*, 24(3):293–318, 1992.

15. P. Buchholz, G. Ciardo, S. Donatelli, and P. Kemper. Complexity of memory-efficient Kronecker operations with applications to the solution of Markov models. *INFORMS J. Comp.*, 12(3):203–222, SUMMER 2000.

16. P. Buchholz and P. Kemper. Numerical analysis of stochastic marked graphs. In *6th Int. Workshop on Petri Nets and Performance Models (PNPM'95)*, pages 32–41, Durham, NC, October 1995. IEEE Comp. Soc. Press.

17. P. Buchholz and P. Kemper. Compact representations of probability distributions in the analysis of superposed GSPNs. In R. German and B. Haverkort, editors, *Proc. 9th International Workshop on Petri Nets and Performance Models (PNPM'01)*, pages 81–90. IEEE Computer Society Press, 2001.

18. P. Buchholz and P. Kemper. Kronecker based matrix representations for large Markov models. This proceedings, 2003.

19. J. R. Burch, E. M. Clarke, K. L. McMillan, D. L. Dill, and J. Hwang. Symbolic model checking: 10^{20} states and beyond. In *Proc. 5th Annual IEEE Symposium on Logic in Computer Science (LICS'90)*, pages 428–439. IEEE Computer Society Press, 1990.

20. G. Ciardo, G. Luettgen, and R. Siminiceanu. Efficient symbolic state-space construction for asynchronous systems. In Mogens Nielsen and Dan Simpson, editors, *Application and Theory of Petri Nets 2000 (Proc. 21st Int. Conf. on Applications and Theory of Petri Nets)*, LNCS 1825, pages 103–122, Aarhus, Denmark, June 2000. Springer-Verlag.

21. G. Ciardo, G. Luettgen, and R. Siminiceanu. Saturation: an efficient iteration strategy for symbolic state space generation. In Tiziana Margaria and Wang Yi, editors, *Proc. Tools and Algorithms for the Construction and Analysis of Systems (TACAS)*, LNCS 2031, pages 328–342, Genova, Italy, April 2001. Springer-Verlag.

22. G. Ciardo and A. Miner. SMART: Simulation and Markovian analyser for reliability and timing. In *Proc. 2nd International Computer Performance and Dependability Symposium (IPDS'96)*, page 60. IEEE Computer Society Press, 1996.

23. G. Ciardo and A. Miner. Storage alternatives for large structured state spaces. In R. Marie, B. Plateau, M. Calzarossa, and G. Rubino, editors, *Proc. 9th International Conference on Modelling Techniques and Tools for Computer Performance Evaluation*, volume 1245 of *LNCS*, pages 44–57. Springer, 1997.

24. G. Ciardo and A. Miner. A data structure for the efficient Kronecker solution of GSPNs. In P. Buchholz and M. Silva, editors, *Proc. 8th International Workshop on Petri Nets and Performance Models (PNPM'99)*, pages 22–31. IEEE Computer Society Press, 1999.

25. G. Ciardo and R. Siminiceanu. Using edge-valued decision diagrams for symbolic generation of shortest paths. In Mark D. Aagaard and John W. O'Leary, editors, *Proc. Fourth International Conference on Formal Methods in Computer-Aided Design (FMCAD)*, LNCS 2517, pages 256–273, Portland, OR, USA, November 2002. Springer.

26. F. Ciesinski and F. Grössner. On probabilistic computation tree logic. This proceedings, 2003.
27. E. Clarke, M. Fujita, P. McGeer, K. McMillan, J. Yang, and X. Zhao. Multi-terminal binary decision diagrams: An efficient data structure for matrix representation. In *Proc. International Workshop on Logic Synthesis (IWLS'93)*, pages 1–15, 1993. Also available in *Formal Methods in System Design*, 10(2/3):149–169, 1997.
28. E. Clarke, K. McMillan, X. Zhao, M. Fujita, and J. Yang. Spectral transforms for large Boolean functions with applications to technology mapping. In *Proc. 30th Design Automation Conference (DAC'93)*, pages 54–60. ACM Press, 1993. Also available in *Formal Methods in System Design*, 10(2/3):137–148, 1997.
29. L. Cloth. Specification and verification of Markov reward models. This proceedings, 2003.
30. O. Coudert, C. Berthet, and J. C. Madre. Verification of synchronous sequential machines based on symbolic execution. In J. Sifakis, editor, *Proc. International Workshop on Automatic Verification Methods for Finite State Systems*, volume 407 of *LNCS*, pages 365–373. Springer, 1989.
31. P. D'Argenio, B. Jeannet, H. Jensen, and K. Larsen. Reachability analysis of probabilistic systems by successive refinements. In L. de Alfaro and S. Gilmore, editors, *Proc. Joint International Workshop on Process Algebra and Probabilistic Methods, Performance Modeling and Verification (PAPM/PROBMIV'01)*, volume 2165 of *LNCS*, pages 39–56. Springer, 2001.
32. I. Davies, W. Knottenbelt, and P. Kritzinger. Symbolic methods for the state space exploration of GSPN models. In T. Field, P. Harrison, J. Bradley, and U. Harder, editors, *Proc. 12th International Conference on Modelling Techniques and Tools for Computer Performance Evaluation (TOOLS'02)*, volume 2324 of *LNCS*, pages 188–199. Springer, 2002.
33. M. Davio. Kronecker products and shuffle algebra. *IEEE Transactions on Computers*, C-30(2):116–125, February 1981.
34. L. de Alfaro, M. Kwiatkowska, G. Norman, D. Parker, and R. Segala. Symbolic model checking of concurrent probabilistic processes using MTBDDs and the Kronecker representation. In S. Graf and M. Schwartzbach, editors, *Proc. 6th International Conference on Tools and Algorithms for the Construction and Analysis of Systems (TACAS'00)*, volume 1785 of *LNCS*, pages 395–410. Springer, 2000.
35. S. Donatelli. Superposed stochastic automata: A class of stochastic Petri nets amenable to parallel solution. *Performance Evaluation*, 18:21–36, 1993.
36. R. Enders, T. Filkorn, and D. Taubner. Generating BDDs for symbolic model checking in CCS. In K. Larsen and A. Skou, editors, *Proc. 3rd International Workshop on Computer Aided Verification (CAV'91)*, volume 575 of *LNCS*, pages 203–213. Springer, 1991.
37. P. Fernandes, B. Plateau, and W. Stewart. Efficient descriptor-vector multiplication in stochastic automata networks. *Journal of the ACM*, 45(3):381–414, 1998.
38. M. Fujita, Y. Matsunaga, and T. Kakuda. On the variable ordering of binary decision diagrams for the application of multi-level synthesis. In *European conference on Design automation*, pages 50–54, 1991.
39. G. Hachtel, E. Macii, A. Pardo, and F. Somenzi. Probabilistic analysis of large finite state machines. In *Proc. 31st Design Automation Conference (DAC'94)*, pages 270–275. ACM Press, 1994.
40. G. Hachtel, E. Macii, A. Pardo, and F. Somenzi. Markovian analysis of large finite state machines. *IEEE Trans. on CAD*, 15(12):1479–1493, 1996.
41. H. Hansson and B. Jonsson. A logic for reasoning about time and probability. *Formal Aspects of Computing*, 6(5):512–535, 1994.

42. V. Hartonas-Garmhausen. *Probabilistic Symbolic Model Checking with Engineering Models and Applications.* PhD thesis, Carnegie Mellon University, 1998.

43. H. Hermanns, M. Kwiatkowska, G. Norman, D. Parker, and M. Siegle. On the use of MTBDDs for performability analysis and verification of stochastic systems. *Journal of Logic and Algebraic Programming: Special Issue on Probabilistic Techniques for the Design and Analysis of Systems,* 56(1-2):23–67, 2003.

44. H. Hermanns, J. Meyer-Kayser, and M. Siegle. Multi terminal binary decision diagrams to represent and analyse continuous time Markov chains. In B. Plateau, W. Stewart, and M. Silva, editors, *Proc. 3rd International Workshop on Numerical Solution of Markov Chains (NSMC'99),* pages 188–207. Prensas Universitarias de Zaragoza, 1999.

45. H. Hermanns and M. Siegle. Bisimulation algorithms for stochastic process algebras and their BDD-based implementation. In J.-P. Katoen, editor, *Proc. 5th International AMAST Workshop on Real-Time and Probabilistic Systems (ARTS'99),* volume 1601 of *LNCS,* pages 244–264. Springer, 1999.

46. J. Hillston and L. Kloul. An efficient Kronecker representation for PEPA models. In *Proc. Joint International Workshop on Process Algebra and Probabilistic Methods, Performance Modeling and Verification (PAPM/PROBMIV'01),* pages 120–135. Springer-Verlag, September 2001.

47. T. Kam, T. Villa, R.K. Brayton, and A. Sangiovanni-Vincentelli. Multi-valued decision diagrams: theory and applications. *Multiple-Valued Logic,* 4(1–2):9–62, 1998.

48. J.-P. Katoen, M. Kwiatkowska, G. Norman, and D. Parker. Faster and symbolic CTMC model checking. In L. de Alfaro and S. Gilmore, editors, *Proc. Joint International Workshop on Process Algebra and Probabilistic Methods, Performance Modeling and Verification (PAPM/PROBMIV'01),* volume 2165 of *LNCS,* pages 23–38. Springer, 2001.

49. M. Kuntz and M. Siegle. Deriving symbolic representations from stochastic process algebras. In H. Hermanns and R. Segala, editors, *Proc. 2nd Joint International Workshop on Process Algebra and Probabilistic Methods, Performance Modeling and Verification (PAPM/PROBMIV'02),* volume 2399 of *LNCS,* pages 188–206. Springer, 2002.

50. M. Kwiatkowska, R. Mehmood, G. Norman, and D. Parker. A symbolic out-of-core solution method for Markov models. In *Proc. Workshop on Parallel and Distributed Model Checking (PDMC'02),* volume 68.4 of *Electronic Notes in Theoretical Computer Science.* Elsevier, 2002.

51. M. Kwiatkowska, G. Norman, and D. Parker. PRISM: Probabilistic symbolic model checker. In T. Field, P. Harrison, J. Bradley, and U. Harder, editors, *Proc. 12th International Conference on Modelling Techniques and Tools for Computer Performance Evaluation (TOOLS'02),* volume 2324 of *LNCS,* pages 200–204. Springer, 2002.

52. M. Kwiatkowska, G. Norman, and D. Parker. Probabilistic symbolic model checking with PRISM: A hybrid approach. In J.-P. Katoen and P. Stevens, editors, *Proc. 8th International Conference on Tools and Algorithms for the Construction and Analysis of Systems (TACAS'02),* volume 2280 of *LNCS,* pages 52–66. Springer, 2002.

53. M. Kwiatkowska, G. Norman, D. Parker, and R. Segala. Symbolic model checking of concurrent probabilistic systems using MTBDDs and Simplex. Technical Report CSR-99-1, School of Computer Science, University of Birmingham, 1999.

54. M. Kwiatkowska, G. Norman, and R. Segala. Automated verification of a randomized distributed consensus protocol using Cadence SMV and PRISM. In G. Berry, H. Comon, and A. Finkel, editors, *Proc. 13th International Conference on Computer Aided Verification (CAV'01)*, volume 2102 of *LNCS*, pages 194–206. Springer, 2001.

55. C. Lee. Representation of switching circuits by binary-decision programs. *Bell System Technical Journal*, 38:985–999, 1959.

56. K. McMillan. *Symbolic Model Checking*. Kluwer Academic Publishers, 1993.

57. A. Miner. *Data Structures for the Analysis of Large Structured Markov Chains*. PhD thesis, Department of Computer Science, College of William & Mary, Virginia, 2000.

58. A. Miner. Efficient solution of GSPNs using canonical matrix diagrams. In R. German and B. Haverkort, editors, *Proc. 9th International Workshop on Petri Nets and Performance Models (PNPM'01)*, pages 101–110. IEEE Computer Society Press, 2001.

59. A. Miner. Efficient state space generation of GSPNs using decision diagrams. In *Proc. 2002 Int. Conf. on Dependable Systems and Networks (DSN 2002)*, pages 637–646, Washington, DC, June 2002.

60. A. Miner and G. Ciardo. Efficient reachability set generation and storage using decision diagrams. In S. Donatelli and J. Kleijn, editors, *Proc. 20th International Conference on Application and Theory of Petri Nets (ICATPN'99)*, volume 1639 of *LNCS*, pages 6–25. Springer, 1999.

61. A. Miner, G. Ciardo, and S. Donatelli. Using the exact state space of a Markov model to compute approximate stationary measures. In *Proc. 2000 ACM SIGMETRICS Conf. on Measurement and Modeling of Computer Systems*, pages 207–216, Santa Clara, CA, June 2000.

62. D. Parker. *Implementation of Symbolic Model Checking for Probabilistic Systems*. PhD thesis, University of Birmingham, 2002.

63. E. Pastor and J. Cortadella. Efficient encoding schemes for symbolic analysis of Petri nets. In *Design Automation and Test in Europe (DATE'98)*, Paris, February 1998.

64. E. Pastor, J. Cortadella, and M. Peña. Structural methods to improve the symbolic analysis of Petri nets. In *20th International Conference on Application and Theory of Petri Nets*, Williamsburg, June 1999.

65. B. Plateau. On the stochastic structure of parallelism and synchronisation models for distributed algorithms. In *Proc. 1985 ACM SIGMETRICS Conference on Measurement and Modeling of Computer Systems*, volume 13(2) of *Performance Evaluation Review*, pages 147–153, 1985.

66. R. Rudell. Dynamic variable ordering for ordered binary decision diagrams. In *Proc. International Conference on Computer-Aided Design (ICCAD'93)*, pages 42–47, 1993.

67. M. Siegle. Compositional representation and reduction of stochastic labelled transition systems based on decision node BDDs. In D. Baum, N. Mueller, and R. Roedler, editors, *Proc. Messung, Modellierung und Bewertung von Rechen- und Kommunikationssystemen (MMB'99)*, pages 173–185. VDE Verlag, 1999.

68. M. Siegle. Behaviour analysis of communication systems: Compositional modelling, compact representation and analysis of performability properties. Habilitation thesis, Universität Erlangen-Nürnberg, 2002.

69. W. J. Stewart. *Introduction to the Numerical Solution of Markov Chains*. Princeton, 1994.

338 A. Miner and D. Parker

70. A. Xie and A. Beerel. Symbolic techniques for performance analysis of timed circuits based on average time separation of events. In *Proc. 3rd International Symposium on Advanced Research in Asynchronous Circuits and Systems (ASYNC'97)*, pages 64–75. IEEE Computer Society Press, 1997.
71. D. Zampunièris. *The Sharing Tree Data Structure, Theory and Applications in Formal Verification*. PhD thesis, Department of Computer Science, University of Namur, Belgium, 1997.

Probabilistic Methods in State Space Analysis

Matthias Kuntz and Kai Lampka

Friedrich-Alexander-Universität Erlangen-Nürnberg, Institut für Informatik
{mskuntz,kilampka}@informatik.uni-erlangen.de

Abstract. Several methods have been developed to validate the correctness and performance of hard- and software systems. One way to do this is to model the system and carry out a state space exploration in order to detect all possible states. In this paper, a survey of existing probabilistic state space exploration methods is given. The paper starts with a thorough review and analysis of bitstate hashing, as introduced by Holzmann. The main idea of this initial approach is the mapping of each state onto a specific bit within an array by employing a hash function. Thus a state is represented by a single bit, rather than by a full descriptor. Bitstate hashing is efficient concerning memory and runtime, but it is hampered by the non deterministic omission of states. The resulting positive probability of producing wrong results is due to the fact that the mapping of full state descriptors onto much smaller representatives is not injective. – The rest of the paper is devoted to the presentation, analysis, and comparison of improvements of bitstate hashing, which were introduced in order to lower the probability of producing a wrong result, but maintaining the memory and runtime efficiency. These improvements can be mainly grouped into two categories: The approaches of the first group, the so called multiple hashing schemes, employ multiple hash functions on either a single or on multiple arrays. The approaches of the remaining category follow the idea of hash compaction. I.e. the diverse schemes of this category store a hash value for each detected state, rather than associating a single or multiple bit positions with it, leading to persuasive reductions of the probability of error if compared to the original bitstate hashing scheme.

1 Introduction

It is commonplace that distributed, concurrent hard- and software systems have become part of our daily life. Because of our high dependency on these systems, it becomes more and more important to assert that they are working correctly and that they meet high performance requirements. Several methods have been developed to validate the correctness and performance of hard- and software systems. One way to do this is to model the system and generate the corresponding state space.

From some high-level description formalism like (stochastic) Petri nets, (stochastic) process algebras, etc., low level representations are derived. The (low-level) model consists of all possible states a system can be in and the set of all

C. Baier et al. (Eds.): Validation of Stochastic Systems, LNCS 2925, pp. 339–383, 2004.

possible transitions between these states. The concurrency of the modelled system that can be expressed by high-level constructs must be made explicit when deriving such a state-transition model. That means that the number of states grows exponentially in the number of parallel components of the system model. Thus, a severe problem of this approach is the huge state space of concurrent system models. This problem is commonly known as *state space explosion problem*. Thus the size and complexity of systems that can be analysed is restricted, since memory space is limited.

There exist many methods to approach this problem, i.e. to accommodate very large state spaces. The existing methods can roughly be divided into two classes: Approaches that perform exhaustive state space search and methods that perform only partial state exploration. The class of approaches that perform exhaustive state space search, can be divided into approaches that utilise mass storage and/or distributed hardware e.g. [8, 16] and those that represent the state space in an efficient and compact way e.g. [1, 9, 24]. The methods in the second class are the so-called partial analysis methods. There exist many ways to organise a partial analysis. The two most prominent representatives of this class are partial order and probabilistic methods. Partial order techniques exploit the fact that the major contribution to the state space explosion stems from the huge number of possible interleavings of execution traces of concurrent processes. It can be observed that not all interleavings are relevant for system analysis. Generally spoken, partial order methods aim to prune away such interleavings. This can be achieved for example, by defining an equivalence relation on system behaviour [3].

It should be noted that in contrast to probabilistic methods partial order methods still guarantee full coverage of the state space to be analysed.

Using probabilistic methods, it is possible that only a fraction of all reachable states of a system being modelled is analysed, thus the probability that states are omitted is greater than zero. This paper concentrates on probabilistic methods. Probabilistic methods have been proven to be very successful in formal verification of large systems like communication protocols or cache coherency protocols. There exist several tools that employ probabilistic state space exploration. The two most prominent tools are SPIN, a model checker for functional verification [13] and DNAmaca a performance analysis tool [18]. The applicability of probabilistic methods was tested on several examples. Among the tested protocols are the IEEE/ANSI standard protocol for the scalable coherent interface (SCI) protocol, cache coherence protocols for the Stanford DASH Multiprocessor [25], the Data Transfer Protocol (DTP) and the Promela File Transfer Protocol (PFTP) [12, 14, 29]. Probabilistic methods were also applied to model flexible manufacturing systems [18].

The paper is organised as follows: In section 2 we will briefly review hashing. In section 3 state space generation in general will be covered. Section 4 introduces and analyses Holzmann's bitstate hashing, the basis of all probabilistic analysis methods. The following sections present newer variants and improvements of bitstate hashing. A thorough analysis of omission probabilities for the exploration methods presented in sections 9, 10 and 11 can be found in the technical report version of this paper [21].

2   Review of Hashing

In the sequel some principles of hashing will be reviewed. The presentation is based on [20, 22]. We briefly discuss the selection of "good" hashing functions, then we introduce and analyse some hashing methods that are employed by the verification methods that are presented in this paper.

2.1   Basic Terminology

Hashing is a scatter storage technique, where records that are to be stored are distributed over a prescribed address range. Hashing aims to retrieve a stored data record by computing the position where the record might be found. For this purpose a hashing function is applied. The parameter k of the hash function is called "key". The set of all keys is denoted $\mathcal{K}$, a key is a natural number such that $\mathcal{K}$ is a subset of the natural numbers. A hash table is an array with indices $0 \ldots m - 1$, where m is the size of the hash table. A hash function h is defined as follows:

$$h : \mathcal{K} \mapsto \{0, \ldots, m - 1\}$$

The value $h(k)$ is called hash address or hash value. Usually, a hash function is not injective. If two keys k and k', map to the same hash value, i.e. $h(k) = h(k')$, a so-called (address) collision occurs. Such keys are also called "synonyms".

2.2   Selection of "good" Hash Functions

Requirements: There are several requirements that a "good" hash function has to fulfill:

- Efficiency: It must be possible to compute the hash function easily and quickly, because it has to be computed millions of times.
- Distribution: To avoid a large number of address collisions the hash values should distribute evenly over the hash table.

Division-with-rest Method: The simplest method to generate a hash address $h(k)$ for given k in a range between 0 and $m - 1$ is to take the remainder of k on integer division by m:

$$h(k) = k \bmod m$$

The choice of m is decisive for the quality of the hash function. If m is even (odd), then $h(k)$ is even (odd), if k is. If m is a power of the number system in which the keys are represented, i.e. $m = r^i$, if r is the basis of the number system, then $h(k)$ has as outcome only the i least significant digits of k. A "good" choice for m is a prime, such that none of the numbers $r^i \pm j$, where r is the basis of the keys' number system, i and j small positive integers, is divisible by m [20, 22].

Multiplicative Method: The given key is multiplied by an irrational number φ. The integer part of the result is cut off. So, for different keys, different values between 0 and 1 are obtained. Ideally, for keys $1, 2, 3, \ldots n$ these values are equally distributed on the interval $[0, 1)$.

$$h(k) = \lfloor m \cdot (k \cdot \varphi - \lfloor k \cdot \varphi \rfloor) \rfloor$$

Universal Hashing: For hashing to produce a "good" result it is important that the keys are equally and uniformly distributed over the set of all possible keys. This assumption is somewhat problematic, as keys are often not uniformly distributed. So, instead of making critical assumptions about the distribution of the keys, assumptions about the hash method are made, that are less problematic. Using universal hashing, one hash function is picked in a uniform and random manner out of a set $\mathcal{H}$ of hash functions. The randomisation in the selection of the hash functions shall guarantee that an unevenly distributed set of keys does not lead to many hash collisions. Note that a single function h of $\mathcal{H}$ might still lead to many collisions but the average over all functions of $\mathcal{H}$ diminishes this.

If $\mathcal{H}$ is a collection of hash functions, with $h : \mathcal{K} \mapsto \{0, \ldots, m - 1\}$, for all $h \in \mathcal{H}$ then $\mathcal{H}$ is called "universal" if for any two keys k, k', $k \neq k'$

$$\frac{|\{h \in \mathcal{H} \mid h(k) = h(k')\}|}{|\mathcal{H}|} \leq \frac{1}{m}$$

holds, i.e. if at most the m-th part of the hash functions leads to a hash collision for the keys.

2.3   Collision Resolution by Chaining

This section introduces chaining and open addressing as collision resolution methods.

Assume that a hash table already contains a key k, at which point another key k' is to be inserted, such that $h(k) = h(k')$, i.e. a collision occurs. The key k' has to be stored somewhere else, not on position $h(k')$. Chaining stores keys that collide with keys that are already in the hash table outside the table. For that purpose a dynamic data structure, a linked list, is associated with each slot of the hash table that contains the keys k' that collide with k.

There are two ways to realise hashing with chaining:

Separate Chaining of Collided Keys: Each element of the hash table is the first element of a linked list containing the collided elements.

Direct Chaining of Collided Keys: Each element of the hash table is a pointer to a linked list. That means that the real collided elements start with the second entry of the linked list. For a more detailed description of chaining cf. for example [20, 22].

2.4   Collision Resolution with Open Addressing:

In contrast to chaining, open addressing stores all keys inside the hash table. If upon insertion of key k' another key k is already stored at $h(k')$, then according to a fixed rule another slot has to be searched on which k' may be accommodated. For that purpose for each key a fixed order in which slots are inspected is defined. As soon as an empty slot is found, the key is stored at this address. The sequence of slots that are inspected during the search is called "probe sequence". Such methods are called "open addressing schemes". There are many variants of open addressing of which only those are briefly introduced that are relevant for this paper. For a more thorough overview we refer to [22] and [20]. and the literature referred to there.

Clustering: Clustering is a phenomenon that may occur with open addressing. Two types of clustering can be distinguished:

- Primary clustering: If adjacent slots of the hash table are filled then primary clustering leads to the effect that these areas tend to become larger.
- Secondary clustering: Two synonyms k and k' are going through the same probe sequence, i.e. interfere with each other on insertion.

Open Addressing Schemes: Assuming a hash table with m slots, open addressing applies a probe sequence that is a permutation of all m possible hash values.

The permutation depends only on the actual key.

- **Uniform Probing:** Under the assumption of uniform probing, each of the $m!$ possible permutations should be equally likely. Practical realisation of uniform probing is very difficult, such that alternatively a scheme called "random probing" is applied, which is almost as efficient as uniform probing [22].
- **Double Hashing:** Here, address collisions are resolved by applying a second hash function h', which ideally should be independent of the primary hash function h. h' has to be chosen in such a way that the probing sequence forms a permutation of the hash addresses, i.e. it is required that $h'(k) \neq 0$ and $h'(k) \not| m$, if m is prime this can be granted. The function h' is used to compute the probe sequence, if $h(k)$ produces a collision. The chosen probe sequence is given as follows:

$$h(k) \bmod m, (h(k) + h'(k)) \bmod m, \ldots, (h(k) + i \cdot h'(k)) \bmod m, \ldots,$$
$$(h(k) + (m - 1) \cdot h'(k)) \bmod m$$

- **Ordered Hashing:** Ordered Hashing is a method that aims to speed up during unsuccessful searches. To reach that goal, ordered hashing simulates ordered insertion, i.e. the hash keys are inserted in an ordered fashion. This method is explained by means of an example:

Assume that items 12 and 19 have been inserted into a hash table. Let the elements 12 and 19 be synonyms that have been inserted in that order. Now, assume that key 5 is searched for, which is synonymous to 12 and 19. The search for 5 starts at address $h(5)$, which is equal to $h(12)$, at this position 12 can be found. Without ordered hashing we would have to continue probing until we can decide that 5 is not in the hash table. Using ordered hashing, we could stop probing after the first probe, which yielded item 12 at position $h(5)$, as $12 > 5$.

Analysis of Uniform Probing: We now turn to the analysis of the efficiency of uniform probing, which is slightly better than random probing. It is assumed that the hash table has m slots, of which $m - n$ are free. Under the assumption of uniform probing, each of the $\binom{m}{n}$ possible assignments of m slots to n hash values is equally probable. The probability, that exactly i slots have to be probed when inserting a new element is computed as follows [22]:

$$p_i = \frac{\binom{m-i}{n-i+1}}{\binom{m}{n}} \tag{1}$$

Eq. 1 stems from the fact, that apart from the $i-1$ fixed keys also $n-(i-1)$ keys that are still to be inserted have to be distributed on the $m - i$ free slots. The average number of probed slots on insertion, C'_n, is then computed as follows:

$$
\begin{aligned}
C'_n \quad &= \quad \sum_{i=1}^{m} i \cdot p_i \\[2mm]
\overset{\sum_{i=1}^{m} p_i = 1}{=} \quad & m + 1 - \sum_{i=1}^{m}(m+1) \cdot p_i - \sum_{i=1}^{m}(-i \cdot p_i) \\[2mm]
&= \quad m + 1 - \sum_{i=1}^{m}(m+1-i) \cdot p_i \\[2mm]
\overset{\binom{a}{b}=\binom{a}{a-b}}{=} \quad & m + 1 - \sum_{i=1}^{m}(m+1-i) \cdot \binom{m-i}{m-n-1} / \binom{m}{n} \\[2mm]
\overset{\binom{a}{b}=\frac{b+1}{a+1}\cdot\binom{a+1}{b+1}}{=} \quad & m + 1 - \sum_{i=1}^{m}(m+1-i) \cdot \frac{m-n}{m-i+1}\binom{m-i+1}{m-n} / \binom{m}{n} \\[2mm]
&= \quad m + 1 - (m-n) \cdot \sum_{i=1}^{m}\binom{m-i+1}{m-n} / \binom{m}{n} \\[2mm]
\overset{\sum_{c=1}^{a}\binom{c}{b}=\binom{a+1}{b+1}}{=} \quad & m + 1 - (m-n) \cdot \binom{m+1}{m-n+1} / \binom{m}{n} \\[2mm]
&= \quad m + 1 - (m-n) \cdot \frac{m+1}{m-n+1} \\[2mm]
&= \quad \frac{m+1}{m-n+1} \tag{2}
\end{aligned}
$$

The average number of probed slots at successful searches using open addressing, C_n is given below:

$$C_n \quad = \quad \frac{1}{n}\sum_{i=0}^{n-1} C_i'$$

$$= \quad \frac{1}{n}\sum_{i=0}^{n-1} \frac{m+1}{m-i+1}$$

$$= \quad \frac{m+1}{n} \cdot (\frac{1}{m+1} + \frac{1}{m} + \dots \frac{1}{m-n+2})$$

$$\underset{H_n=\sum_{i=1}^{n}\frac{1}{i}}{=} \frac{m+1}{n} \cdot (H_{m+1} - H_{m-n+1})$$

This equation can be justified as follows:

Searching a key k follows the same probe sequence as it was applied, when k was inserted. If k is the $(i+1)$st element that was inserted into the hash table, then the expected number of probes, searching for k equals $\frac{m+1}{m-i+1}$. The average of all n keys in the hash table yields the factor C_n.

3   State Space Exploration

As already mentioned in the introduction, the first step in functional or temporal performance analysis is the generation of the state graph (low-level model description). Therefore a short introduction to the state space (SSp) generation will follow now. The section will be rounded by motivating and introducing coarsely probabilistic SSp exploration and clarifying the basic terminology.

3.1   Exhaustive, Explicit State Space Generation

The technique of generating all reachable states of a modeled system is called *exhaustive* SSp exploration. If every state descriptor is explicitly and fully represented by a record, one speaks of *explicit* SSp representation. The data structure required for an exhaustive explicit SSp exploration scheme is a buffer (state buffer) and a large somehow structured storage space (state table). The former holds the detected but yet not explored states, where the latter is used for storing the already visited states. Usually a state is considered to be explored once all its successor states have been determined. Depending on how the state buffer is accessed a depth-first search (dfs) or a breadth-first search (bfs) on the SSp is implemented. The size of the state buffer is bounded by the depth, breadth respectively of the explored state graph and the number of bits required for storing the state descriptors. Since the buffer is accessed in a structured manner, the currently not used parts can be swapped onto secondary storage device (e.g. HDD). Consequently the state buffer is not the bottleneck of an exhaustive and explicit SSp exploration as far as access time and memory consumption are concerned.

The state table holds the already visited states, thus it serves as a database, whose single purpose is an effective look-up of a given state. Common layouts for the state table as used under exhaustive and explicit SSp-exploration differ concerning allocation strategy (dynamic or static) and structure. Their commonness is usually the use of a hashing scheme, where the state descriptors serve as keys and where collision resolution (sec. 2.3) needs to be applied. One may note, that even though nowadays hard disk drives (HDDs) are fast and employ large buffers and block-wise access patterns, the state table needs to be kept in total in the memory. This is due to the fact, that the accesses to the state table are not structured and therefore buffer strategies may not be applied (most recently read...). Concerning the large HDD-buffers one may consider, that they can be viewed as additional memory space. This extensional RAM will only put back the boundaries, but will be easily absorbed by SSps of one or two magnitudes larger, which are not a curiosity when it comes to the modeling of large and highly complex systems. As a consequence and due to the limitation of memory the number of states to be explored and thus the size and complexity of models to be analyzed is restricted.

Figure 1 shows how an exhaustive exploration of a SSp proceeds in principle: The state buffer is here organized as stack $Stack$, thus the algorithm follows a dfs-scheme. The state table is organized as a hash table (H). Whether the algorithm follows a $direct$ - or $separate$ chaining, or a $probing$ approach for handling collided state descriptors is hereby transparent, since a state descriptor s is inserted into the hash table by calling the function $insert(s, H)$.

```
(1)      Stack := ∅, H := ∅, s₀ := initial state
(2)      push(s₀, Stack)
(3)      insert(s, H)
(4)      DFS()
(5)      DFS() {
(6)           while (Stack ≠ ∅)
(7)                s = pop(Stack)
(8)                forall successors s′ of s do
(9)                     if s′ ∉ H
(10)                         push(s′, Stack)
(11)                         insert(s′, H)
(12)                    endif
(13)               od
(14)          endwhile
(15)   }
```

Fig. 1. Algorithm for Exhaustive State Space Search

It is now assumed that *direct chaining* is used for collision resolution and that the number of reachable states (n) is greater than the number of slots m in the hash table. Thus the average of n/m states will obtain the same hash value. If the table is full an average of n/m comparisons is necessary to decide if a state is new. With increasing number of comparisons the time efficiency of the search decreases. If an address pointer requires p bytes, then the hash table consumes $m{\cdot}p$ bytes of memory. Under chaining each entry of the hash table needs $mem(s)$ bytes and an additional p bytes for its referencing. To accommodate the largest possible fraction of the SSp m should be small to keep the memory consumption as small as possible. On the other hand, m should be large to minimise the number of hash conflicts to keep the time efficiency of the SSp exploration high. The memory requirement for administering n states O, if chaining for collision resolution is employed, is given by:

$$O = m \cdot p + (mem(s) + p) \cdot n \qquad (3)$$

3.2   Introduction to Probabilistic State Space Exploration

For exemplification think now of a bounded Petri net (PN) consisting of 1000 places. Let the maximum number of tokens per place be less than 256, yielding a memory consumption of 1 KByte per state descriptor under a non-efficient straight forward encoding scheme. Thus a SSp of $n = 10^6$ states requires 1000 MByte for storing the 10^6 distinct state descriptors only (eq. 3: $mem(s) = $ 1KByte). Furthermore one would need to spend additional memory space for the administration structure required for resolving collisions. I.e. by employing direct chaining and a non-optimal 64-bit addressing scheme and a maximal number of slots ($m = 10^6$) another 16 MByte of memory is required (eq. 3: $p = 8$ MByte). Even under such a worst-case scenario concerning memory space, the memory requirement of the administrative structure for collision resolution is still less than 2% of the overall required memory space (eq. 3: $10^6 \cdot 8\,\text{Byte} + $ ($1000\,\text{Byte} + 8\,\text{Byte}$) $\cdot 10^6 = 1016$ MByte). Thus the space complexity of the administration structure is not the bottleneck of exhaustive and explicit SSp exploration. –Alternatively one may think now of swapping parts of the SSp on to secondary storage devices. But the occurrence of states is mostly not structured, consequently an expensive look-up for each swapped state must be taken into account. Assuming a memory size of $\approx$ 500 MByte RAM, in the above illustrated scenario, one would need to swap $5 \cdot 10^5$ states onto HDD. As a consequence one has to spend $\approx 3hours$ only to reload $5 \cdot 10^5$ states twice into memory, assuming a HDD access time of 10 ms per read operation. Thus the employment of secondary storage devices is for its time complexity not applicable and the number of states to be explored bounded by the size of the available RAM (cf. section 3.1). For coping with the problem of limited memory on one hand and the SSp explosion problem on the other hand many approaches have been developed, targeting a decrease in memory consumption of large SSps and keep the runtime at moderate levels.

A number of approaches, which have been applied successfully (mainly in the tools SPIN [11] and DNAmaca [15]), are based on the idea of not storing the full state descriptors of visited states, but storing a smaller representative for each instead. However since the mapping of the keys (state descriptors) onto the representatives (hash values) is not injective, different state descriptors may be mapped on the same representative[1]. Therefore the exploration algorithm may consider states as identical, which are in fact not. Consequently the later reached states will be misleadingly dropped and not explored, denoted as *state omission*. Furthermore their successor states might not be explored, if they are only reachable through omitted states. Thus the state table may not contain all reachable states. Consequently the modeler takes into account, that the evaluated functional properties and / or performance results might be wrong[2]. I.e. the analysis can produce a positive result (e.g. deadlock free), which is in fact wrong, denoted as the production of *false positives*. Since the omission of states and therefore the production of wrong results is not predictable, nor one does know which portion of the SSp will be visited, such methods having this disadvantage are referred to as *probabilistic SSp exploration* methods.

Since the probabilistic exploration techniques achieve tremendous savings in memory consumption and thus runtime and since they do neither depend on a specific high-level model description method, nor on structural properties of a model, they might produce results, where otherwise an exhaustive, explicit SSp exploration and thus an analysis of a system will fail. Consequently the flaw of producing false positives, might be accepted, where otherwise no analysis could be carried out. Therefore the consumption of memory at a fixed level of error, as well as their time complexity seems to be the most adequate benchmarks for comparing the different probabilistic approaches, where in the sequel of this survey paper the former will be emphasized.

The chance of error, which is induced by the omission of states, is denoted as probability of omission of states (p_{om}). It describes the probability that states might be skipped during SSp generation. Thus one speaks of SSp coverage, if referring to the probability of visiting all reachable states, which is given by $p_{cov} = 1 - p_{om}$. The omission of a specific state depends hereby on the probability of its own collision as well as on the probability that its predecessors were omitted. For simplicity one often presumes, that each state is reachable on different paths (well connected components), thus the condition of a predecessors exploration can be dropped. As a consequence the SSp coverage is equal to the probability that no state collided ($p_{cov} = p_{no}$). But one should be aware, that this consideration does often not coincide with the reality, especially in the field of reliability analysis, since faulty behaviours leading to disastrous outcomes often only occur under very specific conditions, i.e. they are only reachable via few specific paths in the SSp. A placement of the probabilistic approaches within the different methods of SSp exploration is given in figure 2, where each method presented in this paper is briefly characterized.

---

[1] One should note, that the state descriptors may not be uniformingly distributed, thus it might be difficult determining a posteriori an adequate hash function, which will produce uniformingly distributed hash values.

[2] In case of a model's functional behaviour, this means for example, that a model might be considered of being deadlock free, which it is in fact not, due to the coincidental omission of the deadlock states during SSp exploration.

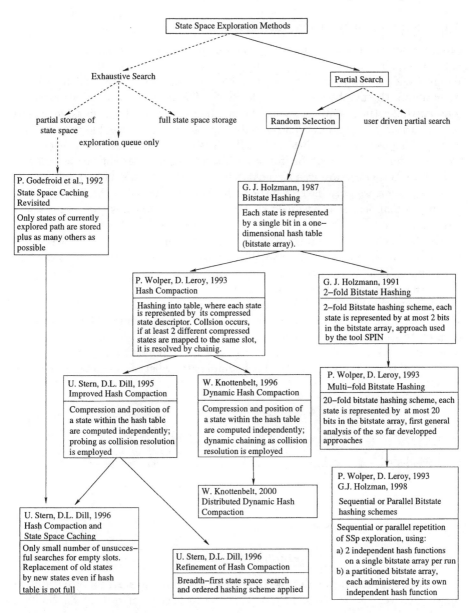

Fig. 2. Classification of probabilistic SSp exploration techniques

4   Bitstate Hashing

In this section the principles of bitstate hashing as it was introduced by Holzmann in [10–12] shall be presented in greater detail. Bitstate hashing can be considered the father of several probabilistic state space exploration methods that will be presented in this paper.

4.1 Functionality of Bitstate Hashing

From section 3 we know that the standard method to maintain a state space is hashing. This traditional approach has the main drawback, that the memory requirements for explicitly storing a single state are very high, which yields a reduction in the number of explorable states. The idea of bitstate hashing is as follows: Under the assumption that the number of slots in the hash table is high in comparison to the number of reachable states, then it is possible to identify a state by its position in the hash table. If this is possible, then it is not necessary to store a full state descriptor in the hash table, a single bit suffices in order to indicate, whether a state was already visited or not. I.e. to decide if a state s is already present we only have to inspect the value of the bit at the index $h(s)$ to which s is mapped. It is assumed that bit value "1" indicates that state s was already found and bit value "0" means that s was not found yet.

At the beginning of the state space search all bits in all slots are set to zero. If M bytes of memory are available, we can use $8 \cdot M$ bits to maintain the state space. That means, the hash table consists of a one-dimensional bit array (bitstate array), whose indices range from $0 \ldots 8 \cdot (M-1)$, i.e. $m = 8 \cdot M$. From that follows, that up to m states can be analysed. The idea of bitstate hashing is illustrated in fig. 3. From a given representation of the system to be analysed, for each of its states a state descriptor s is derived. For s a hash value is computed that lies between 0 and $m-1$. The hash value corresponds to the address of the hash table, where the value of the bit flag indicates whether s is already present or not.

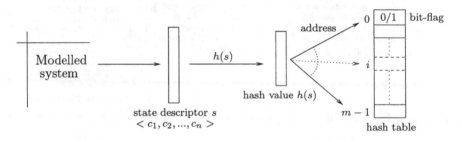

Fig. 3. Bitstate hashing

Ideally, $h(s)$ should be defined in such a way that the states are distributed uniformly over the m slots. Fig. 4 gives the respective algorithm for bitstate hashing, where it is assumed that the basic state space traversal is done by a depth first search.

Bitstate hashing can only be recommended if the size of the state space, and the memory requirements to represent all these states explicitly render an exhaustive search infeasible. As the states are not stored explicitly, hash conflicts cannot be resolved. That means, if s and s' are mapped onto the same position in the hash table, s' and its successors will be ignored, because it is assumed that s' has already been analysed. As $n/m \to 0$ the expected number of hash conflicts approaches zero. From that follows, that m should be as large as possible.

```
(1)   Stack := ∅, i := 0 /*Lines (1) to (5) initialisation of hash table
(2)   while i < m do
(3)          H[i] := 0
(4)          i := i + 1
(5)   od
(6)          j := h(s₀)
(7)          H[j] := 1
(8)          push(s₀, Stack)
(9)          DFS();
(10)  DFS() {
(11)  while (Stack ≠ ∅)
(12)         s = pop(Stack)
(13)         forall successors s' of s do
(14)             k := h(s')
(15)             if H[k] = 0
(16)                 H[k] := 1
(17)                 push(s', Stack)
(18)             endif
(19)         od
(20)  endwhile
(21)  }
```

Fig. 4. Algorithm for Bitstate Hashing

Number of Analysable States: For exhaustive state space search, the maximum number of analysable states given certain memory constraints, is $M/mem(s)$, given $mem(s)$ bytes are needed for storing a single state and M is the available memory space in bytes. In contrast, using bitstate hashing this number can be incremented up to $8 \cdot M$, where the hash conflicts scatter the set of analysed states over the whole set of reachable states.

When comparing the coverage of traditional exhaustive search with that of bitstate hashing, we can distinguish two cases:

1. $n \leq M/mem(s)$: In this case the traditional approach performs better than bitstate hashing, as the latter cannot guarantee a coverage of 100%.
2. $n > M/mem(s)$: In this case, an exhaustive search deteriorates into a partial search, for which the coverage might be lower than for bitstate hashing.

Finally, we compare the memory requirements of bitstate hashing with traditional hashing. For traditional hashing, the memory consumption was assessed in eq. 3 Now, O reduces to

$$O = \frac{m}{8} \tag{4}$$

bytes.

5   Analysis and Criticism of Bitstate Hashing

In this section we will derive some efficiency measures for bitstate hashing, as it was done in [14]. These measures are the collision probability, the loss probability and the coverage ratio.

5.1   Analysis

Let n be the number of reachable states, $mem(s)$ the number of bytes necessary to store a single state. Finally, let M be the size of memory in bytes that is available for state space exploration. That means, that the size of the hash table m is equal to $8 \cdot M$.

For the analysis of the bitstate hashing method we compute the average collision probability if n states are inserted into the hash table. It should be noted that each collision in bitstate hashing leads to the omission of at least one state. The state that collides and all its successor states are omitted. The omission of the successor states is not too severe as *it is assumed that the transition graph is well connected*, i.e. several paths lead to each state such that it is likely that the state omitted due to a collision will be explored anyway. The hashing function is assumed to distribute states uniformly over the m slots of the hash table. We distinguish now the follwing two cases:

1. $n \leq m$: The collision probability for the first state to be inserted is clearly zero, The last state to be inserted has collision probability at most $\frac{n-1}{m}$, if no collision has occurred during the insertion of the other $n - 1$ states. On average, this consideration yields the following estimation for the collision or equivalently omission probability $\tilde{p}_{om}$:

$$\tilde{p}_{om} \leq \frac{1}{n} \sum_{i=0}^{n-1} \frac{i}{m} = \frac{1}{n \cdot m} \sum_{i=0}^{n-1} i = \frac{1}{n \cdot m} \cdot \frac{n \cdot (n-1)}{m} = \frac{n-1}{2m} \qquad (5)$$

Note carefully that the approximation in eq. 5 says only that the omission probability for state i is on average $\tilde{p}_{om}$. The coverage ratio using the approximation of eq. 5 is $1 - \tilde{p}_{om}$. (Since it is assumed, that each collision leads to the omission of only one state, we can identify coverage ratio and probability of no omissions.) This estimation is problematic because of the fact that $\tilde{p}_{om}$ only approximates an average omission probability for a particular state. A more accurate approximation of the probability of no ommissions will therefore be

$$(1 - \tilde{p}_{om})^n, \qquad (6)$$

where the probability of no omissions is now taken over all n reachable states. In [14] it seems that, when comparing the coverage ratio of bitstate hashing with that of a traditional exhaustive search the approximation of eq. 6 was applied, even if never stated.

2. $n > m$: There are more reachable states than slots in the hash table. That means $n - m > 0$ is the minimal number of collisions during insertion. From that follows that the probability that collisions occur at all is certainly one. For a particular state the collision probability is normally less than one. This probability can be estimated as follows:

$$\tilde{p}_{om} = 1 - \frac{m}{n} \tag{7}$$

$\frac{m}{n}$: Fraction of states that are mapped onto different slots.

Finally, we estimate the omission probability using an exhaustive state space exploration method. The omission probability is greater than zero if the number of reachable states is larger than the memory available for state space search:

$$\tilde{p}_{om} = 1 - \frac{M}{n \cdot mem(s)} \tag{8}$$

Recall, M is the available memory in bytes. For each state $mem(s)$ bytes are required, such that the overall memory requirements for storing all n states are equal to $n \cdot mem(s)$. The percentage of states, that can be stored without omissions is thus $\frac{M}{n \cdot mem(s)}$.

5.2   Approximating the Probability of State Omissions

In contrast to Holzmann [14], Leroy and Wolper [29] argued, that the average probability of a state collision is not an appropriate measure for estimating the SSp coverage rate. Their estimation for the SSp coverage is based on the idea, that the SSp is completely explored, if no collision of states occurred. –In the following we consider n states to be explored and a bitstate array consisting of $m > n$ bit positions.

The first state has clearly the chance of $(1 - 0/m)$ of being mapped onto an occupied position, since no bit is marked as occupied at this stage. Assuming that i states have been entered collision-free so far,i.e. i bit positions are occupied, the $(i + 1)$'st state has then the chance of $1 - (i/m)$ of being mapped onto an occupied bit position. Analogous the n'th state to be entered has the chance of $1 - ((n-1)/m)$ of being mapped onto an occupied bit position and thus of being omitted. Since the SSp exploration is exhaustive only, if each position has got at most one state after all n states are explored (= no collision has occurred), the probability of no collision p_{no} is given by the conjunction of the probabilities of each individual state being inserted collision-free into the bitstate array. Since well-connected SSps are assumed, the individual probabilities are independent, yielding:

$$p_{no} = \left(1 - \frac{0}{m}\right) \times \cdots \times \left(1 - \frac{i-1}{m}\right) \times \cdots \times \left(1 - \frac{n-1}{m}\right) = \prod_{i=0}^{n-1} 1 - \frac{i}{m} \tag{9}$$

Since in nowadays hardware configurations one can allocate easily bitstate arrays of some billion bits[3], it can be assumed that the number of available bit positions m is not only very large, but exceeds the number of states n by far, i.e. $m \gg n$. Thus $\frac{(n-1)}{m} \ll 1$ and $1 - \frac{(n-1)}{m}$ can be approximated by $e^{-\frac{(n-1)}{m}}$. This approximation togehter with $\sum_{i=0}^{n-1} \frac{i}{m} = \frac{n \cdot (n-1)}{2m} \approx \frac{n^2}{2m}$ for $n \gg 1$ can be employed in eq. 9, yielding the following approximation:

$$p_{no} \approx \prod_{i=0}^{n-1} e^{-\frac{i}{m}} = e^{-\sum_{i=0}^{n-1} \frac{i}{m}} = e^{-\frac{n \cdot (n-1)}{2m}} \approx e^{-\frac{n^2}{2m}} , \qquad (10)$$

which can be interpreted as the SSp coverage p_{cov}. Since eq. 9 is indeed the same as the one found in [30, 15, 19, 17, 18]:

$$p_{no} = \frac{1}{m^n} \cdot \frac{m!}{(m-n)!} = \prod_{i=0}^{n-1} 1 - \frac{i}{m}, \qquad (11)$$

the more intuitive explanation discussed above yields the same approximation for the probability of no collision. However one should note, that Leroy et al. [30, 17], as well as Knottenbelt et al. [18] developed a more pessimistic approximation. Instead of employing $1 - \frac{(n-1)}{m} \approx e^{-\frac{(n-1)}{m}}$ the authors apply the Stirling approximation for factorials on eq. 11, yielding $p_{no} \approx e^{-n^2/m}$.

Following Knottenbelt [19], one may conclude, that eq. 10 is a lower bound in cases, where $\frac{2}{3}n^2 < m$ holds. This condition is tolerable in all cases of practical interest, since for a high probability of no collision $p_{no} \approx 1$: $n^2 \ll 2m$. Furthermore one should note, that p_{no} can be interpreted as the minimum probability of collision, if at most n states are explored. I.e. in case more states are explored, one would need to readjust m in order to maintain the SSp coverage. This gives one the maximum probability of omission $P_{om}^{max} = 1 - P_{no}^{min}$. This means in a nutshell, that the SSp exploration needs to be aborted, if more than $n = \sqrt{-2m \cdot \ln(p_{no})}$ (derived from eq. 10) states are visited, since otherwise P_{om}^{max} can not be guaranteed. The notion of P_{no}^{min} and P_{om}^{max} will become important if the hash table oriented schemes are compared, since they allow the storage of at most n states only.

5.3   Criticism of Bitstate Hashing

In the sequel the need for new approaches in probabilistic state space exploration is justified by presenting some reasonable criticism of bitstate hashing and its efficiency analysis, as it was reported in this section and in section 4.

In [14] Holzmann derived some approximate formulae to compute the expected coverage of bitstate hashing. This factor is equal to the average probability

---

[3] Todays 32-bit architecture restricts the size of a state table to the size of 2^{32}, i.e. $m \leq 8 \cdot 2^{32}$.

that a particular state is not omitted in analysis. This average is computed over all reachable states. In [14] it is assumed that one hash collision leads in average to the omission of only one successor state. This assumption is in general not valid. In addition this omission probability does not constitute a bound on the omission probability of a particular state. This is essential, since not all states have the same omission probability, states that are only reachable via one or few long paths have a much higher omission probability. It is possible that such states are error states, such that we can not give a bound of the probability that a verifier generates false positives. A further assumption that is also questionable is the assumption that the reachability graph is well-connected. This assumption reduces the omission probability only if the paths that lead to a certain state have only few states in common, but this is doubtable.

From eq. 10 one can conclude, that bitstate hashing requires a huge number of unused positions in the bitstate array in order to maintain a high level of SSp coverage. Since for achieving a high probability of no collision (≈ 1), $n^2/2m$ needs to be close to zero ($2m \gg n^2$). I.e. employing eq. 10 on the memory utilisation rate ($\mu_{mem} = n/m$) and for levels of SSp coverage toward 100% one obtains:

$$\mu_{mem} \approx \sqrt{\frac{-2 \cdot \ln(p_{no})}{m}} \quad \overset{p_{no} \to 1}{=} \quad 0,$$

Thus the major disadvantage of bitstate hashing, is the waste of memory, when it comes to SSp coverage rates close to 1. For exemplification one may assume a SSp of size $n = 10^8$. In order to maintain a $p_{no} = 1 - 10^{-6}$ one would need to allocate a bitstate array of size 625 EByte ($8 \cdot 625 \cdot 10^{18}$ Bit). Such memory space is truly beyond even tomorrows hardware configuration, since even an advanced 64-bit architecture is limited to the use of bitstate arrays of 18.5 EByte ($8 \cdot 2^{64}$ bits = 147.5 Ebit). The other way round, i.e. considering a todays (tomorrows) hardware configuration, say 500 MByte (10 GByte) of RAM, bitstate hashing allows one to explore only $\approx 29,032$ ($\approx 129,837$) states at a poor level of p_{no} of 0.9. Therefore Holzmann [12, 14] suggested to executed k runs (SSp explorations), each with its own independent hash function, augmenting p_{no} to $1 - p_{om}^k$, where $p_{om} < 1$.

6   Multiple Hashing Schemes

The original probabilistic SSp exploration scheme introduced by Holzmann [10] has the disadvantage, that the number of bits m, required to maintain a low probability of state omission p_{om}, needs to be much larger than the number of states to be explored n. In order to increase SSp coverage and minimise the probability of a wrong result some proposals have been made. The suggestions range from the sequential use of 2 independent hash functions up to the repeated execution of SSp generation by employing either k bitstate hashing schemes in parallel or a multi-fold hashing scheme sequentially. This section will focus only on the two main approaches as proposed in [29]. However, before we proceed one should note that in nowadays' hardware configurations one can easily allocate bitstate arrays of some billion bits, consequently in the following only the case $n \ll m$ will be considered.

6.1   Multiple Hashing on a Single Bitstate Array

Multiple hashing is based on the application of k independent hash functions, where each hash function determines a specific bit position, which has to be checked for determining a state's first or repeated occurrence. I.e. a state is considered to be already visited, if and only if all k bit positions in the bitstate array are occupied (flagged with 1), i.e.

if $H[h_1(s)] = 1 \wedge H[h_2(s)] = 1 \wedge \ldots \wedge H[h_k(s)] = 1$

then state s is assumed to be already visited.

Thus each state is now represented by at most k bits, where each bit to be checked is determined by applying one of the k independent hash functions to the state descriptor. In contrast to the subsequently discussed alternative scheme, the multi-fold hashing scheme analyzed below operates on only *one* bitstate array consisting of m bits.

Holzmann [12] introduced a 2-fold, Wolper and Leroy [29] a 20-fold hashing scheme, carried out on a single bitstate array of size m. For shortening the discussion, we will now directly consider the general case of a k-fold hashing scheme, where the analysis is based on the assumption that each state inserted will occupy exactly k bits in the bitstate array. Since the hash functions are independent there are m^k ways of selecting k out of m positions. Due to the above assumption, i states will occupy $i \cdot k$ of the m positions in the bitstate array. Applying k hash functions, $(i \cdot h)^k$ combinations consisting of occupied positions only can be generated. I.e. after inserting state i into the table $m^k - (i \cdot h)^k$ combinations using at least one free bit position are available. This gives the probability of inserting state $(i + 1)$ into the bitstate array without colliding on all bit positions:

$$p_{no}^{i+1} = \frac{m^k - (i \cdot k)^k}{m^k}$$

Thus the probability of no collision for a k-fold scheme can be described by [29]:

$$p_{no} = \prod_{i=0}^{n-1} \frac{m^k - i^k \cdot k^k}{m^k} = \prod_{i=0}^{n-1} 1 - \left(\frac{i \cdot k}{m} \right)^k , \tag{12}$$

As a consequence of the assumption, that each state occupies k bits eq. 12 is pessimistic, since in reality most states will collide at least on some bit positions and even some of the k independent hash function may map a specific state descriptor onto the same bit position. In the same manner as eq. 10 was derived, one obtains as approximation for the probability of no collision under the k-fold hashing scheme for $\frac{k \cdot (n-1)}{m} \ll 1$:

$$p_{no} = \prod_{i=0}^{n-1} 1 - \left(\frac{i \cdot k}{m}\right)^k \approx \prod_{i=0}^{n-1} e^{-\frac{i^k \cdot k^k}{m^k}} = e^{-\frac{k^k}{m^k} \sum_{i=0}^{n-1} i^k}$$

For $\sum_{i=0}^{n} i^k = n^{k+1}/(k+1) + O(n^k)$ [7, 30] and $n \gg 1$ one obtains:

$$p_{no} \approx e^{-\frac{n^{k+1} \cdot k^k}{(k+1) \cdot m^k}} \tag{13}$$

Thus a 2-fold hashing scheme employed on a SSp of 10^8 states requires a memory space of $\approx 144 \cdot 10^3$ GByte in order to maintain a SSp coverage rate of $1 - 10^{-6}$, which is much better than the original bitstate hashing scheme (see example at end of sec. 5.3, p. 354), but still not satisfactory. After assessing various scenarios differing in the probability of no collision and size of available memory, Wolper and Leroy [29] came to the conclusion, that the use of 20 hash function is the scheme of choice. The decision for $k = 20$ is hereby influenced by the fact, that beyond a certain level of k, the bitstate array will fill up too quickly and thus have a negative impact on the probability of no collision, whereas a small number of hash functions will require a huge bitstate array in order to achieve high coverage rates. However, as illustrated in fig. 5 and fig. 6, where in contrast to Wolper and Leroy [29] higher levels of SSp coverage and larger sizes of memory space have been considered, a choice of $k = 20$ seems only appropriate in cases where lower levels of SSp coverage ($p_{om} < 10^{-6}$) are sufficient and a large number of states ($> 8 \cdot 10^7$) needs to be explored. I.e. in nowadays hardware configurations a choice of $k = 30$ seems more appropriate, in order to keep the probability of state omission low.

Let us now consider a scenario, where a bitstate array of 500 MByte ($4000 \cdot 10^6$ Bit) is available. Under the suggested 30-fold hashing approach, one would be able to explore 52.168 (65.191) million states, at a level of 99.9999% (99.99%) of SSp coverage. In case of Holzmann's 2-fold hashing scheme one would be able to explore only $22,894$ ($106,268$) states. It is interesting to note that in this setting, assuming a use of 30 bits per state, the 30-fold hashing scheme yields a memory utilisation rate of $\mu_{mem} \approx 38\%$ ($\approx 49\%$). Thus 47 (31) bits per state are then unused, since for achieving the above given high level of SSp coverage $4000/52.168 \approx 77$ ($4000/65.191 \approx 61$) bits per state need to be allocated in the bitstate array.

6.2   Multiple Hashing on a Partitioned Bitstate Array

As an alternative to the k-fold hashing scheme, Wolper and Leroy [29] suggested a partitioning of the bitstate array into p disjoint parts. Each sub-array H_i is hereby administered by its own independent hash function h_i. Thus a state s is considered as being already visited, if all p bit positions in the p sub-arrays (H_i) are occupied ($H_i[h_i(s)] = 1; 1 \le i \le p$), which may lead to an omission.

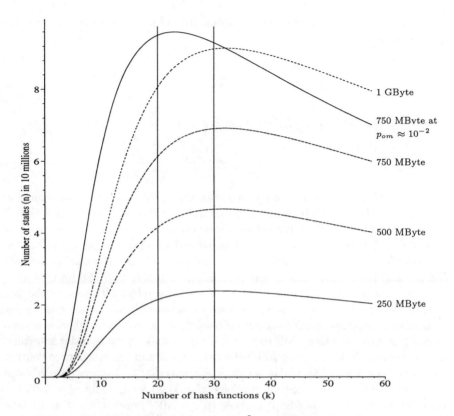

Fig. 5. Fixed SSp coverage ($p_{om} = 10^{-6}$) and varying size of memory

If one employs p independent bitstate arrays, each of size $\frac{m}{p}$ and administered by its own independent hash function h_i, there are $(\frac{m}{p})^p - i^p$ different ways of entering the $i+1$ visited state into the p bitstate arrays, yielding [29]:

$$p_{no} = \prod_{i=0}^{n-1} \frac{\left(\frac{m}{p}\right)^p - i^p}{\left(\frac{m}{p}\right)^p} = \prod_{i=0}^{n-1} 1 - \left(\frac{i \cdot p}{m}\right)^p . \tag{14}$$

This is the same result as obtained for the *multi-fold hashing scheme on a single* bitstate array (k = p). This is not surprising, since under both approaches the occurrence of collisions is identical and in both cases it was assumed that each state inserted into the bitstate array(s) will occupy exactly $p = k$ bits. This assumption turns out to lead to a more pessimistic estimation for the probability of no collision in case only *one* bitstate array is employed. In case of partitioning the bitstate array into p independent parts, this assumption will at least hold for the first state to be entered. Thus one may conclude, following Wolper and Leroy [29] that a multi-fold hashing on a non-partitioned bitstate array will consume fewer bits per state, and thus have a lower probability of error and therefore must to be preferred.

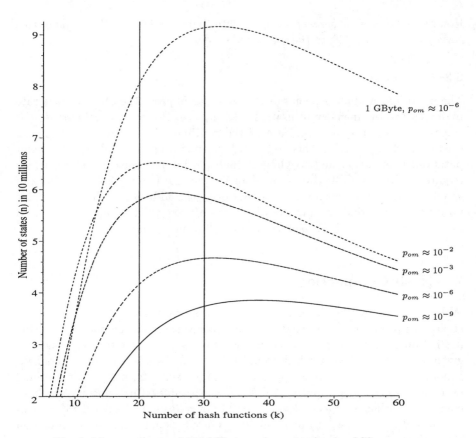

Fig. 6. Memory Space of 500 MByte and varying levels of SSp coverage

One may interpret the hashing scheme on p independent bitstate arrays as the parallel execution of a bitstate hashing on a single bitstate array of size $\frac{m}{p}$, where the SSp exploration is carried out p times and where for each run an independent hash function is used. However doing so would give an overall probability of state omission of:

$$p_{om} = (1 - p_{no})^p \ ,$$

where p_{no} is the probability of no collision per run. The latter is identical for each run, since independent hash functions and bitstate arrays of equal sizes $\left(\frac{m}{p}\right)$ are employed. As a consequence, one can use eq. 9, where the size of the bitstate array m needs now to be adjusted to $\frac{m}{p}$, yielding:

$$p_{om} = (1 - p_{no})^p = \left(1 - \prod_{i=0}^{n-1} 1 - \frac{i \cdot p}{m}\right)^p \ .$$

However further analysis seems appropriate, but goes beyond the scope of this survey paper and is left to the interested reader.

6.3   Summary

The multi-fold hashing schemes discussed so far improve the SSp coverage rate by increasing the memory utilisation rate (μ_{mem}). However, this improvement comes with a substantial increase of the runtime, if compared to the initial bitstate hashing scheme. This additional runtime overhead is due to the computation, comparison and toggling of multiple bit positions per state, which requires additional CPU time. Therefore Wolper and Leroy [29] had the idea of simulating a bitstate hashing scheme by storing the addresses of the occupied bit positions, rather than storing the whole bitstate array. This approach is called hash compaction and is the subject of the following section.

7   Hash Compaction

Wolper and Leroy [29] developed the idea of storing the addresses of the occupied bit positions in the bitstate array, rather than storing the whole array itself. Consequently, such a procedure would simulate a bitstate hashing at a given level of no collision, which would be beyond nowadays hardware configurations. Furthermore, such an approach would not suffer from the computational overhead induced by computing h different hash values. Consequently, the hash compaction scheme targets a low probability of collision by increasing the memory utilisation rate (μ_{mem}) and avoiding computational overhead as in case of the multiple hashing schemes.

7.1   Functionality of the Hash Compaction Approach

Wolper and Leroy [29] suggested that each state descriptor is mapped onto a string of b bits by applying a hash function (h_c). This procedure can be viewed as a compression of state descriptors. The compressed state descriptors are then stored in a hash table, which is assumed to consist of as many slots as states to be explored (n). The position (slot index) of a compressed state descriptor within the hash table is hereby determined by employing another hash function (h_p). Unfortunately Wolper and Leroy [29] do not explain how the slot index for positioning a compressed state descriptor into the hash table is generated. But from a later work of Leroy, Wolper et. al. [30] one can deduce that a state descriptor's position in the hash table is not generated independently of its compressed value (see discussion below). Thus two states are wrongly considered to be identical, if their compressed descriptors are equal, since both will be mapped onto the same position within the hash table. I.e. from $h_c(s_i) = h_c(s_j)$ one concludes $s_i = s_j$, even though $s_i \neq s_j$, since for $h_c(s_i) = h_c(s_j)$ hash function h_p will generate the same slot index ($h_p(s_i) = h_p(s_j)$), due to the dependency of h_p from h_c. In cases, where states collide on slots ($h_p(s_i) = h_p(s_j)$), but for which

$h_c(s_i) \neq h_c(s_j)$, a direct chaining of the overflows for collision resolution is applied. According to Wolper and Leroy [29] the imposed overhead for chaining of overflows is negligible, since the number of slots is n and thus a collision on slots will be rarely the case. However one should note, that the collision resolution is mandatory for the analysis carried out below.

In a nutshell the hash compaction scheme can be understood as an adaptation of the common bitstate hashing approach as discussed in sec. 3.1. Where the original bitstate scheme interpreted the hash values as position within the bitstate array, the hash compaction scheme stores the obtained hash values in a hash table consisting of n slots. Due to the non-injective compression function h_c, states may be wrongly considered as identical and thus the SSp may not fully be covered.

7.2   Analysis of Hash Compaction

It is intuitively clear that the hash compaction scheme has a lower time complexity if compared with the 30-fold hashing, since only 2 hash functions are employed (compression and slot indexing only). The main question to be answered is the performance concerning space complexity. The analysis is hereby straightforward.

The state descriptors are compressed to b bits, where each bit string can be interpreted as positions within a bitstate array. Since the compressed descriptors are mapped onto hash values $\in [0, 2^b - 1]$ the hash compaction method simulates a bitstate hashing scheme on a bitstate array of size $m = 2^b$. Since the probability of no collision for the latter was approximated by $p_{no} \approx e^{\frac{-n^2}{2m}}$, p_{no} under the hash compaction scheme can be approximated by substituting $2m$ by 2^{b+1}, yielding:

$$p_{no} \approx e^{\frac{-n^2}{2^{b+1}}}. \tag{15}$$

From Knottenbelt's work [19] one may conclude, that the above approximation is a lower bound if $2^b > 2/3n^2$. From eq. 15 one can derive the number of bits required for each compressed descriptor in order to explore a SSp of a given size and at a fixed probability of no collision:

$$b = \log_2 \left(\frac{-n^2}{2 \ln(p_{no})} \right). \tag{16}$$

Since eq. 15 is based on the fact that only states are omitted which map onto the same hash value, it is mandatory that a resolution scheme for collisions on slot indices must be employed. Otherwise the approx. of eq. 15 had to be corrected, leading to a lower p_{no}. However it is also clear, that an independent generation of the slot indices will improve the SSp coverage rate, since here a collision only occurs, if slot indices, as well as the compressed state descriptors of two distinct states collide. Therefore it is not clear why Wolper and Leroy [29] did not suggest an independent generation of slot indices. It is intuitively clear that in such a case one would obtain a lower probability of omitting a state, since the probability of colliding on a slot and a compressed descriptor is given by their product, where the individual probabilities are < 1.

Under the hash compaction scheme the number of states (n) to be explored is limited by $\frac{M}{b+1}$, where M is the *size of memory in bits*, where b is the number of bits per compressed state descriptor and where an additional bit, indicating a slots status (occupied or not), must be spent. This flagging is hereby only required under the original scheme, since if one considers a mapping of the state descriptor onto the interval $[0, 2^b - 2]$, one value would be left for indicating a slot's emptiness. The administrative overhead for indicating a slots status is then reduced to b bits and p_{no} is almost not affected from doing so, because $2^b \gg 1$. But for the time being the original approach will be considered, where each state to be stored requires $b + 1$ bits and following Wolper and Leroy [29] the overhead for chaining overflows will be neglected as well.

The size of the compressed state descriptors b influences on the one hand the number of states to be stored and on the other hand the SSp coverage. From the probability of collision one can derive to choose b as large as possible for achieving a high level of SSp coverage ($\frac{n^2}{2^{b+1}} \to 0$). However $n = \frac{M}{(b+1)}$, which is the number of states to be stored in the memory, implies a small value for b. The probability of no collision (eq. 15) together with the required space in bits M for storing n states ($M = n \cdot (b + 1)$) yields:

$$M \approx (b + 1) \cdot \sqrt{-2^{b+1} \cdot \ln(p_{no})}. \qquad (17)$$

i.e. if one uses the above eq. 17 for determining M for a given b and a fixed p_{no}, the latter can be interpreted as the minimum probability of no collision or vice versa it gives us the maximum probability of omission ($P_{om}^{max} = 1 - P_{om}^{min}$). The interpretation as minimum or maximum probability of omission is based on the following view:

1. In case the available memory M_{avail} is less than M, the number of states which can be stored n_s is smaller than $\sqrt{-2^{b+1} \cdot \ln(p_{no})}$. I.e. after the $n_s + 1$'st state is reached, the exploration must be aborted since p_{no} is truly zero, as the $n_s + 1$'st state will produce a memory overflow.
2. In case $M_{avail} > M$ one obtains $n_s > n = \sqrt{-2^{b+1} \cdot \ln(p_{no})}$. But as soon as the $n + 1$'st state is reached, the exploration must be aborted as well, since p_{no} is violated. In such a situation one needs to recalculate b in order to reach the desired level of P_{no}^{min} and restart the exploration routine.

Fig. 7 shows the relation between memory space and size of compressed state descriptors at $P_{no}^{min} = 1 - 10^{-6}$. The graph can be interpreted as follows:

A 8 to 9 Byte compression of each state descriptor requires a memory size of 49.35 MByte up to 1 GByte in order to explore a SSp consisting of 6.074002 up to 97.184040 million states at $P_{no}^{min} = 1 - 10^{-6}$. If one explores fewer states, the actual p_{no} will be larger. In case one explores more states, the exploration routine will run out of memory, yielding $p_{no} = 0$. This means that one obtains the number of states to be explored just by dividing M_{avail} by $b + 1$, where b

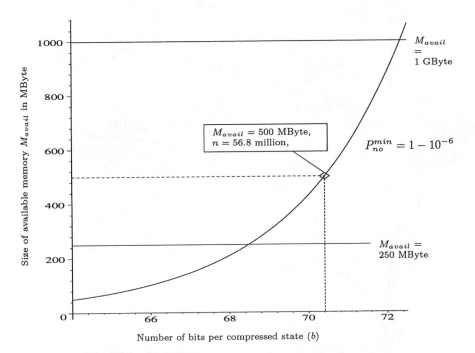

Fig. 7. Relation between size of memory space and size of hash key

depends on the square of the number of states n and the natural logarithm of p_{no}. Thus a 8 to 9 Byte compression of each state descriptor seems to be in the reach of today's average hardware configurations (50 MByte - 1 GByte of RAM) in order to guarantee a minimum coverage rate of 99.9999% and explore SSp between ~ 6 and ~ 97 million states.

For the purpose of illustration one should now imagine, that 500 MByte of memory are available. According to eq. 17 and as illustrated in fig. 7 a compression of each state descriptor onto $b = 70.41$ bits would be optimal in order to reach $P_{no}^{min} = 10^{-6}$.[4] I.e. under a 71-bit compression the hash compaction scheme enables one to store 55.5×10^6 states at a coverage of $0.9999993 < P_{no}^{min}$.[5] In contrast the 2-fold (30-fold) bitstate hashing scheme allows one to explore here $20,398$ (51.57×10^6) states. Thus the hash compaction method outperforms the 2-fold bitstate hashing by far, but the advantage over the 30-fold hashing scheme is not that obvious. For a more detailed analysis one just need to compare the memory requirements of the multiple hashing scheme and the hash compaction at the same levels of SSp coverage and not at the same levels of P_{min}^{no}. This is done in fig. 8, where the memory space of the multiple hashing is considered to

---

[4] One should note that a hash value of 64 (70) bit length, as suggested in [29] does not suffice, since compressed state descriptors of such a size allow one only the exploration of 4.29 (48.6) million states at a minimum coverage rate (P_{cov}^{min}) of $1 - 10^{-6}$.

[5] Each slot (state) requires the 71 bits for storing the compressed descriptor plus an additional status bit, yielding 72 bits per state.

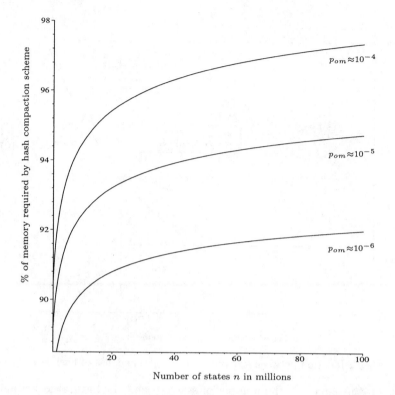

Fig. 8. Mem. requirement of HC compared to Multiple Hashing

be 100% and the ordinate shows the percentage required by the hash compaction. Since the latter is clearly below 100%, the lower space complexity of the hash compaction scheme is obvious. Furthermore the better run-time behaviour speaks to the advantage of the hash compaction method, since the latter only requires the application of two hash functions, where under the multiple hashing approach 30 hash values per states need to be computed.

8   Dynamic Hash Compaction

The dynamic hash compaction scheme is implemented in the performance analysis tool DNAmaca [15]. In [19, 17] and [18] the basic probabilistic SSp exploration scheme was extended for being used in a distributed computing environment. For the purpose of this survey paper we will focus on the non-distributed version only, since the *probabilistic* state space exploration aspect is sufficiently covert there. The interested reader may refer to [21].

8.1   Functionality of Dynamic Hash Compaction

The dynamic hash compaction scheme is based on a hash table and direct chaining of overflows. But rather than storing the full state descriptors it stores compressed state descriptors, generated by applying a second hash function, $h_c : S \mapsto \{0, \ldots, 2^d - 1\}$. The first hash function $h_s : S \mapsto \{0, \ldots, r - 1\}$ determines the slot in which the compressed state descriptor will be stored. Collisions of different compressed descriptors on slots are resolved by chaining.[6] As a consequence two different states ($s_i \neq s_j$) are considered equal, if they collide on *both*, indexing and compression:

$$\textbf{if } h_s(s_i) = h_s(s_j) \wedge h_c(s_i) = h_c(s_j) \textbf{ then } s_i = s_j.$$

Since h_s and h_c are not injective, this conclusion might be wrong, which results in the omission of states.[7]

8.2   Analysis of the Dynamic Hash Compaction

The main improvement of the dynamic hash compaction scheme, in contrast to the original hash compaction approach is, that the calculation of slot indices of the state descriptors is done *independently* from the their compression. Since the hash function h_s maps the n state descriptors onto r different slots, where the independently generated compressed descriptors are stored, the dynamic approach can be understood as the r-fold application of a hash compaction scheme. As one may recall from the hash compaction scheme, a compression of state descriptors onto bit strings of length b simulates a bitstate hashing with an array of size 2^b. Consequently the dynamic hash compaction simulates a bitstate hashing onto an array of size $m = r \cdot 2^d$. I.e. by simply replacing $2m$ in eq. 10 by $r \cdot 2^{d+1}$ one obtains:

$$p_{no} \approx e^{\frac{-n^2}{r \cdot 2^{d+1}}} \tag{18}$$

Knottenbelt [19] has shown that eq. 18 is actually a lower bound, in cases where $2/3n^2 < r \cdot 2^d$. This later criterion is normally given in practice, since a high level of state space coverage requires $\frac{n^2}{r2^{d+1}} \to 0$ and thus $n^2 \ll r2^{d+1}$ [19]. Similar to the analysis of the hash compaction p_{no} can be interpreted as the minimum probability of collision ($P_{om}^{max} = 1 - P_{no}^{min}$), if sufficient memory for storing the n states is supplied. In case the available memory is larger than required, one would be enabled to store more than n states. However, this would violate p_{no} and in such cases one would need to recalculate r and d in order to maintain the fixed p_{no}, which would induce a higher memory requirement.

Concerning the memory consumption, one should note, that the dynamic hash compaction scheme stores the slot overflows not in lists but makes use

---

[6] Examples of appropriate hash functions can be found in [17].

[7] For the sake of completeness it should be mentioned that for generating a transition matrix, the states are consecutively numbered and that this number is stored with the compressed descriptors. However this paper focuses on the generation of different states and not on the subsequent analysis, thus the linear overhead induced by the element counter is neglected.

of dynamic allocated arrays. Since each array is referenced by a pointer and equipped with a counter in order to track the number of overflows per slot individually the space complexity of the dynamic hash compaction is described by:

$$M = n \cdot d + \rho \cdot r, \tag{19}$$

where ρ is the number of bits per slot (row) for administering the at most r different dynamic allocated arrays. This gives an administration overhead of $\frac{\rho \cdot r}{n}$ bits per state. Since $r \ll n$, collision on slots will occur quite frequently, i.e. a substantial number of overflows may be stored per slot. As a consequence the dynamic hashing scheme can be understood as a dynamic 2-*dimensional* hashing scheme, where h_s determines the row index and where h_c computes the column index for each state descriptor to be inserted. Therefore slots are in the following referred to as rows.

The next question to be dealt with, is about the choice for good pairs of r and d, in order to minimize the memory requirement at a given level of state omission and size of SSp.

For simplification let r be described by a power of 2, i.e. $r = 2^k$ and thus $p_{no} \approx e^{\frac{-n^2}{2^{k+d+1}}}$. If the sum of k and d equals the number of bits used under the hash compaction scheme for storing a compressed state descriptors, one cannot notice any difference concerning SSp coverage. From formula 19 one can easily obtain, that an increment of d yields an additional memory consumption of n bits. However an increment of k comes along with an additional memory consumption of

$$\rho \cdot (2^{k+\epsilon} - 2^k) = \rho \cdot 2^k(2^\epsilon - 1),$$

per increase of k by ϵ. In order to keep the probability of collision now constant, it is clear, that an increment of k requires a decrement of d and vice versa. The idea is now, to increase k and decrease d as long as the additional row overhead per state is ≤ 1, since otherwise an increment of d, in order to maintain the probability of no collision and minimize the additional memory space required ought to be favored. Since d can only be decremented by whole numbers, the smallest change of k is given by $\epsilon = 1$. Together with the bound of 1 bit increase in memory consumption per state one obtains:

$$\frac{\rho \cdot 2^k}{n} \leq 1$$

$$=> k \leq \log_2\left(\frac{n}{\rho}\right). \tag{20}$$

Together with formula 18 one finally obtains:

$$d \approx \log_2\left(\frac{-n \cdot \rho}{2 \cdot \ln(p_{no})}\right) \tag{21}$$

Thus for a given number of states n and bits per row ρ, one obtains a value of k, where another doubling of r (increment of k by 1) would lead to an additional memory consumption of more than one bit per state. I.e. in order to minimize memory consumption and maximize SSp coverage, it would be then more sensible to increase the number of bits per compressed state descriptor d by one instead. –One should note, that Knottenbelt et al. [19, 17, 18] make use of a different approach. In order to minimize M they deploy eq. 18 in eq. 19 and derive the resulting term with respect to d. Since a minimum for M is targeted, the resulting eq. can be set equal to 0 and finally resolved for d. This procedure yields almost the same result as the straight forward approximation followed here.

For exemplification one should now consider the exploration of 10^8 states at a coverage rate of $1 - 10^{-6}$. Under the original hash compaction method one has to use $b = 72.08$ per compressed state descriptors in order to maintain the required coverage rate. As a consequence the original hash compaction method requires 912.5 MByte of memory in order to explore 10^8 states and maintain a SSp coverage rate of 99.9999%.[8] According to eq. 20 and 21 the dynamic hash compaction scheme requires here only the use of $d = 51$ bit. The additional allocation of $r = 2^{21}$ rows guarantees the desired level of SSp coverage. Thus the dynamic hashing scheme requires only 625 MByte in order to store 10^8 different states at a coverage rate of $P_{cov}^{min} = 1 - 10^{-6}$. This reduction of memory space is quite obvious if one compares eq. 16 and 21, which discloses $d \leq b$ and considers that the allocation of r rows maintains the desired level of P_{no}^{min} for less memory space as required for the additional $(b-d)$ bits per state descriptor as required in case of the original hash compaction scheme. Furthermore, one should note that the overhead for collision resolution on slots under the hash compaction scheme was neglected so far, which speaks once more for the dynamic hash compaction scheme.

9   Improvement of the Hash Compaction Scheme

In this section a new scheme, introduced by Stern and Dill in [26] will be presented. The method is inspired by Wolper's and Leroy's [29] hash compaction method.

9.1   Improved Hash Compaction Scheme

In general terms, this method proceeds as follows:

1. From a given state descriptor s, using a compression function h_c, a compressed value, $h_c(s)$, which is b bits long, is computed.
2. From the same state descriptor as in 1, applying a hash function h_p, a probing sequence $h_p(s)$ is generated.
3. The compressed value $h_c(s)$ from 1 is inserted at the address, computed in 2.

---

[8] A use of 73 bits per state was assumed, i.e. the status bit as required under the original approach was considered, but $P_{no}^{min} = 1 - 10^{-6}$ is therefore violated slightly.

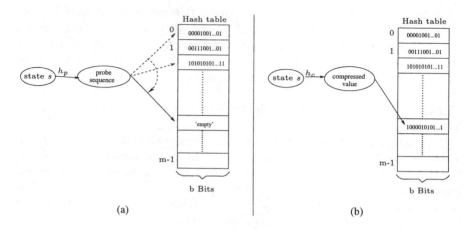

Fig. 9. Improved hash compaction

Note, that the order of 1 and 2 is irrelevant, i.e. 1 and 2 are independent. In fig. 9 the functionality of the improved hash compaction method is shown pictorially. Fig. 9 (a) shows step 2, fig. 9 (b) shows step 3 of the description above.

The compression of the state descriptor was already applied in Wolper's and Leroy's original hash compaction method (cf. section 7). The differences between the original hash compaction method and Stern and Dill's improved scheme are as follows:

- In the improved scheme open addressing instead of chaining is employed to resolve collisions, which is the difference to Knottenbelt's dynamic scheme. The advantage of open addressing is a lower memory consumption, because chaining requires the additional storage of pointers, whose size corresponds roughly to the size of a compressed entity that is to be stored in the hash table. The improved hash compaction scheme requires approximately only half of the memory that is needed by the method presented in section 7.
- In the new scheme, the hash value that determines the address at which the state is to be inserted and the compressed value are computed independently. That means that the hash value is computed from the uncompressed state descriptor. This has the advantage that the same compressed value can appear at several different places in the hash table. As different states may be mapped onto the same compressed value and the same address this might result in less omissions than in Wolper's and Leroy's original hash compaction scheme.

Hashing and Compression:

- The hashing scheme employed here, is uniform probing. A description of uniform probing was given in section 2. The employment of uniform probing is an assumption only used for the analysis. In the implementation some approximation of uniform probing has to be used.

- In improved hash compaction, the compressed state descriptor $h_c(s)$ is stored in the hash table. We assume that the state descriptor s is compressed down to b bits. Thus, the compression function can take $l = 2^b$ different values, so, h_c has a range from 0 to $l - 1$.

Of particular interest is the notion of uniform compression. In uniform compression each state s is equally likely to be one of the values $0, \ldots, l - 1$. To achieve this goal, universal hashing can be employed. If the compression function h_c, is selected as described for universal hashing (cf. section 2) then it is guaranteed that the probability for two states s_1 and s_2 to be mapped onto the same compressed value is not greater than $1/l$:

$$\forall s_1, s_2 \in S, \ s_1 \neq s_2 : \ Pr(h_c(s_1)) = h_c(s_2)) \leq \frac{1}{l}$$

State Insertion: When inserting a new state into the hash table, three cases can be distinguished:

1. The probed slot is empty, the state is inserted.
2. The probed slot is not empty and contains a state whose compressed value is different from the one that is to be inserted (= collision). In this case the probing has to be continued.
3. The probed slot contains a compressed value that is equal to the one that is to be inserted. In this case it is concluded that the two uncompressed states are equal. If this conclusion is wrong, this leads to the omission of the probed state and possibly also to the omission of one or more successor states.

9.2   Analysis

As usual, n is the number of reachable states, m is the number of slots in the hash table.

Omission Probability: The omission probability of improved hash compaction can be approximated as follows:

$$p_{om} = Pr(\text{even one omission during state enumeration})$$

$$\approx 1 - \left(\frac{l-1}{l}\right)^C \tag{22}$$

where

$$C = (m + 1)[H_{m+1} - H_{m-n+1}] - n$$

The approximation 22 has the following explanation: C is equal to the total expected number of hash collisions on h_c. As we employ uniform compression, the probability that two compressed values are not equal is $(l - 1)/l$, assuming that the comparisons are independent, the probability of zero omissions, i.e. zero collisions can be approximated as $[(l - 1)/l]^C$.

The upper bound $\hat{p}_{om}$ on the omission probability is

$$p_{om} \le \hat{p}_{om} = \frac{1}{l}C \qquad (23)$$

Maximum Omission Probability: An approximation of the maximum omission probability in the case, when the hash table fills up is given, using the approximation of eq. 22:
We assume that $l \gg 1$, then, using that for small values of x, $1 - x \approx e^{-x}$, $(l-1)/l = 1 - 1/l \approx e^{-\frac{1}{l}}$. This yields:

$$p_{om} \approx 1 - e^{-\frac{1}{l}\cdot C} \qquad (24)$$

H_n, which appears in the definition of C can be approximated by $\ln(n) + \gamma$, where γ ist the Euler-constant ($\gamma \approx 0.57722$), e^x is approximated by $1 + x$, so we obtain by assuming $m = n$, $l \gg 1$ and also $n \gg 1$:

$$
\begin{aligned}
C &= (m+1) \cdot [H_{m+1} - H_{m-n+1}] - n \stackrel{m=n}{=} (n+1) \cdot [H_{n+1} - H_{n-n+1}] - n \\
&= (n+1) \cdot [H_{n+1} - H_1] - n \\
&= (n+1) \cdot [\ln(n+1) + \gamma - 1] - n = (n+1) \cdot (\ln(n+1) + \gamma) - n - 1 - n \\
&= (n+1) \cdot (\ln(n+1)) + \gamma \cdot (n+1) - 2n - 1 \stackrel{n \gg 1}{\approx} n \cdot \ln(n)) + \gamma \cdot n - 2n \\
&\approx n \cdot (\ln(n) - 1.4)
\end{aligned}
$$

Using eq. 24 and the approximation of the natural exponential function from above this yields:

$$P_{om}^{max} \approx \frac{1}{l}n(\ln(n) - 1.4) \qquad (25)$$

Again, note that this approximation is only applicable for $m = n$.

Number of Bits per State: The question, how many bits for the compressed value is addressed now, where a 4-byte and 5-byte compression is investigated in greater detail:

In the sequel it is assumed that 400 MByte of memory space is available for storing n states during state space search. Using a 5-byte compression scheme, the hash table can accommodate $m = 80$ million states, whereas assuming a 4-byte compression scheme the table can hold up to 100 million states ($m = n$). Fig. 10 depicts the omission probabilities for the different sizes of the compressed state descriptors.

9.3   Comparison of Improved Hash Compaction with Hash Compaction and Dynamic Hash Compaction

In this subsection we will compare Stern's and Dill's improved hash compaction scheme with that of Wolper's and Leroy's original hash compaction and

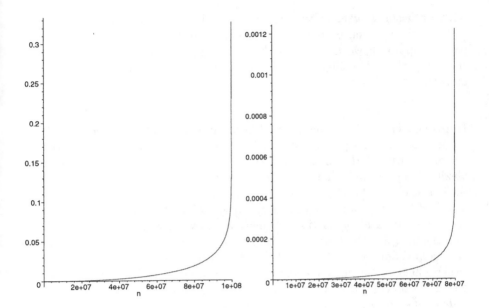

Fig. 10. Omission probabilities, using a 4 and 5 byte compression scheme

Knottenbelt's dynamic hash compaction scheme. As dynamic hash compaction realizes compression and positioning of a state within the hash table independently, it is very interesting whether it performs better or not in comparison to improved hash compaction, both with regard to memory requirements and omission probability.

Hash Compaction and Improved Hash Compaction: It is assumed that 10^8 states shall be analysed. The size of the compressed state will vary between 40 and 72 bits. From that follows, that the memory requirements lie between $\approx$ 500MByte for a 40-bit compression and $\approx$ 1 GByte for a 72-bit compression. From table 1 it becomes obvious that the improved hash compaction scheme

Table 1. Probability of no omissions for hash compaction and improved HC

Bit key length	40	48	56	64	72
HC	$0.11 \cdot 10^{-1974}$	$0.19 \cdot 10^{-7}$	0.9329	0.9997	0.99999
IHC	0.99845	0.99999	0.9999999	0.9999999999	0.999999999999

outperforms Leroy's and Wolper's original scheme by orders of magnitude. This is clear from the fact that the omission probability in the improved scheme is the product of the collision probability on compression and thecollision probability on hashing, whereas the omission probability in the original approach is equal to the collision probability on compression. For the improved hash compaction approach we have taken the maximum omission probability (P_{om}^{max}), such that the probability of no omissions is the minimum probability of no omissions.

Improved Hash Compaction and Dynamic Hash Compaction: In fig. 11 we see the omission probabilities of improved and dynamic hash compaction, where in part (A) the slight disadvantage of the latter concerning space complexity was neglected. Under dynamic hash compaction one can allocate a hash table of size $r = 2220445$ slots and map the state descriptors onto hash values of length of 51 bit, leading to a memory consumption of ≈ 645 MByte. In case of the improved hash compaction scheme a memory consumption of 637.5 MByte was assumed, i.e. the hash table consists of 10^8 slots and the state descriptors are mapped onto 50 or 51 bits. The different compressions are due to the use of either a flagging bit per slot or a dedicated flagging value. I.e. in case one allows only $2^b - 1$ different hash values the overhead for flagging can be reduced to b bits instead of employing n flagging bits. This reduction of the memory requirement enables one to use an additional bit for the compression $(50 + 1)$, where the loss of one hash value is negligible, since $2^b \gg 1$. As a consequence the omission probability, even when it comes to fill-ups close to the size of the employed hash table, is clearly smaller, i.e. the improved hashing scheme performs better than the dynamic scheme concerning omission probability. However in cases were the flagging bit per slot is considered, p_{om} and P_{om}^{max} increase, leading to an advantage of the dynamic scheme (fig. 11(A)). This result is very interesting, since in Knottenbelt's thesis [19] only this less advantageous case of flagging each slot was considered. But we cannot follow this since assuming $2^b - 1$ different hash values, the overhead for flagging is reduced to b bits and a compression onto 51 bits is possible. I.e. assuming an appropriate flagging scheme the improved hash compaction scheme outperforms the dynamic hashing scheme clearly concerning p_{om} and P_{om}^{max}. If we go one step further and compare both approaches at the same memory size, say $M_{avail} \approx 645$ MByte, the maximum number of states which can be stored in the memory varies slightly between the approaches. As a consequence the advantage of the improved hash compaction scheme over the dynamic one is clearly not to be neglected, whether one uses a flagging bit or not (fig. 11(B)). This effect is due to the circumstance, that the hash table under the improved scheme consists of $(645 \cdot 10^8 \cdot 8)/51 = 1.011764706 \cdot 10^8$ slots. I.e. one is actually enabled to store more than 10^8 states in the memory, however the SSp consits at most of 10^8 states. Thus the hash table won't be filled up totally, improving P_{om}^{max} considerably. However, one should note that the length of the probing sequences in case of the improved scheme is problematic and may increase the runtime dramatically, where as in case of the dynamic hash compaction the comparison is bounded by the number of compressed state descriptors per slot, which is on average $\frac{n}{r}$ states per slot.

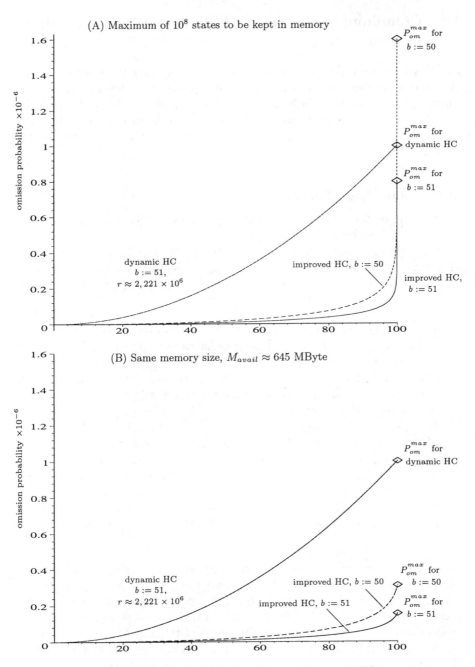

Fig. 11. Comparison of dynamic - and improved HC

10   Combining State Space Caching and Improved Hash Compaction

In this section we will show how an exhaustive state space exploration method, state space caching, can be combined with a probabilistic scheme, namely the improved hash compaction scheme, introduced in section 9. This is done in order to overcome some problems and inefficiencies of both state space caching and improved hash compaction. This approach was first reported by Stern and Dill in [28].

10.1   State Space Caching

We briefly introduce another state space exploration method, namely state space caching, which was investigated in [5].

State space caching performs an exhaustive exploration of all reachable states of a system that is to be analysed. It is a kind of depth first search. We will distinguish two "extreme" positions in performing a state space search as a depth first search:

- A traditional depth first search employs two data structures, a state buffer and a state table. This kind of analysis has very high memory requirements and the size of state spaces that can be analysed is limited by the memory space that is available for the state table (sec. 3.1). If the size of the state space exceeds the memory space the search might deteriorate into an uncontrolled partial search.
- It is possible to perform an exhaustive search by storing only the states of the path that is currently under investigation. This "store nothing" approach [5] has the advantage of an enormous decrease in memory requirements but comes along with a blow up in run time. The reason for the run time blow up is, redundant work has to be done, as the already visited states are not stored, each state and consequently the subgraph associated with it has to be explored as many times as different paths lead to that particular state.

As a compromise between these two extreme positions, state space caching takes the following approach: The stack stores only the states of the path that is currently analysed. In the state table or cache only as many states are stored as memory is available. If the cache fills up, old states must be replaced by newly reached ones. This method pushes the limits of the size of protocols that are verifiable but still has to deal with the problem of an increase of the run time. To overcome the run time problem a partial order reduction method is employed that fully avoids redundant work, but still guarantees full state space coverage. So the method of state space caching is combined with sleep sets cf. [2, 6]. This approach has been successfully used in the model checker "VeriSoft" [4]. The use of sleep set does not affect the analysis of the scheme, so due to lack of space it will be omitted here, and introduced in the technical report version [21] of this paper.

10.2 Combination of State Space Caching and Improved Hash Compaction

The replacement strategy of state space caching, replacing old states by new ones, when the hash table is full, is not very efficient, since a long probe sequence might be necessary before it can be decided whether a newly reached state is in the table. To overcome this inefficiency one now probes at most t slots before, if necessary at all, the new state is inserted into the table. If all probed t slots are non-empty and if all entries at these t slots are different from the state to be inserted then one value of any of the t slots is chosen (at random) to be replaced by the new value. Stern and Dill [28] call this approach t-limited scheme. It is assumed that uniform hashing for the state insertion and uniform compression is employed.

This combination of state space caching and improved hash compaction has the following consequences:

- Replacement now might also take place if the table is not full.
- State space caching itself performs an exhaustive search of the state space, the combination with the improved hash compaction method as presented in this section, turns state space caching into a partial search: Assume two states s and s', $s \neq s'$ are given, then it might be the case that the two states are hashed onto the same slot and also the compressed values for s and s' are equal, in this case it might be erroneously concluded that $s = s'$, such that an omission with probability greater zero occurs, which makes the combination of hash compaction and state space caching a probabilistic verification method.

Analysis: For the analysis, there is one important point to be noticed:
In case of state space caching it is only possible to give the number of different states that were inserted in the table, this number, n', is greater or equal to the real number of reachable states. This is obvious, since every reachable state may be reached by more than one path and is so visited at least once when state space caching is employed. From 22 we have the approximation of the omission probability of state space caching.

$$p_{om} \approx 1 - \left(\frac{l-1}{l} \right)^{C} \qquad (26)$$

with

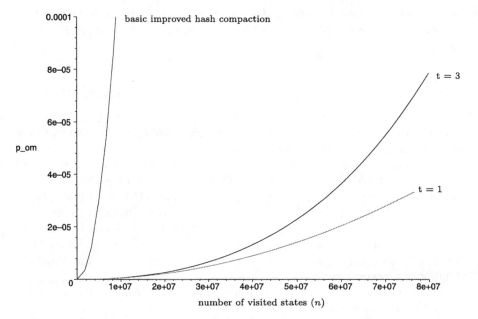

Fig. 12. Omission probabilities, using a 4 byte compression and $t = 1$ and $t = 3$

$l = 2^b$, the number of different compressed keys, each b bits long.

$$C = \begin{cases} \frac{t}{t+1}n' \left(\frac{n'}{m}\right)^t + \sum_{j=0}^{t-1} j \left(\frac{n'}{m}\right)^{j+1} \left(\frac{m}{j+1} - \frac{n'}{j+2}\right) & \text{if } n' \leq m \\ (H_{t+1} - 1)m + t(n' - m) & \text{if } n' > m. \end{cases}$$

The improved hash compaction scheme experienced the problem that the omission probability increased sharply, when the hash table is nearly full, as it can be seen in fig. 10. The combination with state space caching overcomes this problem. The effect of state space caching is shown pictorially. Again it is assumed that the state table is of size 400 MByte. The compressed values are of size 4 bytes (fig. 12) resp. 5 bytes (fig. 13). The t limited scheme is employed for $t = 1$ and $t = 3$.

11   Refinement of Improved Hash Compaction

In this section the hash compaction method that was introduced in section 9 is further refined. The material presented in this section is based on [27] by Stern and Dill. The refinement consists of changing the following points in comparison with improved hash compaction from section 9: Firstly breadth first search is employed and secondly another type of hashing is applied. Both changes will be explained in detail in the next section. The goal of this refinement is to obtain

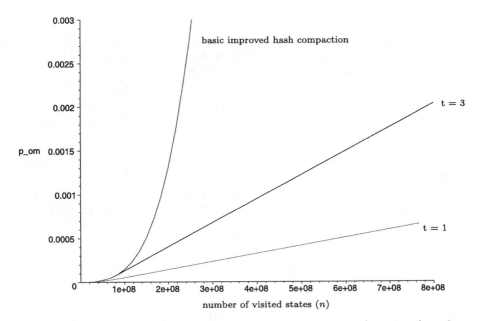

Fig. 13. Omission probabilities, using a 5 byte compression and $t = 1$ and $t = 3$

a tighter bound for the probability of false positives. As a result of this new approach a further reduction of memory requirements is achieved in comparison with the improved hash compaction method of section 9, i.e. given a maximum omission probability, the same value can now be obtained by spending less bits for the compressed state descriptor than the improved hash compaction scheme.

11.1   Refined Method

In the sequel we will describe the differences between this new approach and the improved hash compaction method from section 9.

- Here breadth first search (BFS) for the state space traversal is employed. The omission probability is computed over a longest BFS-path. Let d be the diameter of the reachability graph, i.e. the longest BFS-path from the initial state to any reachable state.
- In the refined method ordered hashing instead of uniform hashing for the state insertion algorithm is employed. Ordered hashing aims to reduce the average number of probes during an unsuccessful search (cf. section 2). Thus, the omission probability can be reduced, as it is dependent on the length of the probe sequence.

11.2   Analysis

Let d be the diameter of the reachability graph. Let s_d be the state on level d on a BFS-path. State s_d is omitted, if any of the states on the path $s_0, \ldots, s_d$ is omitted, thus the omission probability for s_d is bounded by the path's omission probability. To derive an upper bound for s_d's omission probability resp. a lower bound for the probability that s_d is not omitted, we proceed as follows:

Let N_i be the event that s_i is not omitted during its search. State s_i can be omitted for two reasons:

1. A state being necessary for s_i's reachability has been omitted, i.e. s_i is not reached at all.
2. s_i can be reached but is omitted, due to a hash collision and the fact that the compressed values of s_i and the state that is already present are equal.

Thus, N_i can formally be defined as follows:

$$N_0' \Leftrightarrow N_0$$
$$N_i' \wedge N_{i-1} \Rightarrow N_i \quad 0 < i \leq d,$$

where N_i' is the event that s_i is not omitted during its insertion. For the probability of the events N_i and N_i' this yields:

$$Pr(N_0') = Pr(N_0)$$
$$Pr(N_i) \geq Pr(N_i' \wedge N_{i-1}) = Pr(N_i' \mid N_{i-1}) \cdot Pr(N_{i-1})$$

Applying this recursively yields:

$$Pr(N_d) \geq Pr(N_0') \cdot \prod_{i=1}^{d} Pr(N_i' \mid N_{i-1}) \tag{27}$$

It is obvious that the probabilities on the right hand side of eq. 27 depend on the hashing method that is applied. $Pr(N_i' \mid N_{i-1})$ can be assessed for ordered hashing as follows:

Let p_k be the probability that, given k states already in the hash table, during insertion of a new state no omission occurs.

$$p_k = 1 - \frac{2}{l}(H_{m+1} - H_{m-k}) + \frac{2m + k(m-k)}{ml(m-k+1)} \tag{28}$$

A derivation of eq. 28 can be found in [25]
p_k can be used to approximate $Pr(N_i' \mid N_{i-1})$:
Let k_i be the total number of states in the hash table if breadth first search has completed level i. As k_i is taken over all paths of length at most i, k_i is at least as large as the number of states on a single path of length i, thus

$$Pr(N_0') \geq p_{k_0 - 1}$$
$$Pr(N_i' \mid N_{i-1}) \geq p_{k_i - 1}$$

This approximation is justified by the observation that with increasing number of entries in a hash table also the omission probability increases. Using eq. 27 one obtains for $Pr(N_d)$:

$$Pr(N_d) \geq \prod_{i=0}^{d} p_{k_i - 1} \qquad (29)$$

Note carefully that the last level analysed is not necessarily the actual last level, as it might be the case that states on higher/deeper levels are all omitted. In this case the bound for the probability of no omissions is too optimistic.

12   Conclusion

The discussion so far has shown, that the historic development, as far as space complexity and SSp coverage are considered, has led to a step wise improvement of the probabilistic SSp exploration schemes. However concerning run-time, it is clear, that the original bitstate hashing scheme outperforms the latter developments. But its enormous memory requirements in case of high SSp coverage levels contradict its practical usefulness. Considering the developments from bitstate hashing to the approaches where chaining of slot overflows was employed, the dynamic hash compaction seems to be the most advantageous scheme. However if the latter is compared to the improved hash compaction approaches, which were introduced slightly earlier than the dynamic scheme and which are in contrast based on probing for resolving collisions on slots, a different pictures has to be drawn.

The four inner columns of table 2 show the memory requirements of the various approaches, assuming a fixed number of 10^8 states and different levels of omission probability. In contrast to that, the leftmost column shows a varying number of states (n), where a fixed omission probability of $p_{om} \approx 10^{-3}$ and a fixed size of memory of 500 MByte was considered. The respective values were obtained by using the equation introduced in the analysis section of each respective approach. From the table one can obtain, that the improved hash compaction scheme is the most space efficient approach. In table 3 the so far presented schemes are listed and assessed concerning run-time and space complexity. As shown there, the (extended)improved hash compaction approaches as well as the dynamic hash compaction scheme are the most appropriate for being employed in practice, where a slight advantage of the former methods concerning space complexity is given. However it should be noted that in case of the improved hash compaction the length of the probe sequence heavily influences the runtime behaviour of the approach. This dependency may turn out to be a clear disadvantage in comparison to the dynamic hash compaction scheme. Since in cases, where it comes to high fill-up rates of the hash table the runtime of the improved scheme may increase dramatically, where the run time of the dynamic

Table 2. Comparison of probabilistic SSp exploration schemes

Scheme	$p_{om} \approx 10^{-9}$	$p_{om} \approx 10^{-6}$	$p_{om} \approx 10^{-4}$	$p_{om} \approx 10^{-2}$	unit	n
BSH	$624 \cdot 10^3$	625	6.25	$62.17 \cdot 10^{-3}$	EByte	2829
2-fold BSH	$14.5 \cdot 10^3$	144	14.44	1.4	TByte	$228,981$
30-fold BSH	1230	980	840	721	MByte	$1.676 \cdot 10^6$
HC	1087	912.5	825	737	MByte	$67,93 \cdot 10^{6\,a}$
dynamic HC^b	770	645	559	487	MByte	$96 \cdot 10^{8\,c}$
improved HC	762.5	637.5	550	475	MByte	$97.5 \cdot 10^{8\,d}$

^{a} $b = 61$
^{b} $\rho = 28$ bit
^{c} $r = 2,272,599, d = 41$ bits
^{d} $b = 41$ bits

Table 3. Ranking of probabilistic SSp exploration schemes

Scheme	Complexity of		Remark
	Runtime	Memory Space^a	
BSH	++	− − −	very low mem. utilisation (μ_{mem})
2-fold BSH	+	− −	slight improved μ_{mem}
30-fold BSH	− −	−	requires h independent hash function, improved μ_{mem}
HC	+	+	no independent slot indexing
dynamic HC	+	++	search on arrays of overflows is not efficient
improved HC (IHC)	− −	++	probing sequences interfere with runtime
SSp Caching and IHC	−	+ +	reduced P_{om}^{max} compared to IHC
BFS and IHC	−	++	reduced memory req. for same P_{om}^{max} compared to IHC

^{a} Refers to the theoretical memory requirement concerning high-levels of SSp coverage hash compaction scheme is bounded by the maximum number of overflows per slot. Concerning the work of Stern and Dill [28, 26] in combination with breadth first search a comparison is problematic. On one hand the effectiveness of the t-limited scheme cannot be assessed correctly, since one does not have a bound on the number of states that are visited and inserted more than once. On the other hand a similar problem arises in the case of the improved hash compaction scheme. Its efficiency depends on the diameter of the reachability graph, which might be much smaller than the number of states, but cannot be assessed a priori.

References

1. P. Buchholz. Structured Analysis Approaches for Large Markov Chains. *Applied Numerical Mathematics*, 31:375–404, 1999.

2. P. Godefroid. Using Partial Orders to Improve Automatic Verification Methods. In Edmund M. Clarke and Robert P. Kurshan, editors, *Proceedings of 2nd International Workshop on Computer Aided Verification (CAV'90)*, volume 531 of *LNCS*, pages 176–185. Springer, 1990.

3. P. Godefroid. *Partial-Order Methods for the Verification of Concurrent Systems. An Approach to the State-Explosion Problem.* PhD thesis, Université de Liege, 1995.

4. P. Godefroid. Model checking for programming languages using VeriSoft. In *Proceedings of the 24th ACM Symposium on Principles of Programming Languages*, pages 174–186, 1997.

5. P. Godefroid, G. Holzmann, and D. Pirottin. State Space Caching Revisited. In *Proceedings of CAV'92*, volume 663, pages 178–191. Springer, LNCS, 1992.

6. P. Godefroid and P. Wolper. Using Partial Orders for the Efficient Verification of Deadlock Freedom and Safety Properties. *Formal Methods in System Design*, 2(2):149–164, April 1993.

7. R. Graham, D. Knuth, and O. Patashnik. *Concrete Mathematics*. Addison-Wesley, 2 edition, 1994.

8. B.R. Haverkort, A. Bell, and H. Bohnenkamp. On the Efficient Sequential and Distributed Generation of very Large Markov Chains from Stochastic Petri Nets. In *Proc. of IEEE Petri Nets and Performance Models*, pages 12–21, 1999.

9. H. Hermanns, J. Meyer-Kayser, and M. Siegle. Multi Terminal Binary Decision Diagrams to Represent and Analyse Continuous Time Markov Chains. In Plateau et al. [23], pages 188–207. Proc. of the 3rd Int. Workshop on the Numerical Solution of Markov Chains (NSMC'99).

10. G. Holzmann. On Limits and Possibilities of Automated Protocol Analysis. In *Proc. 7th IFIP WG6.1 Int. Workshop on Protocol Specification, Testing, and Verification*, pages 339–344. North-Holland Publishers, 1987.

11. G. Holzmann. An Improved Protocol Reachability Analysis Technique. *Software, Practice and Experience*, 18(2):137–161, 1988.

12. G. Holzmann. *Design and Validation of Computer Protocols*. Prentice-Hall, 1991.

13. G. Holzmann. The Model Checker SPIN. *IEEE Transactions on Software Engineering*, 23(5):1–17, 1997.
14. G. Holzmann. An Analysis of Bitstate Hashing. *Formal Methods in System Design*, 13(3):289–307, 1998.
15. W. Knottenbelt. Generalized Markovian Analysis of Timed Transition Systems, 1996. Master's Thesis, University of Cape Town.
16. W. Knottenbelt and P. Harrison. Distributed Disk-based Solution Techniques for Large Markov Models. In Plateau et al. [23], pages 58–75. Proc. of the 3rd Int. Workshop on the Numerical Solution of Markov Chains (NSMC'99).
17. W. Knottenbelt, P. Harrison, M. Mestern, and P. Kritzinger. Probability, Parallelism and the State Space Exploration Problem. In R. Puigjaner, N. Savino, and B. Serra, editors, *Computer Performance Evaluation: Modelling Techniques and Tools*, volume 1469 of *LNCS*, pages 165–179. Springer, 1998. Proc. of the 10th International Conference, Tools '98, Palma de Mallorca, Spain, September 14-18, 1998.
18. W. Knottenbelt, P. Harrison, M. Mestern, and P. Kritzinger. A Probabilistic Dynamic Technique for the Distributed Generation of Very Large State Spaces. *Performance Evaluation Journal*, 39(1–4):127–148, February 2000.
19. W.J. Knottenbelt. *Parallel Performance Analysis of Large Markov Models*. PhD thesis, University of London, Imperial College, Dept. of Computing, 1999.
20. D. Knuth. *The Art of Computer Programming: Vol 3: Sorting and Searching*. Addison-Wesley, 2 edition, 1998.
21. M. Kuntz and K. Lampka. Probabilistic Methods in State Space Analysis. Technical Report 7 07/02, Universität Erlangen-Nürnberg, Institut für Informatik 7, 2002.
22. Th. Ottman and P. Widmayer. *Algorithmen und Datenstrukturen*. Spektrum Akademischer Verlag, 3. revised edition, 1996.
23. B. Plateau, W.J. Stewart, and M. Silva, editors. *Numerical Solution of Markov Chains*. Prensas Universitarias de Zaragoza, 1999. Proc. of the 3rd Int. Workshop on the Numerical Solution of Markov Chains (NSMC'99).
24. M. Siegle. Behaviour analysis of communication systems: Compositional modelling, compact representation and analysis of performability properties. Habilitationsschrift, Institut für Informatik, Friederich-Alexander-Universität Erlangen-Nürnberg, 2002.
25. U. Stern. *Algorithmic Techniques in Verification by Explicit State Enumeration*. PhD thesis, Technische Universität München, 1997.
26. U. Stern and D. L. Dill. Improved Probabilistic Verification by Hash Compaction. In *Proceedings of CHARME '95*, volume 987, pages 206–224. Springer, LNCS, 1995.
27. U. Stern and D. L. Dill. A New Scheme for Memory-Efficient Probabilistic Verification. In Reinhard Gotzhein and Jan Bredereke, editors, *Formal Description Techniques IX: Theory, application and tools*, volume 96 of *IFIP Conference Proceedings*, pages 333–348. Kluwer, 1996. Proc. of IFIP TC6 WG6.1 International Conference on Formal Description Techniques IX / Protocol Specification, Testing and Verification XVI, Kaiserslautern Germany, October 8–11.
28. U. Stern and D. L. Dill. Combining State Space Caching and Hash Compaction. In *Methoden des Entwurfs und der Verifikation digitaler Systeme, 4. GI/ITG/GME Workshop*, pages 81–90, 1996.

29. P. Wolper and D. Leroy. Reliable Hashing Without Collision Detection. In *Proceedings of the 5th International Computer Aided editor = Costas Courcoubetis, Verication Conference (CAV'93)*, volume 697 of *LNCS*, pages 59–70. Springer, 1993.
30. P. Wolper, U. Stern, D. Leroy, and D. L. Dill. Reliable probabilistic verification using hash compaction. Submitted for publication, downloaded from http://www.citeseer.nj.nec.com/32915.html on 23.08.2002.

Analysing Randomized Distributed Algorithms*

Gethin Norman

School of Computer Science, University of Birmingham,
Birmingham B15 2TT, United Kingdom
G.Norman@cs.bham.ac.uk

Abstract. Randomization is of paramount importance in practical applications and randomized algorithms are used widely, for example in co-ordinating distributed computer networks, message routing and cache management. The appeal of randomized algorithms is their simplicity and elegance. However, this comes at a cost: the analysis of such systems become very complex, particularly in the context of distributed computation. This arises through the interplay between probability and nondeterminism. To prove a randomized distributed algorithm correct one usually involves two levels: classical, assertion-based reasoning, and a probabilistic analysis based on a suitable probability space on computations. In this paper we describe a number of approaches which allows us to verify the correctness of randomized distributed algorithms.

1 Introduction

Distributed algorithms [66] are designed to run on hardware consisting of many interconnected processors. The term 'distributed' originates from algorithms that used to run over a geographically large network, but now applies equally well to shared memory multiprocessors. A *randomized* algorithm is one which contains an assignment to a variable based on the outcome of tossing a fair coin or a random number generator. Since the seminal paper by Michael Rabin [85] randomized algorithms have won universal approval, basically for two reasons: simplicity and speed. Randomized distributed algorithms include a number of theoretical algorithmic schemes, for example the dining philosophers protocol and the consensus of Aspnes and Herlihy [5], as well as the practical real-world protocols for Byzantine agreement due to Kursawe et al. [11] and IEEE 1394 FireWire root contention protocol [46].

A necessary consequence of using randomization is the fact that the correctness statements must combine probabilistic analysis, typically based on some appropriate probability space on computation paths, with classical, assertion-based reasoning. Examples include: "termination occurs *with probability* 1", "delivery of a video frame is guaranteed within time t with probability *at least* 0.98" and "the expected number of rounds until a leader is elected is at most 2^k". The

---

* Supported in part by the EPSRC grant GR/N22960.

C. Baier et al. (Eds.): Validation of Stochastic Systems, LNCS 2925, pp. 384–418, 2004.

analysis of randomized distributed algorithms becomes very complex, sometimes leading to errors.

In a distributed environment, where each of the concurrent processors must make decisions in a highly nondeterministic context, it can be shown that there are *no* (symmetric) deterministic solutions to certain network co-ordination problems [63, 33]. Randomization offers a powerful tool for *symmetry breaking*, in addition leading to faster solutions. It has been applied to a range of algorithms, for example the dining philosophers protocol [63], Byzantine agreement [8], self stabilization [41], and shared-memory consensus [5]. It should be emphasised that the standard liveness properties, for example termination, become probabilistic (hold with probability 1) in the context of randomized algorithms. This applies to the so called *Monte Carlo* algorithms, which are the focus of this paper, but not to the *Las Vegas* algorithms such as the randomized quicksort [77]. In Las Vegas algorithms termination can only be ensured with some appropriately high probability, and so non-termination simply becomes unlikely, albeit still possible.

Randomization is of paramount importance in practical applications: randomized algorithms are used widely, for example in co-ordinating distributed computer networks [66], message routing [1], data structures, cache management, and graph and geometric algorithms [77]. The appeal of randomized algorithms is their simplicity and elegance. However, this comes at a cost: the probability space, and with it the probabilistic analysis, become very complex, particularly in the context of distributed computation. This is caused by both the complex interplay between probabilistic and nondeterministic choices and by the introduction of *dependencies* between random variables that can easily be overlooked. To prove a randomized distributed algorithm correct one usually involves two levels: classical, assertion-based reasoning, and a probabilistic analysis based on a suitable probability space on computations, see e.g. [67, 82, 1]. A correctness argument often crucially relies on the statement "*If* the random variables are independent *then* the algorithm produces the correct answer with probability p". The simplicity of this statement obscures potential dangers, which can arise if the variables turn out *not* to be independent, since in such cases estimated probabilities can differ from the *actual* probabilities by a wide margin.

For example, an error in a mutual exclusion algorithm due to Rabin [84], which was pointed out by Saias [90] and later corrected by Rabin [52], can be explained in terms of an inadvertently introduced dependence of the probability of a process winning the draw on the total number of processes in the network, and not only on those taking part in the draw; when exploited by a malicious adversary, this has the effect of making the actual probability of winning smaller than it should be. In a different but no less critical situation – probabilistic modelling of nuclear power stations – an error can be traced to the incorrect assumption of events being independent, which resulted in the actual probabilities becoming unsatisfactorily large [97]. Thus, satisfactory correctness guarantees are essential supporting evidence when implementing a new randomized algorithm.

2   Background

In this section we give an overview of areas related to analyzing randomized distributed algorithms, and then outline the remainder of the paper.

2.1   Modelling Randomized Distributed Algorithms

The standard modelling paradigm for systems exhibiting probabilistic behaviour is based on Markov chains. Such models often provide sufficient information to assess the probability of events occurring during execution of a sequential system, as each such execution can be represented as a sequence of "steps" made according to the probability distribution. We note that, for such models it is straightforward to define a probability measure of the set of executions (paths) of the model. However, this must be extended to take account of the distributed scenario, in which concurrently active processors handle a great deal of unspecified nondeterministic behaviour exhibited by their environment, as well as make probabilistic choices. As argued in [91], randomized distributed algorithms require models *with nondeterministic choice between probability distributions*. The central idea is to allow a *set* of probability distributions in each state. The choice between these distributions is made *externally* and *nondeterministically*, either by a scheduler that decides which sequential subprocess takes the next "step" (as in e.g. the concurrent Markov chains [101]), or by an *adversary* that has the power to influence and potentially confuse the system [9, 7], as is the case with Byzantine failure. A number of essential equivalent models exist following this paradigm and including *probabilistic automata* [91], *Markov Decision Processes* [26], *probabilistic-nondeterministic systems* [9] and *concurrent probabilistic systems* [7].

Probabilistic choices are *internal* to the process and made according to the selected distribution. It should be emphasised that the complexity introduced through allowing nondeterminism affects the probabilistic modelling: a probability space can be associated with the space of computations *only if* nondeterministic choices have been pre-determined, which is typically achieved through the choice of an adversary (which chooses at most one distribution in a given state).

As in the non-probabilistic setting, we may have to impose *fairness constraints* in order to ensure that liveness properties can be verified. In a distributed environment fairness corresponds to a requirement such as each concurrent component to progress whenever possible. Without fairness, certain liveness properties may trivially fail to hold in the presence of simultaneously enabled transitions of a concurrent component. A number of fairness notions have been introduced [101, 20, 7] for systems containing probabilistic and nondeterministic behaviour, for example in [7] an adversary is called *fair* if any choice of transitions that becomes enabled infinitely often along a computation path is taken infinitely often, whereas in [22] an adversary is called fair if there exists $\varepsilon > 0$ such that all nondeterministic choices are taken with probability at least ε The interested reader is referred to [7, 22] for more information on the subject.

To allow the construction of complex probabilistic systems it is straightforward to extend the definition of *parallel composition* in standard labelled transition systems to this probabilistic setting. Alternatively, to specify complex randomized distributed algorithms one can employ the higher-level modelling languages developed in the field of concurrency. Much of this work is based on replacing the (nondeterministic) choice operator with a probabilistic variant, and hence is not suitable for modelling randomized distributed algorithms where nondeterminism and probabilistic behaviour coexist. Of process calculi developed to include both a nondeterministic and probabilistic choice operator, we mention [36, 35, 104] which are based on CSS [73], and [64, 65, 76] where languages based on CSP [88] are considered.

2.2   Specifying Properties of Randomized Distributed Algorithms

Often the specification of a randomized distributed algorithm can be written in a simple unambiguous form, for example *"eventually the protocol terminates"* or *"the algorithm terminates in at most $\mathcal{O}(n)$ rounds"*, and hence there is no need to use any specification language. However, one must be careful as without a formal specification language it is easy to make errors. An example of this, mentioned earlier, is in [84] where there is an error in the proof of correctness which can be attributed to the fact that the properties of the protocol were not stated correctly. For further details on this error and methods for correctly specifying properties see [90, 89].

Probabilistic logics are a natural choice of specification formalisms to use for stating correctness of randomized distributed algorithms. Such a logic can be obtained from a temporal logic, e.g. CTL [12], in two ways: either by adding probability operators with *thresholds* for truth to the syntax [37] (within this we include 'with probability 1' [101, 79]), or by re-interpreting the formulae to denote (an estimate of) *the probability* instead of the usual truth values [45, 69]. In the threshold approach, path formulae ϕ are annotated with thresholds $[\cdot]_{\geq p}$ for $p \in [0, 1]$, with the formula $[\phi]_{\geq p}$ interpreted as "the probability assigned to the set of paths satisfying the path formula ϕ in the classical sense is *at least* p". The fact that this semantics is well defined follows from the observation that in logics such as CTL path formulae determine measurable sets [101]. For models with nondeterminism quantification over paths A (E) can be added, meaning "for all (some) schedulers"; this models the *worst* (*best*) probabilistic outcome.

2.3   Verification Techniques for Distributed Algorithms

Model checking [12] has become an established industry standard for use in ensuring correctness of systems, and has been successful in exposing flaws in existing designs, to mention e.g. the tools SMV, SPIN and FDR, none of which can handle probability. The process of model checking involves building a (finite-state) *model* of the system under consideration, typically built from components composed in parallel. Then the model checking tool reads the description, builds a model, and for each specification (in temporal/modal logic) attempts to exhaustively check whether it holds in this model; if not, then a diagnostic trace leading to error is output.

One problem with the model checking approach, which makes model checking complex systems infeasible, is the well know *state space explosion problem*: the number of states grows exponentially with the number of components in the design. Furthermore, often the system under consideration is either infinite state or parameterized, for example by the number of processes, and we would like to prove that the system is correct for *any* value of the parameter. In such cases we cannot use finite state model checking, however, one approach to overcome these difficulties and allow for the analysis of both parameterized and infinite state systems is to develop a *verification methodology* which reduces the verification problem for large/infinite state systems to *tractable finite-state* subgoals that can be discharged by a conventional model checker. In the non-probabilistic setting *abstraction* techniques have been used to reduce the complexity of the system under study by constructing a simpler abstract system which *weakly* preserves temporal properties: if a property holds in the abstract system then it holds true in the concrete system, while the converse does not hold. Other techniques developed in the non-probabilistic case include *compositional/refinement* methods [40, 72], *abstract interpretation* [15], *temporal case splitting* [71] and *data independent induction* [16].

For example, *data type reduction*, a specific instance of abstract interpretation, can be used to reduce large or *infinite* types to small finite types, and temporal case splitting breaks the proof into *cases* based on the value of a given variable. Combining data type reduction and temporal case splitting can reduce a complex proof to checking only a small number of simple subcases, thus achieving significant space savings.

We should note that the above techniques can no longer be described as fully automated, it is often up to the user to decide on the abstraction or how the problem should be decomposed. This may not be any easy task as it often requires a detailed understanding of both how the system being modelled works and the verification methodology.

2.4   Outline of Paper

In the next two sections we review techniques for verifying randomized distributed algorithms and give examples applying these methods. In Section 3 we consider methods for verifying *qualitative* properties, that is, properties which require satisfaction with probability 1. Whereas, in Section 4 we consider the case of the more general *quantitative* properties such as "the probability that a leader is elected within t steps is at least 0.75". In Section 5 we outline the complete verification methodology for verifying a complex randomized distributed algorithm. Finally in Section 6 we summarise and consider future research directions.

3   Qualitative Analysis

In this section we outline several techniques used to verify correctness with probability 1, called P-validity in [51, 105]. Examples of such correctness statements include *"the protocol eventually terminates"*, *"eventually a leader is elected"* and *"eventually the resource gets allocated"*.

3.1   Verifying Finite-State Systems

In the case of finite state problems, the verification of probability 1 properties reduces to graph based analysis, i.e. conventional model checking. This result was first established in [38] in the case of termination of finite state systems. In [78] this is extended by introducing deductive proof rules for proving that termination properties of (finite state) probabilistic systems hold with probability 1. While in [79, 102] a sound and complete methodology for establishing correctness with probability 1 for general temporal logic properties are introduced, as well as model checking procedures. We also mention approaches based on the branching time framework [20, 7] where model checking algorithms for probability 1 properties under different notions of fairness are introduced. In all of the above techniques the verification is performed over a non-probabilistic version of the system, in which the probabilistic choices are replaced with nondeterminism with fairness constraints.

Applying such techniques has lead to the verification of a number of randomized distributed algorithms. In particular by using *symbolic techniques*, for example BDDs, one can verify very large randomized distributed algorithms. For example, using the probabilistic symbolic model checker PRISM [54, 83] qualitative analysis of randomized algorithms with over 10^{30} states has been performed. However, in general randomized distributed protocols are often too complex to apply these methods. Furthermore, often the protocols include parameters which are unbounded (for example, the number of rounds), or are genuinely infinite state. In such cases, we may apply model checking techniques to prove that certain instances of the protocol are correct, for example when there are 6 processes in the ring, but we cannot say that the protocol is correct for all instances, for example that the protocol is correct for a ring of *any* size. In the next section, we will consider alternative verification techniques which allow us to prove correctness in such cases.

3.2   Verifying Complex, Parameterized and Infinite-State Systems

As mentioned above, many randomized distributed algorithms are parameterized by, for example the number of processes or nodes in the network, or are infinite state, for example real-time protocols, in such cases the verification problem becomes harder. In this section we will consider methods for establishing qualitative (probability 1) properties of such systems (in Section 4.2 we will consider checking quantitative properties).

In certain cases, for example when considering real-time randomized protocols, although the protocol has an infinite state space one can reduce the verification problem to checking a finite-state quotient of the system, for example [2] present a method for probabilistic real-time systems, using the region graph approach [3]. In this work it is demonstrated that, verifying whether (infinite state) probabilistic real time systems satisfy branching time temporal properties with probability 1, reduces to analyzing a finite-state nondeterministic system.

In addition, abstraction/refinement methods can be applied to simplify the verification of probability 1 properties. For example, in [100] a sequence of stepwise abstractions are used in the verification of the (real-time) IEEE 1394 FireWire root contention protocol [46]. In each step of the refinement process one aspect of the protocol is abstracted, for example one step concerns abstracting timing information, where the correctness of each abstraction is validation through probabilistic simulations [94]. After the final step of the refinement processes one is left with a simple system which is straightforward to analyse.

In the non-probabilistic case, as mentioned in Section 2, much progress has been made in methods for the verification of parameterized and infinite state systems. For certain protocol and properties, after replacing the probabilistic behaviour with nondeterminism, one can directly use these techniques for the verification of randomized distributed algorithms. For example, using Cadence SMV [72], a proof assistant which allows the verification of large, complex, systems by reducing the verification problem to small subproblems that can be solved automatically by model checking, certain properties of the randomized consensus protocol of Aspnes and Herlihy [5] and the Byzantine agreement protocol of Cachin, Kursawe and Shoup [11] have been verified in [55] and [53] respectively. Note that, both these protocols have *unboundedly* many states and the proofs are for *any* number of processes.

In general however, to verify probability 1 properties as in the finite state case one must add fairness constraints when replacing the probabilistic behaviour with nondeterminism. The first example of verifying a non-trivial (parameterized) randomized distributed algorithm appeared in [78], where both a simplified version of Lehmann and Rabin's dining philosophers protocol [63] and a mutual exclusion algorithm are proved correct using deductive proof rules. The proofs presented are complex, require a detailed knowledge of the protocol and are constructed by hand.

The above technique has since been extended [51, 105] to allow one to employ "automated" methods when verifying correctness with probability 1 for parameterized probabilistic systems. This new method is based on the *network invariant* approach developed for proving liveness properties of non-probabilistic systems, see for example [50]. The advance in this work on the network invariant method, over the previous work in this area, which allows for the extension to probability 1 statements, is that in [50] fairness constraints of the system are taken into account. The idea behind the network invariant approach when proving correctness of a parameterized system with N processes $P_1, \ldots, P_N$, is to construct an abstraction of $N-1$ of the processes and the remaining process P_1.

Then show that: (a) composing in parallel two copies of the abstraction behave like the abstraction and (b) the property of interest holds over the abstraction composed with the single process P_1. The technique is not fully automated, although verifying the properties (a) and (b) can be performed mechanically, the construction of a suitable/correct abstraction must be performed manually and may require ingenuity and a deep understanding of the protocol under study. Furthermore, this approach is dependent on using *local arguments*, that is arguments that do not depend on the global state space since $N - 1$ of the processes are abstracted, and therefore is only applicable when correctness can be proved using such arguments.

An alternative method for proving probability 1 properties of randomized distributed algorithms is presented in [29]. In this paper the techniques are based on results in Markov theory and in particular *recurrent* states and *0-1 laws*. The techniques developed are useful for proving "convergence" type properties of systems, for example *eventually a leader is elected* and *the protocol reaches a stable state*. The fundamental steps of this approach are finding a non-increasing measure over the states of the protocol and showing that, with some (non-negligible) probability, from any state this measure strictly decreases within a finite number of steps. This approach can be seen as a type of abstraction, where the set of states of the concrete protocol with the same measure are mapped to a single state in the abstract version. The main effort in using this method is finding the "correct" measure which depends on the protocol under consideration, however as the measure is defined on the states of the protocol on how the states are modelled can also influence this process. We will give a simple example illustrating this method in the next section.

3.3   Example: Verification of a Self-Stabilizing Protocol

In this section we consider the self stabilization algorithm of Israeli and Jalfon [47]. A self-stabilising protocol for a network of processes is a protocol which, when started from some possibly illegal start state, returns to a legal/stable state without any outside intervention within some finite number of steps. When the network is a ring of identical processes, the stable states are those where there is exactly one process designated as "privileged" (has a token). A further requirement in such cases is that the privilege (token) should be passed around the ring forever in a fair manner.

In the protocol of Israeli and Jalfon the network is a ring of N identical processes, $P_0, P_1, \ldots, P_{N-1}$. Each process P_i has a boolean variable q_i which represents the fact that the process has a token. A process is active if it has a token and *only* active processes can be scheduled. When an active process is scheduled, it makes a (uniform) random choice as to whether to move the token to its left (to process P_{i-1}) or right (to process P_{i+1}), and when a process ends up with two tokens, these tokens are merged into a single token. The stable states are those states where there is one token in the ring.

First, in any stable state (where there is only token), the privilege (token) is passed randomly around the ring, and hence is passed around the ring in a fair manner. Next we consider whether from any unstable state with probability 1 (under any scheduler) a stable state is reached. For simplicity we consider the case when in the unstable state there are two tokens in the ring, however the ring is of an arbitrary size. The correctness proof sketched below is based on that presented in [29] and is an example of a proof based on 0-1 laws and ergodic theory.

In any unstable state of the protocol (any state where there are two processes with tokens), the scheduler can only choose between the active processes who moves next, that is, between the two processes which have tokens. Without loss of generality, we can suppose that the two processes are i and j and that the minimum distance between the tokens is $d > 0$ as shown in Figure 1.

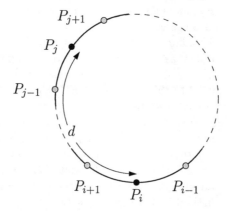

Fig. 1. Ring with two tokens

Now suppose that the scheduler chooses process i to move next, then with probability 1/2 the minimum distance between the two tokens will decrease by 1 (when P_i chooses to pass the token to P_{i+1}). On the other hand, if the scheduler chooses process P_j to move then again with probability 1/2 the minimum distance between the tokens will decrease by 1 (when P_j chooses to pass the token to P_{j-1}). Therefore, since i and j where arbitrary, in any unstable state, no matter what the scheduler decides – since it can only choose between the two processes which have tokens – with probability at least 1/2, the distance between the two tokens decreases by 1.

Letting $m = \lfloor N/2 \rfloor$, we have that the minimum distance between the tokens is less than or equal to m, and hence it follows that from *any* unstable state within m steps (and with probability greater than or equal to $1/2^m$) the distance between the tokens becomes 0 for any scheduler, that is, a stable state is reached (since when the minimum distance between the tokens is zero the tokens are merged into a single token). Using this and the fact that once a stable state is reached it is never left (since new tokens are never created), we can apply standard results from Markov theory to show that, for any adversary and from any unstable state, with probability 1, a stable state is reached.

For the complete formal proof for an arbitrary number of tokens see [29]. The interested reader is also referred to [30], where this method has been applied to a more complex setting: proving the correctness of a variant of Lehmann and Rabin's solution to the dining philosophers problem [63].

4   Quantitative Analysis

In this section we consider approaches and techniques for proving the correctness of quantitative properties. The verification of such properties is more complex than the probability 1 properties considered in the previous section. Examples of such properties include *"the probability of electing a leader within time t is at least 99%"* and *"the expected time until consensus is $\mathcal{O}(k)$"*. In the following section we will consider the verification of finite state systems, then in Section 4.2 consider the case for parameterized and infinite state system and finally in Section 4.3 we give a simple example demonstrating some of these techniques.

4.1   Verifying Finite-State Protocols

Much has been published on quantitative probabilistic model checking, see for example algorithms for probabilistic variants of CTL/CTL* [35, 9, 7], LTL [14] and the mu-calculus [45], and real-time probabilistic systems [56]. The fundamental method for assessing the probability of a path formula holding is through a reduction to solving a system of linear equations or linear optimization problem in the context of models with nondeterminism.

There are two approaches to extending temporal logics to the probabilistic setting. Using the threshold approach, probabilistic variants based on CTL* have been derived, see for example [9, 7], where minimum/maximum probability estimates replace exact probabilities. Furthermore, [20, 21, 23] extends the above logics to allowing the specification (and verification) of *expected time* and *long run average* properties using results from the field of Markov Decision Processes [26]. The alternative, *quantitative* interpretation, where formulae are maps from the set of states to the [0, 1] interval instead of truth and falsity, has been considered in, for example [45]. We also mention the qualitative approach, based on probabilistic predicate transformers [75], presented in [69].

Model checking algorithms for the logic PCTL [9, 7] (an extension of CTL) have been implemented in the probabilistic model checker PRISM [54, 83] which has been used to verify a number of randomized distributed algorithms. For example, in [95] the probabilistic model checker PRISM has been used to verify the Crowds protocol [87], a randomized protocol designed to provide users with a mechanism for anonymous Web browsing. Further case studies can be found through the PRISM web page [83].

In [74] an alternative logic framework for reasoning about randomized distributed algorithm based on the program logic of "weakest preconditions" for Dijkstra's guarded command language *GCL* [27] is considered. The authors consider a probabilistic extension of *GCL*, called *pGCL* and the logic of "weakest precondition" is replaced with "greatest pre-expectation". More formally, in the non-probabilistic case [27], if P is a predicate over final states and *prog* is a (non-probabilistic) program, the "weakest precondition" predicate *wp.prog.P* is defined over initial states and holds only in the initial states from which the program *prog* is guaranteed to reach P. The authors' extension to the probabilistic framework is based on "greatest expectation": now a predicate returns for any initial state the maximum probability which the program *prog* is guaranteed to reach P from the initial state. In [68] it is shown how to formulate expected time directly in the logic, and the author demonstrates this by analysing Lehmann and Rabin's dining philosophers protocol [63].

As in the qualitative case, in general randomized distributed protocols are either too complex to apply probabilistic model checking, include parameters which are unbounded, for example the number of processes or the number of rounds, or are genuinely infinite state to apply these techniques. In the next section, we will consider alternative verification techniques capable of verifying such protocols.

4.2   Verifying Complex, Parameterized and Infinite-State Protocols

Methods for verifying complex randomized distributed algorithms include the approach developed in [103, 99, 98] which allow certain performance measures of I/O automata extended with probabilistic and real time behaviour to be computed compositionally. That is, the performance measures are computed component-wise and eliminate the need to explicitly construct the global state space, and hence combat the state explosion problem. We also mention the compositional, trace based, approach developed in [25] to allow for *assume-guarantee* type reasoning.

Alternatively, D'Argenio et al. [17, 18] present refinement strategies for checking quantitative reachability properties. The approach is based on constructing smaller *abstract* models of the system under study, which preserve reachability probabilities, in the sense that the probabilities obtained on the abstract system are upper and lower bounds on reachability probabilities of the concrete system. Furthermore, automated techniques are developed for refining the abstraction when the results obtained are inconclusive (i.e. when the bounds are too coarse). The authors have implemented these refinement strategies in the tool RAPTURE [86].

In [70] data refinement is extended to the probabilistic setting. Data refinement is a generalization of program refinement. Using the probabilistic programming language *pGCL* [74], the authors study the data refinement method in the probabilistic setting, and use this method to perform quantitative analysis of a probabilistic steam boiler.

In general to verify complex randomized distributed algorithms one can use equivalence relations to reduce the state space, and hence the complexity of the verification problem. Examples, include probabilistic simulation [48, 94], probabilistic bisimulation [61] and testing based equivalences [13].

As in the qualitative case, when studying infinite state systems, for example real-time randomized protocols, in certain cases one can reduce the verification problem to analyzing a finite-state quotient of the system. For general infinite state probabilistic systems methods for calculating *maximum reachability probabilities* have been developed in [57]. In the case of real-time systems, approaches have been developed in [56] for verifying properties of a probabilistic variant of the branching time logic TCTL [39]. The efficient algorithm for calculating bounds on the maximal reachability probability of probabilistic real-time automata presented in [56] has subsequently been implemented in [19]. For case studies concerning the qualitative verification of real-time randomized distributed algorithms protocols see [19, 59, 58].

An alternative approach is to use a theorem proving framework, such as the recent HOL support for the verification of probabilistic protocols [43, 44]. However, this technique has yet to be applied to *distributed* algorithms. We also mention the *approximate techniques* [62] which use Monte-Carlo algorithms to approximate the probability that a temporal formula is true. The advantage of this approach over model checking is that one does not need to construct the state space of the system, and hence reduce the state space explosion problem.

The final type of techniques we consider are those that try to separate the probabilistic and nondeterministic behaviour, in an attempt to isolate the probabilistic arguments required in the proof of correctness. The advantage of this *decomposition* of the verification problem is that simpler non-probabilistic methods can be used in the majority of the analysis, while the more complex probabilistic verification techniques need only be applied to some small isolated part of the protocol. Such approaches applicable to quantitative analysis include: *complexity statements* [91], *coin lemmas* [92, 34] and *scheduler luck games* [28]. We will now consider each of these methods in turn.

Probabilistic Complexity Statements [91, 67, 81, 93] are used to give time or complexity bounds for randomized distributed algorithms. A probabilistic complexity statement has the form:

$$U \xrightarrow{\phi \leq c}_p U'$$

where U and U' are sets of states, ϕ is a complexity measure, c is a nonnegative real number and $p \in [0, 1]$. Informally the above probabilistic complexity statement means that:

> *whenever the protocol is in a state of U, under any adversary, the probability of reaching a state in U' within complexity c is at least p where the complexity is measured according to ϕ.*

A complexity measure is used to determine the complexity of an execution of a system and examples of such measures include: the elapsed time, the number of updates of a variable, number of coin flips, and the number of rounds of a protocol. The key property of complexity statements is *compositionality*, that is complexity statements can be combined in the following way:

$$\text{if } U \xrightarrow{\phi \leq c}_p U' \text{ and } U' \xrightarrow{\phi \leq c'}_{p'} U'' \text{ then } U \xrightarrow{\phi \leq c+c'}_{p \cdot p'} U'' .$$

For this compositionality to hold there are certain conditions on the adversary that can be considered, however it has been shown that this compositionality property of complexity statements holds for fair adversaries [91]. This compositionality result can then be used to simplify the proof of correctness of randomized distributed algorithms into the verification of a number of smaller simpler problems. Furthermore, in [91] it is shown how using complexity statements one can derive upper bounds on the worst case performance of randomized distributed algorithms.

For correctness proofs employing probabilistic complexity statements, which also demonstrates how complexity statements can be used to verify expected time properties, see [81, 67, 82]. Additionally we will give an example of applying complexity statements in Section 5.

Coin Lemmas [92, 34, 93] are a tool for separating the probabilistic and nondeterministic arguments in the analysis of distributed probabilistic systems. They are used for proving upper and lower bounds on the probability of events. Their advantage lies in the reduction of probabilistic analysis to non-probabilistic steps and also force the user into a certain well-defined probabilistic scenario, drawing his or her attention to the possible interference between probability and nondeterministic choices, which has the effect of thus reducing the chance for making errors due to underestimating the complexity of the actual system execution scenario. We motivate the need for coin lemmas through the following example taken from [34].

> Consider the experiment of rolling a die. We know that the probability of rolling any number between 1 and 6 is 1/6. Now consider a simple protocol which can roll the die and sends a beep signal whenever the outcome is an even number. We could then say that the probability of the beep signal being sent is 1/2. However, it may not be the case that in every execution the protocol does roll the die, therefore the correct statement would be that in each execution the probability of either not rolling the die or observing a beep signal is at least 1/2.

The observation in this simple example that one needs to take into the account whether the die is rolled or not is the basis for formulating coin lemmas. It may seem trivial in this simple example but as the complexity of the protocol grows and the number of experiments increases the result is no longer obvious. By adding the event "die is rolled" to the protocol, a coin lemma for this simple protocol is: with probability 1/2 either the event "die is rolled" is *not* observed or a beep signal is observed.

We now illustrate the use of coin lemmas through the following strategy for proving the correctness of probabilistic complexity statements. Suppose that we want to prove the correctness of the probabilistic complexity statement:

$$U \xrightarrow{\phi \le c}_p U'$$

where the probability p arises through the probability that some designated random choices have a certain outcome. By fixing the outcome of these random choices and replacing all remaining random choices to nondeterministic choices we are left with a non-probabilistic system and given certain coin lemmas, by proving that *all* paths of this non-probabilistic system starting from any state in U reach a state in U' while ϕ increases by at most c, it follows that with probability p it holds in the original probabilistic system.

Scheduler Luck Games are introduced in [28], which can be seen as an instance of coin lemmas that can, in certain cases, provide a simpler and more concise correctness proof. For a given randomized distributed algorithm a game is set up between two players: *scheduler* and *luck*. The player *scheduler* decides how the nondeterminism is resolved in an attempt to disprove the correctness of the protocol, while the player *luck* chooses the outcome of some of the probabilistic choices in an attempt to verify the protocol as correct. The player *luck* is said to have a k winning strategy if by fixing at most k coin flips it is ensured that the protocol is correct. Intuitively, when the player has a k winning strategy, it follows that the protocol is correct with probability at least $1/2^k$.

In the cases of protocols with rounds, the authors show how such games can be used to infer expected time properties of randomized distributed algorithms. In particular, if luck has a winning strategy for the game in an expected number of at most r rounds with at most k interventions, (i.e. fixing at most k coin flips), then the protocol terminates within $r \cdot 2^k$ expected number of rounds.

4.3   Example: Verification of a Self Stabilizing Protocol

In this section we return to the self stabilizing protocol of Israeli and Jalfon [47] given in Section 3.3 and establish an upper bound on the expected time until a stable state is reached for an arbitrary sized ring. Again for simplicity restricting attention to when there are two tokens in the ring. The method is based on that given in [29].

We again consider the minimum distance between the two tokens in a ring. Now supposing the ring is of size N we have that the minimum distance between the two tokens can range from 0 to m, where $m = N/2$ if N is even and $(N-1)/2$ if N is odd. If we consider the behaviour of the adversary when the distance between the tokens is $0 < d < m$, then no matter which of the active processes (the two processes with tokens) the adversary chooses to scheduler, with probability $1/2$ in the next state the distance is $d - 1$ and with probability $1/2$ the distance is $d + 1$. If $d = 0$, then there is only one token (since the tokens are merged), and hence we have reached a stable state. On the other hand, if $d = m$, then when N is even with probability 1 in the next state the distance is $d - 1$, on the other hand if N is odd with probability $1/2$ in the next state the distance is $d - 1$ and with probability $1/2$ the distance is d. Intuitively, we can consider the protocol as a random walk with barriers $[0, m]$ (where 0 is absorbing) as shown in Figure 2.

More formally, we can show that there exists a *probabilistic simulation* [94] between the random walk and the protocol (by relating any unstable state of the protocol whose minimum distance between the tokens is d to the state of

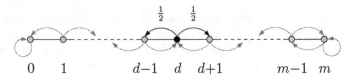

Fig. 2. Random walk with barriers 0 and m where 0 is absorbing

random walk which corresponds to being at d and any stable state of the protocol to being at 0). Then, using the fact that probabilistic simulation is sound with respect to trace inclusion [91], it follows that the expected time to reach a stable state is less than or equal to the maximum expected time to reach the barrier 0 from any state of the random walk. Hence, using random walk theory [32] it follows that the expected time to reach a stable state is bounded above by $\mathcal{O}(m^2) = \mathcal{O}(N^2)$.

5   Case Study: Byzantine Agreement

In this section we describe an approach to the formal verification of Cachin, Kursawe and Shoup's randomized Asynchronous Binary Byzantine Agreement protocol (ABBA) [11], which uses techniques for the verification of non-probabilistic parameterized protocols, (finite-state) probabilistic model checking and probabilistic complexity statements. Further details concerning the models we have constructed and the proof of correctness can be found at the PRISM web page [83]. The results presented in this section first appeared in [53].

Agreement problems arise in many distributed domains, for example, when it is necessary to agree whether to commit or abort a transaction in a distributed database. A *distributed agreement protocol* is an algorithm for ensuring that a collection of distributed parties, which start with some initial value (0 or 1) supplied by an environment, eventually terminate agreeing on the same value. The requirements for a randomized agreement protocol are:

Validity: If all parties have the same initial value, then any party that decides must decide on this value.

Agreement: Any two parties that decide must decide on the same value.

Probabilistic Termination: Under the assumption that all messages between non-corrupted parties eventually get delivered, *with probability 1*, all initialized and non-corrupted parties eventually decide.

5.1   The Protocol

The ABBA protocol is set in a completely asynchronous environment, allows the maximum number of corrupted parties and makes use of cryptography and randomization. There are n parties, an adversary which is allowed to corrupt at most t of them (where $t < n/3$), and a trusted dealer. The parties proceed through possibly unboundedly many rounds: in each round, they attempt to agree by casting votes based on the votes of other parties. In addition to **Validity** and **Agreement**, the protocol guarantees **Probabilistic Termination** in a constant expected number of rounds which is validated through the following property:

Fast Convergence: The probability that an honest party advances by more than $2r + 1$ rounds is bounded above by $2^{-r} + \varepsilon$ where ε is a negligible function in the security parameter.

The Model and Cryptographic Primitives. The ABBA protocol is set in the *static* corruption model: the adversary must decide whom to corrupt at the very beginning of the execution of the protocol. Once the adversary has decided on the corrupted parties these are then simply absorbed into the adversary. The adversary also has complete control over the network: it can schedule and deliver the messages that it desires. The honest parties can therefore be considered as passive: they just respond to requests made by the adversary and do not change state in between such requests. Thus, the adversary can delay messages for an arbitrary length of time, except that it must eventually deliver each message.

The protocol uses two classes of cryptographic primitives. The first are *threshold random-access coin-tossing schemes*. Such a scheme models an unpredictable function F, of which each party holds a *share*, that maps the name of a coin for each round r to its value $F(r) \in \{0, 1\}$. Each party can generate a share of each coin, where $n - t$ shares are both necessary and sufficient to construct the value of a particular coin. The second class of cryptographic primitives the protocol uses are *non-interactive threshold signature schemes*. These schemes are used to prevent the adversary from forging or modifying messages. In [11, 96] it has been shown that the cryptographic primitives have efficient implementations and are proved secure in the random oracle model. We therefore consider a version of the protocol which assumes the correctness of the cryptographic elements.

The Protocol. The protocol is given in Figure 3. Each party's initial value is sent to it by a trusted dealer. The parties proceed in rounds, casting pre-votes and main-votes. A party constructs both the *signature share* and *justification* for each pre-vote and main-vote it casts using threshold signature schemes. The justification for each vote is the *signature* obtained by combining the signature shares of the messages that the party used as the basis for this vote. For example, if a party casts a main-vote for 1 in round r, then the corresponding justification is the signature obtained through combining the signature shares present in the $n - t$ messages which contain pre-votes for 1 in round r that the party must have received. For further details on these justifications see [11].

Observe that the power of the adversary is limited by the requirement that all votes carry a signature share and a justification, and the assumption that the threshold signature scheme is secure (the adversary cannot forge either signature shares or signatures). The presence of the signature shares and this assumption implies that the adversary cannot forge any messages of the honest parties, that is, cannot send a message in which it pretends to be one of the honest parties.

Protocol ABBA for party i with initial value v_i.

0. PRE-PROCESSING. Generate a signature share on the message

$$(\texttt{pre-process}, v_i)$$

and send all parties a message of the form

$$(\texttt{pre-process}, v_i, signature\ share).$$

Collect $2t + 1$ pre-processing messages.

Repeat the following steps 1-4 for rounds $r = 1, 2, \ldots$

1. PRE-VOTE. If $r = 1$, let v be the majority of the received pre-processing votes. Else, select $n - t$ justified main-votes from round $r - 1$ and let:

$$v = \begin{cases} 0 & \text{if there is a main-vote for 0} \\ 1 & \text{if there is a main-vote for 1} \\ F(r-1) & \text{if all main-votes are } \texttt{abstain}. \end{cases}$$

Produce a signature share on the message $(\texttt{pre-vote}, r, v)$ and the corresponding justification, then send all parties a message of the form

$$(\texttt{pre-vote}, r, v, justification, signature\ share).$$

2. MAIN-VOTE. Collect $n - t$ properly justified round r pre-vote messages, and let

$$v = \begin{cases} 0 & \text{if there are } n - t \text{ pre-votes for 0} \\ 1 & \text{if there are } n - t \text{ pre-votes for 1} \\ \texttt{abstain} & \text{if there are pre-votes for 0 and 1}. \end{cases}$$

Produce a signature share on the message $(\texttt{main-vote}, r, v)$
and the corresponding justification, then send all parties a message of the form

$$(\texttt{main-vote}, r, v, justification, signature\ share).$$

3. CHECK FOR DECISION. Collect $n - t$ justified main-votes of round r. If all these are main-votes for $v \in \{0, 1\}$, then decide on v, and continue for one more round. Otherwise simply proceed.

4. COIN. Generate a *coin share* of the coin for round r and send all parties a message of the form

$$(r, coin\ share).$$

Collect $n - t$ shares of the coin for round r and combine to get the value of $F(r) \in \{0, 1\}$.

Fig. 3. Asynchronous binary Byzantine agreement protocol ABBA [11]

The adversary can make one honest party believe that the initial vote of a corrupted party is 0, while another honest party believes it is 1, since these messages do not require justification. However, since all the remaining votes need justification, the adversary cannot just make up the pre-votes and main-votes of the corrupted parties. For example, if in round r there are at least $n - t$ pre-votes for 0 and between 1 and $n - t - 1$ pre-votes for 1 (all of which carry proper justification), then there is justification in round r for both a main-vote for 0 and for abstain, but not for 1. Thus, the adversary can make one honest party believe a corrupted party has a main-vote for 0 in round r, while another honest party believes that the same corrupted party has a main-vote for abstain.

Assumptions. Recall that, to verify the ABBA protocol correct, we need to establish the properties of *Validity*, *Agreement* and *Fast Convergence*. A number of assumptions were needed in order to perform the verification. These include the correctness of the cryptographic primitives; for example, we assume the following properties of the threshold coin-tossing scheme:

Robustness For any round r it is computationally infeasible for an adversary to produce $n - t$ valid shares of the coin for round r such that the output of the share combining algorithm is not $F(r)$.

Unpredictability An adversary's advantage in the following game is negligible. The adversary interacts with the honest parties and receives less than $n - 2t$ shares of the coin for round r from honest parties, then at the end of the interaction outputs a bit $v \in \{0, 1\}$. The adversaries advantage is defined as the distance from $1/2$ of the probability that $F(r) = v$.

These assumptions are implicit in the models we construct, in that they restrict the power of the adversary. For example, the adversary cannot forge messages or make up any of the votes of the corrupted parties which require justification.

The remaining assumptions concern fairness statements which correspond to the fact that the adversary must eventually send all messages (which ensure that parties eventually cast votes). For example, we assume that

Proposition 1. *For any party that enters round $r + 1$:*

(a) *if the party does not decide by round r, then the coin for round r is tossed;*
(b) *if the party does not decide by round $r + 1$, then the coin in round $r + 1$ is tossed.*

5.2   Agreement and Validity

Both these arguments are independent of the actual probability values, and hence can be verified by conventional model checking methods. Below we give a brief outline of the arguments based on two lemmas.

Lemma 1. *If in round r there are main-votes for v, then there are none for $\neg v$.*

Lemma 2. *If party i decides on v in round r, then there are less than $n - 2t$ main-votes for abstain in round r from honest parties.*

Validity: We prove that if all honest parties have the same initial preference, then all honest parties decide on this value in the initial round. Suppose all honest parties have the same initial value v, then in round 1 the pre-votes of all parties will be v, since all will see a majority of pre-processing votes for v (a majority of pre-processing votes requires at least $t+1$ votes, that is, at least one vote from an honest party). It then follows that all parties will have a main-vote for v in round 1, and hence all decide on v in the first round.

Agreement: We prove that if the first honest party to decide decides on v in round r, then all honest parties decide on v either in round r or round $r+1$. Therefore, suppose party i is the first honest party to decide and it decides on v in round r. Then i must have received an honest main-vote for v, and hence, by Lemma 1, there are no main-votes for $\neg v$ in round r. Therefore, any party that decides in round r must decide on v. Now, by Lemma 2, there are less than $n - 2t$ honest main-votes for abstain, and since a party reads at least $n - 2t$ honest main-votes, a party must receive an honest main-vote for something other than abstain in round r and Lemma 1 implies this must be for v. Putting this together, all honest parties receive a main-vote for v and none for $\neg v$ in round r, thus all have a pre-vote for v in round $r+1$. It follows that all will have a main-vote for v in round $r+1$, and hence all will decide on v in round $r+1$.

In [83, 53] fully automated proofs of these properties, for all values of n, have been given using the Cadence SMV [72] a proof assistant.

5.3   Fast Convergence: High Level Proof

Before we give an outline of the proof of *Fast Convergence* we need to introduce the following notation. Let $\rho_r \in \{0,1\}$ be the value (when it exists) which at least $n - 2t$ honest parties cast pre-votes for in round r. Since $2(n-2t) > n-t$ (see Lemma 1), in any round, there cannot be $n - 2t$ honest pre-votes for v and for $\neg v$. Using this notation we introduce the following lemma.

Lemma 3. *If an honest party receives a justified main-vote for* abstain *in round $r+1$, then $\rho_r = 1 - F(r)$.*

Proof. Suppose an honest party receives a justified main-vote for abstain in round $r+1$. Then the justification must include both a pre-vote for 0 and for 1 in round $r+1$. However, as in any round there cannot be main-votes for 0 and 1 (see Lemma 1), one of these pre-votes must be justified by the value of the coin $(F(r))$ and the other by a main-vote (which must be for $1 - F(r)$) in round r. We also have that if there is a main-vote in round r for v, then there are at least $n - t$ pre-votes for v in round r, and hence at least $n - 2t$ honest pre-votes for v in round r. That is, we have $v = \rho_r$. Putting this together we have $\rho_r = 1 - F(r)$ as required.   □

Lemma 3 implies that if ρ_r can be calculated before the coin $F(r)$ "is tossed", then all parties will accept only main-votes for ρ_r in round $r+1$ with probability

1/2 (the probability that $\rho_r = F(r)$). However, if ρ_r is not determined until after the coin is tossed, then the adversary may be able to set $\rho_r = 1 - F(r)$, and hence stop parties from agreeing. Note that, if ρ_r is undefined, then all parties will vote for the value of the coin in round $r + 1$ and agreement is reached.

Consider the state where the $(n - 2t)$th honest party is about to reveal its share of the coin $F(r)$ and suppose that S is the set of parties which have finished round r. All the parties in S have already collected main-votes for round r, and hence their pre-votes for round $r + 1$ are already determined. We have two cases to consider:

- There exists a party in S which is going to vote for a determined $v \in \{0, 1\}$ (not for the coin). Then this party has received at least one main-vote for v in round r. This means that ρ_r is determined and equals v, and hence the value of ρ_r is determined before the coin $F(r)$ is revealed.
- All parties in S will base their pre-vote in round $r + 1$ on the value of the coin $F(r)$. Now, since there are at least $n - 2t$ honest parties in S, there will be at least $n - 2t$ pre-votes for $F(r)$ in round $r + 1$. In this case, the only possible value for ρ_{r+1} is $F(r)$, and therefore ρ_{r+1} is determined before the coin $F(r + 1)$ is revealed.

Therefore, the probability of agreement within two rounds from any $r > 1$ is at least 1/2, and hence the probability that an honest party advances by more than $2r + 1$ rounds is bounded above by 2^{-r}.

The proof of **Fast Convergence** clearly depends on the probability values, and hence cannot be verified using conventional model checking methods. However, from the proof we see that establishing this property reduces to analysing the probabilistic aspects of the protocol over two arbitrary rounds. That is, the complexity of having possibly unboundedly many rounds is removed from the verification. In Section 5.4 we describe how we formulate an abstraction by considering only two arbitrary rounds, verify the correctness of the abstraction, and prove **Fast Convergence** for finite configurations using the probabilistic model checker PRISM [54, 83]. In Section 5.5 we give an alternative proof of **Fast Convergence** for an arbitrary number of parties, which is automated except for one high-level inductive argument involving probabilistic reasoning.

5.4   Fast Convergence: Automated Verification

In this section we use PRISM, to verify **Fast Convergence** for finite configurations $(n = 4, \ldots, 20)$. Based on the high level proof of **Fast Convergence**, for a fixed round $r > 1$ we construct an abstract protocol considering *only* the main-votes for $r - 1$, all votes for round r and the pre-votes for $r + 1$. To initialise the abstract protocol we consider any possible combination of main-votes for round $r - 1$ which satisfies the condition: there cannot be a main-vote for 0 and a main-vote for 1. This restriction holds for the full protocol (see Lemma 1). Furthermore, we suppose that no party has decided in round r, that is, all parties take part in round r and round $r + 1$ (if a party has decided in an earlier round $r' < r$, then by **Agreement** it follows that all honest parties will decide by round $r' + 1 \leq r$). We only explicitly define the main-votes and pre-votes of the honest parties, and express the votes of the corrupted parties in terms of the honest parties votes using the assumptions we have made.

The times at which the coins of rounds $r - 1$ and r are flipped, that is, when the $(n - t)$th share of the coin for rounds $r - 1$ and r are released, are also abstracted. We suppose that this can happen any time after at least $n - 2t$ honest parties have collected the main-votes for round $r - 1$ that they require to cast their pre-vote in round r. The fact that this condition is sufficient for the coin in round $r - 1$ follows from the fact that $n - t$ parties must give their share of the coin to work out the value of the coin, and honest parties do not work out their share until they have collected these main-votes from round $r - 1$. Since clearly the coin in round r cannot be tossed before the coin in round $r - 1$, requiring the same condition for the coin in round r is also sufficient. Note that, although this means that the coins for round $r - 1$ and r may be tossed earlier than would be possible in the actual protocol, and hence give the adversary more power, these requirements are sufficient for proving **Fast Convergence**.

Abstract Protocol. We now introduce the abstract protocol using the PRISM description language which is a variant of reactive modules [4]. The basic components of the language are *modules* and *variables*. A system is constructed as a number of modules which can interact with each other. A module contains a number of variables which express the state of the module, and its behaviour is given by a set of guarded commands of the form:

$$[] \text{ <guard>} \rightarrow \text{<command>};$$

The guard is a predicate over the variables of the system and the command describes a transition which the module can make if the guard is true (using primed variables to denote the next values of variables). If a transition is probabilistic, then the command is specified as:

$$\text{<prob>} : \text{<command>} + \cdots + \text{<prob>} : \text{<command>}$$

To construct the abstract protocol, we define a module for the adversary which decides on the main-votes in round $r - 1$, modules for the coins in rounds $r - 1$ and r, and modules for each honest party. The abstract protocol is then defined as the asynchronous parallel composition of these modules:

$$adversary \ ||| \ coin_{r-1} \ ||| \ coin_r \ ||| \ party_1 \ ||| \ \cdots \ ||| \ party_{n-t}.$$

We let $N = n - t$, the number of honest parties, and $M = n - 2t$, the minimum number of main-votes (pre-votes) from honest parties required before a pre-vote (main-vote) can be made. Also, in the construction of the protocol, we define certain variables which can be updated by any module, which for example count the number of parties that have made a certain pre-vote. We achieve this by using *global* variables.

Module for the Adversary. The adversary decides on the main-votes in round $r - 1$ and the only restriction we impose is that there cannot be votes for both

0 and 1. We suppose that there are always main-votes for abstain, and honest parties can decide on pre-votes after reading any. Instead boolean variables are used to denote whether there are main-votes for 0 and 1. The adversary has the following structure:

> module *adversary*
>
> $m_0^{r-1} : [0..1];$ // *main-vote for 0 in round r − 1*
> $m_1^{r-1} : [0..1];$ // *main-vote for 1 in round r − 1*
>
> [] $(m_0^{r-1}{=}0) \wedge (m_1^{r-1}{=}0) \rightarrow (m_0^{r-1'}{=}1);$ // *choose 0*
> [] $(m_0^{r-1}{=}0) \wedge (m_1^{r-1}{=}0) \rightarrow (m_1^{r-1'}{=}1);$ // *choose 1*
>
> endmodule

Note that, before the adversary makes this choice, we suppose there are only abstain votes in round $r - 1$.

Modules for the Coins. The coin for round $r - 1$ can be be tossed once $n - 2t$ honest parties have decided on their pre-vote for round r and use the (global) variable n_r with range $[0..M]$ to count the number of parties who have decided on their pre-vote in round r (n_r is updated by honest parties when they have decided on their pre-vote for round r).

module $coin_{r-1}$

$c_{r-1} : [0..1];$ // *local state of the coin (0 not tossed and 1 tossed)*
$v_{r-1} : [0..1];$ // *value of the coin*

// *wait until $n_r \geq M$ before tossing the coin*
[] $(c_{r-1}{=}0) \wedge (n_r{\geq}M) \rightarrow 0.5 : (v_{r-1}'{=}0) \wedge (c_{r-1}'{=}1){+}0.5 : (v_{r-1}'{=}1) \wedge (c_{r-1}'{=}1);$

endmodule

The coin for round r is similar and is constructed by renaming $coin_{r-1}$ as follows:

> module $coin_r = coin_{r-1}[c_{r-1} = c_r, v_{r-1} = v_r]$ endmodule

Modules for the Parties. The local state of this party i is represented by the variable $s_i \in \{0, \dots, 9\}$ with following interpretation:

> 0 - read main-votes in round $r - 1$ and decide on a pre-vote for round r;
> 1 - cast pre-vote for 0 in round r;
> 2 - cast pre-vote for 1 in round r;
> 3 - cast pre-vote for coin in round r;
> 4 - read pre-votes and cast main-vote in round r;
> 5 - read main-votes in round r and decide on a pre-vote for round $r + 1$;
> 6 - cast pre-vote for coin in round $r + 1$;
> 7 - cast pre-vote for 0 in round $r + 1$;

8 - cast pre-vote for 1 in round $r + 1$;
9 - finished.

The module of party i has the form:

```
module party_i

    s_i : [0..9]; // local state of party i

    [] ...

    :   :

endmodule
```

where the transitions are dependent upon the local state of the party, what votes have been cast and the values of the coins. We now consider each state of the party in turn.

Read main-votes in round r–1 and decide on a pre-vote for round r: Recall that, before the adversary has chosen which of 0 and 1 is a possible main-vote for round $r - 1$, there are only main votes for abstain, and hence the only pre-vote the party can have is for the coin. However, once the adversary has chosen between 0 and 1 there can be pre-votes for either this value or the coin (since we suppose there are always main-votes for abstain). Since we do not restrict the number of main-votes the party must read before choosing its pre-vote in round r, the transition rules for reading main-votes in round $r - 1$ are given by:

$$[] \ (s_i{=}0) \wedge (m_0^{r-1}{=}1) \rightarrow (s_i'{=}1) \wedge (n_r'{=}\min(M, n_r{+}1)) \ // \textit{pre-vote for 0}$$
$$[] \ (s_i{=}0) \wedge (m_1^{r-1}{=}1) \rightarrow (s_i'{=}2) \wedge (n_r'{=}\min(M, n_r{+}1)) \ // \textit{pre-vote for 1}$$
$$[] \ (s_i{=}0) \rightarrow (s_i'{=}3) \wedge (n_r'{=}\min(M, n_r{+}1)) \ // \textit{pre-vote for coin}$$

Note that, the party increments the variable n_r once it has finished reading main-votes in round $r - 1$ and decided on a pre-vote for round r.

Cast pre-votes in round r: In this state a party has either already decided on its pre-vote, or it will be based on the value of the coin, and hence it must wait for the coin to be tossed. We introduce the variables (which will be needed for parties to decide on their main-votes in round r) p_v^r, for $v = 0, 1$, which count the number of pre-vote for v in round r. The transition rules of the party in this state are given by:

$$[] \ (s_i{=}1) \rightarrow (s_i'{=}5) \rightarrow (p_0^{r'}{=}p_0^r{+}1); \ // \textit{cast pre-vote for 0}$$
$$[] \ (s_i{=}2) \rightarrow (s_i'{=}5) \rightarrow (p_1^{r'}{=}p_1^r{+}1); \ // \textit{cast pre-vote for 1}$$
$$// \textit{cast pre-vote for the coin}$$
$$[] \ (s_i{=}3) \wedge (c_{r-1}{=}1) \wedge (v_{r-1}{=}0) \rightarrow (s_i'{=}5) \wedge (p_0^{r'}{=}p_0^r{+}1);$$
$$[] \ (s_i{=}3) \wedge (c_{r-1}{=}1) \wedge (v_{r-1}{=}1) \rightarrow (s_i'{=}5) \wedge (p_1^{r'}{=}p_1^r{+}1);$$

Note that the global variables p_0^r and p_1^r are incremented when the party casts its vote.

Read pre-votes and cast main-vote in round r: A party must wait until sufficiently many $(n-t)$ pre-votes in round r have been cast, and hence until $n-2t$ honest parties have cast their pre-vote, that is, $p_0^r + p_1^r \geq M$. For the party to cast a main-vote for abstain, it must receive a pre-vote for 0 and for 1 which can be from either an honest or corrupted party. To receive an honest vote for $v \in \{0,1\}$, an honest party must have voted for this value, that is, $p_v^r > 0$. On the other hand, to receive a corrupted vote for v, either this is the value of the coin, or the corrupted party received a main-vote in the previous round for v, that is, $v_v = 0$ or $m_v^{r-1} = 1$. To cast a main-vote for $v \in \{0,1\}$, the party must at least have received at least $n-2t$ honest pre-votes for v. Before we give the transition rules we need the following boolean global variables: m_v^r for $v \in \{0, 1, abstain\}$ to indicate what main-votes have been made. Note that, again, we only record if there is a main-vote for a value as opposed to the total number of votes. The transition rules are then given by:

$$// \text{ main-vote for abstain}$$
$$[] \ (s_i{=}4) \wedge (p_0^r{+}p_1^r \geq M) \wedge ((p_0^r > 0) \vee (v_1{=}0) \vee (m_0^{r-1}{=}1)) \wedge$$
$$((p_1^r > 0) \vee (v_1{=}1) \vee (m_1^{r-1}{=}1)) \rightarrow (s_i'{=}5) \wedge (m_{abs}^{r'}{=}1)$$
$$// \text{ main-vote for } 0$$
$$[] \ (s_i{=}4) \wedge (p_0^r{+}p_1^r \geq M) \wedge (p_0^r \geq M) \rightarrow (s_i'{=}5) \wedge (m_0^{r'}{=}1)$$
$$// \text{ main-vote for } 1$$
$$[] \ (s_i{=}4) \wedge (p_0^r{+}p_1^r \geq M) \wedge (p_1^r \geq M) \rightarrow (s_i'{=}5) \wedge (m_1^{r'}{=}1)$$

The global variables m_0^r, m_1^r and m_{abs}^r are updated when the party decides.

Read main-votes in round r and decide on a pre-vote for round $r+1$: To vote for the coin, the party must have received abstain votes from an honest party (again this is a requirement but is not sufficient in the actual protocol). To vote for $v \in \{0,1\}$ the party needs at least one vote for v from either an honest or corrupted party. To get such a vote from a corrupted party there needs to be at least $n-2t$ honest main-votes for v in round r. The transition rules for reading the main-votes are therefore given by:

$$[] \ (s_i{=}5) \wedge (m_{abs}^r{=}1) \wedge (m_0^r{=}0) \wedge (m_1^r{=}0) \rightarrow (s_i'{=}6) \ // \text{ pre-vote for coin}$$
$$[] \ (s_i{=}5) \wedge ((m_0^r{=}1) \vee (p_0^r \geq M)) \rightarrow (s_i'{=}7) \ // \text{ pre-vote for } 0$$
$$[] \ (s_i{=}5) \wedge ((m_1^r{=}1) \vee (p_1^r \geq M)) \rightarrow (s_i'{=}8) \ // \text{ pre-vote for } 1$$

Cast pre-votes in round $r+1$: Since we are only concerned with finding whether all pre-votes are the same or not, we introduce just (global) boolean variables indicating that a pre-vote for this value has been cast or not, that is p_v^{r+1} for $v = 0, 1$. The transition rules then follow similarly the cases for $s_i = 1, 2, 3$ above:

$$[] \ (s_i{=}7) \wedge (c_r{=}1) \rightarrow (s_i'{=}9) \wedge (p_0^{r+1'}{=}1); \ // \text{ cast pre-vote for } 0$$
$$[] \ (s_i{=}8) \wedge (c_r{=}1) \rightarrow (s_i'{=}9) \wedge (p_1^{r+1'}{=}1); \ // \text{ cast pre-vote for } 1$$
$$// \text{ cast pre-vote for the coin}$$
$$[] \ (s_i{=}6) \wedge (c_r{=}1) \wedge (v_r{=}0) \rightarrow (s_i'{=}9) \wedge (p_0^{r+1'}{=}1);$$
$$[] \ (s_i{=}6) \wedge (c_r{=}1) \wedge (v_r{=}1) \rightarrow (s_i'{=}9) \wedge (p_1^{r+1'}{=}1);$$

This completes the possible transitions of party i. To construct further parties we use renaming, for example:

$$\texttt{module } party_j = party_i[s_i = s_j] \texttt{ endmodule}$$

Correctness of the Abstract Protocol. To prove the correctness of the abstract model constructed in PRISM, we follow the method presented in [58] for timed probabilistic systems. This method reduces the verification of the correctness of the abstraction to constructing non-probabilistic variants of the abstract and concrete models and checking *trace refinement* between these systems. The method is reliant on encoding the probabilistic information and choices of the adversary in *actions* during model construction. Since the Cadence SMV language does not support actions, we use the process algebra CSP [88] and the model checker FDR [31].

More formally, we hand-translate both the abstract protocol and the full protocol (restricted to two arbitrary rounds) into CSP, encoding both the probabilistic choices and the possible non-deterministic choices of the adversary into the actions of the CSP processes. Using the tool FDR we were then able to show that the concrete protocol is a trace refinement of the abstract protocol, and hence the correctness of our abstraction. Note that we were only able to do this for finite configurations. For further details and the FDR code see the PRISM web page [83].

Model Checking Results. The property we wish to verify is that from the initial state, with probability at least 0.5, all honest parties have the same pre-vote in round $r + 1$, and hence decide by round $r + 1$. This property can be expressed by the PCTL formula:

$$\mathcal{P}_{\geq 0.5}\left[\texttt{true } \mathcal{U} \left(\bigwedge_{i=1}^{N} s_i{=}9\right) \wedge \left((p_0^{r+1}{=}1 \wedge p_1^{r+1}{=}0) \vee (p_0^{r+1}{=}0 \wedge p_1^{r+1}{=}1)\right)\right].$$

On all models constructed this property does indeed hold. A summary of the model checking results obtained for the abstract protocol in PRISM is included in Figure 4, where all experiments were run on a 440 MHz SUN Ultra 10 workstation with 512 Mb memory under the Solaris 2.7 operating system. Further details of the experiments can be found at the PRISM web page [83].

5.5   Fast Convergence: Parametric Verification

In this section we give a proof of **Fast Convergence** for any number of parties. The proof is fully automated except for one high-level inductive argument involving probabilistic reasoning. The proof demonstrates how to separate the probabilistic and nondeterministic behaviour and isolate the probabilistic arguments in the proof of correctness.

The high-level probabilistic argument is based on a number of properties (**P1** – **P6**) that can be proved by non-probabilistic reasoning, and have been proved

n	t	number of states	construction time (sec)	model checking time (sec)	**minimum probability**
4	1	16,468	3.00	1.49	**0.5**
5	1	99,772	5.41	4.86	**0.5**
6	1	567,632	8.53	7.81	**0.5**
7	2	1,303,136	10.9	16.0	**0.5**
8	2	8,197,138	20.0	24.4	**0.5**
9	2	5.002e+7	27.0	36.5	**0.5**
10	3	9.820e+7	33.8	58.9	**0.5**
11	3	6.403e+8	62.4	85.1	**0.5**
12	3	4.089e+9	75.4	114	**0.5**
13	4	7.247e+9	98.3	167	**0.5**
14	4	4.856e+10	157	282	**0.5**
15	4	3.199e+11	194	470	**0.5**
16	5	5.273e+11	241	651	**0.5**
17	5	3.605e+12	363	987	**0.5**
18	5	2.429e+13	610	1,318	**0.5**
19	6	3.792e+13	694	1,805	**0.5**
20	6	2.632e+14	1,079	2,726	**0.5**

Fig. 4. Model checking results for PRISM

in the Cadence SMV proof assistant for an arbitrary number of parties. Here we state them in English; for the corresponding formal statements and Cadence SMV proofs see [83]. First we proved that, for any party that enters round $r + 1$ and does not decide in round r:

P1 If before the coin in round r is tossed there is a concrete pre-vote (i.e. a vote *not* based on the value of the coin) for v in round $r + 1$ and after the coin in round r is tossed it equals v, then the party decides in round $r + 1$.

P2 If before the coin in round r is tossed there are no concrete pre-votes in round $r + 1$, then either the party decides in round $r + 1$, or if after the coins in round r and round $r + 1$ are tossed they are equal, then the party decides in round $r + 2$.

In addition, we proved that the following properties hold.

P3 If the coin in round r has not been tossed, then neither has the coin in round $r + 1$.

P4 In any round r there cannot be concrete pre-votes for 0 and 1.

P5 In any round r, if there is a concrete pre-vote for $v \in \{0, 1\}$, then in all future states there is a concrete pre-vote for v.

P6 Each coin is only tossed once.

We complete the proof of **Fast Convergence** with a simple manual proof based on the following classification of protocol states:

- let $Undec(r)$ be the set of states in which the coin in round $r - 1$ is not tossed and there are no concrete pre-votes in round r;

- for $v \in \{0, 1\}$, let *Pre-vote*(r, v) be the set of states where the coin in round $r - 1$ is not tossed and there is a concrete pre-vote for v in round r.

It follows from **P4** that these sets are pairwise disjoint and *any* state where the coin in round $r - 1$ is not tossed is a member of one of these sets. The following proposition is crucial to establishing the efficiency of the protocol.

Proposition 2. *In an idealised system, where the values of the coins in rounds* $1, 2, \ldots, 2r - 1$ *are truly random, the probability of a party advancing by more than* $2r + 1$ *rounds is bounded above by* 2^{-r}.

Proof. We prove the proposition by induction on $r \in \mathbb{N}$. The case when $r = 0$ is trivial since the probability bound is 1. Now suppose that the proposition holds for some $r \in \mathbb{N}$ and suppose a party enters round $2r + 1$. If a party decides in round $2r$, then by **Agreement** all parties will decide by round $2r + 1$, and hence the probability that a party enters round $2r + 3$ given a party enters round $2r + 1$ is bounded above by 0. On the other hand, if no party decides in round $2r$, then by Proposition 1(a) the coin for round $2r$ is tossed. For any state s reached just before the coin is tossed we have two cases to consider:

- $s \in$ *Pre-vote*$(2r + 1, v)$ for some $v \in \{0, 1\}$: by **P1**, if the coin in round $2r$ equals v, any party which enters round $2r + 1$ decides in round $2r + 1$, and hence using **P6** it follows that the probability of a party advancing more than $2r + 3$ rounds given that a party advances more than $2r + 1$ rounds is bounded above by $1/2$.
- $s \in$ *Undec*$(2r + 1)$: using **P5**, there are no concrete pre-votes in round $2r + 1$ before the coin in round $2r + 1$ is tossed, and hence by **P2** any party either decides in round $2r + 1$, or, if the coins in round $2r$ and $2r + 1$ are equal, it decides in round $2r + 2$. Now, since in s the coin for round $2r$ has not been tossed, by **P3** neither has the coin for round $2r + 1$. Therefore, using Proposition 1 and **P6** it follows that the probability of a party advancing more than $2r + 3$ rounds given that a party advances more than $2r + 1$ rounds is bounded above by $1/2$.

Putting this together and since $\mathbf{P}(A \cap B) = \mathbf{P}(A|B) \cdot \mathbf{P}(B)$, we have

$\mathbf{P}$(a party advances $>2r+3$ rounds)

$= \mathbf{P}$(a party advances $>2r+3$ rounds and a party advances $>2r+1$ rounds)

$= \mathbf{P}$(a party advances $>2r+3$ rounds | a party advances $>2r+1$ rounds) $\cdot$

$\quad \mathbf{P}$(a party advances $>2r+1$ rounds)

$\leq 1/2 \cdot 2^{-r} = 2^{-(r+1)}$

as required.   □

It can be argued that in a real system the probability of a party advancing by more than $2r+1$ rounds is bounded above by $2^{-r} + \varepsilon$, where ε is negligible. This follows from the **Unpredictability** property of the coin tossing scheme and **P6**; for more details see [10].

In addition to **Fast Convergence** we can directly prove that the protocol guarantees **Probabilistic Termination** in a constant expected number of rounds by using probabilistic complexity statements. As in [82], the complexity measure of interest corresponds to the increase in the maximum round number among all the parties. We now sketch the argument for proving that the protocol guarantees **Probabilistic Termination** in a constant expected number of rounds using the probabilistic complexity statements. First, let $\phi_{MaxRound}$ be the complexity measure that corresponds to the increase in the maximum round number among all the parties, and define the following sets of states:

- $\mathcal{R}$, the set of reachable states of the protocol;
- $\mathcal{D}$, the set of reachable states of the protocol in which all parties have decided;
- *Undec*, the set of states in which the coin in round $r_{\max} - 1$ is not tossed and there are no concrete pre-votes in round $r_{\max}$, where $r_{\max}$ is the current maximum round number among all parties;
- *Pre-vote*(v), the set of states where the coin in round $r_{\max} - 1$ is not tossed and there is a concrete pre-vote for v in round $r_{\max}$, where $r_{\max}$ is the current maximum round number among all parties.

Next we require the following property (which is straightforward to prove in Cadence SMV): from any state (under any fair scheduling of the non-determinism) the maximum round increases by at most one before we reach a state where either all parties have decided or the coin in the maximum round has not been tossed, which can be expressed as the following probabilistic complexity statement:

$$\mathcal{R} \xrightarrow{\phi_{MaxRound} \leq 1}_1 \mathcal{D} \cup Undec \cup Pre\text{-}vote(0) \cup Pre\text{-}vote(1).$$

Applying similar arguments to those given in Proposition 2 we can show that the following probabilistic complexity statements hold:

$$Undec \xrightarrow{\phi_{MaxRound} \leq 2}_{\frac{1}{2}} \mathcal{D} \quad \text{and} \quad Pre\text{-}vote(v) \xrightarrow{\phi_{MaxRound} \leq 2}_{\frac{1}{2}} \mathcal{D} \quad \text{for } v \in \{0,1\}.$$

Alternatively, one could use *coin lemmas* [92, 34] to validate these probabilistic complexity statements. Then, using the compositionality result of complexity statements [91] and the fact that the sets *Undec*, *Pre-vote*(0) and *Pre-vote*(1) are disjoint, the above complexity statements can be combined to give:

$$\mathcal{R} \xrightarrow{\phi_{MaxRound} \leq 2+1}_{1 \cdot \frac{1}{2}} \mathcal{D},$$

that is, from any state of the protocol the probability of reaching a state where all parties have decided while the maximum round increases by at most 3 is at least 1/2. Finally, again using results presented in [91], it follows that from any state of the protocol all parties decide within at most $O(1)$ rounds.

6   Discussion and Conclusions

We have presented a number of techniques that can be applied to the analysis of randomized distributed algorithms. The main problem in verifying such algorithms is correctly dealing with the interplay between probability and non-determinism. A number of approaches exist however, the lesson to be learnt when dealing with complex randomized distributed algorithms is to first try and separate the probabilistic reasoning to a small isolated part of the protocol. This then simplifies the probabilistic arguments and allows one to use standard non-probabilistic techniques in the majority of the verification.

The differing techniques have been used to verify a number of different randomized distributed algorithms and it can be seen that both the structure of the protocol under study and the type of property being verified has influence over the applicability of each approaches and how easy it is to apply.

It may be beneficial to consider new methods for verification for example based on the techniques developed in the areas of performance analysis and the extensive literature concerning dynamic programming and Markov Decision Processes. In particular, it would be useful to examine how methods using *rewards* can be incorporated in these approaches, for example to compute the expected number of rounds.

With regards to computer-aided verification, state-of-the-art probabilistic model checking tools are applicable to only complete, finite state models. On the other hand, there exist non-probabilistic model checkers which can deal with parametric and infinite state programs, however they do not support probabilistic reasoning. A fully automated proof of correctness of randomized distributed algorithms could feasibly be derived using a theorem prover e.g. [43]. An alternative goal is to develop proof techniques for probabilistic systems in the style of for example Cadence SMV, incorporating both those used in Cadence SMV and the proof rules for probabilistic complexity statements following [91]. Given an implementation of such rules as a layer on top of, for example, the PRISM model checking tool, one may be able to *fully automate* the proof of correctness of complex randomized distributed algorithms.

References

1. S. Aggarwal and S. Kutten. Time-optimal self stabilizing spanning tree algorithms. In R. Shyamasundar, editor, *Proc. Foundations of Software Technology and Theoretical Computer Science*, volume 761 of *LNCS*, pages 15–17. Springer, 1993.
2. R. Alur, C. Courcoubetis, and D. Dill. Model-checking for probabilistic real-time systems. In J. Albert, B. Monien, and M. Rodrf8guez-Artalejo, editors, *Proc. Int. Col. Automata, Languages and Programming (ICALP'91)*, volume 510 of *LNCS*, pages 115–136. Springer, 1991.
3. R. Alur, C. Courcoubetis, and D.L. Dill. Model-checking in dense real-time. *Information and Computation*, 104(1):2–34, 1993.
4. R. Alur and T. Henzinger. Reactive modules. *Formal Methods in System Design*, 15:7–48, 1999.
5. J. Aspnes and M. Herlihy. Fast randomized consensus using shared memory. *Journal of Algorithms*, 11(3):441–460, 1990.

6. C. Baier, M. Huth, M. Kwiatkowska, and M. Ryan, editors. *Proc. Int. Workshop Probabilistic Methods in Verification (PROBMIV'98)*, volume 22 of *ENTCS*. Elsevier Science, 1998.

7. C. Baier and M. Kwiatkowska. Model checking for a probabilistic branching time logic with fairness. *Distributed Computing*, 11:125–155, 1998.

8. M. Ben-Or. Another advantage of free choice: Completely asynchronous agreement protocols. In *Proc. Symp. Principles of Distributed Computing (PODC'83)*, pages 27–30. ACM Press, 1983.

9. A. Bianco and L. de Alfaro. Model checking of probabilistic and nondeterministic systems. In P. Thiagarajan, editor, *Proc. Foundations of Software Technology and Theoretical Computer Science (FSTTCS'95)*, volume 1026 of *LNCS*, pages 499–513. Springer, 1995.

10. C. Cachin, K. Kursawe, F. Petzold, and V. Shoup. Secure and efficient asynchronous broadcast protocols (extended abstract). In J. Kilian, editor, *Proc. Advances in Cryptology - CRYPTO 2001*, volume 2139 of *LNCS*, pages 524–541. Springer, 2001.

11. C. Cachin, K. Kursawe, and V. Shoup. Random oracles in Constantinople: practical asynchronous Byzantine agreement using cryptography (extended abstract). In *Proc. Symp. Principles of Distributed Computing (PODC'00)*, pages 123–132. ACM Press, 2000.

12. E. Clarke, O. Grumberg, and D. Peled. *Model Checking*. MIT Press, 1999.

13. R. Cleaveland, S. Smolka, and A. Zwarico. Testing preorders for probabilistic processes. In W. Kuich, editor, *Proc. Int. Col. Automata, Languages and Programming (ICALP'92)*, volume 623 of *LNCS*, pages 708–719. Springer, 1992.

14. C. Courcoubetis and M. Yannakakis. The complexity of probabilistic verification. *Journal of the ACM*, 42(4):857–907, 1995.

15. P. Cousot and R. Cousot. Abstract interpretation: a unified lattice model for static analysis of programs by construction or approximation of fixpoints. In *Proc. Symp. Principles of Programming Languages (POPL'77)*, pages 238–252. ACM Press, 1977.

16. S. Creese and A. Roscoe. Data independent induction over structured networks. In H. Arabnia, editor, *Proc. Int. Conf. Parallel and Distributed Processing Techniques and Applications (PDPTA'00)*, volume II. CSREA Press, 2000.

17. P. D'Argenio, B. Jeannet, H. Jensen, and K. Larsen. Reachability analysis of probabilistic systems by successive refinements. In de Alfaro and Gilmore [24], pages 39–56.

18. P. D'Argenio, B. Jeannet, H. Jensen, and K. Larsen. Reduction and refinement strategies for probabilistic anylsis. In Hermanns and Segala [42], pages 57–76.

19. C. Daws, M. Kwiatkowska, and G. Norman. Automatic verification of the IEEE 1394 root contention protocol with KRONOS and PRISM. In *Proc. Int. Workshop Formal Methods for Industrial Critical Systems (FMICS'02)*, volume 66(2) of *ENTCS*. Elsevier Science, 2002.

20. L. de Alfaro. *Formal Verification of Probabilistic Systems*. PhD thesis, Stanford University, 1997.

21. L. de Alfaro. Temporal logics for the specification of performance and reliability. In R. Reischuk and M. Morvan, editors, *Proc. Symp. Theoretical Aspects of Computer Science (STACS'97)*, volume 1200 of *LNCS*, pages 165–176. Springer, 1997.

22. L. de Alfaro. From fairness to chance. In Baier et al. [6].

414   G. Norman

23. L. de Alfaro. How to specify and verify the long-run average behaviour of probabilistic systems. In *Proc. Symp. Logic in Computer Science (LICS'98)*, pages 454–465. IEEE Computer Society Press, 1998.
24. L. de Alfaro and S. Gilmore, editors. *Proc. Int. Workshop Process Algebra and Probabilistic Methods, Performance Modeling and Verification (PAPM/PROB-MIV'01)*, volume 2165 of *LNCS*. Springer, 2001.
25. L. de Alfaro, T. Henzinger, and R. Jhala. Compositional methods for probabilistic systems. In Larsen and Nielsen [60], pages 351–365.
26. C. Derman. *Finite-State Markovian Decision Processes*. New York: Academic Press, 1970.
27. E. Dijkstra. *A Discipine of Programming*. Prenticec Hall International, 1976.
28. S. Dolev, A. Israeli, and S. Moran. Analyzing expected time by scheduler-luck games. *IEEE Transactions on Software Engineering*, 21(5):429–439, 1995.
29. M. Duflot, L. Fribourg, and C. Picaronny. Randomized finite-state distributed algorithms as Markov chains. In J. Welch, editor, *Proc. Distributed Computing (DISC'2001)*, volume 2180 of *LNCS*, pages 240–254. Springer, 2001.
30. M. Duflot, L. Fribourg, and C. Picaronny. Randomized dining philosophers without fairness assumption. In *Proc. IFIP Int. Conf. Theoretical Computer Science (TCS'02)*, volume 223 of *IFIP Conference Proceedings*, pages 169–180. Kluwer Academic, 2002.
31. Failures divergence refinement (FDR2). Formal Systems (Europe) Limited, http://www.formal.demon.co.uk/FDR2.html.
32. W. Feller. *An Introduction to Probability Theory and its Applications*, volume 1. John Wiley & Sons, 1950.
33. M. Fischer, N. Lynch, and M. Paterson. Impossibility of distributed consensus with one faulty process. *Journal of the ACM*, 32(5):374–382, 1985.
34. K. Folegati and R. Segala. Coin lemmas with random variables. In de Alfaro and Gilmore [24], pages 71–86.
35. H. Hansson. *Time and Probability in Formal Design of Distributed Systems*. Elsevier, 1994.
36. H. Hansson and B. Jonsson. A calculus for communicating systems with time and probabilities. In *Proc. Real-Time Systems Symposium*, pages 278–287. IEEE Computer Society Press, 1990.
37. H. Hansson and B. Jonsson. A logic for reasoning about time and reliability. *Formal Aspects of Computing*, 6(4):512–535, 1994.
38. S. Hart, M. Sharir, and A. Pnueli. Termination of probabilistic concurrent programs. *ACM Transactions on Programming Languages and Systems*, 5(3):356–380, 1983. A preliminary version appeared in Proc. ACM Symp. Principles of Programming Languages, pages 1–6, 1982.
39. T. Henzinger, X. Nicollin, J. Sifakis, and S. Yovine. Symbolic model checking for real-time systems. In *Proc. Symp. Logic in Computer Science (LICS'98)*, pages 394–406. IEEE Computer Society Press, 1992.
40. T. Henzinger, S. Qadeer, and S. Rajamani. You assume, we guarantee: Methodology and case studies. In A. Hu and M. Vardi, editors, *Proc. Computer-aided Verification (CAV'98)*, volume 1427 of *LNCS*, pages 440–451. Springer, 1998.
41. T. Herman. Probabilistic self stabilisation. *Information Processing Letters*, 35(2):63–67, 1990.
42. H. Hermanns and R. Segala, editors. *Proc. Int. Workshop Process Algebra and Probabilistic Methods, Performance Modeling and Verification (PAPM/PROB-MIV'02)*, volume 2399 of *LNCS*. Springer, 2002.

43. Joe Hurd. Verification of the Miller-Rabin probabilistic primality test. In R. Boulton and P. Jackson, editors, *TPHOLs 2001: Supplemental Proceedings*, number EDI-INF-RR-0046 in Informatics Report Series, pages 223–238. Division of Informatics, University of Edinburgh, 2001.

44. Joe Hurd. *Formal Verification of Probabilistic Algorithms*. PhD thesis, University of Cambridge, 2002.

45. Michael Huth and Marta Kwiatkowska. Quantitative analysis and model checking. In *Proc. Symp. Logic in Computer Science (LICS'97)*, pages 111–122. IEEE Computer Society Press, 1997.

46. IEEE Computer Society. IEEE Standard for a High Performance Serial Bus. Std 1394-1995. August 1996.

47. A. Israeli and M. Jalfon. Token management schemes and random walks yield self-stabilizing mutual exclusion. In *Proc. Symp. Principles of Distributed Computing (PODC'90)*, pages 119–131. ACM Press, 1990.

48. B. Jonnsson and K.G. Larsen. Specification and refinement of probabilistic processes. In *Proc. Symp. Logic in Computer Science (LICS'91)*, pages 266–277. IEEE Computer Society Press, 1991.

49. B. Jonsson and J. Parrow, editors. *Proc. Int. Conf. Concurrency Theory (CONCUR'94)*, volume 836 of *LNCS*. Springer, 1994.

50. Y. Kesten and A. Pnueli. Control and data abstraction: the cornerstone of practical formal verification. *Software Tools for Technology Transfer*, 4(2):328–342, 2000.

51. Y. Kesten, A. Pnueli, E. Shahar, and L. Zuck. Network invariants in action. In L. Brim, P. Jancar, M. Kretinsky, and A. Kucera, editors, *Proc. CONCUR'02 – Concurrency Theory*, volume 2421 of *LNCS*, pages 101–115. Springer, 2002.

52. E. Kushilevitz and M. Rabin. Randomized mutual exclusion algorithm revisited. In PODC92 [80], pages 275–284.

53. M. Kwiatkowska and G. Norman. Verifying randomized Byzantine agreement. In D. Peled and M. Vardi, editors, *Proc. Formal Techniques for Networked and Distributed Systems (FORTE'02)*, volume 2529 of *LNCS*, pages 194–209. Springer, 2002.

54. M. Kwiatkowska, G. Norman, and D. Parker. PRISM: Probabilistic symbolic model checker. In T. Field, P. Harrison, J. Bradley, and U. Harder, editors, *Proc. Modelling Techniques and Tools for Computer Performance Evaluation (TOOLS'02)*, volume 2324 of *LNCS*, pages 200–204. Springer, April 2002.

55. M. Kwiatkowska, G. Norman, and R. Segala. Automated verification of a randomized distributed consensus protocol using Cadence SMV and PRISM. In G. Berry, H. Comon, and A. Finkel, editors, *Proc. Int. Conf. Computer Aided Verification (CAV'01)*, volume 2102 of *LNCS*, pages 194–206. Springer, 2001.

56. M. Kwiatkowska, G. Norman, R. Segala, and J. Sproston. Automatic verification of real-time systems with discrete probability distributions. *Theoretical Computer Science*, 282:101–150, 2002.

57. M. Kwiatkowska, G. Norman, and J. Sproston. Symbolic computation of maximal probabilistic reachability. In Larsen and Nielsen [60], pages 169–183.

58. M. Kwiatkowska, G. Norman, and J. Sproston. Probabilistic model checking of deadline properties in the IEEE 1394 FireWire root contention protocol. *Special Issue of Formal Aspects of Computing*, 2002. To appear.

59. M. Kwiatkowska, G. Norman, and J. Sproston. Probabilistic model checking of the IEEE 802.11 wireless local area network protocol. In Hermanns and Segala [42], pages 169–187.

60. K. Larsen and M. Nielsen, editors. *Proc. CONCUR'01: Concurrency Theory*, volume 2154 of *LNCS*. Springer, 2001.
61. K. Larsen and A. Skou. Bisimulation through probabilistic testing. *Information and Computation*, 94(1):1–28, 1991. Preliminary version of this paper appeared in Proc. 16th Annual ACM Symposium on Principles of Programming Languages, pages 134-352, 1989.
62. R. Lassaigne and S. Peyronnet. Approximate verification of probabilistic systems. In Hermanns and Segala [42], pages 213–214.
63. D. Lehmann and M. Rabin. On the advantage of free choice: A symmetric and fully distributed solution to the dining philosophers problem (extended abstract). In *Proc. Symp. on Principles of Programming Languages (POPL'81)*, pages 133–138. ACM Press, 1981.
64. G. Lowe. *Probabilities and Priorities in Timed CSP*. PhD thesis, Oxford University Computing Laboratory, 1993.
65. G. Lowe. Representing nondeterministic and probabilistic behaviour in reactive processes. Technical Report PRG-TR-11-93, Oxford University Computing Laboratory, 1993.
66. N. Lynch. *Distributed Algorithms*. Morgan Kaufmann, 1996.
67. N. Lynch, I. Saias, and R. Segala. Proving time bounds for randomized distributed algorithms. In *Proc. Symp. Principles of Distributed Computing (PODC'94)*, pages 314-323. ACM Press, 1994.
68. A. McIver. Quantitative program logic and expected time bounds in probabilistic distributed algorithms. *Theoretical Computer Science*, 282:191–219, 2002.
69. A. McIver and C. Morgan. An expectation-based model for probabilistic temporal logic. *Logic Journal of the IGPL*, 7(6):779–804, 1999.
70. A. McIver, C. Morgan, and E. Troubitsyna. The probabilistic steam boiler: a case study in probabilistic data refinement. In J. Grundy, M. Schwenke, and T. Vickers, editors, *Proc. Int. Refinement Workshop and Formal Methods Pacific 1998*, Discrete Mathematics and Theoretical Computer Science, pages 250–265. Springer, 1998.
71. K. McMillan. Verification of infinite state systems by compositional model checking. In L. Pierre and T. Kropf, editors, *Proc. Advanced Research Working Conference on Correct Hardware Design and Verification Methods (CHARME'99)*, volume 1703 of *LNCS*, pages 219–233. Springer, 1999.
72. K. McMillan. A methodology for hardware verification using compositional model checking. *Science of Computer Programming*, 37(1–3):279–309, 2000.
73. R. Milner. *Communication and Concurrency*. Prentice Hall, 1989.
74. C. Morgan and A. McIver. *pGL*: Formal reasoning for randomized distributed algorithms. *South African Computer Journal*, pages 14–27, 1999.
75. C. Morgan, A. McIver, and K. Seidel. Probabilistic predicate transformers. *ACM Transactions on Programming Languages and Systems*, 18(3):325–353, 1996.
76. C. Morgan, A. McIver, K. Seidel, and J. Sanders. Refinement-oriented probability for CSP. *Formal Aspects of Computing*, 8(6):617–647, 1996.
77. R. Motwani and P. Raghavan. *Randomized Algorithms*. Cambridge University Press, 1995.
78. A. Pnueli and L. Zuck. Verification of multiprocess probabilistic protocols. *Distributed Computing*, 1(1):53–72, 1986.
79. A. Pnueli and L. Zuck. Probabilistic verification. *Information and Computation*, 103:1–29, 1993.
80. *Proc. Symp. Principles of Distributed Computing (PODC'92)*. ACM Press, 1992.

81. A. Pogosyants and R. Segala. Formal verification of timed properties of randomized distributed algorithms. In *Proc. Symp. Principles of Distributed Computing (PODC'95)*, pages 174–183. ACM Press, 1995.
82. A. Pogosyants, R. Segala, and N. Lynch. Verification of the randomized consensus algorithm of Aspnes and Herlihy: a case study. *Distributed Computing*, 13(3):155–186, 2000.
83. PRISM web page. http://www.cs.bham.ac.uk/~dxp/prism/.
84. M. Rabin. N-process mutual exclusion with bounded waiting by $4\log_2 N$-valued shared variable. *Journal of Computer and System Sciences*, 25(1):66–75, 1982.
85. M. O. Rabin. Probabilistic algorithms. In J. Traub, editor, *Algorithms and Complexity: New Directions and Recent Results*, pages 21–39. Academic Press, New York, 1976.
86. RAPTURE web page. http://www.irisa.fr/prive/bjeannet/prob/prob.html.
87. M. Reiter and A. Rubin. Crowds: Anonymity for web transactions. *ACM Transactions on Information and System Security (TISSEC)*, 1(1):66–92, 1998.
88. A. Roscoe. *The Theory and Practice of Concurrency*. Prentice-Hall, 1997.
89. A. Saias. *Randomness versus Non-Determinism in Distributed Computing*. PhD thesis, Massachusetts Institute of Technology, 1994.
90. I. Saias. Proving probabilistic correctness statements: the case of Rabin's algorithm for mutual exclusion. In PODC92 [80], pages 263–274.
91. R. Segala. *Modelling and Verification of Randomized Distributed Real-Time Systems*. PhD thesis, Massachusetts Institute of Technology, 1995.
92. R. Segala. The essence of coin lemmas. In Baier et al. [6].
93. R. Segala. Verification of randomized distributed algorithms. In E. Brinksma, H. Hermanns, and J.-P. Katoen, editors, *Lectures on Formal Methods and Performance Analysis (First EEF/Euro Summer School on Trends in Computer Science)*, volume 2090 of *LNCS*, pages 232–260. Springer, 2001.
94. R. Segala and N. Lynch. Probabilistic simulations for probabilistic processes. In Jonsson and Parrow [49], pages 481–496.
95. V. Shmatikov. Probabilistic analysis of anonymity. In *Proc. Computer Security Foundations Workshop (CSFW'02)*, pages 119–128. IEEE Computer Society Press, 2002.
96. V. Shoup. Practical threshold signatures. In B. Preneel, editor, *Proc. Advances in Cryptology - EUROCRYPT 2000*, volume 1807 of *LNCS*, pages 207–220. Springer, 2000.
97. T. Speed. Probabilistic risk assessment in the nuclear industry : WASH-1400 and beyond. In L. LeCam and R. Olshen, editors, *Proc. Berkeley Conference in honour of Jerzy Neyman and Jack Kiefer*. Wadsworth Inc., 1985.
98. E. Stark and G. Pemmasani. Implementation of a compositional performance analysis algorithm for probabilistic I/O automata. In J. Hillston and M. Silva, editors, *Proc. Int. Workshop Process Algebra and Performance Modelling (PAPM'99)*, pages 3–24. Prensas Universitarias de Zaragoza, 1999.
99. E. Stark and S. Smolka. Compositional analysis of expected delays in networks of probabilistic I/O automata. In *Proc. Symp. Logic in Computer Science (LICS'98)*, pages 466–477. IEEE Computer Society Press, 1988.
100. M. Stoelinga and F. Vaandrager. Root contention in IEEE 1394. In J.-P. Katoen, editor, *Proc. AMAST Workshop on Real-Time and Probabilistic Systems (ARTS'99)*, volume 1601 of *LNCS*, pages 53–74. Springer, 1999.
101. M. Vardi. Automatic verification of probabilistic concurrent finite state programs. In *Proc. Symp. Foundations of Computer Science (FOCS'85)*, pages 327–338. IEEE Computer Society Press, 1985.

102. M. Vardi and P. Wolper. An automata-theoretic approach to automatic program verification. In *Proc. Symp. Logic in Computer Science (LICS'86)*, pages 332–344. IEEE Computer Society Press, 1986.
103. S-H. Wu, S.A. Smolka, and E.W. Stark. Composition and behaviour of probabilistic I/O automata. In Jonsson and Parrow [49], pages 513–528.
104. Wang Yi and K.G. Larsen. Testing probabilistic and non-deterministic processes. In R. Linn Jr. and M. Ümit Uyar, editors, *Protocol Specification, Testing and Verification*, volume C-8 of *IFIP Transactions*, pages 47–61. North-Holland, 1992.
105. L. Zuck, A. Pnueli, and Y. Kesten. Automatic verification of probabilistic free choice. In A. Cortesi, editor, *Proc. Verification, Model Checking, and Abstract Interpretation, Third International Workshop (VMCAI 2002)*, volume 2294 of *LNCS*, pages 208–224. Springer, 2002.

An Abstraction Framework for Mixed Non-deterministic and Probabilistic Systems

Michael Huth

Department of Computing, South Kensington Campus,
Imperial College London, London, SW7 2AZ, United Kingdom,
M.Huth@doc.imperial.ac.uk

Abstract. We study abstraction techniques for model checking systems that combine non-deterministic with probabilistic behavior, emphasizing the discrete case. Existing work on abstraction offers a host of isolated techniques which we discuss uniformly through the formulation of abstracted model-checking problems (MCPs). Although this conceptualization is primarily meant to be a useful focal point for surveying the literature on abstraction-based model checking even beyond such combined systems, it also opens up new research opportunities and challenges for abstract model checking of mixed systems. In particular, we sketch how quantitative domain theory may be used to specify the precision of answers obtained from abstract model checks.

1   Motivation and Objectives

This survey deals with the abstraction of dynamical systems whose computations are guided by a mix of non-deterministic and probabilistic (or quantitative) behavior. For sake of brevity, this paper uses the term "mixed models" as a general reference to models of such systems. E.g. a Markov decision process is a mixed model, since its dynamics alternates between non-deterministic and probabilistic state transitions. Concurrent or under-specified randomized algorithms are another example of mixed models and we will meet more kinds of mixed models in this survey. Although the proper design of models is a central concern in engineering at large, this paper will focus on the analysis of mixed models through abstraction. Ideally, analysis activities lead to an incremental and validated design of a computing artifact. Not surprisingly, formal analysis is mandatory for safety-critical systems and becomes increasingly important in business-critical information systems as well. The objectives of this survey are to

- motivate the use of mixed models for the design and analysis of computational systems;
- motivate the need for abstraction techniques in the design and analysis of mixed models;
- present the design and analysis of mixed models as model-checking problems (MCPs);

C. Baier et al. (Eds.): Validation of Stochastic Systems, LNCS 2925, pp. 419–444, 2004.
© Springer-Verlag Berlin Heidelberg 2004

- propose a unified theory of abstraction for MCPs;
- study main abstraction techniques for mixed models as instances of this unified theory; and
- formulate research challenges for abstracting mixed models and their MCPs.

Outline of paper In Section 2, we develop a general and quite flexible framework for specifying model-checking problems as the formal topic of our discourse and present an important model-checking problem for Markov decision processes in Section 3. The need for having abstract versions of model-checking problems is discussed in Section 4. In Section 5, we survey the various aspects of model-checking problems that can be subjected to abstractions. Model-checking problems give rise to query equivalences which leads to a discussion of bisimulations in Section 6. In Section 7, we define the approximation problem for abstract model-checking problems and point out connections to work in quantitative domain theory. Section 8 reflects on the need of diagnostics as supporting evidence for answers to model-checking instances and Section 9 lists the caveats for effective use of abstract model-checking problems. Although soundness is at the heart of abstraction-based model checking, Section 10 states the main reasons and circumstances in which one wants to abandon such soundness. In Section 11, we present a subjective list of research challenges in the area of abstraction-based model checking of mixed models and Section 12 concludes.

2   Model-Checking Problems

The design of a computational system requires a model of this system. Such a model is necessarily an abstraction of the real system. E.g. a hardware design may start out as a mathematical model of a boolean circuit, but its realization needs to take into account cost and physical constraints. Although models abstract from real systems, their abstractions tend to leave out "irrelevant" details only. Models, unlike real systems, can be represented in a computer and may therefore be subjected to formal and automatic analysis. In this paper, we focus on the latter advantage of models over real systems.

Similar systems will have similar kinds of models and queries about those models. We may axiomatize such a "kind" as

- a class $\mathcal{M}$ of models ($M \in \mathcal{M}$);
- a class Φ of queries about such models ($\phi \in \Phi$);
- a partial order Ans of possible answers to such queries ($a \in$ Ans) — thus also allowing answer domains used for performance modeling (e.g. [43, 45]); and
- a specification $\models: \mathcal{M} \times \Phi \to$ Ans of how to compute answers to queries over models.

It is customary to use infix notation for applications of $\models$:

$$(M \models \phi) \in \text{Ans}. \tag{1}$$

An Abstraction Framework   421

In this paper, the syntax $M \models \phi$ denotes a *model-checking instance* whose evaluation (1) is referred to as a *model check*. With a set of data as above, we call the tuple

$$\mathcal{P} = (\mathcal{M}, \Phi, \text{Ans}, \models) \tag{2}$$

a *model-checking problem* (MCP). An example of a MCP is SMV [58], where $\mathcal{M}$ is determined by the class of SMV programs — which describe Kripke structures — Φ is computation-tree logic (CTL) [17], Ans = {ff < tt}, and $\models$ is the standard semantics of CTL over Kripke structures [17].

Queries need not have logical structure. For example, models may be computer programs, queries may be questions such as "Which portions of the code will never be reached?" (dead code detection, e.g. [13]), and $\models$ may be the specification of a family of program analyses [82].

For an example of a MCP with mixed models, consider $\mathcal{M}$ as the class of Markov decision processes and Φ as linear-time temporal logic (LTL) [17]. A suitable partial order of answers is the interval domain [94]

$$\mathcal{I} = \{[r, s] \mid 0 \leq r \leq s \leq 1\} \tag{3}$$

ordered by $[r, s] \leq [r', s']$ iff $r \leq r'$ and $s' \leq s$. The answer $[r, s]$ to a model check $M \models \phi$ is specified to represent

- r: the worst-case probability of M satisfying the LTL formula ϕ;
- s: the best-case probability of M satisfying the LTL formula ϕ.

We denote this MCP by MDP subsequently.

3   Markov Decision Processes

MDP is a natural choice for modeling probabilistic algorithms. E.g. for the Java method in Figure 1, based on [76], we may want to know the best-case/worst-case probabilities of satisfying the postcondition, given the precondition. If we are interested in computing the interval $(M \models (0 \leq x \leq 2) \rightarrow \mathsf{G}(x < 3))$ — where G denotes "globally" or "at all future points" — then a natural model M of this method is depicted in Figure 2. Non-determinism results from the

```
int aRandomUpdate(x : int) {// precondition: 0 <= x <= 2
  for (i = 0; i < 5; i++) {
    // coin_flip() returns 0 or 1 with probability 0.5
    x = x + coin_flip();
  }
  return x;
}// postcondition: x < 3
```

Fig. 1. A probabilistic method with return type int and qualitative preconditions and postconditions

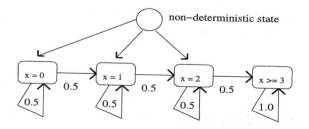

Fig. 2. A mixed model of the method `aRandomUpdate` of Figure 1. Non-deterministic nodes and probabilistic nodes may not alternate strictly. The choice of model was informed by the structure of the program and the query

under-specification of the parametric method input x. Probabilistic behavior is caused by the method's coin flip. The model in Figure 2 is an abstraction, since it already assumes the precondition in its non-deterministic choice and abstracts $x \geq 3$ into a single state.

The models M of a MCP $(\mathcal{M}, \Phi, \text{Ans}, \models)$ are typically represented symbolically in a programming language (e.g. Java), a specification language (e.g. process algebras [46, 70, 2, 42]) or through data structures that implement model checks (e.g. binary decision diagrams [9], their multiple-valued variants MTBBDs [16], and ADDs [92]). A structural operational semantics [90] then maps such syntactic structures to full mathematical models. Abstraction and its soundness are typically defined and proved for those mathematical models, e.g. the abstract model in Figure 2.

In our example, the interval $(M \models (0 \leq x \leq 2) \to \mathsf{G}(x < 3))$ equals $[0, 0]$ since any non-deterministic choice results in a state that eventually has to increase the value of x. Thus, one needs an abstract interpretation of these checks which restricts the scope of the G modality to the first five states on each path. This is sound since the method's sole for-statement executes exactly five times. We may reduce the computation of this check to the one of $(M \models \mathsf{F}(x >= 3))$ — where F denotes "at some future point" — since the precondition is enforced in the model and since F and G are logical duals. Thus, we now evaluate a bounded-until query

$$\mathsf{tt}\ \mathsf{U}^{\leq 5}\ (x >= 3) \tag{4}$$

whose semantics $[r, s]$ can be computed through fixed-point iterations or linear programming techniques. We leave the computation of these values as an exercise. Neither r nor s are conventional probabilities. For r, a fictitious opponent of the verifier controls the initial value of x with the aim of minimizing that probability, choosing x $==$ 0. Dually, for s the opponent turns into a collaborator who chooses x $==$ 2 in order to maximize that value. The meaning of $\mathsf{G}(x < 3)$ is then $[1 - s, 1 - r]$; this negation swaps the roles of opponent and collaborator, the opponent now chooses x $==$ 2 and the collaborator picks x $==$ 0. These different roles of opponent and collaborator are also reflected in the process algebra CSP in the form of demonic and angelic choice (respectively) [46].

The two numbers r and s are the extremal outcome points of a set of possible opponent strategies. In that sense, the interval $[r, s]$ is already an abstraction of itself. For example, an opponent may put a uniform probability measure on the precondition domain of x, resulting in a form of probabilistic testing which turns the model into a fully probabilistic one, a Markov chain. Since each value of x determines a likelihood $f(x)$ of the postcondition's being true, we obtain a measurable function f from the value domain of x to $[0, 1]$ and the postcondition's expected likelihood is then the integral $\int f d\mu$ of that function f over the probability measure μ on the values of x. In our example, let that measure μ assign $\frac{1}{3}$ to each value 0, 1, and 2. Then the expected likelihood in the worst-case equals the one in the best-case since all non-determinism is resolved by the distribution on the method's input. We compute this likelihood as $\int f d\mu = \frac{1}{3} \cdot p_0 + \frac{1}{3} \cdot p_1 + \frac{1}{3} \cdot p_2$, where p_i is the probability of the trace property in question at state x == i; these are actual probabilities since these states are not, and cannot, reach non-deterministic states. In general, soundness of this interpretation means that the expected likelihood be in between the angelic (s) and demonic (r) one.

The origin of these semantic ideas can be traced back to Kozen's work [59, 60] which pioneered the semantics of deterministic programs as transformers of measures or measurable functions for deterministic programs; we highly recommend reading these papers. This line of work has been extended by the Probabilistic Systems Group at Oxford University to abstraction and the presence of non-determinism; see e.g. [80]. In their work, mathematical models treat non-determinism as an abstraction of probabilities such that qualitative non-determinism can be refined by any probabilities. For example, consider the process-algebra specification

$$(\text{heads} \oplus_{.25} \text{tails}) \sqcap (\text{heads} \oplus_{.75} \text{tails}) \tag{5}$$

where $P \oplus_p Q$ is the process that, with probability p behaves like P, and, with probability $1 - p$ behaves as Q; and $P \sqcap Q$ models non-deterministic choice between P and Q — controlled by some unspecified environment. If $P \oplus_{[r,s]} Q$ denotes a process that can act as $P \oplus_p Q$ for any p in the interval $[r, s] \in \mathcal{I}$, then their semantics equates heads $\oplus_{[.25,.75]}$ tails with the specification in (5). Such "convex closure" semantics occur frequently in models that mix probabilities and non-deterministic choice — be it explicitly in the modeling formalism or implicitly by forming the convex closure of models or analysis results.

Pre/postcondition validation of aRandomUpdate can be reduced to a *probabilistic reachability problem* on the Markov decision process in Figure 2: "What is the likelihood of reaching state x >= 3 from the initial state?" Probabilistic LTL checking [100] is tailored for answering such questions. Our example also suggests that *pre/postconditions* (as well as object invariants and control-flow constraints of the program) *guide us in determining how to abstract a model at the source level.*

Bounded model checking [15] abstracts $\models$ by truncating the computation of least fixed points in model checks for a branching-time temporal logic and qualitative models (Kripke structures). This idea applies to LTL and our probabilistic models as well: approximate reachability probabilities by putting bounds on the length of computation traces, thereby enforcing a *finite horizon*. In Figure 2, the for-statement in Figure 2 supplies us with such a bound that is even guaranteed not to lose any precision.

E.g. for ϕ being $\mathsf{F}\,p$, we may abstract ϕ with ϕ' iff ϕ' equals p or $p \vee \mathsf{X}\,p$ or $p \vee \mathsf{X}\,(p \vee \mathsf{X}\,p)$ or ... Without knowing the implementation of a model checker, it is difficult to say whether this really abstracts queries or just models. Queries $\mathsf{G}\,p$ ("Globally, p holds.") may be rewritten as $p \wedge \mathsf{X}\,\mathsf{G}\,p$, but all finite approximations $p \wedge \mathsf{X}\,(p \wedge \mathsf{X}(p \wedge \ldots))$ are unsound in general since paths can be infinite. The semantic equivalence of $\mathsf{G}\,p$ and $\neg\mathsf{F}\,\neg p$, therefore illustrates the delicate role of negation in abstraction-based model checking.

Cousot & Cousot [21] developed abstract interpretation as a theory for systematically designing such sound program abstractions; this theory was originally developed for qualitative computation and has been extended to Markov decision processes by Monniaux in his doctoral thesis [77] and subsequent papers [72–76]. These papers are *mandatory* reading on abstraction of Markov decision processes. Implementing `coin_flip` is another, potentially unsound, source of abstraction.

4   Why Abstraction?

Since the specification and computation of abstractions incur costs, we need to justify the need for constructing abstractions for the purpose of analysis.

4.1   Decidability

The specification of $\models$ may not be computable, e.g. if we were to extend the MCP of SMV to infinite-state Kripke structures, we could encode the Halting problem as a model check. Since Markov decision processes faithfully subsume such qualitative systems, the MCP of infinite-state Markov decision processes and CTL is undecidable as well. Note that models of under-specified or concurrent probabilistic programs are potentially infinite state.

The answers to individual model-checking instances $M \models \phi$ may be very hard to determine and such a determination may be closely tied to finding a 'magic' abstraction. E.g. the non-probabilistic method `Collatz` in Figure 3 determines an infinite-state deterministic Kripke structure M in x and c. The answer to the LTL check $(M \models (1 \leq x) \rightarrow \mathsf{F}(c = 1))$ is unknown at the time of writing. One could conceivably prove that its answer is tt by (1) specifying a well-founded order and a variant; and (2) showing that each execution of the method's while-statement decreases the variant in that well-founded order. It is plausible to believe that such a feat is tantamount to specifying a suitable abstraction of M that does not possess infinite computation paths. Since the MCP of infinite-state Kripke structures with LTL "embeds" into MDP, our example of Figure 3 is relevant to this discussion.

```
int Collatz(x : int) {// precondition: 1 <= x
  int c = x;
  while (c != 1) {
    if (c%2 == 0) { c = c/2; } // %2 is ''modulo 2''
    else { c = 3*c + 1; }
  }
  return c;
}// postcondition: c = 1.
```

Fig. 3. A non-probabilistic method which determines an infinite-state Kripke structure M in x and c. The answer to the LTL check $(M \models (1 \leq x) \rightarrow F(c = 1))$ is unknown

4.2   Complexity

For MCPs with decidable specification $\models: \mathcal{M} \times \Phi \rightarrow \mathsf{Ans}$, the complexity of computing its instances may be too high. Such assessments depend on the type of complexity (e.g. worst-case, average-case or randomized [84]) as well as on its type of bound (lower or upper bounds). For example, the decision problem STOCHASTIC SCHEDULING [83], a scheduling problem of tasks with completion time determined by a Poisson distribution, is PSPACE-complete; and so is the problem of deciding whether the synchronous composition of processes can reach a deadlocked state. For mixed models in MDP, the worst-case upper-bound time complexity of $M \models \phi$ is doubly exponential in the size of ϕ and polynomial in the size of M if M is finite [20]. The complexity 'bottleneck' resides in the semantics of conjunction $\phi_1 \wedge \phi_2$ which may contain conditional probabilities, a non-compositional notion. Baier et al. [5] therefore propose to approximate this semantics, an interval $[r, s]$, in a compositional manner. The approximation is computed via the function $c: [0, 1] \times [0, 1] \rightarrow [0, 1]$ given by

$$c([r_1, s_1], [r_2, s_2]) = [\max\{0, r_1 + r_2 - 1\}, \min\{s_1, s_2\}]. \tag{6}$$

For $[r', s'] = c([r_1, s_1], [r_2, s_2])$ soundness of this approach is guaranteed since $r' \leq r_1, r_2$ and $s_1, s_2 \leq s'$ follow from the modularity axiom of measures [37]. This new interpretation of conjunction modifies $\models$ to $\models'$ and makes the time complexity single exponential in the size of the formula and has a linear space complexity [5]. The function c is built from t-norms which have been used for similar abstractions for Φ being a probabilistic mu-calculus [50]. The first component of c in (6) is well known and applied in artificial intelligence as it computes the best lower bound on the joint probability of two events, without knowing anything about the independence of these events; that t-norm also forms an integral part of a characterization of a modal algebra for quantitative temporal logics [79].

If M is a finite-state Kripke structure and ϕ a CTL formula, then checks are linear in the sizes of M and ϕ respectively [10]. Although the latter complexity is as good as one could hope for, model checks cannot be conducted in practice whenever M is too large. Alas, this is often observed due to the state-space-explosion problem: the composition of n components often results in a model whose size is of the order 2^n. Of course, what applies to the linear bound for

SMV becomes even more so true for the polynomial bound of MDP. There is a reasonable consensus in the formal-methods community that, in light of the known complexity of model checks, scalability is only achievable through the use of aggressive abstraction techniques.

It is worth noting that the computation of $\models$ may be approximated efficiently with randomized algorithms (see e.g. [81]) if the average complexity of model checks is feasible. In that case, one obtains a modification of $\models$ to some $\models'$ which may compromise soundness. Essentially, such approaches replace formal model checks with a form of probabilistic testing [76, 89, 65]. A similar phenomenon can be observed in the complexity of decision problems. For example, trace equivalence in labeled transition systems is PSPACE complete [57], but trace equivalence in fully probabilistic systems is time polynomial [52]; thus, randomization can improve a problem's complexity.

4.3   Environments

If models M are to interact with the world in which they are embedded, they need to be *open*. The method of Figure 1 has an open model, since it interacts with its environment through its formal parameter x. Open models under-specify their environment. In our example, we only know the range for the value of x and this under-specification is represented through non-deterministic choice. In open models, we may also abstract from the interaction with such an environment. The code for method aRandomUpdate does not reveal the semantics of parameter binding, e.g. call-by-value, call-by-name etc. Our model, however, assumes call-by-value.

The formal analysis of models typically requires that models be *closed* in the sense that they are fully specified. In our example, non-determinism in the choice of value for x closes the model. There are many techniques for closing open models. For example, one may encode the interaction with an environment as an LTL formula over a set of communication actions and then synthesize source code from such a formula for the abstraction of a driver [86]. One may also refine such a closure by replacing a check $M \models \phi$ with a check $M \models \psi \rightarrow \phi$ where ψ encodes an additional assumption about the interaction with an environment, provided that Φ supports a sound abstract interpretation of implications — as is the case for LTL [31]. We mention that these techniques seem similar to the ones being applied when systems are enriched with features [34].

If possible, one may also randomize the interaction with an environment, e.g. by replacing the non-deterministic choice for parameter values in methods with a suitable probability distribution. In this manner, the relational semantics of qualitative programs turns into probabilistic predicate transformers [59, 80].

More generally, the probabilities of sets of traces are a function of what constitutes a legal trace in a model, i.e. a legal strategy of resolving non-determinism. For example, in deciding her next non-deterministic move an adversary may have access to [77]

- the current state only; this Markov condition of history-free strategy is used in computing answers to queries for models in MDP;

- the execution trace from the initial to the present state (history-dependent strategy); or
- may see all the system's random choices during its unfolding, e.g. this is discussed in [19]; Monte Carlo methods such as [76] are based on this.

For a more detailed discussion of the role of adversaries, we refer to chapter 5 of Segala's thesis [95]. Clearly, designers of model-checking frameworks and their users need to be aware of the expressive power of strategies for the generation of trace sets. For example, it is well known that the probabilistic semantics of safety properties over Markov decision processes does not change if one replaces history-free with history-dependent strategies; this is no longer true for liveness properties (see e.g. page 105 in [77]).

4.4   Composition

Connecting interfaces of components makes models M more abstract through the hiding of internal communication, as can be seen in the process algebras CCS [70], CSP [46], ACP [2], the π-calculus [71], and probabilistic process algebras [42] where components connect via typed channels. Such hiding of internal computation results in weak bisimulation as a natural model equivalence [70]. Since the complexity of checks is also a significant function of the size of the model, it may therefore be desirable to replace a model with a weakly bisimilar one with small or minimal state set.

4.5   Specifications

We argued that open models allow for the under-specification of an interaction environment. Under-specification in the form of under-determination or uncertainty is also useful in the description of closed models. Models M then abstract *sets* of implementations. Consequently, the verification of checks on such models applies to the entire set of its implementations. This allows flexibility for an implementation without compromising formal analysis. Examples of under-specified models and their checks abound in the literature. We only mention the modal transition systems of Larsen & Thomsen [64, 62, 12], Bruns & Godefroid's model checker for partial systems [7, 8], Sagiv et al.'s heap shape analysis of C programs [93], CSP expressions and refinement checking [32], and Morgan & McIver's quantitative mu-calculus [68]. The latter works over systems that mix non-determinism and probabilistic choice.

There are also good reasons for under-specification in the abstraction-based model checking of *completely* specified mixed models, whenever one requires the soundness of such checks for queries with unrestricted use of negation [49].

4.6   Occam's Razor

For MCPs, Occam's razor specializes to

"Whatever aspects of M, $\models$, and ϕ are not needed in computing $(M \models \phi)$ should be discarded."

As an example, consider software slicing, e.g. [41], which abstracts M — the model of a program — to M' and then checks $M' \models \phi$. The model M' is computed by

- defining the program points C of interest, typically determined by the query ϕ;
- computing all program points C' on which points in C depend; and
- by using C' to compress the model M to some M'.

The soundness of this abstraction rests on the soundness of determining C and C'. For realistic fragments of programming languages, e.g. a concurrent subset of Java — where models are mixed due to probabilistic assignments and compositional reasoning at method boundaries — such soundness proofs can be a monumental task [41]. We don't know of any formal work on slicing probabilistic programs.

4.7   Viewpoints

System design and analysis needs to consider multiple points of view. A telecommunications software may be conceived with the utmost formal rigor, its functional correctness may be proven or tested to complete satisfaction, but millions of Canadian dollars may have been wasted because the system's performance turns out to be too slow by a factor of 100 upon deployment [98]. In Figure 4, we see an adaptation of a figure from [56] that illustrates this point in a trivial system. Qualitative, probabilistic, and cost aspects are all important in the design and analysis of this system. Changing such viewpoints constitutes an abstraction. E.g. in making a transition from the probabilistic to the qualitative view, we may want to identify non-zero probabilities with tt and 0 with ff. We mention some formal work on the soundness of such viewpoint changes [56, 47] and point out that no abstraction framework for such changes of viewpoints exists to date.

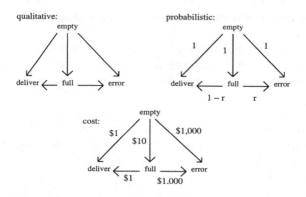

Fig. 4. A system, based on [56], described from three points of view: qualities, probabilities, and costs

5   What to Abstract?

Given a MCP as in (2), we may want to abstract

- models M to M', e.g. through abstract interpretation [21, 77], probabilistic bisimulation [63], the abstraction of an SMV program that describes M [18], the reduction of probabilistic timed automata to timed automata [61] etc;
- queries ϕ to some ϕ', e.g. the approximations of ATL* queries with ATL queries whose model-checking complexity is much more feasible [40]; and
- the answer domain Ans to an abstract answer domain Ans', e.g. by reducing "with probability 1" MCPs over Markov decision processes to model checking over Kripke structures; or by replacing a probabilistic measure of information flow with a possibilistic answer [14].

This suggests to define a signature for abstract MCPs of (2).

Definition 1 (Abstract MCP). *A MCP $\mathcal{P}' = (\mathcal{M}', \Phi', \mathtt{Ans}', \models')$ abstracts the MCP $\mathcal{P} = (\mathcal{M}, \Phi, \mathtt{Ans}, \models)$ with witness (α, β, γ) iff*

1. *$\alpha \subseteq \mathcal{M} \times \mathcal{M}'$ is a relation where $(M, M') \in \alpha$ means that model M' abstracts model M;*
2. *$\beta \subseteq \Phi \times \Phi'$ is a relation where $(\phi, \phi') \in \beta$ means that the query ϕ' abstracts the query ϕ; and*
3. *$\gamma \colon \mathtt{Ans}' \to \mathtt{Ans}$ is a total function such that $\gamma(a')$ re-interprets the abstract answer $a' \in \mathtt{Ans}'$.*

Alternatively, one could define $\gamma \subseteq \mathtt{Ans}' \times \mathtt{Ans}$ as a right-total relation: every abstract answer a' has at least one concrete answer a. However, powerdomains [1] over answer domains are able to express such relations as total functions.

Naturally, abstract answers computed on abstract models or queries should yield meaningful concrete answers for the un-abstracted models and queries. Since $\gamma(M' \models' \phi')$ is the concrete re-interpretation of an abstract check, we need to understand how $\gamma(M' \models' \phi')$ and $(M \models \phi)$ should relate in case that $(M, M') \in \alpha$ and $(\phi, \phi') \in \beta$. Standard MCPs and their soundness of abstraction-based model checking are a reliable guide for determining this relationship.

Sound verification certificates Suppose that $\mathcal{P}'$ abstracts $\mathcal{P}$ with $\mathtt{Ans} = \{\mathtt{ff} < \mathtt{tt}\}$ and witness $(\alpha, \{(\phi, \phi) \mid \phi \in \Phi\}, \gamma = \lambda a.a)$, i.e. only models are being abstracted. E.g. this is the case in the thesis [77] and the papers [72–76]. Given $(M, M') \in \alpha$, we may want that

$$(M' \models \phi) = \mathtt{tt} \quad \Rightarrow \quad (M \models \phi) = \mathtt{tt}, \tag{7}$$

i.e. certified verifications on abstractions apply to abstracted models as well. Using the order $\mathtt{ff} < \mathtt{tt}$ in Ans, we may express this as $\gamma(M' \models \phi) \le (M \models \phi)$. A concrete example of a MCP with sound verification certificates are Markov decision processes with reachability properties as queries and probabilistic simulations [24] as instances of α.

Sound refutation certificates Under the same assumptions on $\mathcal{P}$ as above, we may want that certified refutations on abstractions are refutations of the un-abstracted model:

$$(M, M') \in \alpha \ \& \ (M' \models \phi) = \mathtt{ff} \ \Rightarrow \ (M \models \phi) = \mathtt{ff} \,. \tag{8}$$

Again, $\gamma(M' \models \phi) \leq (M \models \phi)$ guarantees such a sound transfer of results, provided that now $\mathtt{tt} < \mathtt{ff}$. Thus, a combination of sound refutation and verification seems impossible at first sight. Indeed, for probabilistic automata and various forms of (weak) simulations only very limited fragments of PCTL observe such a sound transfer [96]. However, if $\mathtt{tt}$ and $\mathtt{ff}$ are maximal elements in Ans such a combination is indeed possible [23, 7, 93], as demonstrated for Markov decision processes and a branching-time temporal mu-calculus in [49].

Sound [worst,best]-case probabilities Consider MDP as an abstraction of itself, where only α is non-trivial and left unspecified here. Then $\gamma(M' \models \phi) \leq (M \models \phi)$ reads as $[r', s'] \leq [r, s]$ which means $r' \leq r$ and $s \leq s'$. Note that the order of approximation in $\mathcal{I}$ naturally captures the soundness of abstract model checks according to $\gamma(M' \models \phi) \leq (M \models \phi)$: the abstract worst-case probability r' is a conservative abstraction of the concrete one r for the worst-case; similarly, the abstract best-case probability s' is a conservative abstraction of the concrete one s for the best-case.

The general formulation of sound abstract MCPs is now clear.

Definition 2 (Sound Abstract MCP). *Let $\mathcal{P}' = (\mathcal{M}', \Phi', \mathtt{Ans}', \models')$ and $\mathcal{P} = (\mathcal{M}, \Phi, \mathtt{Ans}, \models)$ be MCPs such that $\mathcal{P}'$ abstracts $\mathcal{P}$ with witness (α, β, γ). Then $\mathcal{P}'$ is sound for $\mathcal{P}$ iff for all $M, M' \in \mathcal{M}'$, and $\phi, \phi' \in \Phi'$ we have*

$$(M, M') \in \alpha \ \& \ (\phi, \phi') \in \beta \ \Rightarrow \ \gamma(M' \models' \phi') \leq (M \models \phi) \,. \tag{9}$$

We hasten to point out that the utility of this notion depends on mild assumptions about the witnesses α and β. The former is typically right-total and left-total, the latter should at least be right-total — i.e. every $\phi \in \Phi$ has at least one $\phi' \in \Phi'$ with $(\phi, \phi') \in \beta$ — and will often be a function. The function γ need not enjoy special properties beyond (9), but its monotonicity ensures that the composition of sound abstract MCPs remains sound. In that case, if (α, β, γ) witnesses that $\mathcal{P}'$ abstracts $\mathcal{P}$ and if $(\alpha', \beta', \gamma')$ witnesses that $\mathcal{P}''$ abstracts $\mathcal{P}'$, then $(\alpha; \alpha', \beta; \beta', \gamma'; \gamma)$ witnesses that $\mathcal{P}''$ abstracts $\mathcal{P}$.

Example 1 (Abstract MCPs). We remark that the MCPs for sound refutation certificates, sound verification certificates, and sound [worst,best]-case probabilities presented above are all instances of sound abstract MCPs.

It turns out that there are many kinds of models, queries, and answers if we mix non-deterministic and probabilistic behavior. Models may be Markov decision processes (e.g. [91]), probabilistic automata [96] etc; queries could be formulas of a temporal logic [100, 67], random variables [59] etc; and the legitimate answer types could be truth values [17], real numbers [59], real-valued intervals [51] etc. The query ϕ could even represent another, less specified, model and $(M \models \phi) = \texttt{tt}$ may mean *"M implements ϕ."* This is the basis of the model checker FDR whose mathematical foundation has been extended to probabilistic choice by Morgan et al. [78, 80, 97] and Jifeng et al. [53]. In [78], the process algebra CSP is enriched with a probabilistic choice that distributes through all other operators, non-deterministic choice in particular. A process has a denotational semantics using the probabilistic power domain [55, 54] and the failure/divergence semantics [46] provides a sound notion of process refinement.

Given this abundance of different approaches and interpretations, it comes as no surprise that we make (2) a syntactic leitmotif of this survey since it helps to present and organize the vast literature on this topic in a rather uniform and more coherent manner.

Papers on abstraction in model checking typically focus on one aspect of abstraction at a time:

- simulations [69] and abstract interpretation [21] specify a sound instance of α, where queries support only a limited form of negation such as "for all computation paths, ..."; in that case, $\texttt{Ans} = \{\texttt{ff} < \texttt{tt}\}$, and $\gamma = \lambda a.a$;
- bisimulations [85, 70] specify another sound instance of α, where queries have an unrestricted form of negation, $\gamma = \lambda a.a$, and $\texttt{Ans} = \{\texttt{ff}, \texttt{tt}\}$ in the *discrete* ordering;
- the bounded unfolding of least-fixed point queries specifies a sound instance of β for verification certificates;
- bounded model-checking techniques [15] specify an abstract version $\models'$ of $\models$ and
- possibilistic abstractions, e.g. [14] and [56], specify a *qualitative* abstract answer domain for a probabilistic one.

In practice, several of these abstractions may be employed to abstract a check and such a combination is then encoded in the witness (α, β, γ). In any event, (9) specifies the meaning of soundness for such abstract checks. Although we argue that Definition 2 unifies most abstraction techniques encountered in practice, a single MCP cannot embrace every form of abstraction there is. For example, the annotation of LTL or CTL* path modalities (next, until etc) with probability thresholds [38, 39], as done for LTL in Figure 5, does not give rise to a sound β unless β is a function of the model M under analysis [51], forcing M to be a singleton.

Example 2 (Abstracting answers and queries). Consider a federal officer whose job is to certify the reliability of safety-critical programs. She will primarily be interested in knowing, whether the probability of meeting the safety requirement is above a legally prescribed threshold. Her query set Φ, with atomic observables o, is given in Figure 5; note that her queries ϕ are elements $\phi' \in \Phi' = \text{LTL}$ annotated with threshold values at X and U connectives. The semantics $\models_\geq$ of these thresholds is that the worst-case probability of the underlying trace set be $\geq p$.

In [51], a connection between these two semantics is established: the worst-case and best-case probabilities, computed for path sub-formulas in a bottom-up manner, determine two maps $(\cdot)^\flat, (\cdot)^\sharp \colon \Phi' \to \Phi$ (respectively) such that a positive worst-case probabilities for ϕ' makes $(\phi')^\flat$ true and a best-case probability for ϕ' below 1 makes $(\phi')^\sharp$ false; this is shown for t-norms, such as the one used in [5], as a semantics of conjunction $(\wedge)$.

$$\phi ::= \text{tt} \mid o \mid \neg\phi \mid \phi \wedge \phi \mid [\mathsf{X}\,\phi]_{\geq p} \mid [\phi\,\mathsf{U}\,\phi]_{\geq p} \qquad (10)$$

Fig. 5. Syntax of LTL$_\geq$ with thresholds $p \in [0,1]$

A worst-case abstraction $\alpha^\forall \colon [0,1] \to \{\text{ff} < \text{tt}\}$ identifies ff with $\{0\}$ and tt with $(0,1]$; dually, the best-case abstraction $\alpha^\exists \colon [0,1] \to \{\text{ff} < \text{tt}\}$ maps 1 to tt and $[0,1)$ to ff. These abstractions have a common concretization $\xi \colon \{\text{ff} < \text{tt}\} \to [0,1]$ which maps ff to 0 and tt to 1. The pairs $(\xi, \alpha^\forall)$ and $(\xi, \alpha^\exists)$ constitute Galois embeddings [21,33] where $\alpha^\forall$ and $\alpha^\exists$ are the lower and upper adjoint of ξ (respectively). This mirrors the situation of Galois adjunctions used in [22] for sound abstraction of a temporal logic with unrestricted use of negation, where the concretization function has a universal and existential abstraction function as adjoints.

MDP abstracts the MCP $(\mathcal{M}', \Phi, \text{Ans}, \models)$, where $\models$ equals $\models_\geq \times \models_\geq$, Φ equals LTL$_\geq$ $\times$ LTL$_\geq$, and Ans equals $\{\text{ff} < \text{tt}\} \times \{\text{tt} < \text{ff}\}$. The abstraction relation on models, α, is the identity; the abstraction relation on queries is defined as those pairs $((\psi, \eta), \phi)$ with $\psi = \phi^\flat$ and $\eta = \phi^\sharp$; and $\gamma \colon \text{Ans}' \to \text{Ans}$ maps intervals $[r,s]$ to $(\alpha^\forall(r), \alpha^\exists(s)) \in \text{Ans}$. Soundness (9) follows from the result in [51] stated above. It is the soundness and the definition of γ that determine which of the maps $\alpha^\forall$ and $\alpha^\exists$ have a worst-case and best-case reading (respectively).

Many other people have also realized such connections between answer domains. For example, Jonsson & Larsen [56] relate systems whose transitions are annotated with probability intervals $[r,s]$ to modal transition systems [64, 62] — abstracting the interval domain $\mathcal{I}$ to the flat booleans $\mathbb{B}_\perp$; guaranteed $[r,s]$-transitions are those with $0 < r$; possible $[r,s]$-transitions are those for which $0 < s$. This matches our definitions of $\alpha^\forall$ and $\alpha^\exists$: if $0 < r$, then every $p \in [r,s]$ is mapped to tt via $\alpha^\forall$; if $s < 1$, then $\alpha^\exists(p) = \text{ff}$ for all $p \in [r,s]$. We refer to [47,48] for a more detailed study of such abstractions.

6   Query Equivalence

Occam's razor also applies whenever models cannot be distinguished by checks. Given a MCP $\mathcal{P}$ as in (2), we define for $M, M' \in \mathcal{M}$ the relation

$$M \sim M' \qquad \text{iff} \qquad \lambda\phi\colon \Phi.(M \models \phi) = \lambda\phi\colon \Phi.(M' \models \phi)\,. \qquad (11)$$

The query equivalence $\sim$ partitions $\mathcal{M}$ and induces an abstract MCP $\mathcal{P}'$ for $\mathcal{P}$, where α is $\sim$, $\beta = \{(\phi, \phi) \mid \phi \in \Phi\}$, and Ans$'$ = Ans with $\gamma(a') = a'$ for all $a' \in \text{Ans}'$. As an example, consider the MCP of labeled transition systems with a propositional modal mu-calculus as query language [6]. Then all query-equivalent models are also trace equivalent but not the other way around [70]. Occam's razor suggests that M in $M \models \phi$ be replaced with some M' such that $M \sim M'$ and the computation of $(M' \models \phi)$ have minimal cost — e.g. M' may have minimal state space among all models that are query equivalent to M. It should be apparent that such a reduction is less aggressive than the ones that are driven by a particular query ϕ, where the equality of the functions in (11) is being replaced with the equality of their evaluation at ϕ. We already encountered software slicing as such an example and note that property-driven abstraction is not adequately explored in the study of mixed models.

This approach of model simplification is not always guaranteed to be fruitful. For example, there are continuous-space labeled Markov chains that are not equivalent to any finite-state model for a simple probabilistic modal logic Φ [28].

Bisimulations [85, 70] are a device for establishing instances of the query equivalence $\sim$ for branching-time query languages such as CTL and the modal mu-calculus. Bisimulations relate models in an intuitive operational manner, lead to co-inductive reasoning principles, and have typically efficient decision algorithms [3, 4, 87, 11]. Bisimulations are sound for a query language if bisimilar models are also query equivalent. The converse relationship is called completeness. Bisimulation of labeled transition systems is sound and complete for CTL and the propositional modal mu-calculus.

If internal actions are hidden from the top layer of an operational semantics, then state transitions may be pre- or post-composed with a finite number of internal, silent actions. This results in a new state-transition relation whose bisimulation is called "weak" when applied to the original model [70]. With respect to the original model, such a saturation of state transitions may not be sound (e.g. [36]) or may be sound and complete, as it is the case for Markov decision processes and pCTL* [30]. It is worth noting that branching bisimulation [99] avoids the unsoundness encountered in [36]. A better understanding of the connections between probabilistic bisimulations and branching bisimulations is desired; see [44].

7   Precise Abstract MCPs

Model checking is referred to as a property-verification technique: given a model M, we mean to check one property ϕ at a time. In the context of an abstract MCP $\mathcal{P}' = (\mathcal{M}', \Phi', \mathrm{Ans}', \models')$ for a MCP $\mathcal{P} = (\mathcal{M}, \Phi, \mathrm{Ans}, \models)$ with witness (α, β, γ), the possible sound abstractions of the check $M \models \phi$ that are offered by $\mathcal{P}'$ are all $\gamma(M' \models' \phi')$ with $(M, M') \in \alpha$ and $(\phi, \phi') \in \beta$. Thus, the set

$$A_\phi^M = \{\gamma(M' \models \phi') \mid (M, M') \in \alpha, \ (\phi, \phi') \in \beta\} \tag{12}$$

collects all the information about $(M \models \phi)$ that we could ever infer from performing abstract and sound checks according to $\mathcal{P}'$.

Definition 3 (Precise Abstract MCPs). *Let* $\mathcal{P}' = (\mathcal{M}', \Phi', \mathsf{Ans}', \models')$ *and* $\mathcal{P} = (\mathcal{M}, \Phi, \mathsf{Ans}, \models)$ *be MCPs such that* $\mathcal{P}'$ *abstracts* $\mathcal{P}$ *with witness* (α, β, γ). *For* $\phi \in \Phi$, $\mathcal{P}'$ *is* ϕ-*precise for* $\mathcal{P}$ *iff for all* $M \in \mathcal{M}$, *the set* A_ϕ^M *has* $(M \models \phi)$ *as a least upper bound. We call* $\mathcal{P}'$ *precise for* $\mathcal{P}$ *iff it is* ϕ-*precise for* $\mathcal{P}$ *for all* $\phi \in \Phi$.

Since Ans is a partial order, the statement about A_ϕ^M above includes the assumption that the least upper bound of A_ϕ^M exists. Evidently, it makes sense to stipulate that Ans be a continuous domain [1], a partial order with a rich and formal notion of approximation. This stipulation holds for all answer domains presented in this survey. Although the set A_ϕ^M may not be directed, we expect that its information content leads to directed approximations. E.g. consider MDP. If $[r', s']$ and $[r'', s'']$ are in A_ϕ^M for Ans $= \mathcal{I}$, then $[\max(r', r''), \min(s', s'')]$ is a rational reconstruction of an upper bound in $\mathcal{I}$ that still approximates $(M \models \phi)$ — despite the fact that $[\max(r', r''), \min(s', s'')]$ may not be an element of A_ϕ^M.

Consider $\mathcal{P}$ as the MCP of Kripke structures and CTL, $\mathcal{P}'$ as the same MCP restricted to finite models, and Ans $=$ Ans$' = \{\mathtt{ff} < \mathtt{tt}\}$. We leave α unspecified, whereas $\beta = \{(\phi, \phi) \mid \phi \in \Phi\}$, and $\gamma = \lambda a.a$. Fix a CTL formula ϕ. Then $\mathcal{P}'$ is ϕ-precise iff for each model M with $(M \models \phi) = \mathtt{tt}$ there is some finite model M' with $(M, M') \in \alpha$ such that $(M' \models \phi) = \mathtt{tt}$ as well. Thus, this amounts to saying that all verifications of ϕ have certified verifications in some finite abstraction. Therefore, finite-model properties for model checking are instances of precise abstract MCPs.

If the query language is a logic with the finite model property, then the abstract MCP may indeed be precise for a suitable abstraction relation. As an example, Desharnais et al. [29] showed that each continuous-state labeled Markov process has a sequence of finite acyclic labeled Markov processes as abstractions such that this sequence is precise for a probabilistic modal logic.

Even if one has proven that $\mathcal{P}'$ is precise for $\mathcal{P}$, practical use of abstract model checking requires feasible strategies or heuristics for finding "precise" abstractions. In the case of $\mathtt{ff} < \mathtt{tt}$, at least, one knows what to look for: a positive check of ϕ on an abstraction does the trick. In the [worst,best]-case probability situation, however, single checks cannot be expected to recover the actual concrete value of $(M \models \phi)$: all members of A_ϕ^M may be strictly below it. In essence, practitioners need to specify a desired degree of accuracy and the result $\gamma(M' \models' \phi')$ is then compared to $(M \models \phi)$. Fortunately, most continuous domains have a natural way of measuring such a distance [66, 102]. For example, the distance between the concrete and abstract checks — $[r, s]$ and $[r', s']$ (respectively) — may be given as

$$|r - r'| + |s - s'|. \tag{13}$$

Quantitative domain theory [66, 102] has the potential of supplying such distance notions for a vast class of answer domains.

Definition 4 (The MCP Approximation Problem). *Let us assume that there is some suitable[1] distance function* $d\colon \mathtt{Ans} \times \mathtt{Ans} \to [0, +\infty)$. *Let* $\mathcal{P}' = (\mathcal{M}', \Phi', \mathtt{Ans}', \models')$ *and* $\mathcal{P} = (\mathcal{M}, \Phi, \mathtt{Ans}, \models)$ *be MCPs such that* $\mathcal{P}'$ *abstracts* $\mathcal{P}$ *with witness* (α, β, γ). *Given* $\epsilon > 0$, $M \in \mathcal{M}$, *and* $\phi \in \Phi$ *are there* $(M, M') \in \alpha$ *and* $(\phi, \phi') \in \beta$ *such that*

$$d(\gamma(M' \models' \phi'), (M \models \phi)) < \epsilon ? \tag{14}$$

Although this definition is conceptually pleasing, it begs the question of what value $(M \models \phi)$ is. Measurements [66] $\mu\colon \mathtt{Ans} \to [0, \infty)$, on the other hand, compute a self-distance which can inform us how far away the analysis is from a complete result. E.g. the canonical measurement for $\mathcal{I}$ is simply $\mu[r, s] = |\, r - s \,|$. The MCP Approximation Problem can then be restated:

"Given $\epsilon > 0$, $M \in \mathcal{M}$, and $\phi \in \Phi$ are there $(M, M') \in \alpha$ and $(\phi, \phi') \in \beta$ such that

$$\mu(\gamma(M' \models' \phi')) < \epsilon ?" \tag{15}$$

Unfortunately, if the underlying model M is under-specified or partial in some way, then we cannot uniquely identify the source of such partiality — whether it resides in the incompleteness of our abstraction-based analysis or in the partiality of the abstracted model. In Markov decision processes, for example, this partiality is encoded in the non-determinism which gives rise to genuine intervals $[r, s]$ of [worst,best]-case probabilities. Although this partiality and the resulting blur of sources of partiality in answers is absent in Markov chains, the latter tend to have Markov decision processes as abstractions. This blurring of the source of incompleteness seems to be specific to quantitative model checking. Recall that $\mathtt{ff} < \mathtt{tt}$ for verification certificates; the measurement is then $\mu_{\mathrm{v}}(\mathtt{ff}) = 1$ and $\mu_{\mathrm{v}}(\mathtt{tt}) = 0$. Dually, refutation checks have $\mathtt{tt} < \mathtt{ff}$ and the measurement is $\mu_{\mathrm{r}}(\mathtt{tt}) = 1$ and $\mu_{\mathrm{r}}(\mathtt{ff}) = 0$. In either case, $\epsilon = 0.5$ will then turn (15) into a check for precision.

8   Diagnostics

The specification of model checks typically stipulates that the answer domain $\mathtt{Ans}$ to checks of (2) has an extensional type such as $\{\mathtt{ff} < \mathtt{tt}\}$ for Kripke structures [17] or the unit interval $[0, 1]$ for a probabilistic semantics of LTL [100]. The practical utility of model checking, however, depends on its ability to attach intensional content to such extensional replies. Little may be gained from knowing that a system can reach deadlock from some initial state, unless information about a possible execution trace from some initial to a deadlocked state

---

[1] We require that the distance function be Scott-continuous of type $\mathtt{Ans} \times \mathtt{Ans} \to [0, \infty)^{\mathrm{op}}$ such that $d(a, a') = 0$ implies $a = a'$.

be provided. Without a strict regime on managing negation and path quantifiers, this demand applies to refuted (answer is ff) and verified properties (answer is tt) alike. The utility of abstraction techniques will therefore also depend on their ability to produce such diagnostic information. For systems that mix non-deterministic and probabilistic behavior, this constitutes a major research challenge.

9   Cost Effectiveness

We need to remark that all the wonderful work on abstraction in model checking is of little use if

- the cost of specifying or computing executable abstractions outweighs the savings over executing un-abstracted models; or
- if there are no heuristics or other support that guides one in the choice of an abstraction, even in case that the abstract MCP is precise for the un-abstracted one.

The details of assessing this trade-off are too dependent on the particular MCP at hand to be studied in the general framework presented in this survey. However, the reader should keep in mind that the first objection is mute if the MCP under consideration is undecidable; abstraction then functions as a 'last straw.' Also, there is already some work on the establishment of strategies for the refinement of abstractions of probabilistic transition systems with respect to certain reachability properties [24, 25].

10   Abandoning Soundness

Formal modeling and analysis presupposes that the underlying methodology be sound. We formulated such soundness for abstract MCPs in (9). Yet there are a number of good reasons and situations in which one wishes to abandon (9).

- A model may only be subjected to a limited exploration of its computation paths: testing of software, e.g. with the tool Verisoft [35], is a viable alternative to full verification through model checking, especially in the presence of tight time-to-market constraints.
- Similarly, complexity results on mixed models may suggest not to explore all computation paths but to truncate them at a certain bound, e.g. [65].
- For mixed models, probabilistic testing can result in Monte Carlo methods for the analysis of such models [76, 89]. Not only will probabilistically sampled paths be verified, but one can then often extend this verification to the entire system by specifying a probability for this extension to fail, e.g. *hypothesis testing*. Specifying a measurable extension error is not always possible for qualitative models.
- A MCP may not have a finite model property or may not be subjectable to symbolic analysis. One may then have to do with the ability to run simulations (generate a computation path/tree) of the model in question.

- Query equivalence and the related notions of bisimulation and weak bisimulation are relations. Thus, two models are either related or not. In some practical situations, notably secure information flow in computer systems [26, 27, 101], one cannot hope to ever achieve such equivalences between a model M (the implementation) and M' (the specification as an abstraction of the implementation). Instead, one may want to stipulate that $\gamma(M' \models' \phi')$ and $(M \models \phi)$ have a distance below ϵ. For the distance in (13), such a property won't guarantee nor care for (9). As an example we mention the approach of Di Pierro et al. [88].

11 Research Challenges

The challenges that lie ahead for abstraction-based model checking of mixed models are many, tremendous, and possibly very discouraging. We formulate some key problems that are common to all such MCPs.

- Establish sound formal techniques for a query-driven (ϕ) abstraction M' of a model M such that $(M \models \phi)$ equals $(M' \models \phi)$. Investigate the complexity of computing such an M' that has minimal "cost."
- For classes of queries used in practive, prove or disprove that abstract MCPs are precise for their concrete counterparts.
- Find heuristics or formal techniques for solving the abstract MCP approximation problem. This includes the problem of revising abstractions if their checks or diagnostics are insufficiently informative.
- For MCPs with unrestricted negation of queries, develop diagnostics that capture the evidence of model checking results. For mixed models, further develop probabilistic games as such diagnostics.
- Push established methods and results from qualitative model checking into the model checking of mixed and, more generally, quantitative MCPs.

12 Conclusions

A unified framework for specifying MCPs and abstract MCPs was presented. A partial survey of the literature on mixed models revealed that most work on abstraction can be represented within this framework. We also pointed out possibly fruitful connections to ongoing work in quantitative domain theory. Finally, research issues and connections to ongoing work in verification of mixed models were identified.

Acknowledgments

We gratefully acknowledge Joost-Pieter Katoen for pointing out omissions, typos, and ambiguities in drafts of this tutorial. Annabelle McIver and Carroll Morgan also made many comments that improved this manuscript; they kindly provided the example in (5).

References

1. S. Abramsky and A. Jung. Domain theory. In S. Abramsky, D. M. Gabbay, and T. S. E. Maibaum, editors, *Handbook of Logic in Computer Science*, volume 3, pages 1–168. Oxford Univ. Press, 1994.
2. J. C. M. Baeten and W. P. Weijland. *Process Algebra*, volume 18 of *Cambridge Tracts in Theoretical Computer Science*. Cambridge University Press, 1990.
3. C. Baier, B. Engelen, and M. Majster-Cederbaum. Deciding Bisimilarity and Similarity for Probabilistic Processes. *Journal of Computer and System Sciences*, 60:187–231, 2000.
4. C. Baier and M. I. A. Stoelinga. Norm fuctions for probabilistic bisimulations with delays. In J. Tiuryn, editor, *Proceedings of 3rd International Conference on Foundations of Science and Computation Structures (FOSSACS)*, Berlin, Germany, March 2000, volume 1784 of *Lecture Notes in Computer Science*, pages 1–16, 2000.
5. C. Baier, M. Kwiatkowska, and G. Norman. Computing probability bounds for linear time formulas over concurrent probabilistic systems. In M. Kwiatkowska, C. Baier, M. Huth and M. Ryan, editors, *Electronic Notes in Theoretical Computer Science*, volume 22. Elsevier Science Publishers, 2000.
6. J. C. Bradfield. *Verifying Temporal Properties Of Systems*. Birkhäuser, Boston, Mass., 1991.
7. G. Bruns and P. Godefroid. Model Checking Partial State Spaces with 3-Valued Temporal Logics. In *Proceedings of the 11th Conference on Computer Aided Verification*, volume 1633 of *Lecture Notes in Computer Science*, pages 274–287. Springer Verlag, July 1999.
8. G. Bruns and P. Godefroid. Generalized Model Checking: Reasoning about Partial State Spaces. In *Proceedings of CONCUR'2000 (11th International Conference on Concurrency Theory)*, volume 1877 of *Lecture Notes in Computer Science*, pages 168–182. Springer Verlag, August 2000.
9. R. R. Bryant. Symbolic Boolean Manipulation with Ordered Binary-Decision Diagrams. *ACM Computing Surveys*, 24(3):293–318, September 1992.
10. J. R. Burch, E. M. Clarke, D. L. Dill, K. L. McMillan, and J. Hwang. Symbolic model checking: 10^{20} states and beyond. Proceedings of the Fifth Annual Symposium on Logic in Computer Science, June 1990.
11. S. Cattani and R. Segala. Decision algorithms for probabilistic bisimulation. In *Proceedings of the 13th International Conference on Concurrency Theory (CONCUR'02)*, volume 2421 of *Lecture Notes in Computer Science*, pages 371–385, Brno, Czech Republic, August 2002. Springer Verlag.
12. K. Cerans, J. Chr. Godskesen, and K. G. Larsen. Timed modal specification - theory and tools. In *Computer Aided Verification*, pages 253–267, 1993.
13. Y.-F. Chen, E. R. Gansner, and E. Koutsofios. A C++ data model supporting reachability analysis and dead code detection. In M. Jazayeri and H. Schauer, editors, *Proceedings of the Sixth European Software Engineering Conference (ESEC/FSE 97)*, pages 414–431. Springer–Verlag, 1997.
14. D. Clark, C. Hankin, S. Hunt, and R. Nagarajan. Possibilistic Information Flow is safe for Probabilistic Non-Interference. In *Workshop on Issues in the Theory of Security (WITS '00)*, Geneva, Switzerland, 7-8 July 2000.
15. E. Clarke, A. Biere, R. Raimi, and Y. Zhu. Bounded Model Checking Using Satisfiability Solving. *Formal Methods in System Design*, 19(1):7–34, July 2001.

16. E. M. Clarke, M. Fujita, and X. Zhao. *Representations of discrete functions*, chapter 'Multi-terminal binary decision diagrams and hybrid decision diagrams', pages 93–108. Kluwer academic publishers, 1996.

17. E. M. Clarke, O. Grumberg, and D. A. Peled. *Model Checking*. The MIT Press, January 2000.

18. E. M. Clarke, O. Grumberg, and D. E. Long. Model checking and abstraction. *ACM Transactions on Programming Languages and Systems*, 16(5):1512–1542, 1994.

19. R. Cleaveland, S. A. Smolka, and A. E. Zwarico. Testing preorders for probabilistic processes. In W. Kuich, editor, *Automata, Languages and Programming, 19th International Colloqium*, volume 623 of *Lecture Notes in Computer Science*, pages 708–719, Vienna Austria, 13–17 July 1992. Springer Verlag.

20. C. Courcoubetis and M. Yannakakis. The Complexity of Probabilistic Verification. *Journal of the Association of Computing Machinery*, 42(4):857–907, July 1995.

21. P. Cousot and R. Cousot. Abstract interpretation: a unified lattice model for static analysis of programs. In *Proc. 4th ACM Symp. on Principles of Programming Languages*, pages 238–252. ACM Press, 1977.

22. P. Cousot and R. Cousot. Temporal abstract interpretation. In *Conference Record of the 27th Annual ACM SIGPLAN-SIGACT Symposium on Principles of Programming Languages*, pages 12–25, Boston, Mass., January 2000. ACM Press, New York, NY.

23. D. Dams. *Abstract Interpretation and Partition Refinement for Model Checking*. PhD thesis, Eindhoven University of Technology, P.O. Box 513, 5600 MB Eindhoven, The Netherlands, July 1996.

24. P. R. D'Argenio, B. Jeannet, H. E. Jensen, and K. G. Larsen. Reachability Analysis of Probabilistic Systems by Successive Refinements. In L. de Alfaro and S. Gilmore, editors, *Process Algebra and Probabilistic Methods: Performance Modelling and Verification*, volume 2165 of *Lecture Notes in Computer Science*, pages 39–56, Aachen, Germany, September 12–14 2001. Springer Verlag.

25. P. R. D'Argenio, B. Jeannet, H. E. Jensen, and K. G. Larsen. Reduction and Refinement Strategies for Probabilistic Analysis. In H. Hermanns and R. Segala, editors, *Process Algebra and Probabilistic Methods, Performance Modeling and Verification, Second Joint International Workshop PAPM-PROBMIV 2002*, volume 2399 of *Lecture Notes in Computer Science*, pages 57–76, Copenhagen, Denmark, July 25–26 2002. Springer.

26. D. Denning. A Lattice Model of Secure Information Flow. *Communications of the ACM*, 19(5):236–243, 1976.

27. D. Denning. Certification of Programs for Secure Information Flow. *Communications of the ACM*, 20(7):504–513, 1977.

28. J. Desharnais, A. Edalat, and P. Panangaden. Bisimulation for Labelled Markov Processes. *Journal of Information and Computation*, 179(2):163–193, December 2002.

29. J. Desharnais, V. Gupta, R. Jagadeesan, and P. Panangaden. Approximating Labeled Markov Processes. In *15th Annual IEEE Symposium on Logic in Computer Science (LICS'00)*, Santa Barbara, California, 26–29 June 2000. IEEE Computer Society Press.

30. J. Desharnais, V. Gupta, R. Jagadeesan, and P. Panangaden. Weak Bisimulation is Sound and Complete for PCTL*. In *Proc. 13th Int'l Conference CONCUR 2002 - Concurrency Theory*, volume 2421 of *Lecture Notes in Computer Science*, pages 355–370, Brno, Czech Republic, 20-23 August 2002. Springer Verlag.

31. M. B. Dwyer and D. A. Schmidt. Limiting State Explosion with Filter-Based Refinement. In *Proceedings of the ILPS'97 Workshop on Verification, Model Checking, and Abstraction*, 1997.

32. URL: http://www.fsel.com/fdr2_download.html.

33. G. Gierz, K. H. Hofmann, K. Keimel, J. D. Lawson, M. Mislove, and D. S. Scott. *A Compendium of Continuous Lattices*. Springer Verlag, 1980.

34. S. Gilmore and M. Ryan, editors. *Language Constructs for Describing Features, Proc. of the FIREworks workshop*. Springer Verlag, 2001.

35. P. Godefroid. Model Checking for Programming Languages using VeriSoft. In *Proceedings of the 24th ACM Symposium on Principles of Programming Languages*, pages 174–186, Paris, January 1997.

36. S. Graf and J. Sifakis. Readiness Semantics for Regular Processes with Silent Actions. In *In: Proc. of ICALP'87*, pages 115–125, 1987.

37. P. R. Halmos. *Measure Theory*. Graduate Texts in Mathematics 18. Springer Verlag, 1950.

38. H. Hansson. *Time and Probability in Formal Design of Distributed Systems*. PhD thesis, Department of Computer Science, Uppsala University, Uppsala, Sweden, 1991.

39. H. A. Hansson and B. Jonsson. A logic for reasoning about time and reliability. *Formal Aspects of Computing*, 6(5):512–535, 1994.

40. A. Harding, M. Ryan, and P.-Y. Schobbens. Approximating ATL* in ATL. In *Third International Workshop on Verification, Model Checking and Abstract Interpretation*, volume 2294 of *Lecture Notes in Computer Science*, pages 289–301, Venice, Italy, January 21–22 2002. Springer Verlag.

41. J. Hatcliff, J. C. Corbett, M. B. Dwyer, S. Sokolowski, and H. Zheng. A formal study of slicing for multi-threaded programs with JVM concurrency primitives. In *Static Analysis Symposium*, pages 1–18, 1999.

42. H. Hermanns, U. Herzog, and J.-P. Katoen. Process algebra for performance evaluation. *Theoretical Computer Science*, 274(1–2):43–87, 2002.

43. H. Hermanns and J.-P. Katoen. Performance Evaluation := (Process Algebra + Model Checking) × Markov Chains. In K. G. Larsen and M. Nielson, editors, *12th Int'l Conference CONCUR 2001 - Concurrency Theory*, volume 2154 of *Lecture Notes in Computer Science*, pages 59–81, Aalborg, Denmark, 20–25 August 2001. Springer Verlag.

44. H. Hermanns. *Interactive Markov Chains*, volume 2428 of *Lecture Notes in Computer Science*. Springer Verlag, September 2002.

45. J. Hillston. *A Compositional Approach to Performance Modelling*. Cambridge University Press, 1996.

46. C. A. R. Hoare. *Communicating Sequential Processes*. Prentice-Hall, 1985.

47. M. Huth. A Unifying Framework for Model Checking Labeled Kripke Structures, Modal Transition Systems, and Interval Transition Systems. In *Proceedings of the 19th International Conference on the Foundations of Software Technology & Theoretical Computer Science*, Lecture Notes in Computer Science, pages 369–380, IIT Chennai, India, December 1999. Springer Verlag.

48. M. Huth. *Domains and Processes*, chapter 'Domains of view: a foundation for specification and analysis', pages 183–218. Kluwer Academic Press, December 2001.
49. M. Huth. Possibilistic and Probabilistic Abstraction-Based Model Checking. In H. Hermanns and R. Segala, editors, *Process Algebra and Probabilistic Methods, Performance Modeling and Verification, Second Joint International Workshop PAPM-PROBMIV 2002*, volume 2399 of *Lecture Notes in Computer Science*, pages 115–134, Copenhagen, Denmark, July 25–26 2002. Springer.
50. M. Huth and M.Kwiatkowska. Quantitative analysis and model checking. In *Proceedings of the 12th Annual IEEE Symposium on Logic in Computer Science*, pages 111–122, Warsaw, Poland, 1997. IEEE Computer Society Press.
51. M. Huth. The interval domain: A matchmaker for aCTL and aPCTL. In M. Mislove, R. Cleaveland, and P. Mulry, editors, *Electronic Notes in Theoretical Computer Science*, volume 14. Elsevier Science Publishers, 2000.
52. T. Huynh and L. Tian. On some Equivalence Relations for Probabilistic Processes. *Fundamenta Informaticae*, 17:211–234, 1992.
53. H. Jifeng, K. Seidel, and A. McIver. Probabilistic models for the guarded command language. *Science of Computer Programming*, 28(2–3):171–192, April 1997.
54. C. Jones. *Probabilistic Nondeterminism*. PhD thesis, Laboratory for the Foundations of Computer Science, University of Edinburgh, Edinburgh, Scotland, 1990. Monograph ECS-LFCS-90-105.
55. C. Jones and G. Plotkin. A probabilistic powerdomain of evaluations. In *In: Proceedings of the IEEE 4th Annual Symposium on Logic in Computer Science*, pages 186–195. IEEE Computer Society Press, 1989.
56. B. Jonsson and K. G. Larsen. Specification and Refinement of Probabilistic Processes. In *6th Annual IEEE Symposium on Logic in Computer Science*, pages 266–277, Amsterdam, The Netherlands, 15–18 July 1991. IEEE Computer Society Press.
57. P. Kannelakis and S. Smolka. CCS Expressions, Finite State Processes and Three Problems of Equivalence. *Journal of Information and Computation*, 86:43–68, 1990.
58. K.L. McMillan. *Symbolic Model Checking*. Kluwer Academic Publishers, 1993.
59. D. Kozen. Semantics of Probabilistic Programs. *Computer and System Sciences*, 22:328–350, 1981.
60. D. Kozen. A Probabilistic PDL. *Computer and System Sciences*, 22(2):162–178, 1985.
61. M. Kwiatkowska, G. Norman, and J. Sproston. Probabilistic Model Checking of the IEEE 802.11 Wireless Local Area Network Protocol. In H. Hermanns and R. Segala, editors, *Process Algebra and Probabilistic Methods, Performance Modeling and Verification, Second Joint International Workshop PAPM-PROBMIV 2002*, volume 2399 of *Lecture Notes in Computer Science*, pages 169–187, Copenhagen, Denmark, July 25–26 2002. Springer.
62. K. G. Larsen. Modal Specifications. In J. Sifakis, editor, *Automatic Verification Methods for Finite State Systems*, volume 407 of *Lecture Notes in Computer Science*, pages 232–246. Springer Verlag, June 12–14 1989. International Workshop, Grenoble, France.
63. K. G. Larsen and A. Skou. Bisimulation through probabilistic testing. *Information and Computation*, 94(1):1–28, September 1991.
64. K. G. Larsen and B. Thomsen. A Modal Process Logic. In *Third Annual Symposium on Logic in Computer Science*, pages 203–210. IEEE Computer Society Press, 1988.

65. R. Lassaigne and S. Peyronnet. Approximate Verification of Probabilistic Systems. In H. Hermanns and R. Segala, editors, *Process Algebra and Probabilistic Methods, Performance Modeling and Verification, Second Joint International Workshop PAPM-PROBMIV 2002*, volume 2399 of *Lecture Notes in Computer Science*, pages 213–214, Copenhagen, Denmark, July 25–26 2002. Springer.

66. K. Martin. The measurement process in domain theory. In *Proc. of Automata, Languages and Programming (ICALP'00)*, pages 116–126, 2000.

67. A. McIver and C. Morgan. Almost-certain eventualities and abstract probabilities in quantitative temporal logic. In C. Fidge, editor, *Electronic Notes in Theoretical Computer Science*, volume 42. Elsevier Science Publishers, 2001.

68. A. McIver and C. Morgan. Games, probability and the quantitative μ-calculus qMμ. In *Proc. 9th Int. Conf. on Logic for Programming, Artificial Intelligence and Reasoning, LPAR 2002*, volume 2514 of *Lecture Notes in Artificial Intelligence*, pages 292–310, Tbilisi, Georgia, 2002. Springer Verlag.

69. R. Milner. An algebraic definition of simulation between programs. In *2nd International Joint Conference on Artificial Intelligence*, pages 481–489, London, United Kingdom, 1971. British Computer Society.

70. R. Milner. *Communication and Concurrency*. Prentice-Hall, 1989.

71. R. Milner. *Communicating and Mobile Systems: the π-Calculus*. Cambridge University Press, 1999.

72. D. Monniaux. Abstract interpretation of programs as Markov decision processes. Technical report, Départment d'Informatique, École Normale Supérieure, 45, rue d'Ulm, 75230 Paris cedex 5, France, 2001.

73. D. Monniaux. Backwards abstract interpretation of probabilistic programs. In *European Symposium on Programming Languages and Systems (ESOP '01)*, number 2028 in Lecture Notes in Computer Science. Springer-Verlag, 2001.

74. D. Monniaux. Abstract interpretation of probabilistic semantics. In *Seventh International Static Analysis Symposium (SAS'00)*, number 1824 in Lecture Notes in Computer Science. Springer-Verlag, 2000. Extended version on the author's web site.

75. D. Monniaux. An abstract analysis of the probabilistic termination of programs. In *8th International Static Analysis Symposium (SAS'01)*, number 2126 in Lecture Notes in Computer Science. Springer-Verlag, 2001.

76. D. Monniaux. An abstract Monte-Carlo method for the analysis of probabilistic programs (extended abstract). In *28th Symposium on Principles of Programming Languages (POPL '01)*, pages 93–101. Association for Computer Machinery, 2001.

77. D. Monniaux. *Analyse de programmes probabilistes par interprétation abstraite*. Thèse de doctorat, Université Paris IX Dauphine, 2001. Résumé étendu en français. Contents in English.

78. C. Morgan, A. McIver, K. Seidel, and J. W. Sanders. Refinement-oriented probability for CSP. *Formal Aspects of Computing*, 8(6):617–647, 1996.

79. C. Morgan and A. McIver. An expectation-based model for probabilistic temporal logic. *Logic Journal of the IGPL*, 7(6):779–804, 1999.

80. C. Morgan, A. McIver, and K. Seidel. Probabilistic predicate transformers. *ACM Transactions on Programming Languages and Systems*, 18(3):325–353, May 1996.

81. R. Motvani and P. Raghavan. *Randomized Algorithms*. Cambridge University Press, 1995.

82. F. Nielson, H. R. Nielson, and C. Hankin. *Principles of Program Analysis*. Springer Verlag, 1999.

83. C. H. Papadimitriou. Games against nature. *Journal of Computer and System Sciences*, 31:288–301, 1985.

84. C. H. Papadimitriou. *Computational Complexity*. Addison-Wesley, 1994.

85. D. M. R. Park. Concurrency and automata on infinite sequences. In P. Deussen, editor, *In Proc. of the 5th GI Conference*, volume 104 of *Lecture Notes in Computer Science*, pages 167–183. Springer Verlag, 1989.

86. C. S. Pasareanu. DEOS kernel: Environment modeling using LTL assumptions. Technical Report #NASA-ARC-IC-2000-196, NASA Ames, July 2000.

87. A. Phillipou, I. Lee, and O. Sokolsky. Weak Bisimulation for Probabilistic Systems. In *Proceedings of the 11th International Conference on Concurrency Theory (CONCUR'02)*, volume 1877 of *Lecture Notes in Computer Science*, pages 334–349, University Park, Pennsylvania, August 2000. Springer Verlag.

88. A. Di Pierro, C. Hankin, and H. Wiklicky. Approximate Non-interference. In *CSFW'02 15th IEEE Computer Security Foundation Workshop*, pages 1–15, June 2002.

89. A. Di Pierro and H. Wicklicky. Probabilistic Abstract Interpretation and Statistical Testing. In H. Hermanns and R. Segala, editors, *Process Algebra and Probabilistic Methods, Performance Modeling and Verification, Second Joint International Workshop PAPM-PROBMIV 2002*, volume 2399 of *Lecture Notes in Computer Science*, pages 211–212, Copenhagen, Denmark, July 25–26 2002. Springer.

90. G. D. Plotkin. A Structural Approach to Operational Semantics. Technical Report FN-19, DAIMI, Computer Science Department, Aarhus University, Ny Munkegade, Building 540, DK-8000 Aarhus, Denmark, September 1981. Reprinted April 1991.

91. M. L. Puterman. *Markov decision processes: discrete stochastic dynamic programming*. Wiley Series in Probability and Mathematical Statistics. John Wiley & Sons, 1994.

92. R. I. Bahar, E. A. Frohm, C. M. Gaona, G. D. Hachtel, E. Macii, A. Pardo, and F. Somenzi. Algebraic Decision Diagrams and Their Applications. In *IEEE /ACM International Conference on CAD*, pages 188–191, Santa Clara, California, 1993. IEEE Computer Society Press.

93. M. Sagiv, T. Reps, and R. Wilhelm. Parametric Shape Analysis via 3-Valued Logic. In *Proceedings of the 26th ACM SIGPLAN-SIGACT Symposium on Principles of programming languages*, pages 105–118, January 20-22, San Antonio, Texas 1999.

94. D. Scott. Continuous lattices. In F. W. Lawvere, editor, *Toposes, Algebraic Geometry and Logic*, volume 274 of *Lecture Notes in Mathematics*, pages 97–136. Springer Verlag, 1972.

95. R. Segala. *Modeling and Verification of Randomized Distributed Real-Time Systems*. PhD thesis, Laboratory for Computer Science, Massachusetts Institute of Technology, June 1995. Available as Technical Report MIT/LCS/TR-676.

96. R. Segala and N. Lynch. Probabilistic Simulations for Probabilistic Processes. *Nordic Journal of Computing*, 2(2):250–273, Summer 1995.

97. K. Seidel, C. Morgan, and A. McIver. An introduction to probabilistic predicate transformers. Technical Report PRG-TR-6-96, Programming Research Group, Oxford Computing Laboratory, Wolfson Building, Parks Road, Oxford OX1 3QD, 1996.

98. B. Selic. Physical programming: Beyond mere logic, April 2001. Invited Talk at ETAPS 2001.

99. R. J. van Glabbeek and W. P. Weijland. Branching Time and Abstraction in Bisimulation Semantics. *Journal of the ACM*, 43(3):555–600, May 1996.

100. M. Vardi. Automatic verification of probabilistic concurrent finite-state programs. In *Proc. 26th IEEE Symp. on Foundations of Computer Science*, pages 327–338, Portland, Oregon, October 1985.

101. D. Volpano. Provably secure programming languages for remote evaluation. *ACM Computing Surveys*, 28A(2): electronic, December 1996.

102. P. Waszkiewicz. *Quantitative Continuous Domains*. PhD thesis, School of Computer Science, University of Birmingham, United Kingdom, July 2002.

The Verification of Probabilistic Lossy Channel Systems

Philippe Schnoebelen

Lab. Spécification & Vérification
ENS de Cachan & CNRS UMR 8643
61, av. Pdt. Wilson, 94235 Cachan Cedex France
phs@lsv.ens-cachan.fr

Abstract. Lossy channel systems (LCS's) are systems of finite state automata that communicate via unreliable unbounded fifo channels. Several probabilistic versions of these systems have been proposed in recent years, with the two aims of modeling more faithfully the losses of messages, and circumventing undecidabilities by some kind of randomization. We survey these proposals and the verification techniques they support.

1 Introduction

Channel systems are systems of finite state automata that communicate via asynchronous unbounded fifo channels. An example, S_{exmp}, is depicted in Fig. 1.

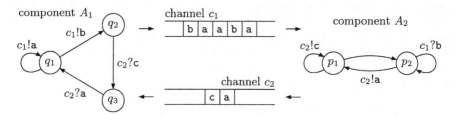

Fig. 1. S_{exmp}: a channel system with two component automata and two channels

They are a natural model for asynchronous communication protocols and, indeed, they form the semantical basis of protocol specification languages such as SDL and Estelle.

The behaviour of a system like S_{exmp} is as expected: component A_1 may move from q_1 to q_2 by sending a b message to channel c_1 where it will be enqueued (channels are fifo buffers). Then A_1 may move from q_2 to q_3 if it can read a c from channel c_2, which is only possible if the channel is not empty and the first available message is a c, in which case the message is dequeued (consumed). The two components, A_1 and A_2, evolve asynchronously.

That channel systems are a *bona fide* model of computation, indeed a Turing-powerful one, was pointed out by Brand and Zafiropulo [11]. This is easy

C. Baier et al. (Eds.): Validation of Stochastic Systems, LNCS 2925, pp. 445–465, 2004.
© Springer-Verlag Berlin Heidelberg 2004

to see: the channels are unbounded and one channel can simulate the work-tape of a Turing machine. One does not have a real scanning and overwriting head that moves back and forth along this "work-tape", but this can be simulated by rotating the contents of the channel to position oneself on any required cell (one keeps track of where the reading head is currently sitting by means of some extra marking symbols).

As an immediate corollary, a Rice Theorem can be stated: *"all nontrivial behavioral properties are undecidable for channel systems"*. Hence fully algorithmic verification (model checking) of arbitrary channel systems cannot be achieved. One is left with investigating restricted methods (that only deal with subclasses of the general model, e.g. systems with a bound on the size of the channels) or approximate methods (that admit false negatives) or semi-algorithmic methods (that may fail to terminate).

Lossy channels. A few years ago Finkel [14] and, independently, Abdulla and Jonsson [5], introduced *lossy channel systems* (LCS's), a very interesting class of channel systems. In lossy systems, messages can be lost while they are in transit, without any notification. These lossy systems are the natural model for fault-tolerant protocols where the communication channels are not supposed to be reliable.

Surprisingly, several verification problems become decidable when one assumes channels are lossy: termination, reachability, safety properties over traces, inevitability properties over states, and several variant problems are decidable for lossy channel systems [14, 6, 12, 5, 18].

This does not mean that lossy channel systems are an artificial model where, since no communication can be fully enforced, everything becomes trivial. To begin with, many important problems are undecidable: recurrent reachability properties are undecidable, so that model checking of liveness properties is undecidable too [4]. Furthermore, boundedness is undecidable [17], as well as all behavioral equivalences [22]. Finally, none of the decidable problems listed in the previous paragraph can be solved in primitive recursive time [23]!

Probabilistic lossy channel systems. It is natural to see message losses as some kind of faults having a probabilistic behaviour. This idea, due to Purushothaman Iyer and Narasimha, led to the introduction of the first Markov chain model for lossy channel systems [20].

There are two different benefits one can expect from investigating probabilistic versions of lossy channel systems: (1) they are a more realistic model, where quantitative information on faults is present, or (2) they are a more tractable model, where randomisation effectively rules out some malicious behaviours.

The verification of probabilistic lossy channel systems is a challenging problem because the underlying objects are countably infinite Markov chains (or Markovian decision processes). Furthermore, these infinite Markov chains are not *bounded*: probabilities of individual transitions can be arbitrarily close to zero.

Below we survey the main existing results on the verification of probabilistic lossy channel systems. The literature we review is still scarce, consisting of [20, 8,

2, 9, 7, 10]. We try to abstract from specific notational or definitional details and present the above works uniformly. As far as verification is concerned, we also try to abstract from specific algorithmic details[1] and extract the main ideas, in the hope that they can be applied to other infinite-state probabilistic verification problems. A recurrent theme is that decidability results rely on the existence of finite attractors.

Plan of this chapter. In section 2, we recall the main definitions and results on classical (non-probabilistic) lossy channel systems. The Markovian models for probabilistic lossy channel systems are given in section 3. We describe qualitative verification in section 4 and quantitative verification in section 5. Finally, in section 6, we present a Markovian decision process model and some results on adversarial verification.

2 Lossy Channel Systems

2.1 Perfect Channel Systems

The simplest way to present lossy channel systems is to start with *perfect*, i.e. not lossy, channel systems.

Definition 1 (Channel System). *A channel system (with m channels) is a tuple $S = \langle Q, C, \Sigma, \Delta, \sigma_0 \rangle$ where*

- $Q = \{r, s, \dots\}$ *is a finite set of* control locations *(or* control states*),*
- $C = \{c_1, \dots, c_m\}$ *is a finite set of m* channels,
- $\Sigma = \{a, b, \dots\}$ *is a finite alphabet of* messages,
- $\Delta \subseteq Q \times C \times \{?, !\} \times \Sigma \times Q$ *is a finite set of* rules.

A rule $\delta \in \Delta$ of the form $(s, c, ?, a, r)$ or $(s, c, !, a, r)$ is written "$s \xrightarrow{c?a} r$" (resp. "$s \xrightarrow{c!a} r$") and means that S can move from control location s to r by reading a from (resp. writing a to) channel c. Reading a is only possible if c is not empty and its first available message is a.

Remark 1. Def. 1 assumes there is only one component automaton in a channel system. This is no loss of generality since several components can be combined into a single one via a classical asynchronous product of automata. □

Remark 2. Def. 1 further assumes that all rules either consume or produce one message. Again, this is no loss of generality and internal rules (no message involved), or rules consuming and producing several messages, or rules testing for emptiness of a channel, etc., could be accounted for. □

---

[1] The papers we survey are mostly theoretical and the algorithms they propose have not yet been implemented and tested on actual examples.

The operational semantics of S is given via a transition system $\mathcal{T}_{\text{perf}}(S) = \langle Conf, \rightarrow_{\text{perf}} \rangle$ where $Conf = Q \times \Sigma^{*C}$ is the set of configurations (with typical elements $\sigma, \theta, \ldots$), and $\rightarrow_{\text{perf}} \subseteq Conf \times Conf$ is the unlabelled transition relation.

A *configuration* of S is a pair $\sigma = \langle r, U \rangle$ where $r \in Q$ is a control location and $U \in \Sigma^{*C}$ is a *channel contents*, i.e. a C-indexed vector of Σ-words: for any $c \in C$, $U(c) = u$ means that c contains u. The void channel contents, where every channel contains the empty word ε, is also denoted ε. The configuration of S_{exmp} in Fig. 1 has channel contents $U = \{c_1 \mapsto \mathbf{abaab}; c_2 \mapsto \mathbf{ca}\}$.

The possible transitions between configurations are given by the rules of S. Formally, for $\sigma, \sigma' \in Conf$, we have $\sigma \rightarrow_{\text{perf}} \sigma'$ iff either

reads: σ is some $\langle s, U \rangle$, there is a rule $s \xrightarrow{c?a} r$ in Δ, $U(c)$ is some $a.u'$, and $\sigma' = \langle r, U\{c \mapsto u'\} \rangle$ (using the standard notation $U\{\_ \mapsto \_\}$ for variants).

writes: σ is some $\langle s, U \rangle$, there is a rule $s \xrightarrow{c!a} r$ in Δ, $U(c)$ is some $u \in \Sigma^*$, and $\sigma' = \langle r, U\{c \mapsto u.a\} \rangle$.

We write $\sigma \xrightarrow{\delta}_{\text{perf}} \sigma'$ when we want to explicit that rule $\delta \in \Delta$ allows the step between σ and σ', and let $En(\sigma)$ denote the set $\{\delta \in \Delta \mid \sigma \xrightarrow{\delta}_{\text{perf}}\}$ of rules that are *enabled* in configuration σ.

2.2   Lossy Channel Systems

Lossy channel systems are channel systems where messages can be lost while they are in transit. A first formal definition was proposed by Finkel (who named them *completely specified protocols*) [14]. Finkel defined them as the subclass of channel systems where it is required that for any control state $s \in Q$, any channel $c \in C$ and any message $a \in \Sigma$, there is a rule $s \xrightarrow{c?a} s$. These rules implement the losses by always allowing the removal, without any modification of the current control state, of whatever message is at the front of any buffer.

Later Abdulla and Jonsson proposed a different definition where, rather than requiring special rules implementing the message losses, we have an altered operational semantics [5]. While this provides for essentially the same behaviour, their definition is much more tractable mathematically (see Remark 3) and this is the one we adopt below.

Formally, given two channel contents U and U', we write $U \sqsubseteq U'$ when U can be obtained from U' by removing an arbitrary number of messages at arbitrary places in U'. Thus $U \sqsubseteq U'$ iff for all $c \in C$, $U(c)$ is a *subword* of $U'(c)$. This extends into a relation between configurations:

$$\langle s, U \rangle \sqsubseteq \langle s', U' \rangle \overset{\text{def}}{\Leftrightarrow} s = s' \wedge U \sqsubseteq U'. \tag{1}$$

Observe that $U \sqsubseteq U$ for all U (by removing no message), and that $\sqsubseteq$ is a partial ordering between channel contents and between configurations. Higman's lemma [16] states it is a well-quasi-ordering (a *wqo*). Thus sets of configurations have a finite number of minimal elements.

It is now possible to define lossy steps, written $\sigma \xrightarrow{\delta}_{\text{loss}} \sigma'$, with

$$\sigma \xrightarrow{\delta}_{\text{loss}} \sigma' \stackrel{\text{def}}{\Leftrightarrow} \sigma \sqsupseteq \theta \xrightarrow{\delta}_{\text{perf}} \theta' \sqsupseteq \sigma' \text{ for some } \theta, \theta' \in Conf. \tag{2}$$

Thus a lossy step is a perfect step possibly preceded and followed by an arbitrary number of message losses.

The *lossy semantics* of a channel system S is the transition system $T_{\text{loss}}(S) = \langle Conf, \rightarrow_{\text{loss}} \rangle$.

Below we omit the "loss" subscript since we consider the lossy semantics by default (we never omit the "perf" subscript). As usual, $\xrightarrow{+}$ and $\xrightarrow{*}$ will denote the transitive and (resp.) reflexive-transitive closures of the one-step $\rightarrow$ relation.

Remark 3. What is mathematically nice in Abdulla and Jonsson's definition is that it induces the following monotonicity property: if $\theta \rightarrow \theta'$ and $\sigma \sqsupseteq \theta$, then $\sigma \rightarrow \theta'$. Similarly, if $\theta' \sqsupseteq \sigma'$, then $\theta \rightarrow \sigma'$. Thus sets of predecessors are upward-closed and sets of successors are downward-closed w.r.t. $\sqsubseteq$. Systems exhibiting such a monotonicity property are said to be *well-structured*, and enjoy general decidability results [3, 15]. □

2.3 Verification of Lossy Channel Systems

Several verification problems are decidable for lossy channel systems. In this survey, we only need the decidability of *reachability* and of *control state reachability*. These problems ask:

Reachability:
  Given: a LCS S and two configurations σ_0 and σ_f,
  Question: does $\sigma_0 \xrightarrow{*} \sigma_f$?

Control State Reachability:
  Given: a LCS S, a configuration σ_0, and a control state $s \in Q$,
  Question: does $\sigma_0 \xrightarrow{*} \langle s, U \rangle$ for some U?

These two problems are equivalent (inter-reducible). Their decidability is shown in [12, 5]. They cannot be solved in primitive recursive time and are thus highly intractable from a worst-case complexity viewpoint [23]. However, there exist clever symbolic methods that can answer reachability questions in many cases [1].

Some other verification problems are undecidable for lossy channel systems. Most notably *Büchi acceptance* and *control state loop*, where one asks:

Büchi Acceptance:
  Given: a LCS S, a configuration σ_0, and a control state $s \in Q$,
  Question: is it possible, starting from σ_0, to visit s infinitely many times, i.e. does $\sigma_0 \xrightarrow{*} \langle s, U_1 \rangle \xrightarrow{+} \langle s, U_2 \rangle \xrightarrow{+} \dots$ for an infinite sequence $U_1, U_2, \dots$?

Control State Loop:
    Given: a LCS S, a configuration σ_0, and a control state $s \in Q$,
    Question: does $\sigma_0 \xrightarrow{*} \langle s, U \rangle \xrightarrow{+} \langle s, U \rangle$ for some U?

These two problems are equivalent (inter-reducible). Their undecidability is proved in [4]. A corollary is that model checking of liveness properties is undecidable for LCS's.

3   Probabilistic Lossy Channel Systems

Purushothaman Iyer and Narasimha were the first to consider probabilistic variants of LCS's [20].

Investigating such variants is rather natural since just saying that "*any message can be lost at any time*" leads to very pessimistic conclusions about the behaviour of channel systems. In reality, many protocols dealing with unreliable channels are designed with the idea that message losses are usually unlikely, so that a bounded number of retries is a sufficient solution in most situations[2]. Capturing these ideas requires models where notions such as "message losses are unlikely" and "most situations" are supported. Hence the introduction of PLCS's, i.e. LCS's where message losses follow some kind of probabilistic distribution.

Definition 2. [20] *A probabilistic lossy channel system (PLCS) is a tuple $S = \langle Q, C, \Sigma, \Delta, \sigma_0, p_{\text{loss}}, D \rangle$ where*

- $\langle Q, C, \Sigma, \Delta, \sigma_0 \rangle$ *is some* underlying channel system,
- $p_{\text{loss}} \in (0, 1)$ *is a* loss probability, *and*
- $D : \Delta \mapsto (0, \infty)$ *is a* weight function *of the rules.*

3.1   The Global-Fault Model

The semantics of a PLCS S is given in terms of a Markov chain $\mathcal{M}_g(S) = \langle Conf, p_g \rangle$ where $Conf$ is the same set of configurations that appears in $\mathcal{T}_{\text{loss}}(S)$ and $p_g : Conf \times Conf \to [0, 1]$ is the transition probability (the g subscript stands for "global", as explained in section 3.2).

Here p_g assigns probabilities to the transitions of $\mathcal{T}_{\text{loss}}(S)$ in accordance with the following principles:

1. A step in $\mathcal{M}_g(S)$ is either a perfect step, or the loss of *one* message.
2. In any given configuration, there is a fixed probability p_{loss} that the next step will be the loss of a message. If there are several messages in σ, each of these messages can be lost with equal probability, so that when $\sigma' \sqsubseteq \sigma$ can

---

[2] The *Alternating Bit Protocol* is not so optimistic. As a result, it is not efficient enough for networks where transmission delays are long (e.g., the Internet).

be obtained by removing one message from σ, we would expect something like

$$p_g(\sigma, \sigma') = \frac{p_{\text{loss}}}{|\sigma|}, \tag{3}$$

assuming $|\langle s, U \rangle|$ denotes $|U|$, i.e. the number $\sum_{c \in C} |U(c)|$ of messages in U.

3. In any given configuration, there is a fixed probability $1 - p_{\text{loss}}$ that the next step will be a perfect step. The probability that $\delta \in \Delta$ will account for the next step is given by its weight $D(\delta)$ after some normalisation against the other enabled rules. If $\sigma \xrightarrow{\delta}_{\text{perf}} \sigma'$ is a step in $\mathcal{T}_{\text{perf}}(S)$, we would expect something like

$$p_g(\sigma, \sigma') = (1 - p_{\text{loss}}) \times \frac{D(\delta)}{\displaystyle\sum_{\delta' \in En(\sigma)} D(\delta')}. \tag{4}$$

Turning these principles into a rigorous definition is a tedious and boring task, that we will not repeat here, leaving it to the reader's imagination. Note that several special cases have to be taken into account[3] or simplified away. Most of what we explain below does not depend on these details. For clarity, we assume there is at least one enabled rule in any configuration σ, so that $\sum_{\delta \in En(\sigma)} D(\delta)$ is nonzero.

Remark 4. Strictly speaking, the transition system underlying $\mathcal{M}_g(S)$ differs from $\mathcal{T}(S)$: a step in $\mathcal{T}(S)$ combines several steps from $\mathcal{M}_g(S)$ (one per message loss plus one for the perfect step), but this is just an unimportant question of granularity, and essentially the same behaviour is exhibited in both cases.   □

We call $\mathcal{M}_g(S)$ the *global-fault* model because it assumes p_{loss} is the probability that the next step is a fault (a message loss) as opposed to a perfect step. A consequence of this assumption is that the fixed fault probability has to be distributed over all messages currently in transit: the more messages are currently in transit, the less likely it is that any single message will be lost in the next step, so that message losses are not independent events.

3.2   The Local-Fault Model

It can be argued that a more realistic model would have p_{loss} applying to any single message, independently of the other messages. This prompted Bertrand and Schnoebelen [9] and, independently, Abdulla and Rabinovich [7], to introduce a new model, here called the *local-fault* model, and denoted $\mathcal{M}_l(S)$.

More precisely, $\mathcal{M}_l(S) = \langle Conf, p_l \rangle$ is defined in accordance with the following principles

---

[3] What about losses when U is empty, and perfect steps when $En(\sigma)$ is empty? What about situations where different message losses account for a same $\sigma \longrightarrow \sigma'$? Or when the reading of a message has the same effect as a loss?

1. A step in $\mathcal{M}_l(S)$ is a perfect step followed by any given number of message losses (possibly zero).
2. In any given configuration, the perfect step is chosen probabilistically, by normalising the weights of the enabled rules.
3. Then, any message is lost with probability p_{loss} (and kept with probability $1 - p_{\text{loss}}$).
4. Thus, if $\sigma \xrightarrow{\delta}_{\text{perf}} \sigma' \sqsupseteq \sigma''$, we would expect something like

$$p_l(\sigma, \sigma'') = \frac{D(\delta)}{\sum_{\delta' \in En(\sigma)} D(\delta')} \times (p_{\text{loss}})^{|\sigma'| - |\sigma''|} \times (1 - p_{\text{loss}})^{|\sigma''|}. \qquad (5)$$

Here too the actual formal definition is more complex because there usually are several ways to reach a same σ'' from a given σ.[4]

3.3   Other Models

It is of course possible to define many other Markov chain models for probabilistic channel systems, e.g. with channel-dependent loss probabilities, etc. The two proposals we just discussed strive for minimality.

In the literature, two other models can be found:

1. The undecidability proof in [2] assumes that messages can only be lost while they are being enqueued in the channel and not later. (We leave it to the reader to write the definition of the associated Markov chain.) This choice mainly aims at simplifying the technical development of the aforementioned paper. However, it underlines the fact that our two earlier models see losses as occurring inside the channels when other possibilities are of course possible.
2. Abdulla and Rabinovich [7] consider different kind of faulty behaviours (not just losses). They allow insertion errors, duplication errors, and corruptions of messages. Such errors were considered in [12] but Abdulla and Rabinovich propose a probabilistic definition of these transmission faults and analyse when decidability is preserved.

4   Qualitative Verification of PLCS's

A Markov chain like $\mathcal{M}_g(S)$ or $\mathcal{M}_l(S)$ comes with a standard probability measure on its set of runs (see e.g. [19]). We let $\mathbb{P}_{\sigma_0}(\Diamond \sigma)$ denote the measure of the set of runs that start from σ_0 and eventually visit σ. More generally, let $\mathbb{P}_{\sigma_0}(\varphi)$ denote the measure of the set of runs from σ_0 that verify some linear-time property φ.

---

[4] However, compared to the global-fault model, the number of special cases is reduced since losses do not clash with perfect steps and since Eq. (5) tolerates $|\sigma| = 0$.

For verification purposes, proving that $\mathcal{M}$ has $\mathbb{P}_{\sigma_0}(\varphi) = 1$ is almost as good as proving $\mathcal{T}, \sigma_0 \models \varphi$ in the classical (non probabilistic) setting, since it shows that the set of runs where φ does not hold is "negligible". However, such negligible sets can make the difference between a decidable and an undecidable problem. Indeed, qualitative verification of PLCS's was investigated as a way to circumvent the undecidability of model checking for LCS's.

Formally, the problems we are interested in here are:

Almost-Sure Inevitability:
  Given: a PLCS S, a configuration σ_0, and some set $W \subseteq Conf$ of configurations,
  Question: does $\mathbb{P}_{\sigma_0}(\Diamond W) = 1$?

Almost-Sure Model Checking:
  Given: a PLCS S, a configuration σ_0, and some LTL−X formula φ,
  Question: does $\mathbb{P}_{\sigma_0}(\varphi) = 1$?

In almost-sure inevitability, we usually only consider sets W given in some finite way, e.g. a finite W, or a W given by specifying the possible control states but putting no restriction on the channel contents. Similarly, we assume that the propositions used in temporal formula φ are simply individual configurations or control states, so that no special extra labelling of $\mathcal{M}(S)$ is required.

Remark 5. [20] and [8] consider formulae in LTL−X, the fragment of LTL that is insensitive to stuttering. This is because $\mathcal{M}_g(S)$ and $\mathcal{T}_{\mathrm{loss}}(S)$ do not have the same granularity (see Remark 4), so that properties sensitive to stuttering may evaluate differently in the two models.   □

Here we follow [24] and assume properties are given under the form of a deterministic Streett automaton $\mathcal{A}_\varphi$. Checking whether $\mathcal{M}_g(S)$ satisfies φ almost surely reduces to checking whether the accepting runs in the product $\mathcal{M}_g(S) \otimes \mathcal{A}_\varphi$ have measure 1. When φ is insensitive to stuttering, we have $\mathcal{M}_g(S) \otimes \mathcal{A}_\varphi \equiv \mathcal{M}_g(S \otimes \mathcal{A}_\varphi)$ [8], and this also holds for $\mathcal{M}_l(S)$. Finally, writing S' for $S \otimes \mathcal{A}_\varphi$, we are left with verifying that a Streett acceptance property, of the form $\alpha = \bigwedge_{i=1}^{n}(\Box\Diamond A_i \Rightarrow \Box\Diamond A_i')$, holds almost surely on $\mathcal{M}_g(S')$.

4.1   The Decidability Results

Baier and Engelen [8] show that almost-sure model checking is decidable for $\mathcal{M}_g(S)$ when $p_{\mathrm{loss}} \geq \frac{1}{2}$ [5]. Bertrand and Schnoebelen [9], and Abdulla and Rabinovich [7], show that almost-sure model checking is decidable for $\mathcal{M}_l(S)$ for any $p_{\mathrm{loss}} \in (0, 1)$. These results apply to almost-sure inevitability since it is a special case of almost-sure model checking.

---

[5] It is believed that this threshold cannot be improved: using a slightly modified model (see section 3.3), Abdulla *et al.* were able to show that the problem is undecidable when $p_{\mathrm{loss}} < \frac{1}{2}$ [2].

The techniques underlying all these decidability results are similar and rely on the existence of finite attractors (see below). Furthermore, in all cases, whether $\mathcal{M}(S)$ almost surely satisfies φ does not depend on the precise value of the fault probability p_{loss}[6], of the weights D, and of the choice of a local vs. global fault model! This is one benefit of qualitative verification: one does not have to worry too much about whether the numerical constants used in the PLCS are realistic or not.

We argued in section 3 that the local-fault model is more realistic than the global-fault model. This is especially true for quantitative properties (dealt with in next section). For qualitative properties, the real superiority of this model is that the existence of finite attractors (entailing decidability) is guaranteed and does not require $p_{\text{loss}} \geq \frac{1}{2}$.

4.2   The Algorithmic Ideas

Verifying that a finite Markov chain almost surely satisfies a Streett property is decidable [13, 24]. However, the techniques involved do not always extend to *infinite* chains, in particular to chains that are not bounded.

It turns out it is possible to adapt these techniques to countable Markov chains *where a finite attractor exists*. We now develop these ideas, basically by simply streamlining the techniques of [8]. As is customary, rather than answering the question whether $\mathbb{P}(\varphi) = 1$, we deal with the dual question of whether $\mathbb{P}(\neg\varphi) > 0$, which allows a simpler technical exposition. More details and full proofs are available in [9].

Below we assume a given Markov chain $\mathcal{M} = \langle Conf, p \rangle$ with underlying transition system $\mathcal{T} = \langle Conf, \rightarrow \rangle$.

We say a non-empty set $W_a \subseteq Conf$ of configurations is an *attractor* when

$$\mathbb{P}_\sigma(\Diamond W_a) = 1 \text{ for all } \sigma \in Conf. \qquad (6)$$

Note that (6) implies $\mathbb{P}_\sigma(\Box\Diamond W_a) = 1$ for all $\sigma \in Conf$. The attractor is *finite* when W_a is.

Observe that an attractor is not the same thing as a set of recurrent configurations (however, an attractor must contain at least one configuration from each recurrent class).

Assume $W_a \subseteq Conf$ is a finite attractor. We define $G(W_a)$ as the finite directed graph $\langle W_a, \rightarrow \rangle$ where the vertices are the configurations from W_a, and where there is an edge from σ to σ' iff σ' is reachable from σ by some nonempty path in $\mathcal{T}$. Observe that the edges in $G(W_a)$ are transitive (but not reflexive in general).

In $G(W_a)$, we have the usual graph-theoretic notion of (maximal) strongly connected components (SCC's), denoted $B, B', \ldots$ These SCC's are ordered by reachability and a minimal SCC (i.e. an SCC B that cannot reach any other

---

[6] Assuming it is $\geq \frac{1}{2}$ in the global-fault model.

SCC) is a *bottom SCC* (a BSCC). A *trivial* SCC is a singleton without the self-loop. Observe that, in $G(W_a)$, a BSCC B cannot be trivial: since W_a is an attractor, one of its configurations must be reachable from B.

Assume a given Streett property $\alpha = \bigwedge_{i=1}^{n}(\Box\Diamond A_i \Rightarrow \Box\Diamond A_i')$. We say a BSSC B of $G(W_a)$ is *correct* for α if, for all $i = 1, ..., n$, either A_i is not reachable from B (in T), or A_i' is. Write $B_1, \ldots , B_k$ for the set of correct BSCC's.

Lemma 1. [21] $\mathbb{P}_\sigma(\alpha) = \mathbb{P}_\sigma(\Diamond(B_1 \cup \cdots \cup B_k))$.

Proof (Idea). Since W_a is a finite attractor, almost all paths eventually visit a BSSC B of $G(W_a)$. These paths almost surely visit B infinitely often, so that they almost surely visit infinitely often any configuration reachable from B and not any other configuration. Thus these paths satisfy α almost surely iff B is correct for α.   □

The key result for our purposes is a reduction of the probabilistic verification of Streett properties of $\mathcal{M}$ to reachability questions on the finite $G(W_a)$.

Corollary 1. $\mathbb{P}_\sigma(\bigwedge_{i=1}^{n}(\Box\Diamond A_i \Rightarrow \Box\Diamond A_i')) > 0$ *iff* $B_1 \cup \cdots \cup B_k$ *is reachable from* σ.

Remark 6. Corollary 1 reduces the question $\mathbb{P}_\sigma(\bigwedge_{i=1}^{n}(\Box\Diamond A_i \Rightarrow \Box\Diamond A_i')) > 0$? to graph-theoretic notions on $G(W_a)$ where the transition probability p of $\mathcal{M}$ does not appear. Similarly, p has no role in the definition of $G(W_a)$. Where p does appear is in making W_a an attractor!   □

Corollary 1 applies if we can find finite attractors in our Markov chains: Let S be a PLCS and let $W_0 \overset{\text{def}}{=} \{\langle q, \varepsilon\rangle \mid q \in Q\}$ be the set of all configurations where the channels are empty.

Lemma 2. W_0 *is a (finite) attractor in* $\mathcal{M}_l(S)$.
Furthermore, if $p_{\text{loss}} \geq \frac{1}{2}$, W_0 *is an attractor in* $\mathcal{M}_g(S)$.

Proof (Idea). The computations proving this are a bit tedious but the idea is easy to understand when one has some familiarity with random walks. We consider $\mathcal{M}_l(S)$ first. Assume $|\sigma| = m$. When $m > \frac{\log(2)}{-\log(1-p_{\text{loss}})}$, and thanks to Eq. (5), it is more probable to see steps $\sigma \to \sigma'$ where σ' exhibits a decrease rather an increase of size (relative to σ). Furthermore increases are at most single increments. Thus the "small" configurations are an attractor, and since W_0 is reachable from any set (thanks to lossiness), W_0 itself is an attractor.

In $\mathcal{M}_g(S)$, the same kind of reasoning explains why the same set W_0 is an attractor when $p_{\text{loss}} \geq \frac{1}{2}$: losses become so likely that the system cannot avoid being attracted to empty configurations.   □

We now have all the necessary ingredients for the proof that almost-sure reachability and almost-sure model checking are decidable for PLCS's: since reachability is decidable in the underlying transition system T, $G(W_a)$ can be built effectively, and it can be decided if a BSSC is correct for α and if it is reachable from σ.

These techniques apply to any extension where reachability remains decidable and where finite attractors can be identified. For example, in [7], Abdulla and Rabinovich investigate an extension of the local-fault model where other forms of channel unreliability exist: messages in transit can be *corrupted* (randomly replaced by some other message), *duplicated* (the duplicate message appears just alongside the original message), and *spurious* messages can appear out of the blue (these are called *insertion errors* in [12]). If corruptions and duplications have a per-message probability[7] of, respectively, p_{corrupt} and p_{dupl}, then W_0 is an attractor if $p_{\text{dupl}} < p_{\text{loss}}$, in which case decidability of almost-sure model-checking can be inferred from the decidability of reachability.

5   Approximate Quantitative Verification of PLCS's

Computing *quantitative* properties, e.g. computing the actual probability $\mathbb{P}(\varphi)$ that φ will be satisfied, is usually harder than deciding qualitative properties (like we did in Section 4). For one thing, and unless it is zero or one, $\mathbb{P}(\varphi)$ will depend on the actual values of the weights and p_{loss}.

Purushothaman Iyer and Narasimha proposed algorithms for the verification of quantitative properties of PLCS's. It turns out these algorithms are flawed (as observed by Rabinovich [21]) in the global-fault model for which they were intended. However, the underlying ideas can be salvaged and made to work for the local-fault model (or the global-fault model if $p_{\text{loss}} \geq \frac{1}{2}$: the required condition is the existence of a finite attractor).

5.1   The Decidability Results

Under the local-fault interpretation, the following problems have effective solutions:

Quantitative Probabilistic Reachability:
  Given: a PLCS S, two configurations σ_0 and σ_f, and a *tolerance* $\nu > 0$,
  Problem: find a p such that $p - \nu \leq \mathbb{P}_{\sigma_0}(\Diamond\sigma) \leq p + \nu$.

Quantitative Probabilistic Model Checking:
  Given: a PLCS S, a configuration σ_0, some LTL$-$X formula φ, and a tolerance $\nu > 0$,
  Problem: find a p such that $p - \nu \leq \mathbb{P}_{\sigma_0}(\varphi) \leq p + \nu$.

We speak of *approximate* quantitative verification because of the tolerance parameter ν (that can be as small as one wishes).

---

[7] Regarding insertion errors, we find it more natural to see them as having a global probability, not a per-message one. But one can model insertion as duplication+corruption going in pairs.

5.2   The Algorithmic Ideas

The methods for answering quantitative probabilistic reachability may be of more general interest. Assume we are given some PLCS S with two configurations σ_0 and σ_f. For $k \in \mathbb{N}$ we let $T_k(\sigma_0)$ denote the tree obtained by unfolding $\mathcal{M}_l(S)$ from σ_0 until depth k. Fig. 2 displays a schematic example for $k = 3$.

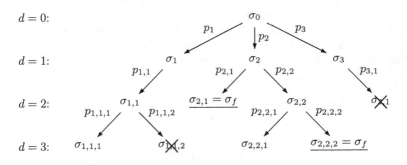

Fig. 2. A tree $T_3(\sigma_0)$ obtained by unfolding some $\mathcal{M}(S)$

Not all paths in $T_k(\sigma_0)$ have length k: paths are not developed beyond σ_f when it is encountered (these leaves are underlined in Fig. 2), or beyond any configuration σ from which σ_f is not reachable (these leaves are crossed).

Observe that it is effectively possible to build $T_k(\sigma_0)$: the crucial part is to cross out the leaves from which σ_f is not reachable, but this can be done since reachability is decidable (recall section 2.3).

A path from the root of $T_k(\sigma_0)$ to some leaf denotes a basic cylinder in the space of runs of $\mathcal{M}_l(S)$. The measure of these cylinders is given by multiplying the individual probabilities along the edges of the path, e.g. the leftmost leaf in our example denotes a set of runs with measure $p_1 \times p_{1,1} \times p_{1,1,1}$. We collect these cylinders in three classes: a path is a $\top$-*path* if it ends in σ_f, a $\bot$-*path* if it ends in a crossed-out leaf, and a ?-*path* otherwise. Thus $T_k(\sigma_0)$ partitions the runs from σ_0 in a finite number of cylinders belonging to three different types. If we now collect the measures of these cylinders according to type we end up with total measure $\mathbb{P}_\top^k$ for the $\top$-paths, $\mathbb{P}_\bot^k$ for the $\bot$-paths, and $\mathbb{P}_?^k$ for the ?-paths, ensuring $\mathbb{P}_\top^k + \mathbb{P}_\bot^k + \mathbb{P}_?^k = 1$ for all $k \in \mathbb{N}$. Obviously, T_{k+1} refines T_k in such a way that $\mathbb{P}_\top^{k+1} \geq \mathbb{P}_\top^k$ and $\mathbb{P}_\bot^{k+1} \geq \mathbb{P}_\bot^k$, entailing $\mathbb{P}_?^{k+1} \leq \mathbb{P}_?^k$.

The value $\mathbb{P}_{\sigma_0}(\Diamond \sigma_f)$ we are looking for satisfies

$$\mathbb{P}_\top^k \leq \mathbb{P}_{\sigma_0}(\Diamond \sigma_f) \leq \mathbb{P}_\top^k + \mathbb{P}_?^k \tag{7}$$

Lemma 3. $\lim_{k\to\infty} \mathbb{P}_?^k = 0$.

Proof. $\mathbb{P}_?^k$ is the probability that, in its first k steps, a run only visits configurations from which σ_f is reachable (called *good configurations*) without actually visiting σ_f itself. Thus $\lim_{k\to\infty} \mathbb{P}_?^k$, denoted $\mathbb{P}_?^\omega$, is the probability that an infinite path only visits good configurations but not σ_f.

Now, $\mathcal{M}_l(S)$ has a finite attractor W_0 (Lemma 2) and an infinite run almost surely visits W_0 infinitely often. Hence $\mathbb{P}_?^\omega$ is also the probability that an infinite run only visits good configurations, visits infinitely often some configuration $\sigma_a \in W_0$, and never visits σ_f. But if σ_a is visited infinitely often, and σ_f is reachable from σ_a, then σ_f is visited almost surely.   □

A possible algorithm for evaluating $\mathbb{P}_{\sigma_0}(\Diamond\sigma_f)$ within ν is to build $T_k(\sigma_0)$, use it to evaluate $\mathbb{P}_\top^k$ and $\mathbb{P}_?^k$, and check if the margin provided by Eq. (7) is low enough for ν, i.e. if $\mathbb{P}_?^k \le 2\nu$. If this is not the case, we retry with a larger k: Lemma 3 ensures that eventually the margin will be as small as needed.

The same method can be used for approximating the value of some $\mathbb{P}_{\sigma_0}(\bigwedge_{i=1}^n(\Box\Diamond A_i \Rightarrow \Box\Diamond A_i'))$: Lemma 1 equates this to some $\mathbb{P}_{\sigma_0}(\Diamond(B_1\cup\cdots\cup B_k))$.

Remark 7. In the global-fault model, Lemma 3 does not always hold and $\mathbb{P}_?^\omega$ can be strictly positive. This is what was missed in [20]. For example, consider a simple system with only the rule $q \xrightarrow{c!a} q$. $\mathcal{M}_g(S)$ is essentially a random walk on $\mathbb{N}$ with a reflecting barrier in 0: in configuration $\langle q, a^n\rangle$ with $n > 0$ messages, one moves to $\langle q, a^{n-1}\rangle$ with probability p_{loss} and to $\langle q, a^{n+1}\rangle$ with probability $1 - p_{\text{loss}}$. It is well-known that if $p_{\text{loss}} < \frac{1}{2}$ then there is a non-zero probability that a run from some nonempty configuration will never visit the empty configuration. This non-zero probability coincides with $\mathbb{P}_?^\omega$.   □

The ideas underlying the above algorithm are quite general and apply to all finitely-branching countable Markov chains with a finite attractor. Effectiveness relies on the fact that ⊥-paths can be identified: the decidability of reachability is an essential ingredient in the algorithm for quantitative probabilistic reachability.

5.3   A First Assessment

The positive results of section 4 and 5 provide methods for verifying channel systems where message losses are seen as some kind of fault obeying probabilistic laws. These results can handle arbitrary LTL formulae (indeed, arbitrary ω-regular properties) and therefore they can be seen as circumventing the undecidability of LTL model checking for standard, non-probabilistic, lossy channel systems.

However there are two main limitations with these results.

1. The value of $\mathbb{P}_{\sigma_0}(\varphi)$ depends on p_{loss} and on the weights D of rules in S. Assigning a fixed fault probability is natural in many situations, and sound reasoning can be carried out even under the assumption of an overly pessimistic fault probability. However, given a distributed protocol modelled as some LCS, it is difficult to meaningfully give values for D. Thus it is hard to make sense of any quantitative evaluation of some $\mathbb{P}_{\sigma_0}(\varphi)$.
2. This difficulty disappears with qualitative verification, since the answers do not depend on the exact values of D or p_{loss}. However, it is still the case that the models we have been considering, $\mathcal{M}_l(S)$ or $\mathcal{M}_g(S)$, see the rules of S as

probabilistic instead of nondeterministic. In practical verification situations, there are at least four possible reasons for nondeterminism in rules:

- Arbitrary interleaving of deterministic but asynchronously coupled components.
- Under-specification, at some early stage of the design process.
- Inclusion of the unknown environment as part of the model for an open system.
- Abstraction of some complex protocol to make it fit the finite-state control paradigm.

The first kind of nondeterminism can perhaps accommodate randomisation of the rules, but the other kinds usually cannot: a branch at a nondeterministic choice point in the abstract model may well never be followed in the actual system, even if the choice point is visited infinitely often.

This difficulty with Markov chains is well-known and its solution usually requires using a richer family of models: the reactive (or concurrent) Markov chains, also known as Markovian decision processes.

6   PLCS's as Markovian Decision Processes

It is very natural to model PLCS's as Markovian decision processes (MDP's) where message losses have some probabilistic behaviour, and where the LCS rules retain their classical nondeterministic meaning. Such a model was first introduced by Bertrand and Schnoebelen [9].

There are several possibilities for associating a MDP with a channel system S. The proposal in [9] adopts the conventions of [24] and elaborates on the ideas underlying the definition of $\mathcal{M}_l(S)$:

1. we have two kinds of configurations: nondeterministic ones and probabilistic ones,
2. steps alternate between firing LCS rules (going from a nondeterministic configuration to probabilistic ones) and losing messages (from a probabilistic configuration to nondeterministic ones), hence the underlying graph is bipartite.

In [9] the losses follow the "local", per-message, interpretation of the p_{loss} parameter, so that we shall write $\mathcal{P}_l(S)$ to denote the MDP associated with S.

As usual, the behaviour of $\mathcal{P}_l(S)$ is defined by means of schedulers (denoted $u, u', \ldots$) that make the nondeterministic choices, based on what happened earlier. When such an scheduler u is provided, the system gives rise to a Markov chain (denoted $\mathcal{P}_l^u(S)$) in the usual way.

Remark 8. [9] introduces one important definitional detail that make technicalities easier: we assume that it is always possible to idle, written $\sigma \xrightarrow{0} \sigma$, instead of firing a rule of Δ. This possibility prevents deadlocks and the definitional difficulties they raise, but it also give more flexibility to schedulers: they can empty the channels by just waiting (idling) until the probabilistic losses do the emptying job, which will eventually happen almost surely.   □

6.1   The Decidability Results

The main verification questions in this framework are

Adversarial Model Checking:
   Given: a PLCS S, a configuration σ_0, and some LTL−X formula φ,
   Question: does $\mathbb{P}_{\sigma_0}(\varphi) = 1$ hold in $\mathcal{P}^u(S)$ for all u?

Cooperative Model Checking:
   Given: a PLCS S, a configuration σ_0, and some LTL−X formula φ,
   Question: does $\mathbb{P}_{\sigma_0}(\varphi) = 1$ hold in $\mathcal{P}^u(S)$ for some u?

Note that, though the two problems are related, they cannot be reduced one to the other. It can be argued that, in verification settings, adversarial model checking is the more natural question.

Bertrand and Schnoebelen [9, 10] show that adversarial and cooperative model checking are decidable when one only considers finite-memory[8] schedulers (i.e. when the quantifications over "all u" is relativized to "all finite-memory u").
   They show that adversarial and cooperative model checking are undecidable when there are no restrictions on the schedulers[9].

6.2   The Algorithmic Ideas

When considering adversarial model checking, we follow the traditional wisdom that it is simpler to reason about questions of the form $\exists u\ \mathbb{P}(\neg \dots) > 0$, rather than of the form $\forall u\ \mathbb{P}(\dots) = 1$, even though the two are dual. Since Streett properties are closed by negation, we can further simplify our framework and consider questions of the form $\exists u\ \mathbb{P}(\alpha) > 0$.
   To begin with, we try to explain how decidability of adversarial model checking depends on the finite-memory assumption. For this we reproduce the proof that the unrestricted problem is undecidable: this provides a lively example of the difference between unrestricted and finite-memory schedulers.

Ideas for undecidability. Let $S = \langle Q, \{c\}, \Sigma, \Delta, \sigma_0 \rangle$ be a single-channel LCS where σ_0 is $\langle r_0, \varepsilon \rangle$.

---

[8] A scheduler is said to be *finite-memory* when it makes its choices as a function of the current configuration and some finite-state information about the history of the computation.

[9] Or on the LTL−X formulae: for example, formulae of the simpler form $\Diamond W$ lead to decidable adversarial or cooperative problems with no restriction on the schedulers [9].

We modify S to obtain S', a new LCS. The construction is illustrated in Fig. 3, where S is copied in the dashed box.

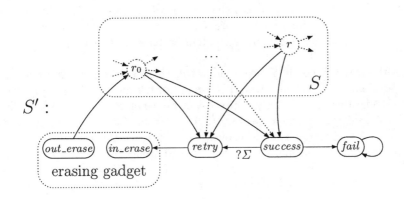

Fig. 3. The NPLCS S' associated with LCS S

S' is obtained by adding three control states (*success*, *retry* and *fail*), some fixed gadget for erasing (that is, emptying) the channel,[10] and rules allowing to jump[11] from any S-state $r \in Q$ to *success* or to *retry*. Jumping from *success* to *fail* is always possible but moving from *success* to *retry* requires that one message be read from the channel: the "?Σ" label is a shorthand for all ?a where $a \in \Sigma$.

Let us now ask ourselves whether we could devise a scheduler u s.t. $\mathbb{P}_{\sigma_0}(\Box\Diamond A) > 0$ in $\mathcal{P}_l^u(S')$ for $A = \uparrow success \setminus \langle success, \varepsilon \rangle$, i.e. whether there exists a scheduler that can make S' visit *success* infinitely often with nonzero probability (and without just idling there with an empty channel).

This seems easy: when we are in the S part of S', we can visit *success* whenever we want by jumping there, then we can erase the channels, jump to σ_0 and start again.

There is a catch however: if the channel is empty when we visit *success*, we will not be allowed to reach the erasing gadget via *retry* (this requires a read) and will end up stuck in *fail*. Now the bad news is that, whenever we decide to jump from some $\langle r, U \rangle$ configuration to *success*, there always is a nonzero probability (a *risk*) that all messages currently in the channel will be lost during this one jump, leaving us in configuration $\langle success, \varepsilon \rangle$ and thwarting our purposes. Precisely, if $|U| = m$, then the risk is $(p_{\text{loss}})^m$.

---

[10] The specification of the gadget is that, from any $\langle in\_erase, U \rangle$ configuration it is always possible to reach $\langle out\_erase, \varepsilon \rangle$ whatever happens (there are no deadlock) and it is impossible to reach $\langle out\_erase, V \rangle$ for $V \neq \varepsilon$. See [10] for a fully detailed solution.

[11] These are internal rules where no reading or writing takes place. Such rules can be simulated by writing to a dummy channel.

Therefore, visiting *success* infinitely many times requires that we make an infinite number of jumps, each of them with a nonzero risk. The product of these risks will only be nonzero if the values are diminishingly small (e.g. the risk associated with the nth jump is less than n^{-2}), which can only be the case if we jump from configurations having larger and larger channels contents. Thus, if S is bounded and such ever larger configurations do not exist, then necessarily $\mathbb{P}_{\sigma_0}(\Box\Diamond A) = 0$ for all u.

On the other hand, if S is not bounded, there exists ever larger reachable configurations in $T_{\mathrm{loss}}(S)$. It would be smart to try and reach these larger and larger configurations, and only jump to *success* when a large enough configuration is reached. Since the losses in $\mathcal{P}(S')$ are probabilistic, we may need some luck to reach these large configurations. However, when the losses do not comply, we can simply have another go (via the rules jumping from S to *retry*) so that a persistent scheduler will eventually reach any (reachable) configuration it wishes. Such a scheduler is clearly not finite-memory: it has to remember what large configuration it is currently aiming at, and there are an infinite number of them.

Finally, in $\mathcal{P}_l(S')$, $\mathbb{P}_{\sigma_0}(\Box\Diamond A) = 0$ for all u iff $T_{\mathrm{loss}}(S)$ is bounded. Since boundedness of $T_{\mathrm{loss}}(S)$ is undecidable [17], we obtain undecidability for adversarial model checking.

Ideas for decidability. It is now possible to hint at why adversarial model checking is decidable when we restrict to finite-memory schedulers.

Assume we want to check whether there is some scheduler u that yields $\mathbb{P}(\alpha) > 0$ for some PLCS $S = \langle Q, C, \Sigma, \Delta, \sigma_0, p_{\mathrm{loss}}, D \rangle$ and some Streett property $\alpha = \bigwedge_{i=1}^{n}(\Box\Diamond A_i \Rightarrow \Box\Diamond A_i')$. We assume the sets $A_i, A_i' \subseteq \mathit{Conf}$ are determined by sets of control states $X_i, X_i' \subseteq Q$.

A possible strategy for such a scheduler is to try to reach a set $X \subseteq Q$ of control states s.t. when one is in X-configurations (that is, configurations with a control state from X) then one can fulfill α without stepping out of X. More formally, we want that X is reachable from σ_0, is non-trivially connected by transitions that do not visit states out of X, and satisfies α: such an X is called a *safe set*.

If a safe X exists, a scheduler exists for $\mathbb{P}(\alpha) > 0$: it tries to reach X (this succeeds with non-zero probability) and, when in X, it just has to fill some $\Box\Diamond A_i'$ obligations. Since from any $\langle x, \varepsilon \rangle$ with $x \in X$, there is a path to A_i' that does not step out of X, it is enough to try that path. When this fails because probabilistic losses did not comply with the path, the scheduler waits until the channels are empty and tries again from the $\langle x', \varepsilon \rangle$ configuration we end up in. This scheme is bound to eventually succeed, and only requires finite memory. Thus, once we are in a safe X, we have $\mathbb{P}(\alpha) = 1$ (the only risk is in whether we can reach X from σ_0, but this has a nonzero probability of success).

The remarkable thing with finite-memory schedulers is that, if there exists a finite-memory scheduler ensuring $\mathbb{P}(\alpha) > 0$, then there exists one that follows the simple "pick a safe $X \subseteq Q$" strategy.

To see this, remember that any scheduler u will make us visit the attractor W_0 infinitely often, and in particular some $\langle x, \varepsilon \rangle$ infinitely often. But a finite-memory scheduler cannot distinguish between all these infinitely many visits to $\langle x, \varepsilon \rangle$ and there is some sequence $(\langle x, \varepsilon \rangle =) \langle x_0, U_0 \rangle \xrightarrow{\delta_1} \langle x_1, U_1 \rangle \xrightarrow{\delta_2} \langle x_2, U_2 \rangle \cdots \langle x_n, U_n \rangle$ of transitions, with x_n in some obligation A_i', that it will try infinitely many times. Trying this infinitely many times means that all the $\langle x_j, \varepsilon \rangle$ for $j = 0, \ldots, n$ are bound to happen infinitely often because of probabilistic losses. Assuming that u ensures α with a nonzero probability entails that the x_j's form a safe set X.

Now that we have equated the existence of a finite-memory u with the existence of a safe set, it remains to check that safe sets can be computed: this is easily done by enumerating the finite number of candidates X and checking each of them using the decidability of reachability in T_{loss}.

6.3   An Assessment of the Adversarial Verification of PLCS's

Modelling PLCS's as Markovian decision processes is a recent idea, and many questions remain unanswered. However, it seems this approach may lead to satisfactory ways of circumventing the undecidability of (classical) LCS model checking: decidability is recovered by simply omitting to take into account exaggeratedly malicious nondeterministic behaviours, be they schedulers that need finite memory or losses that have a zero probability of occurring in real life. One further advantage of this approach is that it relies on reachability questions of the kind that has been shown tractable via symbolic methods.

7   Conclusions and Perspectives

There are two main reasons for moving from LCS's to probabilistic LCS's:

1. obtaining quantitative information about the behaviour of the system (average times, probability that something happens, etc.), and
2. getting rid of exaggeratedly malicious nondeterministic behaviours by imposing some form of fairness.

The results we surveyed are very recent and it is not yet clear what will be the main directions for further research in this area. We just mention the problems that have been left unanswered in our survey:

In the *quantitative* approach:

- Can one compute exactly (not approximately) $\mathbb{P}_{\sigma_0}(\varphi)$ in the Markov chain models?
- Can one compute, or approximate, other numerical values like mean hitting time, etc.?
- Can one compute, or approximate, the extremal values of $\mathbb{P}_{\sigma_0}(\varphi)$ when ranging over all adversaries in the MDP models?

In the *qualitative* approach:

- Is cooperative model checking decidable?
- Do the decidability results extend to more general MDP models of LCS's (e.g. where probabilistic steps are not limited to message losses)?

From a more general perspective, it would be interesting to see how much of the
ideas that have been put forward in the analysis of probabilistic lossy channel
systems can be of use in other infinite-state probabilistic models.

References

1. P. A. Abdulla, A. Annichini, and A. Bouajjani. Symbolic verification of lossy channel systems: Application to the bounded retransmission protocol. In *Proc. 5th Int.
   Conf. Tools and Algorithms for the Construction and Analysis of Systems (TA-
   CAS'99), Amsterdam, The Netherlands, Mar. 1999*, volume 1579 of *Lecture Notes
   in Computer Science*, pages 208–222. Springer, 1999.
2. P. A. Abdulla, C. Baier, S. Purushothaman Iyer, and B. Jonsson. Reasoning about
   probabilistic lossy channel systems. In *Proc. 11th Int. Conf. Concurrency Theory
   (CONCUR'2000), University Park, PA, USA, Aug. 2000*, volume 1877 of *Lecture
   Notes in Computer Science*, pages 320–333. Springer, 2000.
3. P. A. Abdulla, K. Čerāns, B. Jonsson, and Yih-Kuen Tsay. Algorithmic analysis of programs with well quasi-ordered domains. *Information and Computation*,
   160(1/2):109–127, 2000.
4. P. A. Abdulla and B. Jonsson. Undecidable verification problems for programs
   with unreliable channels. *Information and Computation*, 130(1):71–90, 1996.
5. P. A. Abdulla and B. Jonsson. Verifying programs with unreliable channels. *Information and Computation*, 127(2):91–101, 1996.
6. P. A. Abdulla and M. Kindahl. Decidability of simulation and bisimulation between
   lossy channel systems and finite state systems. In *Proc. 6th Int. Conf. Theory of
   Concurrency (CONCUR'95), Philadelphia, PA, USA, Aug. 1995*, volume 962 of
   Lecture Notes in Computer Science, pages 333–347. Springer, 1995.
7. P. A. Abdulla and A. Rabinovich. Verification of probabilistic systems with faulty
   communication. In *Proc. 6th Int. Conf. Foundations of Software Science and Computation Structures (FOSSACS'2003), Warsaw, Poland, Apr. 2003*, volume 2620
   of *Lecture Notes in Computer Science*, pages 39–53. Springer, 2003.
8. C. Baier and B. Engelen. Establishing qualitative properties for probabilistic lossy
   channel systems: An algorithmic approach. In *Proc. 5th Int. AMAST Workshop
   Formal Methods for Real-Time and Probabilistic Systems (ARTS'99), Bamberg,
   Germany, May 1999*, volume 1601 of *Lecture Notes in Computer Science*, pages
   34–52. Springer, 1999.
9. N. Bertrand and Ph. Schnoebelen. Model checking lossy channels systems is prob-
   ably decidable. In *Proc. 6th Int. Conf. Foundations of Software Science and Computation Structures (FOSSACS'2003), Warsaw, Poland, Apr. 2003*, volume 2620
   of *Lecture Notes in Computer Science*, pages 120–135. Springer, 2003.
10. N. Bertrand and Ph. Schnoebelen. Verifying nondeterministic channel systems
    with probabilistic message losses. In *Proc. 3rd Int. Workshop on Automated Ver-
    ification of Infinite-State Systems (AVIS'04), Barcelona, Spain, Apr. 2004*, 2004.
    To appear.
11. D. Brand and P. Zafiropulo. On communicating finite-state machines. *Journal of
    the ACM*, 30(2):323–342, 1983.
12. G. Cécé, A. Finkel, and S. Purushothaman Iyer. Unreliable channels are easier to
    verify than perfect channels. *Information and Computation*, 124(1):20–31, 1996.
13. C. Courcoubetis and M. Yannakakis. The complexity of probabilistic verification.
    Journal of the ACM, 42(4):857–907, 1995.

14. A. Finkel. Decidability of the termination problem for completely specificied protocols. *Distributed Computing*, 7(3):129–135, 1994.
15. A. Finkel and Ph. Schnoebelen. Well structured transition systems everywhere! *Theoretical Computer Science*, 256(1–2):63–92, 2001.
16. G. Higman. Ordering by divisibility in abstract algebras. *Proc. London Math. Soc. (3)*, 2(7):326–336, 1952.
17. R. Mayr. Undecidable problems in unreliable computations. In *Proc. 4th Latin American Symposium on Theoretical Informatics (LATIN'2000), Punta del Este, Uruguay, Apr. 2000*, volume 1776 of *Lecture Notes in Computer Science*, pages 377–386. Springer, 2000.
18. B. Masson and Ph. Schnoebelen. On verifying fair lossy channel systems. In *Proc. 27th Int. Symp. Math. Found. Comp. Sci. (MFCS'2002), Warsaw, Poland, Aug. 2002*, volume 2420 of *Lecture Notes in Computer Science*, pages 543–555. Springer, 2002.
19. P. Panangaden. Measure and probability for concurrency theorists. *Theoretical Computer Science*, 253(2):287–309, 2001.
20. S. Purushothaman Iyer and M. Narasimha. Probabilistic lossy channel systems. In *Proc. 7th Int. Joint Conf. Theory and Practice of Software Development (TAPSOFT'97), Lille, France, Apr. 1997*, volume 1214 of *Lecture Notes in Computer Science*, pages 667–681. Springer, 1997.
21. A. Rabinovich. Quantitative analysis of probabilistic lossy channel systems. In *Proc. 30th Int. Coll. Automata, Languages, and Programming (ICALP'2003), Eindhoven, NL, July 2003*, volume 2719 of *Lecture Notes in Computer Science*, pages 1008–1021. Springer, 2003.
22. Ph. Schnoebelen. Bisimulation and other undecidable equivalences for lossy channel systems. In *Proc. 4th Int. Symp. Theoretical Aspects of Computer Software (TACS'2001), Sendai, Japan, Oct. 2001*, volume 2215 of *Lecture Notes in Computer Science*, pages 385–399. Springer, 2001.
23. Ph. Schnoebelen. Verifying lossy channel systems has nonprimitive recursive complexity. *Information Processing Letters*, 83(5):251–261, 2002.
24. M. Y. Vardi. Probabilistic linear-time model checking: An overview of the automata-theoretic approach. In *Proc. 5th Int. AMAST Workshop Formal Methods for Real-Time and Probabilistic Systems (ARTS'99), Bamberg, Germany, May 1999*, volume 1601 of *Lecture Notes in Computer Science*, pages 265–276. Springer, 1999.

Author Index

Lecture Notes in Computer Science

For information about Vols. 1–3056

please contact your bookseller or Springer

Vol. 3108: H. Wang, J. Pieprzyk, V. Varadharajan (Eds.), Information Security and Privacy. XII, 494 pages. 2004.

Vol. 3107: J. Bosch, C. Krueger (Eds.), Software Reuse: Methods, Techniques and Tools. XI, 339 pages. 2004.

Vol. 3106: K.-Y. Chwa, J.I. Munro (Eds.), Computing and Combinatorics. XIII, 474 pages. 2004.

Vol. 3105: S. Göbel, U. Spierling, A. Hoffmann, I. Iurgel, O. Schneider, J. Dechau, A. Feix (Eds.), Technologies for Interactive Digital Storytelling and Entertainment. XVI, 304 pages. 2004.

Vol. 3104: R. Kralovic, O. Sykora (Eds.), Structural Information and Communication Complexity. X, 303 pages. 2004.

Vol. 3103: K. Deb, e. al. (Eds.), Genetic and Evolutionary Computation – GECCO 2004. XLIX, 1439 pages. 2004.

Vol. 3102: K. Deb, e. al. (Eds.), Genetic and Evolutionary Computation – GECCO 2004. L, 1445 pages. 2004.

Vol. 3101: M. Masoodian, S. Jones, B. Rogers (Eds.), Computer Human Interaction. XIV, 694 pages. 2004.

Vol. 3100: J.F. Peters, A. Skowron, J.W. Grzymała-Busse, B. Kostek, R.W. Świniarski, M.S. Szczuka (Eds.), Transactions on Rough Sets I. X, 405 pages. 2004.

Vol. 3099: J. Cortadella, W. Reisig (Eds.), Applications and Theory of Petri Nets 2004. XI, 505 pages. 2004.

Vol. 3098: J. Desel, W. Reisig, G. Rozenberg (Eds.), Lectures on Concurrency and Petri Nets. VIII, 849 pages. 2004.

Vol. 3097: D. Basin, M. Rusinowitch (Eds.), Automated Reasoning. XII, 493 pages. 2004. (Subseries LNAI).

Vol. 3096: G. Melnik, H. Holz (Eds.), Advances in Learning Software Organizations. X, 173 pages. 2004.

Vol. 3095: C. Bussler, D. Fensel, M.E. Orlowska, J. Yang (Eds.), Web Services, E-Business, and the Semantic Web. X, 147 pages. 2004.

Vol. 3094: A. Nürnberger, M. Detyniecki (Eds.), Adaptive Multimedia Retrieval. VIII, 229 pages. 2004.

Vol. 3093: S.K. Katsikas, S. Gritzalis, J. Lopez (Eds.), Public Key Infrastructure. XIII, 380 pages. 2004.

Vol. 3092: J. Eckstein, H. Baumeister (Eds.), Extreme Programming and Agile Processes in Software Engineering. XVI, 358 pages. 2004.

Vol. 3091: V. van Oostrom (Ed.), Rewriting Techniques and Applications. X, 313 pages. 2004.

Vol. 3089: M. Jakobsson, M. Yung, J. Zhou (Eds.), Applied Cryptography and Network Security. XIV, 510 pages. 2004.

Vol. 3087: D. Maltoni, A.K. Jain (Eds.), Biometric Authentication. XIII, 343 pages. 2004.

Vol. 3086: M. Odersky (Ed.), ECOOP 2004 – Object-Oriented Programming. XIII, 611 pages. 2004.

Vol. 3085: S. Berardi, M. Coppo, F. Damiani (Eds.), Types for Proofs and Programs. X, 409 pages. 2004.

Vol. 3084: A. Persson, J. Stirna (Eds.), Advanced Information Systems Engineering. XIV, 596 pages. 2004.

Vol. 3083: W. Emmerich, A.L. Wolf (Eds.), Component Deployment. X, 249 pages. 2004.

Vol. 3080: J. Desel, B. Pernici, M. Weske (Eds.), Business Process Management. X, 307 pages. 2004.

Vol. 3079: Z. Mammeri, P. Lorenz (Eds.), High Speed Networks and Multimedia Communications. XVIII, 1103 pages. 2004.

Vol. 3078: S. Cotin, D.N. Metaxas (Eds.), Medical Simulation. XVI, 296 pages. 2004.

Vol. 3077: F. Roli, J. Kittler, T. Windeatt (Eds.), Multiple Classifier Systems. XII, 386 pages. 2004.

Vol. 3076: D. Buell (Ed.), Algorithmic Number Theory. XI, 451 pages. 2004.

Vol. 3075: W. Lenski, Logic versus Approximation. VIII, 205 pages. 2004.

Vol. 3074: B. Kuijpers, P. Revesz (Eds.), Constraint Databases and Applications. XII, 181 pages. 2004.

Vol. 3073: H. Chen, R. Moore, D.D. Zeng, J. Leavitt (Eds.), Intelligence and Security Informatics. XV, 536 pages. 2004.

Vol. 3072: D. Zhang, A.K. Jain (Eds.), Biometric Authentication. XVII, 800 pages. 2004.

Vol. 3071: A. Omicini, P. Petta, J. Pitt (Eds.), Engineering Societies in the Agents World. XIII, 409 pages. 2004. (Subseries LNAI).

Vol. 3070: L. Rutkowski, J. Siekmann, R. Tadeusiewicz, L.A. Zadeh (Eds.), Artificial Intelligence and Soft Computing - ICAISC 2004. XXV, 1208 pages. 2004. (Subseries LNAI).

Vol. 3068: E. André, L. Dybkjær, W. Minker, P. Heisterkamp (Eds.), Affective Dialogue Systems. XII, 324 pages. 2004. (Subseries LNAI).

Vol. 3067: M. Dastani, J. Dix, A. El Fallah-Seghrouchni (Eds.), Programming Multi-Agent Systems. X, 221 pages. 2004. (Subseries LNAI).

Vol. 3066: S. Tsumoto, R. Słowiński, J. Komorowski, J.W. Grzymała-Busse (Eds.), Rough Sets and Current Trends in Computing. XX, 853 pages. 2004. (Subseries LNAI).

Vol. 3065: A. Lomuscio, D. Nute (Eds.), Deontic Logic in Computer Science. X, 275 pages. 2004. (Subseries LNAI).

Vol. 3064: D. Bienstock, G. Nemhauser (Eds.), Integer Programming and Combinatorial Optimization. XI, 445 pages. 2004.

Vol. 3063: A. Llamosí, A. Strohmeier (Eds.), Reliable Software Technologies - Ada-Europe 2004. XIII, 333 pages. 2004.

Vol. 3062: J.L. Pfaltz, M. Nagl, B. Böhlen (Eds.), Applications of Graph Transformations with Industrial Relevance. XV, 500 pages. 2004.

Vol. 3061: F.F. Ramos, H. Unger, V. Larios (Eds.), Advanced Distributed Systems. VIII, 285 pages. 2004.

Vol. 3060: A.Y. Tawfik, S.D. Goodwin (Eds.), Advances in Artificial Intelligence. XIII, 582 pages. 2004. (Subseries LNAI).

Vol. 3059: C.C. Ribeiro, S.L. Martins (Eds.), Experimental and Efficient Algorithms. X, 586 pages. 2004.

Vol. 3058: N. Sebe, M.S. Lew, T.S. Huang (Eds.), Computer Vision in Human-Computer Interaction. X, 233 pages. 2004.

Vol. 3057: B. Jayaraman (Ed.), Practical Aspects of Declarative Languages. VIII, 255 pages. 2004.

Lecture Notes in Computer Science 2925

Edited by G. Goos, J. Hartmanis, and J. van Leeuwen

Lecture Notes in Computer Science
Edited by G. Goos, J. Hartmanis and J. van Leeuwen